Lecture Notes in Physics

Founding Editors: W. Beiglböck, J. Ehlers, K. Hepp, H. Weidenmüller

The Lecture Notes in Physics

The series Lecture Notes in Physics (LNP), founded in 1969, reports new developments in physics research and teaching – quickly and informally, but with a high quality and the explicit aim to summarize and communicate current knowledge in an accessible way. Books published in this series are conceived as bridging material between advanced graduate textbooks and the forefront of research and to serve three purposes:

- to be a compact and modern up-to-date source of reference on a well-defined topic

- to serve as an accessible introduction to the field to postgraduate students and nonspecialist researchers from related areas

- to be a source of advanced teaching material for specialized seminars, courses and schools

Both monographs and multi-author volumes will be considered for publication. Edited volumes should, however, consist of a very limited number of contributions only. Proceedings will not be considered for LNP.

Volumes published in LNP are disseminated both in print and in electronic formats, the electronic archive being available at springerlink.com. The series content is indexed, abstracted and referenced by many abstracting and information services, bibliographic networks, subscription agencies, library networks, and consortia.

Proposals should be sent to a member of the Editorial Board, or directly to the managing editor at Springer:

Christian Caron
Springer Heidelberg
Physics Editorial Department I
Tiergartenstrasse 17
69121 Heidelberg / Germany
christian.caron@springer.com

P. Kopietz
L. Bartosch
F. Schütz

Introduction to the Functional Renormalization Group

 Springer

Peter Kopietz
Goethe-Universität Frankfurt
Institut für Theoretische Physik
Max-von-Laue-Str. 1
60438 Frankfurt
Germany
pk@itp.uni-frankfurt.de

Lorenz Bartosch
Goethe-Universität Frankfurt
Institut für Theoretische Physik
Max-von-Laue-Str. 1
60438 Frankfurt
Germany
lb@itp.uni-frankfurt.de

Florian Schütz
Goethe-Universität Frankfurt
Institut für Theoretische Physik
Max-von-Laue-Str. 1
60438 Frankfurt
Germany

Kopietz, P. et al., *Introduction to the Functional Renormalization Group*, Lect. Notes Phys. 798 (Springer, Berlin Heidelberg 2010), DOI 10.1007/978-3-642-05094-7

Lecture Notes in Physics ISSN 0075-8450   e-ISSN 1616-6361
ISBN 978-3-642-26325-5   ISBN 978-3-642-05094-7(eBook)
DOI 10.1007/978-3-642-05094-7
Springer Heidelberg Dordrecht London New York

Cover design: Integra Software Services Pvt. Ltd., Pondicherry

Printed on acid-free paper

Springer is part of Springer Science+Business Media (www.springer.com)

For Julius, Jannis, Jona, and Cornelia,
for Zoë, Elias, and Claudia,
and for Wencke

Preface

The renormalization group (RG) has nowadays achieved the status of a *meta-theory*, which is a theory about theories. The theory of the RG consists of a set of concepts and methods which can be used to understand phenomena in many different fields of physics, ranging from quantum field theory over classical statistical mechanics to nonequilibrium phenomena. RG methods are particularly useful to understand phenomena where fluctuations involving many different length or time scales lead to the emergence of new collective behavior in complex many-body systems. In view of the diversity of fields where RG methods have been successfully applied, it is not surprising that a variety of apparently different implementations of the RG idea have been proposed. Unfortunately, this makes it somewhat difficult for beginners to learn this technique. For example, the field-theoretical formulation of the RG idea looks at the first sight rather different from the RG approach pioneered by Wilson, the latter being based on the concept of the effective action which is iteratively calculated by successive elimination of the high-energy degrees of freedom. Moreover, the Wilsonian RG idea has been implemented in many different ways, depending on the particular problem at hand, and there seems to be no canonical way of setting up the RG procedure for a given problem. Fortunately, in the last decade the development of the so-called functional renormalization group (FRG) method has somewhat unified the field by providing a mathematically elegant and yet simple way of expressing Wilson's idea of successive mode elimination in terms of a formally exact functional differential equation for the suitably defined generating functionals of a given theory. While the basic ideas of the Wilsonian RG as well as the field-theoretical RG are explained in many excellent textbooks, a pedagogic introduction to the Wilsonian RG using its modern formulation in terms of the FRG seems not to exist in the literature. It is the purpose of this book to fill this gap.

The book is subdivided into three parts. In Part I, which consists of the first five chapters, we introduce the reader to the basic concepts of the RG. This part is elementary and requires only previous knowledge of some introductory equilibrium statistical mechanics. In the four chapters of Part II we then give a self-contained introduction to the FRG. Since we are aiming at applications of the FRG to nonrelativistic quantum many-body systems, we start in Chap. 6 with an introduction to functional methods, defining various types of generating functionals and vertex functions. With these preparations, we derive in the central Chap. 7 of this book formally exact FRG flow equations of general field theories involving fermions, bosons, or mixtures thereof. We also discuss in detail how to include the emergence of finite vacuum expectation values of some of the field components into the FRG. In the following two chapters we discuss the two most common truncation strategies of the FRG flow equations, namely the vertex expansion in Chap. 8, and the derivative expansion in Chap. 9. Finally, in Part III of this book we apply the FRG method to nonrelativistic fermions. This part consists of three chapters: we first discuss the purely fermionic FRG in Chap. 10, and then present in Chaps. 11 and 12 partially bosonized FRG flow equations for interacting Fermi systems, where certain types of interaction processes are represented by suitable bosonic fields. The selected topics of Part III reflect our own research. We would like to emphasize, however, that the formulation of the FRG method developed in Chap. 7 is rather general and should be useful beyond the limited scope of our own research interest.

This book is based on a special topics course taught by one of us (P.K.) at the Goethe-Universität Frankfurt during the summer semesters 2006 and 2008. The course consisted of two 90-min lectures and one two-hour tutorial each week. The complete material in Part I and the first two chapters of Part II can be covered in 13 weeks provided the audience is familiar with the functional integral formulation of quantum many-body theory as developed on the first 100 pages of the textbook by Negele and Orland. The exercises at the end of Chaps. 1–7 are sometimes nontrivial and should be solved by the students at home. A complete discussion of the solutions of all exercises requires 11 or 12 two-hour tutorials.

We would like to thank several people who, in one way or the other, helped us to complete this book. First of all, we are grateful to Andreas Kreisel, who skillfully used the open source vector graphics editor *inkscape* to create a large part of the figures presented in this book and helped us to optimize our presentation. We also thank our collaborators on topics related to the functional renormalization group: Alvaro Ferraz, Hermann Freire, Nils Hasselmann, Thomas Kloss, Sascha Ledowski, and Andreas Sinner. In particular, the long-term collaborations with Sascha Ledowski and Nils Hasselmann influenced some of the presentations in Part II and Part III of this book. We have also profited from many useful comments and suggestions from some of the students who attended the courses on the renormalization group taught by one of us at the Goethe-Universität Frankfurt; we especially thank Christopher Eichler, who made several useful suggestions. Finally, we would like to thank Nicolas Dupuis, Holger Gies, Carsten Honerkamp, Christoph Kopper, Brad Marston, Jan Martin Pawlowski, Oliver Rosten, and Manfred Salmhofer for illuminating discussions on the functional renormalization group. We are extremely grateful to Manfred Salmhofer for his comments on Chaps. 6 and 10.

Our greatest intellectual debt is to Sudip Chakravarty, Konstantin Efetov, Subir Sachdev, and Kurt Schönhammer, who have been our teachers and mentors. Many of their questions, suggestions and ideas have found their way into this book.

Last but not least, we thank the *Sonderforschungsbereich SFB/TRR49* and the *DAAD/CAPES PROBRAL* program for financial support, and Christine Dinges for patiently producing the first LATEX version of the manuscript from the original hand-written lecture notes. We will maintain a web page

http://itp.uni-frankfurt.de/~rgbook/

where we will list all errors and points of confusion which will undoubtedly come to our attention.

Frankfurt am Main

July 2009

Peter Kopietz
Lorenz Bartosch
Florian Schütz

Contents

Part I

Foundations of the Renormalization Group

In Part I of this book, we introduce the basic concepts of the renormalization group (RG) idea. We start in Chap. 1 with a short review of the phenomenology of phase transitions and the concept of scaling, which motivated the development of the Wilsonian RG. We then give in Chap. 2 a self-contained introduction to the mean-field approximation and the leading fluctuation correction to mean-field theory, the so-called Gaussian approximation. In the central Chap. 3 of this Part I, we introduce the basic ideas and concepts of the Wilsonian RG. Of course, there exist already excellent textbooks where these topics are covered and our presentation is influenced by the books by Ma (1976), Goldenfeld (1992), and Cardy (1996). However, our presentation does not strictly follow any of these works; for example, in contrast to Cardy and Goldenfeld, we introduce diagrammatic perturbation theory from the very beginning in order to prepare the reader for the functional RG introduced in Part II of this book. For completeness we also give in Chap. 5 a brief introduction to the field theoretical RG, using the classical φ^4-theory describing the Ising universality class as an example.

Historically, the RG idea has first been developed in the context of quantum field theory. The so-called field-theoretical RG goes back to the work by Stueckelberg and Petermann (1953) and by Gell-Mann and Low (1954), who studied the ambiguities in the regularization procedure of the divergencies encountered in perturbative quantum field theory. They realized that in the so-called *renormalizable* field theories, where all divergencies encountered in perturbation theory can be absorbed in a finite number of renormalized coupling constants, the ambiguities in the energy scale which one is forced to introduce in the renormalization procedure can be used to relate the behavior of the theory at small and large energies. However, renormalizability is a rather strong assumption about the structure of a theory. In fact, the lattice models used to describe condensed matter systems are often not renormalizable. The lattice spacing itself provides a natural cutoff and there is no reason why one should assume renormalizability. Often, no divergencies appear and it makes perfect sense that even an infinite number of couplings in an effective field theory can depend on the cutoff, such as the lattice spacing.

Motivated by the challenge to understand the experimentally established universality of continuous phase transitions, in the 1970s another version of the RG idea was developed by Wilson (1969, 1971a,b, 1972, 1975), and others (Wilson and Fisher 1972, Wilson and Kogut 1974, Ma 1973, 1976, Fisher 1974, 1983, Wegner 1976, Di Castro and Jona-Lasinio 1976). In 1982 the physics Nobel Price was awarded to Wilson for his contributions to the development of this method; we shall therefore call this method the Wilsonian RG. For a recent review of the foundations of the Wilsonian RG with interesting historical remarks, see Fisher (1998). The Wilsonian formulation of the RG idea is based on the concept of a scale-dependent effective action, which is obtained by eliminating the high-energy degrees of freedom with energies above the cutoff scale Λ. Such a procedure can be carried out without making any strong assumptions about the structure of a theory (such as renormalizability), so that the Wilsonian RG is more general than the field-theoretical RG. In fact, in many applications of the RG in statistical mechanics and condensed matter, there are no infinities and the RG serves as a nonperturbative method for treating strongly interacting many-body systems. Moreover, generalizations of the Wilsonian RG idea turn out to be useful to study the dynamics of systems far from equilibrium (Goldenfeld 1992, McComb 2004).

References

Cardy, J. L. (1996), *Scaling and Renormalization in Statistical Physics*, Cambridge University Press, Cambridge.

Di Castro, C. and G. Jona-Lasinio (1976), *The Renormalization Group Approach to Critical Phenomena*, in C. Domb and M. S. Green, editors, *Phase Transitions and Critical Phenomena*, volume 6, Academic Press, London.

Fisher, M. E. (1974), *The renormalization group in the theory of critical behavior*, Rev. Mod. Phys. **46**, 597.

Fisher, M. E. (1983), *Scaling, Universality and Renormalization Group Theory*, in F. J. W. Hahne, editor, *Lecture Notes in Physics*, volume 186, Springer, Berlin.

Fisher, M. E. (1998), *Renormalization group theory: Its basis and formulation in statistical physics*, Rev. Mod. Phys. **70**, 653.

Gell-Mann, M. and F. E. Low (1954), *Quantum electrodynamics at small distances*, Phys. Rev. **95**, 1300.

Goldenfeld, N. (1992), *Lectures on Phase Transitions and the Renormalization Group*, Addison-Wesley, Reading.

Ma, S. K. (1973), *Introduction to the renormalization group*, Rev. Mod. Phys. **45**, 589.

Ma, S. K. (1976), *Modern Theory of Critical Phenomena*, Benjamin/Cummings, Reading.

McComb, W. (2004), *Renormalization Methods: A Guide for Beginners*, Oxford University Press, Oxford.

Stueckelberg, E. C. G. and A. Petermann (1953), *La normalisation des constantes dans la théorie des quanta*, Helv. Phys. Acta **26**, 499.

Wegner, F. J. (1976), *The Critical State, General Aspects*, in C. Domb and M. S. Green, editors, *Phase Transitions and Critical Phenomena*, volume 6, Academic Press, London.

Wilson, K. G. (1969), *Non-Lagrangian models of current algebra*, Phys. Rev. **179**, 1499.

Wilson, K. G. (1971a), *The renormalization group and critical phenomena I: Renormalization group and the Kadanoff scaling picture*, Phys. Rev. B **4**, 3174.

Wilson, K. G. (1971b), *The renormalization group and critical phenomena II: Phase-space cell analysis of critical behavior*, Phys. Rev. B **4**, 3184.

Wilson, K. G. (1972), *Feynman-graph expansion for critical exponents*, Phys. Rev. Lett. **28**, 548.

Wilson, K. G. (1975), *The renormalization group: Critical phenomena and the Kondo problem*, Rev. Mod. Phys. **47**, 773.

Wilson, K. G. and M. E. Fisher (1972), *Critical exponents in 3.99 dimensions*, Phys. Rev. Lett. **28**, 240.

Wilson, K. G. and J. Kogut (1974), *The renormalization group and the ϵ expansion*, Phys. Rep. **12**, 75.

Chapter 1
Phase Transitions and the Scaling Hypothesis

In the vicinity of continuous phase transitions, thermodynamic quantities and correlation functions typically behave as power laws characterized by universal exponents, which are independent of microscopic parameters of a system. The development of the Wilsonian RG in the 1970s was driven by the desire to gain a microscopic understanding of this universality. In this introductory chapter we briefly review the phenomenology of phase transitions, define the critical exponents, and discuss the relevant scaling laws. For a more detailed discussion of these topics, we refer the reader to other reviews and textbooks (Fisher 1983, Binney et al. 1992, Goldenfeld 1992, Ivanchenko and Lisyansky 1995, Cardy 1996, McComb 2004, Kardar 2007).

1.1 Classification of Phase Transitions

In thermal equilibrium the thermodynamic behavior of a macroscopic system can be derived from the relevant thermodynamic potential. For a quantum system with Hamiltonian $\hat{H}$ that is coupled to a heat bath with temperature T and a particle reservoir characterized by a chemical potential μ, one should calculate the grand canonical partition function $\mathcal{Z}(T, \mu)$ and the corresponding grand canonical potential $\Omega(T, \mu)$,

$$\mathcal{Z}(T, \mu) = e^{-\Omega(T,\mu)/T} = \mathrm{Tr}[e^{-(\hat{H}-\mu\hat{N})/T}] \,, \tag{1.1}$$

where $\hat{N}$ is the particle number operator and the trace is over the relevant Fock space containing an arbitrary number of particles. In the thermodynamic limit, where the volume V approaches infinity while the average density $n = \langle\hat{N}\rangle/V$ is held constant, we expect that the (generalized) free energy $f(T, \mu) = \Omega(T, \mu)/V$ approaches a finite limit independent of V. More generally, for a system whose macroscopic state is characterized by a set of k coupling constants $g_1, \ldots, g_k$, the partition function in the thermodynamic limit is expected to be of the form

$$\mathcal{Z}(g_1, \ldots, g_k) = e^{-Vf(g_1,\ldots,g_k)/T} \,. \tag{1.2}$$

Kopietz, P. et al.: *Phase Transitions and the Scaling Hypothesis.* Lect. Notes Phys. **798**, 5–22 (2010)
DOI 10.1007/978-3-642-05094-7_1

For example, for a system of electrons subject to an external magnetic field with magnitude h in the z-direction, one should add to the Hamiltonian the magnetic energy due to the spin degrees of freedom and thus replace $\hat{H} - \mu\hat{N} \rightarrow \hat{H} - \mu\hat{N} - h\hat{M}$ in the exponent of Eq. (1.1). Here, $\hat{M}$ is the operator representing the component of the total magnetization in the direction of the magnetic field.[1] In this case $k = 3$ with $g_1 = T$, $g_2 = \mu$, and $g_3 = h$.

In the k-dimensional coupling space spanned by $g_1, \ldots, g_k$, the generalized free energy $f(g_1, \ldots, g_k)$ is almost everywhere analytic. However, in the thermodynamic limit it can happen that there are points, lines, or other manifolds in coupling space with dimension smaller than k where $f(g_1, \ldots, g_k)$ exhibits some kind of nonanalyticity. These are the phase boundaries separating different phases of the system. The domains in coupling space where $f(g_1, \ldots, g_k)$ is analytic are called the phases of the system. Depending on the type of nonanalyticity at the phase boundaries, one distinguishes different types of phase transitions. In general, $f(g_1, \ldots, g_k)$ is continuous at the phase boundaries, but its derivatives with respect to the g_i can be discontinuous. A phase transition is called discontinuous (or first order) if at least one of the partial derivatives $\partial f/\partial g_i$ is discontinuous at the phase boundary. On the other hand, if all $\partial f/\partial g_i$, $i = 1, \ldots, k$, are continuous at the phase boundary, the phase transition is called continuous (or second order). Different phases are often (but not always) characterized by different symmetries. One usually tries to characterize phases with lower symmetry quantitatively in terms of a suitable order parameter, which is chosen such that it is nonzero only in the phase with lower symmetry. Let us illustrate these concepts with two examples:

Example 1: Paramagnet–Ferromagnet Transition

The phase transition in some magnetic insulators from a paramagnetic state to a state with ferromagnetic long-range order in the absence of an external magnetic field is an example for a continuous phase transition. The thermodynamic state of the system in a magnetic field can be described in terms of the two relevant parameters T and h. The order parameter is the spontaneous magnetization, which is defined in terms of the expectation value $\langle\hat{M}\rangle$ of the magnetization operator per unit volume in the thermodynamic limit,

$$m = -\lim_{h \to 0} \frac{\partial f(T, h)}{\partial h} = \lim_{h \to 0}\left[\lim_{V \to \infty} \frac{\langle\hat{M}\rangle}{V}\right] . \tag{1.3}$$

---

[1] Throughout this work we measure the temperature and the magnetic field in units of energy, which amounts to formally setting the Boltzmann constant and the Bohr magneton equal to unity. h is then the Zeemann energy and the magnetization operator $\hat{M}$ is dimensionless and gives the component of the total angular momentum in the direction of the field in units of $\hbar$.

If m is finite, the state of the system has lower symmetry than the Hamiltonian, which for $h = 0$ is at least invariant under the reversal of the direction of all spins. Since the magnetization remains finite for $T < T_c$ even without external magnetic field, this phenomenon is called spontaneous symmetry breaking. The order of limits in Eq. (1.3) is important: Only if one first takes the thermodynamic limit $V \to \infty$, the zero-field limit can be finite. If the magnetic system can be described by an Ising model whose one-dimensional version is introduced in Exercise 1.1, then the ferromagnetic state breaks the up–down (Z_2) symmetry of the Hamiltonian. The typical behavior of the spontaneous magnetization curve $m(T)$ as a function of temperature is shown in Fig. 1.1. In the vicinity of the critical temperature T_c where m first becomes finite, the magnetization curve follows a universal power law,

$$m(T) \propto (T_c - T)^\beta, \quad T \leq T_c , \tag{1.4}$$

where the magnetization exponent β is universal in the sense that it has the same value for a whole class of systems which is characterized by rather general properties such as symmetry and dimensionality. While the critical behavior described above is quite general, there are exceptions to this rule in low-dimensional systems (such as the one-dimensional Ising model introduced in Exercise 1.1), where the effect of fluctuations can prohibit spontaneous symmetry breaking at any finite temperature.

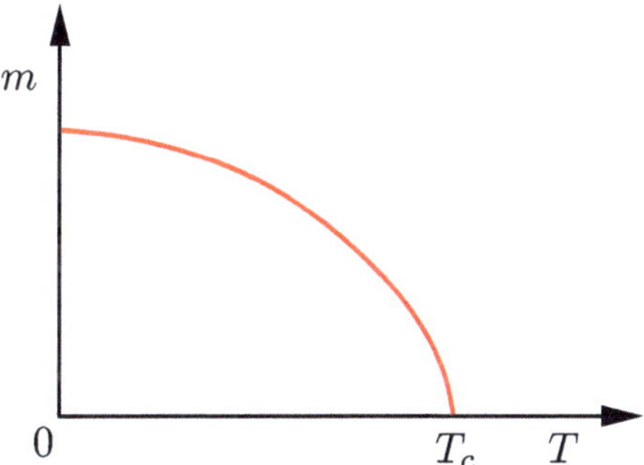

Fig. 1.1 Typical behavior of the spontaneous magnetization of a system exhibiting spontaneous ferromagnetism for temperatures T below a critical temperature T_c

Example 2: Liquid–Gas Transition

The phenomenon that upon heating a liquid begins to boil at a certain transition temperature and transforms into a gas is familiar to everybody. This is an example of a first-order phase transition between two phases with the same symmetry. If one plots the density $n = -\partial f(T, \mu)/\partial \mu$ as a function of temperature for different values of the pressure p, one obtains the curves shown in Fig. 1.2. As we lower the temperature keeping the pressure $p < p_c$ fixed at a value smaller than a certain critical pressure p_c, there is a critical temperature T_c below which gas and liquid can coexist in a certain regime of densities. At the *critical end point* (T_c, p_c, n_c) shown in Fig. 1.2, the density interval where coexistence of gas and liquid is possible collapses to a single point. The shape of the coexistence curve for densities $|n - n_c| \ll n_c$ is again characterized by a universal power law, with the same exponent as observed in the ferromagnet–paramagnet transition in uniaxial ferromagnets (which can be described by an Ising model). For $p < p_c$ the quantity $n - n_c$ is thus analogous to the magnetization in Example 1 and plays the role of an order parameter for the liquid–gas transition. For a nice discussion of similarities and differences between the paramagnet–ferromagnet transition and the liquid–gas transition, see, for example, the review by Fisher (1983). In Exercise 1.2 the coexistence curve of the liquid–gas transition marked in Fig. 1.2 will be calculated approximately within the van der Waals theory, which provides a simple mean-field description for the liquid–gas transition.

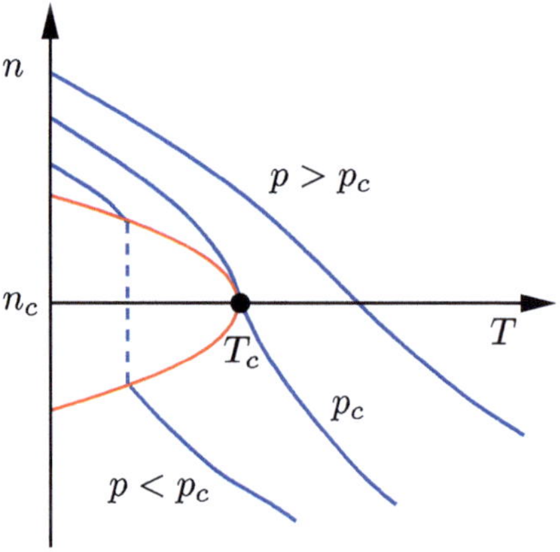

Fig. 1.2 Curves of constant pressure (isobars) in the temperature-density plane of a simple fluid exhibiting a liquid–gas transition. The critical point is denoted by a *black dot* and the *red curve* marks the boundary of the coexistence regime. In the vicinity of the critical point the shape of the coexistence curve can be described by the same power law as the magnetization close to T_c in Fig. 1.1, see Eq. (1.4). Note that in general the coexistence curve is not symmetric with respect to $n - n_c \to -(n - n_c)$, corresponding to reflection on the temperature axis

In the two examples above the phase transition is driven by thermal fluctuations and occurs at a finite temperature. In quantum systems one encounters sometimes phase transitions at zero temperature which are triggered by varying some nonthermal control parameter such as the density or an external magnetic field (see, e.g., Sachdev 1999, 2000). These so-called quantum phase transitions can be classified analogously into first order and continuous ones. We shall give some examples for quantum phase transitions in Sect. 1.4.

1.2 Critical Exponents and Universality of Continuous Phase Transitions

As the temperature approaches the critical temperature T_c of a continuous phase transition, an increasing number of microscopic degrees of freedom are coupled to each other and effectively act as a single entity. The correlation length ξ denotes the typical length scale of the regions where the degrees of freedom are strongly coupled. For example, in the vicinity of the paramagnet–ferromagnet transition (with the temperature slightly larger than the critical temperature) microscopic spins within regions of linear size ξ tend to point in the same direction, while spins belonging to different regions whose distance is large compared with ξ remain uncorrelated. At the critical point associated with a continuous phase transition, the correlation length ξ is infinite and there are fluctuations on all length scales, so that the system is scale invariant. As a consequence, thermodynamic observables are homogeneous functions of the relevant thermodynamic variables so that they exhibit power-law behavior. The exponents which characterize the leading behavior of thermodynamic observables for $T \rightarrow T_c$ are called *critical exponents*. For simplicity, consider the paramagnet–ferromagnet transition. It is convenient to measure the distance from the critical point on the temperature axis in terms of the reduced temperature

$$t = \frac{T - T_c}{T_c} \ . \tag{1.5}$$

Historically, one introduces the critical exponents α, β, and γ to characterize the asymptotic behavior of the following observables for $t \rightarrow 0$,

$$
\begin{array}{lll}
\text{Specific heat:} & C(t) \propto |t|^{-\alpha} \ , & \\
\text{Spontaneous magnetization:} & m(t) \propto (-t)^{\beta} \ , & t \leq 0 \ , \\
\text{Magnetic susceptibility:} & \chi(t) \propto |t|^{-\gamma} \ . &
\end{array}
\tag{1.6}
$$

Moreover, the shape of the critical isotherm giving the magnetic field dependence of the magnetization $m(h)$ at $T = T_c$ for a small magnetic field h defines another critical exponent δ,

$$\text{Critical isotherm:} \quad m(h) \propto |h|^{1/\delta} \operatorname{sgn}(h) \ , \quad t = 0 \ . \tag{1.7}$$

According to the scaling hypothesis to be discussed in Chap. 1.3, the above thermodynamic exponents α, β, γ, and δ can all be related to the scaling behavior of the singular part of the free energy for small t and h.

In addition to these thermodynamic exponents, one usually introduces two more exponents ν and η via the behavior of the order-parameter correlation function $G(r - r')$ for large distances $r - r'$. For a magnetic phase transition with Ising symmetry, $G(r - r')$ is defined as follows. Let us denote by $\hat{m}(r)$ the operator representing the local magnetization density at space point r. By translational invariance the thermal average

$$m = \langle \hat{m}(r) \rangle \equiv \frac{\operatorname{Tr} e^{-\hat{H}/T} \hat{m}(r)}{\operatorname{Tr} e^{-\hat{H}/T}} \tag{1.8}$$

is then independent of r, so that $\delta \hat{m}(r) = \hat{m}(r) - \langle \hat{m}(r) \rangle = \hat{m}(r) - m$. The order-parameter correlation function is then defined by the thermal average

$$\begin{aligned} G(r - r') &= \langle \hat{m}(r)\hat{m}(r') \rangle - \langle \hat{m}(r) \rangle \langle \hat{m}(r') \rangle \\ &= \langle \delta \hat{m}(r) \delta \hat{m}(r') \rangle \;, \end{aligned} \tag{1.9}$$

which depends only on the difference $r - r'$. It is sometimes more convenient to Fourier transform $G(r - r')$ to wave vector space,

$$G(k) = \int d^D r\, e^{-i k \cdot r} G(r) \;. \tag{1.10}$$

Typically, one finds for the asymptotic behavior of $G(r)$ for distances large compared with the correlation length ξ,

$$G(r) \propto \frac{e^{-|r|/\xi}}{\sqrt{\xi^{D-3}|r|^{D-1}}} \;, \quad |r| \gg \xi \;. \tag{1.11}$$

For $T \to T_c$ the correlation length ξ diverges with a power law,

$$\boxed{\xi \propto |t|^{-\nu} \;,} \tag{1.12}$$

where ν is called correlation length exponent. When the system approaches the critical point, the correlation length diverges and the regime of validity of Eq. (1.11) is pushed to infinity. Precisely at the critical point ξ is infinite and the order-parameter correlation function decays with a power that is different from the power in the prefactor in Eq. (1.11); in D dimensions we write

$$G(r) \propto \frac{1}{|r|^{D-2+\eta}} \;, \quad T = T_c \;. \tag{1.13}$$

To motivate the above definition of the correlation function exponent η, note that in wave vector space Eq. (1.13) implies for $k \to 0$,

$$\boxed{G(k) \propto |k|^{-2+\eta} \;, \quad T = T_c \;.} \tag{1.14}$$

We shall show in Chap. 2.3 that for systems whose critical fluctuations are in some sense weakly interacting,[2] the exponent η vanishes so that $G(k) \propto k^{-2}$. Hence, a finite exponent η indicates strongly interacting critical fluctuations. Because in Eq. (1.13) the exponent η can be viewed as a correction to the physical dimension D of the system, η is sometimes called an *anomalous dimension*.

It turns out that the numerical values of the critical exponents are not simply given by rational numbers as one might have expected on the basis of dimensional analysis. Moreover, the exponents are universal in the sense that entire classes of materials consisting of very different microscopic constituents can have the same exponents. As a consequence of this, a uniaxial ferromagnet and a simple fluid for example are believed to have exactly the same critical exponents. All materials can thus be divided into *universality classes*, which are characterized by the same critical exponents. The universality class in turn is determined by some rather general properties of a system, such as its dimensionality, the symmetry of its order parameter, or the range of the interaction. For example, in Table 1.1 we list the critical exponents for the Ising universality class in two and three dimensions and for the XY and Heisenberg universality classes in $D = 3$. Although experimental evidence for the universality of the critical exponents had already emerged in the 1930s, a microscopic understanding of this universality was only achieved in the 1970s with the help of the renormalization group, which also provided a systematic method for explicitly calculating critical exponents in cases where mean-field theory fails.

Table 1.1 Critical exponents of the Ising, XY, and Heisenberg universality classes. The corresponding symmetry groups of the order parameter are Z_2 for the Ising universality class, $O(2)$ for the XY universality class, and $O(3)$ for the Heisenberg universality class. The small subscripts in the first line denote the dimensionality. While the exponents of the two-dimensional Ising universality class are exact, the exponents in three dimensions are only known approximately. The numbers for Ising$_3$ and the error estimates are from the review by Pelissetto and Vicari (2002). For XY$_3$ we give rounded values for α, γ, ν, and η up to two significant figures, as compiled in Pelissetto and Vicari (2002, Table 19). The values for β and δ are obtained using the scaling relations (1.33b) and (1.33d). For Heisenberg$_3$ we quote the results by Holm and Janke (1993)

Exponent	Ising$_2$	Ising$_3$	XY$_3$	Heisenberg$_3$
α	0 (log)	0.110(1)	−0.015	−0.10
β	1/8	0.3265(3)	0.35	0.36
γ	7/4	1.2372(5)	1.32	1.39
δ	15	4.789(2)	4.78	5.11
ν	1	0.6301(4)	0.67	0.70
η	1/4	0.0364(5)	0.038	0.027

---

[2] More precisely, $\eta = 0$ for systems whose critical behavior can be obtained within the so-called Gaussian approximation, which will be discussed in Chap. 2.3.

1.3 The Scaling Hypothesis

It turns out that under quite general conditions (to be discussed below) only two of the six exponents α, β, γ, δ, ν, and η are independent, so that we can obtain the thermodynamic exponents α, β, γ, and δ from the exponents ν and η related to the scaling of the correlation function using so-called scaling relations. The latter follow from the scaling behavior of the free energy and the order-parameter correlation function in the vicinity of a continuous phase transition. The scaling relations can be obtained microscopically with the help of the renormalization group. Historically, the scaling form of the free energy (Widom 1965) and of the correlation function (Kadanoff 1966) was formulated as a hypothesis before the renormalization group was invented.

Let us first discuss the scaling form of the free energy. For simplicity, consider again a magnet with free energy density $f(t, h)$, which is a function of the reduced temperature and the magnetic field. In the vicinity of a continuous phase transition, we expect that $f(t, h)$ can be decomposed into a singular and a regular part,

$$f(t, h) = f_{\text{sing}}(t, h) + f_{\text{reg}}(t, h). \tag{1.15}$$

Sufficiently close to the critical point, the singular part $f_{\text{sing}}(t, h)$ is assumed to be determined by power laws characteristic of a given critical point. According to the scaling hypothesis for the free energy, its singular part satisfies the following generalized homogeneity relation,

$$\boxed{f_{\text{sing}}(t, h) = b^{-D} f_{\text{sing}}(b^{y_t} t, b^{y_h} h) \,,} \tag{1.16}$$

where b is an arbitrary (dimensionless) scale factor and the exponents y_t and y_h are characteristic for a given universality class. It is now easy to show that with this assumption the four thermodynamic exponents α, β, γ, and δ can all be expressed in terms of y_t, y_h and the dimensionality D of the system. Using the fact that b is arbitrary, we may set $b^{y_t} = 1/|t|$, or equivalently $b = |t|^{-1/y_t}$. Then we obtain from Eq. (1.16),

$$f_{\text{sing}}(t, h) = |t|^{D/y_t} \Phi_{\pm} \left(\frac{h}{|t|^{y_h/y_t}} \right) \,, \tag{1.17}$$

with the scaling functions $\Phi_{\pm}(x) \equiv f_{\text{sing}}(\pm 1, x)$. We obtain the desired relations between the thermodynamic exponents α, β, γ, and y_t, y_h by taking appropriate derivatives of Eq. (1.17) with respect to t and h and then setting $h = 0$, assuming that close to the critical point the derivatives of the free energy density are dominated by its singular part,

$$C \approx T_c^{-1} \left. \frac{\partial^2 f_{\text{sing}}}{\partial t^2} \right|_{h=0} \propto |t|^{\frac{D}{y_t}-2} = |t|^{-\alpha} , \tag{1.18a}$$

$$m \approx - \left. \frac{\partial f_{\text{sing}}}{\partial h} \right|_{h=0} \propto (-t)^{\frac{D-y_h}{y_t}} = (-t)^{\beta} , \tag{1.18b}$$

$$\chi \approx \left. \frac{\partial^2 f_{\text{sing}}}{\partial h^2} \right|_{h=0} \propto |t|^{\frac{D-2y_h}{y_t}} = |t|^{-\gamma} , \tag{1.18c}$$

where we have used the definition (1.6) of the critical exponents α, β, and γ. Hence,

$$\alpha = 2 - \frac{D}{y_t} , \tag{1.19a}$$

$$\beta = \frac{D - y_h}{y_t} , \tag{1.19b}$$

$$\gamma = \frac{2y_h - D}{y_t} . \tag{1.19c}$$

To express the exponent δ associated with the critical isotherm in terms of y_h and D, we consider the derivative of Eq. (1.17) for finite $h > 0$,

$$m(t, h) \approx - \frac{\partial f_{\text{sing}}}{\partial h} = |t|^{\frac{D-y_h}{y_t}} \Phi'_{\pm} \left(\frac{h}{|t|^{y_h/y_t}} \right) , \tag{1.20}$$

where $\Phi'_{\pm}(x) = d\Phi_{\pm}(x)/dx$. If the scaling functions $\Phi_{\pm}(x)$ are known, we may solve Eq. (1.20) for $h = h(t, m)$ to obtain the scaling form of the thermal equation of state. To obtain a finite value of m for $t \to 0$, the scaling function $\Phi'_{\pm}(x)$ must behave as $x^{\frac{D}{y_h}-1}$ for $x \to \infty$. We thus obtain for the critical isotherm

$$m(h) \propto h^{\frac{D}{y_h}-1} = h^{1/\delta} , \tag{1.21}$$

and hence,

$$\delta = \frac{y_h}{D - y_h} . \tag{1.22}$$

We may now eliminate the two variables y_t and y_h from the four equations (1.19a), (1.19b), (1.19c), and (1.22) to obtain two scaling relations involving only the experimentally measurable exponents α, β, γ, and δ,

$$\boxed{\begin{aligned} 2 - \alpha &= 2\beta + \gamma , \\ 2 - \alpha &= \beta(\delta + 1) . \end{aligned}} \tag{1.23} \tag{1.24}$$

In order to express also the exponents ν and η in terms of y_t and y_h, we need another scaling hypothesis involving the correlation function $G(r)$. We assume that sufficiently close to the critical point, $G(r)$ is dominated by the singular part $G_{\text{sing}}(|r|; t, h)$ which satisfies (Kadanoff 1966)

$$G_{\text{sing}}(|r|; t, h) = b^{-2(D-y_h)} G_{\text{sing}}\left(\frac{|r|}{b}; b^{y_t} t, b^{y_h} h\right) . \tag{1.25}$$

For simplicity, consider the case $h = 0$. Setting again $b = |t|^{-1/y_t}$, we obtain

$$G_{\text{sing}}(|r|; t, 0) = |t|^{\frac{2(D-y_h)}{y_t}} \Psi_{\pm}\left(\frac{|r|}{|t|^{-1/y_t}}\right) , \tag{1.26}$$

with $\Psi_{\pm}(x) = G_{\text{sing}}(x; \pm 1, 0)$. For $|t| \neq 0$ and $|r| \to \infty$, we expect $G_{\text{sing}}(|r|) \propto \exp[-|r|/\xi]$, so that Eq. (1.26) implies for the correlation length

$$\xi \propto |t|^{-1/y_t} = |t|^{-\nu} , \tag{1.27}$$

where we have used the definition (1.12) of the correlation length exponent ν. We conclude that

$$\nu = \frac{1}{y_t} . \tag{1.28}$$

Using Eq. (1.19a) to express y_t in terms of the specific heat exponent α, we obtain from Eq. (1.28) the so-called hyperscaling relation

$$2 - \alpha = D\nu . \tag{1.29}$$

Finally, we relate the critical exponent η to y_h and D by observing that for $|t| \to 0$ the function $G_{\text{sing}}(|r|; t, 0)$ can only be finite if $\Psi_{\pm}(x)$ is proportional to $|x|^{-2(D-y_h)}$ for large x. We therefore obtain at the critical point

$$G_{\text{sing}}(|r|; 0, 0) \propto |r|^{-2(D-y_h)} = |r|^{-(D-2+\eta)} , \tag{1.30}$$

where we have used the definition (1.13) of the correlation function exponent η. We therefore identify $2(D - y_h) = D - 2 + \eta$, or equivalently

$$\eta = D + 2 - 2y_h . \tag{1.31}$$

Expressing y_h in terms of the susceptibility exponent γ using Eq. (1.19c), we obtain another hyperscaling relation

$$\gamma = (2 - \eta)\nu . \tag{1.32}$$

Equations (1.29) and (1.32) are called hyperscaling relations because they connect singularities in thermodynamic observables with singularities related to the correlation function. It turns out, however, that the underlying scaling hypothesis (1.25) is only valid if the dimension D of the system is smaller than a certain upper critical dimension D_{up}, which depends on the universality class (for the Ising universality class $D_{up} = 4$). As will be discussed in Sect. 2.3.4, for $D > D_{up}$ the Gaussian approximation is sufficient to calculate the critical behavior of the system. The failure of hyper-scaling for $D > D_{up}$ is closely related so the existence of a so-called *dangerously irrelevant coupling* (Fisher 1983, Appendix D).[3] If hyper-scaling is satisfied, we may combine the two thermodynamic scaling relations (1.23) and (1.24) with the two hyperscaling relations (1.29) and (1.32) to express the four thermodynamic exponents α, β, γ, and δ in terms of the two correlation function exponents,

$$\alpha = 2 - D\nu , \tag{1.33a}$$

$$\beta = \frac{\nu}{2}(D - 2 + \eta) , \tag{1.33b}$$

$$\gamma = \nu(2 - \eta) , \tag{1.33c}$$

$$\delta = \frac{D + 2 - \eta}{D - 2 + \eta} . \tag{1.33d}$$

If one considers not only static but also dynamic (i.e., time-dependent) phenomena in the vicinity of a critical point, one observes that temporal correlations of the order parameter decay slower and slower as one approaches the critical point, a phenomenon which is called *critical slowing down*. The typical decay time of temporal order-parameter fluctuations is called correlation time τ_c. One usually observes that in the vicinity of a critical point, τ_c diverges as a power law

$$\boxed{\tau_c \propto \xi^z \propto |t|^{-\nu z} ,} \tag{1.34}$$

where z is called the dynamic exponent.

1.4 Scaling in the Vicinity of Quantum Critical Points

Although quantum mechanics can be essential to understand the existence of ordered phases in matter (e.g., superconductivity and magnetism are quantum effects), it turns out that quantum mechanics does not influence the asymptotic critical behavior of finite temperature phase transitions. The reason is that sufficiently close to the critical point, classical thermal fluctuations are always dominant. To understand this, note that according to Eq. (1.34) the typical energy scale E_c associated with temporal fluctuations vanishes in the vicinity of a continuous phase transition as a power law,

---

[3] To understand this rather subtle point, the reader should read the following two chapters and carefully do Exercise 3.3.3 at the end of Chap. 3.

$$E_c = \hbar/\tau_c \propto |t|^{\nu z} \propto |T - T_c|^{\nu z} \, . \tag{1.35}$$

As pointed out many years ago by Hertz (1976), in quantum systems static and dynamic fluctuations are not independent, because the Hamiltonian $\hat{H}$ determines not only the partition function, but also the time evolution of any observable $\hat{A}$ via the Heisenberg equation of motion,

$$i\hbar \frac{d\hat{A}}{dt} = [\hat{A}, \hat{H}] \, . \tag{1.36}$$

In quantum systems the energy E_c associated with the correlation time is therefore also the typical fluctuation energy for static fluctuations. Quantum mechanics should be negligible for $E_c \ll T$, because quantum effects are then washed out by thermal excitations. According to Eq. (1.35), this condition is always satisfied sufficiently close to T_c, so that a purely classical description of order-parameter fluctuations is sufficient to calculate the critical exponents.

On the other hand, there are many interesting systems which exhibit phase transitions at zero temperature when some nonthermal control parameter r is fine-tuned to a critical point r_c. These so-called *quantum phase transitions* can be associated with a nonanalyticity of the ground state properties of the system at $r = r_c$. The corresponding point in the relevant parameter space is called a *quantum critical point.* Some of the most interesting phenomena in condensed matter systems are related to quantum critical points, for example:

(a) *Anderson localization:* Electrons in disordered systems undergo a metal–insulator transition as a function of the disorder strength. The nonthermal control parameter r is in this case some measure for the disorder strength, such as the elastic mean free path in units of the inverse Fermi wave vector.

(b) *Quantum Hall effect:* The quantization of the Hall conductance in the two-dimensional electron gas in semiconductor heterostructures for certain values of an external magnetic field is an example of a class of quantum phase transitions where the magnetic field plays the role of the nonthermal control parameter r. Alternatively, one can fix the magnetic field and vary the density, so that r is the density.

(c) *Paramagnet–ferromagnet transition in uniaxial ferromagnets in a transverse magnetic field:* In some magnetic insulators with an easy axis (such as $LiHoF_4$), one can generate a quantum phase transition by applying a transverse magnetic field perpendicular to the easy axis. In this case the transverse magnetic field plays the role of the nonthermal control parameter r. The critical behavior of these systems can be described by using an Ising model in a transverse magnetic field.

(d) *Mott–Hubbard transition:* This metal-insulator transition occurs in strongly correlated electronic systems. The strength of the interaction plays here the role of the nonthermal control parameter r. Experimentally, the strength of the interaction can be controlled by external pressure or by chemical doping.

(e) *Fermi-liquid spin-density wave transition*: In some magnetic conductors (such as the heavy fermion material $CeCu_{6-x}Au_x$), it is possible to induce a quantum phase transition from a Fermi liquid state to a spin-density wave state with long-range magnetic order by varying the strength of the interaction.

For a detailed introduction to the fascinating field of quantum phase transitions, we refer the reader to the book and review articles by Sachdev (1999, 2000, 2006, 2009) and to the review by Vojta (2003). Here, we would only like to motivate a generalized scaling hypothesis for the singular part of the free energy density in the vicinity of a quantum critical point. Therefore, we note that in quantum systems the time acts in some sense as an extra dimension and thus increases the effective dimensionality of the system. This intuitive idea can be made mathematically precise by expressing the partition function as an imaginary time functional integral over a suitably defined field $\Phi(\tau, r)$ representing fluctuations of the order parameter,[4]

$$\mathcal{Z} = \int \mathcal{D}[\Phi] \exp\left\{ -\int_0^{1/T} d\tau \int d^D r \mathcal{L}[\Phi(\tau, r)] \right\} . \tag{1.37}$$

Here, $\mathcal{L}[\Phi(\tau, r)]$ is the Lagrangian density for the order-parameter field. Apart from the spatial point r, the field $\Phi(\tau, r)$ depends on the imaginary time τ which takes values in the interval $[0, 1/T]$. Obviously, the imaginary time direction acts like an extra dimension, which becomes infinite for $T \to 0$. But according to Eq. (1.34), in the vicinity of a quantum critical point characterized by the dynamic exponent z, time scales as (length)^z, so that the scaling properties of the quantum system are identical with those of an effective classical system in dimension $D + z$. Due to the extra dimensions, the homogeneity relations have to be modified for continuous quantum phase transitions. At $T = 0$, we obtain instead of the classical relation (1.16) for the scaling of the singular part of the free energy density,

$$\boxed{f_{\text{sing}}(g, h) = b^{-(D+z)} f_{\text{sing}}(b^{y_g} g, b^{y_h} h) ,} \tag{1.38}$$

where $g = |r - r_c|/r_c$ and $1/y_g = \nu$ can be identified with the correlation length exponent. At finite temperature, Eq. (1.38) should be generalized as follows,

$$\boxed{f_{\text{sing}}(g, h, T) = b^{-(D+z)} f_{\text{sing}}(b^{y_g} g, b^{y_h} h, b^z T) .} \tag{1.39}$$

The interplay between classical and quantum fluctuations in the vicinity of a quantum critical point leads to interesting finite temperature crossovers (Hertz 1976, Chakravarty et al. 1988, 1989, Millis 1993, Sachdev 1999, Continentino 2001). In particular, there is a *quantum critical region* where the physics is dominated by thermal excitations of the quantum critical ground state. In systems which do not exhibit

---

[4] The microscopic origin of Eq. (1.37) will become clear in Chaps. 11 and 12 where we use so-called Hubbard–Stratonovich transformations to describe collective fluctuations and spontaneous symmetry breaking in quantum systems.

any long-range order at finite temperature (such as the (one-dimensional) Ising chain
in a transverse field or two-dimensional systems where the symmetry group of
the order parameter is continuous), there are three different regimes as shown in
Fig. 1.3: a *quantum critical* fan emerging from a quantum critical point, which is
bounded by a *thermally disordered* regime and a *quantum disordered* regime. If the
system exhibits long-range order at finite temperature, the corresponding diagram
contains in addition a regime where classical critical behavior can be observed, as
shown in Fig. 1.4. The finite temperature crossovers of quantum phase transitions
have been studied with the help of conventional Wilsonian RG methods (Hertz 1976,
Chakravarty et al. 1988, 1989, Millis 1993, Continentino 2001). However, these

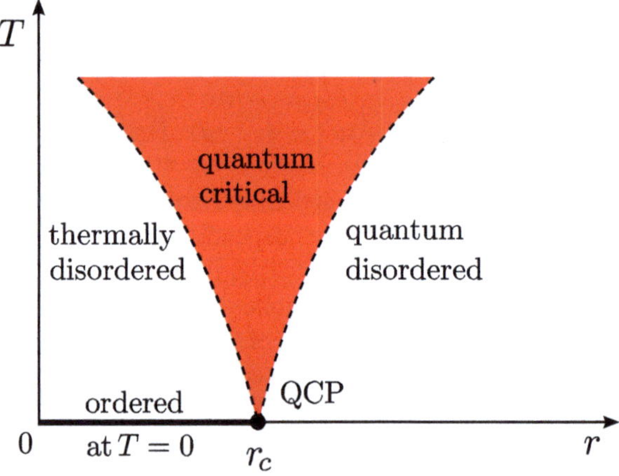

Fig. 1.3 Different regimes in the vicinity of a quantum critical point (QCP) where there is no
long-range order at $T \neq 0$. Here r is some nonthermal control parameter, such as an external
magnetic field or the density. The *dotted curves* indicate the crossover lines at $T = E_c \propto |r - r_c|^{\nu z}$

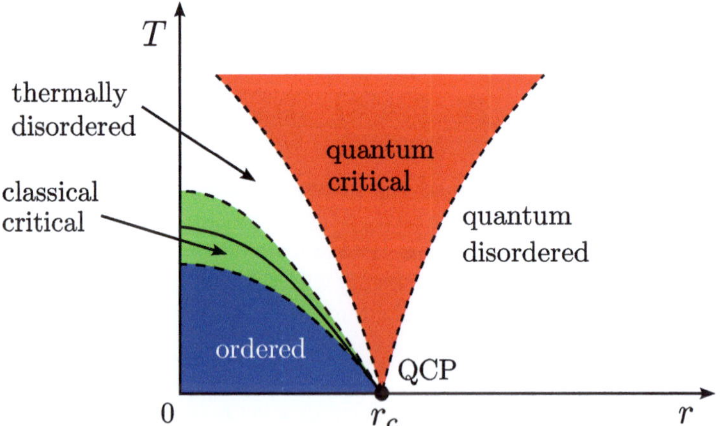

Fig. 1.4 Different regimes in the vicinity of a quantum critical point (QCP) for a system exhibiting
long-range order at finite temperature

works focused on the thermodynamic quantities; the calculation of momentum- and frequency-dependent correlation functions is more difficult and requires functional RG methods (Hasselmann et al. 2007, Sinner et al. 2008). Modern functional RG methods have only very recently been used to study quantum phase transitions (Wetterich 2007, Jakubczyk et al. 2008). One of the main goals of this book is to show that the functional RG is a convenient and powerful tool to calculate momentum- and frequency-dependent correlation functions near quantum critical points.

Exercises

1.1 Exact Solution and Scaling Properties of the One-Dimensional Ising Model

One of the most important model systems in statistical physics is the Ising model which in one dimension allows for a relatively simple exact solution. Let us therefore consider the one-dimensional Ising model of N spins $s_i = \pm 1$ (with $i = 1, \ldots, N$ and periodic boundary conditions) in a magnetic field h. Denoting the coupling constant for ferromagnetic nearest neighbor interactions by J, its Hamiltonian reads

$$H = -J \sum_{i=1}^{N} s_i s_{i+1} - h \sum_{i=1}^{N} s_i .$$

(a) Write the partition function $\mathcal{Z}$ as a product of terms of the form $f(s_i, s_{i+1}) \equiv \exp[(J/T)s_i s_{i+1} + (h/2T)(s_i + s_{i+1})]$ and show that the partition function can also be written as $\mathcal{Z} = \mathrm{Tr}\left[\mathbf{T}^N\right]$, where the *transfer matrix* $\mathbf{T}$ is defined by

$$\mathbf{T} = \begin{pmatrix} e^{(J+h)/T} & e^{-J/T} \\ e^{-J/T} & e^{(J-h)/T} \end{pmatrix} .$$

(b) Diagonalize $\mathbf{T}$ and show that $\mathcal{Z} = \lambda_+^N + \lambda_-^N$, where $\lambda_+ > \lambda_-$ are the two eigenvalues of $\mathbf{T}$. Evaluate λ_+ and λ_- and show that in the thermodynamic limit $N \to \infty$, the free energy per spin, $f \equiv F/N$, is given by

$$f = -T \ln \lambda_+ = -J - T \ln \left[\cosh(h/T) + \sqrt{\sinh^2(h/T) + x} \right] ,$$

where $x = e^{-4J/T}$ is a temperature-like quantity which vanishes for $T \to 0$. Argue that although there is no phase transition at any *finite* temperature, there is a critical point at zero temperature, that is, $x = 0$.

(c) By differentiating f with respect to h, calculate the average magnetization per spin, $m \equiv \frac{1}{N}\left\langle \sum_{i=1}^{N} s_i \right\rangle$. Sketch $m(h)$ for different values of T.

(d) For $h = 0$, calculate and sketch the magnetic susceptibility $\chi(T) = \left(\frac{\partial m}{\partial h}\right)_T$ and the specific heat $C(T) = -T\partial^2 F/\partial T^2$.

(e) Due to its critical point at zero temperature, the critical properties of the one-dimensional Ising model are a little special (Fisher 1983). To discuss scaling properties, it is advantageous to subtract the zero field, zero temperature ground state energy from f and introduce $\tilde{f} = (f + J)/T$. Suitable scaling variables can be chosen to be the temperature-like quantity $x = e^{-4J/T}$ introduced above and $\tilde{h} = h/T$. Determine a scaling function, its scaling eigenvalues y_x and $y_{\tilde{h}}$, and derive the thermodynamic exponents α_x, β_x, γ_x, and δ_x. The index x should remind you of the fact that all thermodynamic exponents are defined with $t = (T - T_c)/T_c$ replaced by x, e.g., $C(x) \propto \partial^2\tilde{f}/\partial x^2 \propto x^{-\alpha_x}$. Discuss your exponents and compare with your results in (d).

(f) Use hyperscaling relations to make a prediction for the critical exponents η and ν_x. Verify your predictions by generalizing the above transfer matrix method and evaluating the spin–spin correlation function $\langle s_i s_j \rangle \equiv \mathcal{Z}^{-1} \sum_{\{s_l = \pm 1\}} s_i s_j \exp(-H/T)$ for $h = 0$ which you should express as $\langle s_i s_j \rangle \propto \exp(-|i - j|/\xi)$. (Hint: For $i < j$ you can write the correlation function as $\langle s_i s_j \rangle = \mathcal{Z}^{-1} \operatorname{Tr}\left[\mathbf{T}^i \sigma^z \mathbf{T}^{j-i} \sigma^z \mathbf{T}^{N-j}\right]$, where σ^z is a usual Pauli matrix.)

1.2 Critical Exponents of the van der Waals Gas

To take into account interaction corrections to an ideal gas of atoms, van der Waals proposed the following equation of state (with $a, b > 0$),

$$(p + a\,(N/V)^2)(V - Nb) = NT \ .$$

Roughly speaking, the coefficient b represents effects due to a hard core repulsive interaction and decreases the effective volume of the system. At larger distances, the interaction between the atoms is attractive and leads to a reduction of the pressure. Because the interaction is always between pairs of molecules, this correction is expected to be proportional to the square of the density $n = N/V$.

(a) Using the van der Waals equation of state, it is possible to describe the discontinuous liquid–gas transition and the critical end point. Sketch $p(V)$ for representative values of T. Rewrite the equation of state as a cubic polynomial in V and argue that at the critical point (p_c, V_c, T_c), the van der Waals equation of state reduces to $(V - V_c)^3 = 0$. Compare coefficients and express a and b through p_c and V_c.

(b) Calculate the free energy $F(T, V)$ by integrating $-p(T, V)$ along an isotherm with respect to V. You may adjust the temperature-dependent additive constant by comparing your result in the limit $V \to \infty$ with that of an ideal gas. Using the relation $C_V = -T\left(\frac{\partial^2 F}{\partial T^2}\right)_V$, calculate the specific heat at constant volume. What do you obtain for the critical exponent α defined by $C_V \propto |t|^{-\alpha}$?

(c) The exponent γ is defined by $\kappa_T \equiv -\frac{1}{V}\left(\frac{\partial V}{\partial p}\right)_T \propto |t|^{-\gamma}$ (with $V = V_c$). The exponent δ is defined at $T = T_c$ by $p - p_c \propto (n - n_c)^\delta \propto (V - V_c)^\delta$. Calculate the critical exponents γ and δ.

(d) As you might have noticed, the van der Waals equation of state predicts regions in the phase diagram where the compressibility $\kappa_T \equiv -\frac{1}{V}\left(\frac{\partial V}{\partial p}\right)_T$ does not satisfy $\kappa_T \geq 0$. This instability is of course unphysical and can be traced back to the fact that the van der Waals equation does not allow for phase separation. A simple remedy is the Maxwell construction: Draw a line parallel to the V-axis which cuts the graph $p(V)$ in such a way that the two areas enclosed by this line and $p(V)$ are equal. In the inner part of the graph (ranging from V_{liquid} to V_{gas}), the pressure $p(V)$ is now replaced by the horizontal line. Justify the Maxwell construction by considering an isotherm of $F(T, V) = -\int p\,dV + \text{const}(T)$. By allowing for phase separation, you can now reduce the free energy $F(T, V)$ for $V_{\text{liquid}} < V < V_{\text{gas}}$, turning the free energy convex. Apply the Maxwell construction to your above sketch and mark the coexistence curve in the p–V diagram (where F is nonanalytic).

(e) To obtain the critical exponent β, write $V = V_c(1 + v)$ and $T = T_c(1 + t)$ and expand $p(T, V)$ for small t and v. You need to keep terms up to order $\mathcal{O}(t, tv, v^3)$. You should justify this later on. Apply the Maxwell construction to obtain the coexistence curve. Go ahead and calculate the coefficient β defined by $V_{\text{gas}} - V_{\text{liquid}} \propto |t|^\beta$ with $t \leq 0$. (You may consult Goldenfeld (1992) if you need more help.)

References

Binney, J. J., N. J. Dowrick, A. J. Fisher, and M. E. J. Newman (1992), *The Theory of Critical Phenomena: An Introduction to the Renormalization Group*, Oxford University Press, Oxford.

Cardy, J. L. (1996), *Scaling and Renormalization in Statistical Physics*, Cambridge University Press, Cambridge.

Chakravarty, S., B. I. Halperin, and D. R. Nelson (1988), *Low-temperature behavior of two-dimensional quantum antiferromagnets*, Phys. Rev. Lett. **60**, 1057.

Chakravarty, S., B. I. Halperin, and D. R. Nelson (1989), *Two-dimensional quantum Heisenberg antiferromagnet at low temperatures*, Phys. Rev. B **39**, 2344.

Continentino, M. A. (2001), *Quantum Scaling in Many-Body Systems*, World Scientific, Singapore.

Fisher, M. E. (1983), *Scaling, Universality and Renormalization Group Theory*, in F. J. W. Hahne, editor, *Lecture Notes in Physics*, volume 186, Springer, Berlin.

Goldenfeld, N. (1992), *Lectures on Phase Transitions and the Renormalization Group*, Addison-Wesley, Reading.

Hasselmann, N., A. Sinner, and P. Kopietz (2007), *Two-parameter scaling of correlation functions near continuous phase transitions*, Phys. Rev. E **76**, 040101.

Hertz, J. A. (1976), *Quantum critical phenomena*, Phys. Rev. B **14**, 1165.

Holm, C. and W. Janke (1993), *Critical exponents of the classical three-dimensional Heisenberg model: A single-cluster Monte Carlo study*, Phys. Rev. B **48**, 936.

Ivanchenko, Y. M. and A. A. Lisyansky (1995), *Physics of Critical Fluctuations*, Springer, New York.

Jakubczyk, P., P. Strack, A. A. Katanin, and W. Metzner (2008), *Renormalization group for phases with broken discrete symmetry near quantum critical points*, Phys. Rev. B **77**, 195120.

Kadanoff, L. P. (1966), *Scaling laws for Ising models near T_c*, Physics **2**, 263.

Kardar, M. (2007), *Statistical Physics of Fields*, Cambridge University Press, Cambridge.

McComb, W. (2004), *Renormalization Methods: A Guide for Beginners*, Oxford University Press, Oxford.

Millis, A. J. (1993), *Effect of a nonzero temperature on quantum critical points in itinerant fermion systems*, Phys. Rev. B **48**, 7183.

Pelissetto, A. and E. Vicari (2002), *Critical phenomena and renormalization group theory*, Phys. Rep. **368**, 549.

Sachdev, S. (1999), *Quantum Phase Transitions*, Cambridge University Press, Cambridge.

Sachdev, S. (2000), *Quantum criticality: Competing ground states in low dimensions*, Science **288**, 475.

Sachdev, S. (2006), *Quantum Phase Transitions*, in G. Fraser, editor, *The New Physics for the Twenty-First Century*, Cambridge University Press.

Sachdev, S. (2009), *Exotic phases and quantum phase transitions: model systems and experiments*, arXiv:0901.4103v6 [cond-mat.str-el].

Sinner, A., N. Hasselmann, and P. Kopietz (2008), *Functional renormalization group in the broken symmetry phase: momentum dependence and two-parameter scaling of the self-energy*, J. Phys.: Condens. Matter **20**, 075208.

Vojta, M. (2003), *Quantum phase transitions*, Rep. Prog. Phys. **66**, 2069.

Wetterich, C. (2007), *Bosonic effective action for interacting fermions*, Phys. Rev. B **75**, 085102.

Widom, B. (1965), *Surface tension and molecular correlations near the critical point*, J. Chem. Phys. **43**, 3892.

Chapter 2
Mean-Field Theory and the Gaussian Approximation

The Wilsonian renormalization group (RG) was invented in order to study the effect of strong fluctuations and the mutual coupling between different degrees of freedom in the vicinity of continuous phase transitions. Before embarking on the theory of the RG, let us in this chapter describe two less sophisticated methods of dealing with this problem, namely the mean-field approximation and the Gaussian approximation. Within the mean-field approximation, fluctuations of the order parameter are completely neglected and the interactions between different degrees of freedom are taken into account in some simple average way. The Gaussian approximation is in some sense the leading fluctuation correction to the mean-field approximation. Although these methods are very general and can also be used to study quantum mechanical many-body systems[1], for our purpose it is sufficient to introduce these methods using the nearest-neighbor Ising model in D dimensions as an example. The Ising model is defined in terms of the following classical Hamiltonian,

$$H = -\frac{1}{2} \sum_{ij=1}^{N} J_{ij} s_i s_j - h \sum_{i=1}^{N} s_i \, . \tag{2.1}$$

Here, i and j label the sites $\boldsymbol{r}_i$ and $\boldsymbol{r}_j$ of a D-dimensional hypercubic lattice with N sites, the variables $s_i = \pm 1$ correspond to the two possible states of the z-components of spins localized at the lattice sites, the J_{ij} denote the exchange interaction between spins localized at sites $\boldsymbol{r}_i$ and $\boldsymbol{r}_j$, and h is the Zeeman energy associated with an external magnetic field in the z-direction. The above model is called classical because it does not involve any noncommuting operators. By simply rotating the magnetic field in the x-direction we obtain a quantum Ising model, as discussed in Exercise 2.1. The quantum mechanical origin of magnetism is hidden in the exchange energies J_{ij}. Due to the exponential decay of localized wave functions, it is often sufficient to assume that the J_{ij} are nonzero only if the sites $\boldsymbol{r}_i$ and $\boldsymbol{r}_j$

---

[1] For example, in quantum many-body systems the *self-consistent Hartree–Fock approximation* can be viewed as a variant of the mean-field approximation, while the Gaussian approximation is usually called *random phase approximation*.

Kopietz, P. et al.: *Mean-Field Theory and the Gaussian Approximation*. Lect. Notes Phys. **798**, 23–52 (2010)
DOI 10.1007/978-3-642-05094-7_2

are nearest neighbors on the lattice. Denoting the nonzero value of J_{ij} for nearest neighbors by J, we obtain from Eq. (2.1) the nearest-neighbor Ising model

$$H = -J \sum_{\langle ij \rangle} s_i s_j - h \sum_{i=1}^{N} s_i \; . \tag{2.2}$$

Here, $\langle ij \rangle$ denotes the summation over all distinct pairs of nearest neighbors. In order to obtain thermodynamic observables, we have to calculate the partition function,

$$\mathcal{Z}(T, h) = \sum_{\{s_i\}} e^{-\beta H} \equiv \sum_{s_1 = \pm 1} \sum_{s_2 = \pm 1} \cdots \sum_{s_N = \pm 1} \exp \left[\beta J \sum_{\langle ij \rangle} s_i s_j + \beta h \sum_i s_i \right] ,$$

$$\tag{2.3}$$

where we have introduced the notation $\beta = 1/T$ for the inverse temperature.[2] While in one dimension it is quite simple to carry out this summation (see Exercise 1.1), the corresponding calculation in $D = 2$ is much more difficult and until now the exact $\mathcal{Z}(T, h)$ for $h \neq 0$ is not known. For $h = 0$ the partition function of the two-dimensional Ising model was first calculated by Onsager (1944), who also presented an exact expression for the spontaneous magnetization at a conference in 1949. A proof for his result was given in 1952 by C. N. Yang. See the textbooks by Wannier (1966), by Huang (1987), or by Mattis (2006) for pedagogical descriptions of the exact solution for $h = 0$. In dimensions $D = 3$ there are no exact results available, so one has to rely on approximations. The simplest is the mean-field approximation discussed in the following section.

2.1 Mean-Field Theory

2.1.1 Landau Function and Free Energy

Mean-field theory is based on the assumption that the fluctuations around the average value of the order parameter are so small that they can be neglected. Let us therefore assume that the system has a finite magnetization,[3]

$$m = \langle s_i \rangle \equiv \frac{\sum_{\{s_j\}} e^{-H/T} s_i}{\sum_{\{s_j\}} e^{-H/T}} \; , \tag{2.4}$$

---

[2] The inverse temperature $\beta = 1/T$ should not be confused with the order-parameter exponent β. Because these notations are both standard we shall adopt them here; usually the meaning of the symbol β is clear from the context.

[3] For lattice models it is convenient to divide the total magnetic moment by the number N of lattice sites and not by the total volume. For simplicity we use here the same symbol as for the magnetization in Eq. (1.3), which has units of inverse volume.

where we have used the fact that by translational invariance the thermal expectation values $\langle s_i \rangle$ are independent of the site label i. Writing $s_i = m + \delta s_i$ with the fluctuation $\delta s_i = s_i - m$, we have

$$s_i s_j = m^2 + m(\delta s_i + \delta s_j) + \delta s_i \delta s_j = -m^2 + m(s_i + s_j) + \delta s_i \delta s_j \; . \tag{2.5}$$

We now assume that the fluctuations are small so that the last term $\delta s_i \delta s_j$ which is quadratic in the fluctuations can be neglected. Within this approximation, the Ising Hamiltonian (2.1) is replaced by the mean-field Hamiltonian

$$\begin{aligned}
H_{\mathrm{MF}} &= \frac{m^2}{2} \sum_{ij} J_{ij} - \sum_i \left(h + \sum_j J_{ij} m \right) s_i \\
&= N \frac{zJ}{2} m^2 - \sum_i (h + zJm) s_i \; ,
\end{aligned} \tag{2.6}$$

where the second line is valid for the nearest-neighbor interactions, and $z = 2D$ is the number of nearest neighbors (coordination number) of a given site of a D-dimensional hypercubic lattice. As the spins in the mean-field Hamiltonian (2.6) are decoupled, the partition function factorizes into a product of N independent terms which are just the partition functions of single spins in an effective magnetic field $h_{\mathrm{eff}} = h + zJm$,

$$\begin{aligned}
\mathcal{Z}_{\mathrm{MF}}(T, h) &= e^{-\beta N zJm^2/2} \sum_{\{s_i\}} e^{\beta(h+zJm)\sum_i s_i} \\
&= e^{-\beta N zJm^2/2} \prod_i \left[e^{\beta(h+zJm)} + e^{-\beta(h+zJm)} \right] \\
&= e^{-\beta N zJm^2/2} \Big[2\cosh[\beta(h + zJm)] \Big]^N \; .
\end{aligned} \tag{2.7}$$

Writing this as

$$\mathcal{Z}_{\mathrm{MF}}(T, h) = e^{-\beta N \mathcal{L}_{\mathrm{MF}}(T,h;m)} \; , \tag{2.8}$$

we obtain in mean-field approximation

$$\boxed{\; \mathcal{L}_{\mathrm{MF}}(T, h; m) = \frac{zJ}{2} m^2 - T \ln\Big[2\cosh[\beta(h + zJm)] \Big] \; . \;} \tag{2.9}$$

As will be explained in more detail below, the function $\mathcal{L}_{\mathrm{MF}}(T, h; m)$ is an example for a Landau function, which describes the probability distribution of the order parameter: the probability density of observing for the order parameter the value m is proportional to $\exp[-\beta N \mathcal{L}_{\mathrm{MF}}(T, h; m)]$. The so far unspecified parameter m is now determined from the condition that the physical value m_0 of the order parameter maximizes its probability distribution, corresponding to the minimum of the Landau function,

$$\left. \frac{\partial \mathcal{L}_{\mathrm{MF}}(T, h; m)}{\partial m} \right|_{m_0} = 0 \, . \tag{2.10}$$

From Eq. (2.9) we then find that the magnetization m_0 in mean-field approximation satisfies the self-consistency condition

$$\boxed{m_0 = \tanh[\beta(h + zJm_0)] \, ,} \tag{2.11}$$

which defines $m_0 = m_0(T, h)$ as a function of T and h. The mean-field result for the free energy per site is thus

$$f_{\mathrm{MF}}(T, h) = \mathcal{L}_{\mathrm{MF}}(T, h; m_0(T, h)) \, . \tag{2.12}$$

The mean-field self-consistency equation (2.11) can easily be solved graphically. As shown in Fig. 2.1, for $h \neq 0$ the global minimum of $\mathcal{L}_{\mathrm{MF}}(T, h; m)$ occurs always at a finite $m_0 \neq 0$. On the other hand, for $h = 0$ the existence of nontrivial solutions with $m_0 \neq 0$ depends on the temperature. In the low-temperature regime $T < zJ$ there are two nontrivial solutions with $m_0 \neq 0$, while at high temperatures $T > zJ$ our self-consistency equation (2.11) has only the trivial solution $m_0 = 0$, see Fig. 2.2. In D dimensions the mean-field estimate for the critical temperature is therefore

$$T_c = zJ = 2DJ \, . \tag{2.13}$$

For $D = 1$ this is certainly wrong, because we know from the exact solution (see Exercise 1.1) that $T_c = 0$ in one dimension. In two dimensions the exact critical temperature of the nearest-neighbor Ising model satisfies $\sinh(2J/T_c) = 1$ (Onsager 1944), which yields $T_c \approx 2.269J$ and is significantly lower than the mean-field prediction of $4J$. As a general rule, in lower dimensions fluctuations are more important and tend to disorder the system or at least reduce the critical temperature.

2.1.2 Thermodynamic Critical Exponents

For temperatures close to T_c and small $\beta|h|$ the value m_0 of the magnetization at the minimum of $\mathcal{L}_{\mathrm{MF}}(T, h; m)$ is small. We may therefore approximate the Landau function (2.9) by expanding the right-hand side of Eq. (2.9) up to fourth order in m and linear order in h. Using

$$\ln[2\cosh x] = \ln 2 + \frac{x^2}{2} - \frac{x^4}{12} + \mathcal{O}(x^6) \, , \tag{2.14}$$

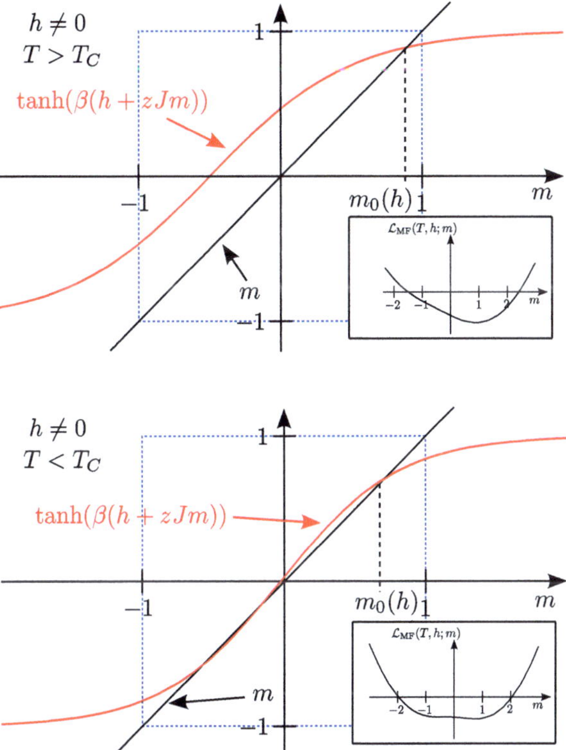

Fig. 2.1 Graphical solution of the mean-field self-consistency equation (2.11) for $h > 0$. The inset shows the behavior of the corresponding Landau function $\mathcal{L}_{\mathrm{MF}}(T, h; m)$ defined in Eq. (2.9). For $T > T_c$ or h sufficiently large the Landau function exhibits only one minimum at finite $m_0 > 0$. For $T < T_c$ and h sufficiently small, however, there is another local minimum at negative m, but the global minimum of $\mathcal{L}_{\mathrm{MF}}(T, h; m)$ is still at $m_0 > 0$

we obtain from Eq. (2.9),

$$\mathcal{L}_{\mathrm{MF}}(T, h; m) = f + \frac{r}{2}m^2 + \frac{u}{4!}m^4 - hm + \dots , \qquad (2.15)$$

with

$$f = -T \ln 2 , \qquad (2.16a)$$

$$r = \frac{zJ}{T}\left(T - zJ\right) \approx T - T_c , \qquad (2.16b)$$

$$u = 2T\left(\frac{zJ}{T}\right)^4 \approx 2T_c , \qquad (2.16c)$$

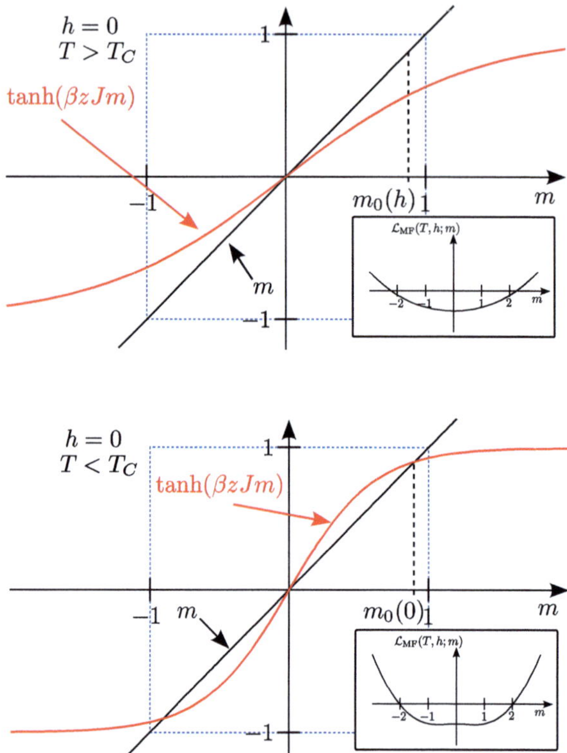

Fig. 2.2 Graphical solution of the mean-field self-consistency equation (2.11) for $h = 0$. The upper figure shows the typical behavior in the disordered phase $T > T_c$, while the lower figure represents the ordered phase $T < T_c$. The behavior of the Landau function is shown in the insets: while for $T > T_c$ it has a global minimum at $m_0 = 0$, it develops for $T < T_c$ two degenerate minima at $\pm|m_0| \neq 0$

where the approximations are valid close to the critical temperature, where $|T - T_c| \ll T_c$ and $zJ/T \approx 1$. Obviously, the sign of the coefficient r changes at $T = T_c$, so that for $h = 0$ the global minimum of $\mathcal{L}_{\mathrm{MF}}(T, h; m)$ for $T > T_c$ evolves into a local maximum for $T < T_c$, and two new minima emerge at finite values of m, as shown in Fig. 2.2. The crucial point is now that for a small reduced temperature $t = (T - T_c)/T_c$ the value of m at the minima of $\mathcal{L}_{\mathrm{MF}}(T, h; m)$ is small compared with unity, so that our expansion (2.15) in powers of m is justified a posteriori. Taking the derivative of Eq. (2.15) with respect to m, it is easy to see that Eq. (2.11) simplifies to

$$\left. \frac{\partial \mathcal{L}_{\mathrm{MF}}(T, h; m)}{\partial m} \right|_{m_0} = rm_0 + \frac{u}{6}m_0^3 - h = 0 \,. \tag{2.17}$$

The behavior of thermodynamic observables close to T_c is now easily obtained:

(a) *Spontaneous magnetization:* Setting $h = 0$ in Eq. (2.17) and solving for m_0, we obtain for $T \leq T_c$ (with $r \leq 0$),

$$m_0 = \sqrt{\frac{-6r}{u}} \propto (-t)^{1/2} . \tag{2.18}$$

Comparing this with the definition (1.6) of the critical exponent β, we conclude that the mean-field approximation predicts for the Ising universality class $\beta = 1/2$, independently of the dimension D.

(b) *Zero-field susceptibility:* To obtain the mean-field result for the susceptibility exponent γ, we note that for small but finite h and $T \geq T_c$ we may neglect the terms of order m_0^3 in Eq. (2.17), so that $m_0(h) \propto h/r$, and hence the zero-field susceptibility behaves for $t \to 0$ as

$$\chi = \left. \frac{\partial m_0(h)}{\partial h} \right|_{h=0} \propto \frac{1}{r} \propto \frac{1}{T - T_c} . \tag{2.19}$$

It is a simple exercise to show that $\chi \propto |T - T_c|^{-1}$ also holds for $T < T_c$. The susceptibility exponent is therefore $\gamma = 1$ within mean-field approximation.

(c) *Critical isotherm:* The equation of state at the critical point can be obtained by setting $r = 0$ in Eq. (2.17), implying

$$m_0(h) \propto \left(\frac{h}{u} \right)^{1/3} , \tag{2.20}$$

and hence the mean-field result $\delta = 3$.

(d) *Specific heat:*   The specific heat C per lattice site can be obtained from the thermodynamic relation

$$C = -T \frac{\partial^2 f_{\mathrm{MF}}(T, h)}{\partial T^2} , \tag{2.21}$$

where the mean-field free energy per site $f_{\mathrm{MF}}(T, h)$ is given in Eq. (2.12). Setting $h = 0$, we find from Eq. (2.17) for $T > T_c$ that $f_{\mathrm{MF}}(t, 0) = f$ because $m_0 = 0$ in this case. On the other hand, for $T < T_c$ we may substitute Eq. (2.18), so that

$$f_{\mathrm{MF}}(t, 0) = f - \frac{3}{2} \frac{r^2}{u} , \quad T < T_c . \tag{2.22}$$

Setting $r \approx T - T_c$ and taking two derivatives with respect to T, we obtain

$$C \approx -T_c \frac{\partial^2 f}{\partial T^2} , \qquad\qquad T > T_c , \tag{2.23a}$$

$$\approx -T_c \frac{\partial^2 f}{\partial T^2} + 3\frac{T_c}{u} , \quad T < T_c . \tag{2.23b}$$

Note that according to Eq. (2.16c) $u \approx 2T_c$ so that $3T_c/u \approx 3/2$. We conclude that within the mean-field approximation the specific heat is discontinuous at T_c, so that $C \propto |t|^0$, implying $\alpha = 0$. Note that the mean-field results are consistent with the scaling relations (1.23) and (1.24), $2 - \alpha = 2\beta + \gamma = \beta(\delta + 1)$.

In order to obtain the exponents ν and η, we need to calculate the correlation function $G(r)$, which we shall do in Sect. 2.3 within the so-called Gaussian approximation. However, a comparison of the mean-field results $\alpha = 0$, $\beta = 1/2$, $\gamma = 1$, and $\delta = 3$ with the correct values given in Table 1.1 at the end of Sect. 1.2 shows that the mean-field approximation is not suitable to obtain quantitatively accurate results, in particular in the physically accessible dimensions $D = 2$ and $D = 3$.

2.2 Ginzburg–Landau Theory

Our ultimate goal will be to develop a systematic method for taking into account the fluctuations neglected in mean-field theory, even if the fluctuations are strong and qualitatively change the mean-field results. As a first step to achieve this ambitious goal, we shall in this chapter derive from the microscopic Ising Hamiltonian (2.1) an effective (classical) field theory whose fields $\varphi(r)$ represent suitably defined spatial averages of the fluctuating magnetization over sufficiently large domains such that $\varphi(r)$ varies only slowly on the scale of the lattice spacing. The concept of an effective field theory representing *coarse-grained* fluctuations averaged over larger and larger length scales lies at the heart of the Wilsonian RG idea. It turns out that in the vicinity of the critical point the form of the effective coarse-grained action $S[\varphi]$ is constrained by symmetry and can be written down on the basis of phenomenological considerations. Such a strategy has been adopted by Ginzburg and Landau (1950) to develop a phenomenological theory of superconductivity which is particularly useful to treat spatial inhomogeneities.

2.2.1 Exact Effective Field Theory

To derive the effective order-parameter field theory for the Ising model, let us write the partition function of the Hamiltonian (2.1) in compact matrix form,

$$\mathcal{Z} = \sum_{\{s_i\}} \exp\left[\frac{\beta}{2} \sum_{ij} J_{ij} s_i s_j + \beta h \sum_i s_i \right] = \sum_{\{s_i\}} \exp\left[\frac{1}{2} s^T \tilde{\mathbf{J}} s + \tilde{\mathbf{h}}^T s \right] ,$$

$$\tag{2.24}$$

where s and $\tilde{\mathbf{h}}$ are N-dimensional column vectors with $[s]_i = s_i$ and $[\tilde{\mathbf{h}}]_i = \beta h$, and $\tilde{\mathbf{J}}$ is an $N \times N$-matrix with matrix elements $[\tilde{\mathbf{J}}]_{ij} = \tilde{J}_{ij} = \beta J_{ij}$. We now use the following mathematical identity for N-dimensional Gaussian integrals, valid for any positive symmetric matrix $\mathbf{A}$,

$$\left(\prod_{i=1}^{N} \int_{-\infty}^{\infty} \frac{dx_i}{\sqrt{2\pi}} \right) e^{-\frac{1}{2}x^T \mathbf{A} x + x^T s} = [\det \mathbf{A}]^{-1/2} e^{\frac{1}{2}s^T \mathbf{A}^{-1} s} , \qquad (2.25)$$

to write the Ising interaction in the following form (we identify $\tilde{\mathbf{J}} = \mathbf{A}^{-1}$)

$$e^{\frac{1}{2}s^T \tilde{\mathbf{J}} s} = \frac{\int \mathcal{D}[x] \exp\left[-\frac{1}{2}x^T \tilde{\mathbf{J}}^{-1} x + x^T s\right]}{\int \mathcal{D}[x] \exp\left[-\frac{1}{2}x^T \tilde{\mathbf{J}}^{-1} x\right]} , \qquad (2.26)$$

where we have introduced the notation

$$\int \mathcal{D}[x] \equiv \prod_{i=1}^{N} \int_{-\infty}^{\infty} \frac{dx_i}{\sqrt{2\pi}} . \qquad (2.27)$$

In fact, the identity (2.26) can be easily proven without using the formula (2.25) by redefining the integration variables in the numerator via the shift $x = x' + \tilde{\mathbf{J}}s$ and completing the squares,

$$-\frac{1}{2}(x' + \tilde{\mathbf{J}}s)^T \tilde{\mathbf{J}}^{-1}(x' + \tilde{\mathbf{J}}s) + (x' + \tilde{\mathbf{J}}s)^T s = -\frac{1}{2}x'^T \tilde{\mathbf{J}}^{-1} x' + \frac{1}{2}s^T \tilde{\mathbf{J}} s . \quad (2.28)$$

In a sense, Eq. (2.26) amounts to reading the Gaussian integration formula from right to left, introducing an auxiliary integration to write the right-hand side in terms of a Gaussian integral. Analogous transformations turn out to be very useful to introduce suitable collective degrees of freedom in quantum mechanical many-body systems; in this context transformations of the type (2.26) are called Hubbard–Stratonovich transformations (Hubbard 1959, Stratonovich 1957, Kopietz 1997). The Ising partition function (2.24) can now be written as

$$\mathcal{Z} = \frac{\int \mathcal{D}[x] \exp\left[-\frac{1}{2}x^T \tilde{\mathbf{J}}^{-1} x\right] \sum_{\{s_i\}} \exp\left[(\tilde{\mathbf{h}} + x)^T s\right]}{\int \mathcal{D}[x] \exp\left[-\frac{1}{2}x^T \tilde{\mathbf{J}}^{-1} x\right]} . \qquad (2.29)$$

For a given configuration of the Hubbard–Stratonovich field x, the spin summation in the numerator factorizes again in a product of independent terms, each describing the partition function of a single spin in a site-dependent magnetic field $h + x_i/\beta$. Therefore, this spin summation can easily be carried out,

$$\sum_{\{s_i\}} \exp\left[(\tilde{\mathbf{h}} + x)^T s\right] = \prod_{i=1}^{N}\left[\sum_{s_i=\pm 1} e^{(\beta h + x_i)s_i}\right] = \prod_{i=1}^{N}\left[2\cosh(\beta h + x_i)\right]$$

$$= \exp\left[\sum_{i=1}^{N} \ln\left[2\cosh(\beta h + x_i)\right]\right] . \tag{2.30}$$

The partition function (2.29) can thus be written as

$$\mathcal{Z} = \frac{\int \mathcal{D}[x]e^{-\tilde{S}[x]}}{\int \mathcal{D}[x]\exp\left[-\frac{1}{2}x^T\tilde{\mathbf{J}}^{-1}x\right]} = \frac{1}{\sqrt{\det \tilde{\mathbf{J}}}}\int \mathcal{D}[x]e^{-\tilde{S}[x]} , \tag{2.31}$$

where

$$\tilde{S}[x] = \frac{1}{2}x^T\tilde{\mathbf{J}}^{-1}x - \sum_{i=1}^{N} \ln\left[2\cosh(\beta h + x_i)\right] . \tag{2.32}$$

To understand the physical meaning of the variables x, we calculate the expectation value of its i-th component

$$\langle x_i \rangle_{\tilde{S}} \equiv \frac{\int \mathcal{D}[x]e^{-\tilde{S}[x]}x_i}{\int \mathcal{D}[x]e^{-\tilde{S}[x]}} . \tag{2.33}$$

Introducing an auxiliary N-component column vector $y = (y_1, \ldots, y_N)^T$ to write

$$x_i = \lim_{y\to 0} \frac{\partial}{\partial y_i} e^{x^T y} , \tag{2.34}$$

and performing the Gaussian integration using Eq. (2.26), we have the following chain of identities,

$$\langle x_i \rangle_{\tilde{S}} = \lim_{y\to 0} \frac{\partial}{\partial y_i} \frac{\int \mathcal{D}[x]\exp\left[-\frac{1}{2}x^T\tilde{\mathbf{J}}^{-1}x\right]\sum_{\{s_i\}}\exp\left[(\tilde{\mathbf{h}} + x)^T s + x^T y\right]}{\int \mathcal{D}[x]e^{-\tilde{S}[x]}}$$

$$= \lim_{y\to 0} \frac{\partial}{\partial y_i} \frac{\sum_{\{s_i\}}\exp\left[\frac{1}{2}(s + y)^T\tilde{\mathbf{J}}(s + y) + \tilde{\mathbf{h}}^T s\right]}{\sum_{\{s_i\}}\exp\left[\frac{1}{2}s^T\tilde{\mathbf{J}}s + \tilde{\mathbf{h}}^T s\right]}$$

$$= \frac{\sum_{\{s_i\}} e^{-\beta H}[\tilde{\mathbf{J}}s]_i}{\sum_{\{s_i\}} e^{-\beta H}} = \langle [\tilde{\mathbf{J}}s]_i \rangle , \tag{2.35}$$

or in vector notation,

$$\langle x \rangle_{\tilde{S}} = \tilde{\mathbf{J}}\langle s \rangle . \tag{2.36}$$

In order to introduce variables φ_i whose expectation value can be identified with the average magnetization $m = \langle s_i \rangle$, we simply define

$$\boldsymbol{\varphi} = \tilde{\mathbf{J}}^{-1}\boldsymbol{x} \, , \qquad (2.37)$$

so that the expectation value of the i-th component of $\boldsymbol{\varphi}$ is

$$\langle \varphi_i \rangle_{\tilde{S}} = [\tilde{\mathbf{J}}^{-1}\langle \boldsymbol{x} \rangle_{\tilde{S}}]_i = \langle s_i \rangle = m \, . \qquad (2.38)$$

Substituting $\boldsymbol{\varphi} = \tilde{\mathbf{J}}^{-1}\boldsymbol{x}$ in Eqs. (2.31) and (2.32), we finally obtain the following exact representation of the partition function,

$$\boxed{\; \mathcal{Z} = \frac{\int \mathcal{D}[\varphi] e^{-S[\varphi]}}{\int \mathcal{D}[\varphi] \exp\left[-\frac{1}{2}\boldsymbol{\varphi}^T \tilde{\mathbf{J}} \boldsymbol{\varphi}\right]} = \sqrt{\det \tilde{\mathbf{J}}} \int \mathcal{D}[\varphi] e^{-S[\varphi]} \, , \;} \qquad (2.39)$$

where the effective action $S[\boldsymbol{\varphi}] \equiv \tilde{S}[\boldsymbol{x} \to \tilde{\mathbf{J}}\boldsymbol{\varphi}]$ is given by

$$\boxed{\; S[\boldsymbol{\varphi}] = \frac{\beta}{2} \sum_{ij} J_{ij}\varphi_i\varphi_j - \sum_{i=1}^{N} \ln\left[2\cosh\left[\beta\left(h + \sum_{j=1}^{N} J_{ij}\varphi_j\right)\right]\right] . \;} \qquad (2.40)$$

Equation (2.39) expresses the partition function of the Ising model in terms of an N-dimensional integral over variables φ_i whose expectation value is simply the magnetization per site. The integration variables φ_i can therefore be interpreted as the fluctuating magnetization. The infinite-dimensional integral obtained from Eq. (2.39) in the limit $N \to \infty$ is an example of a *functional integral*. The resulting effective action $S[\boldsymbol{\varphi}]$ then defines a (classical) effective field theory for the order-parameter fluctuations of the Ising model. We shall refer to the integration variables φ_i as the *fields* of our effective field theory.

It should be noted that Eq. (2.38) does not precisely generalize to higher order correlation functions. If we proceed as above and use the identity

$$x_{i_1} x_{i_2} \ldots x_{i_n} = \lim_{y \to 0} \frac{\partial}{\partial y_{i_1}} \frac{\partial}{\partial y_{i_2}} \cdots \frac{\partial}{\partial y_{i_n}} e^{x^T y} \qquad (2.41)$$

to carry out the same transformations as in Eq. (2.35), we arrive at

$$\langle \varphi_{i_1} \varphi_{i_2} \ldots \varphi_{i_n} \rangle_S = \langle s_{i_1} s_{i_2} \ldots s_{i_n} \rangle + \big\{ \text{averages involving at most}$$
$$n - 2 \text{ of the spins } s_{i_1}, \ldots, s_{i_n} \big\} \, , \qquad (2.42)$$

where the lower order correlation functions involving $n - 2$ and less spin variables arise from the repeated differentiation of the y-dependent terms in the exponent of the second line of Eq. (2.35). Fortunately, for short-range J_{ij} these additional terms can be neglected as long as the distances between the external points $i_1, \ldots, i_n$ are large compared with the range of J_{ij}. For example, for $n = 2$ we have

$$\langle \varphi_i \varphi_j \rangle_S = \langle s_i s_j \rangle - [\tilde{\mathbf{J}}^{-1}]_{ij} \sim \langle s_i s_j \rangle \,. \tag{2.43}$$

We thus conclude that we may also use the effective action $S[\varphi]$ given in Eq. (2.40) to calculate the long distance behavior of correlation functions involving two and more spins.

2.2.2 Truncated Effective Action: φ^4-Theory

The effective action given in Eqs. (2.39) and (2.40) is formally exact but very complicated. In order to make progress, let us assume that the integral (2.39) is dominated by configurations where the integration variables φ_i are in some sense small, so that we may expand the second term in the effective action (2.40) in powers of the φ_i, truncating the expansion at fourth order. Keeping in mind that physically φ_i represents the fluctuating magnetization, we expect that this truncation can only be good in the vicinity of the critical point, where the magnetization is small. Whether or not this truncation is sufficient to obtain quantitatively correct results for the critical exponents is a rather subtle question which will be answered with the help of the RG.[4] With the expansion (2.14) we obtain

$$S[\varphi] = -N \ln 2 + \frac{\beta}{2} \sum_{ij} J_{ij} \varphi_i \varphi_j - \frac{\beta^2}{2} \sum_i \left[h + \sum_j J_{ij} \varphi_j \right]^2$$

$$+ \frac{\beta^4}{12} \sum_i \left[h + \sum_j J_{ij} \varphi_j \right]^4 + \mathcal{O}(\varphi_i^6) \,. \tag{2.44}$$

Since our lattice model has discrete translational invariance, we may simplify Eq. (2.44) by Fourier transforming the variables φ_i to wave vector space, defining

$$\varphi_i = \frac{1}{\sqrt{N}} \sum_k e^{i k \cdot r_i} \varphi_k \,, \tag{2.45}$$

where the wave vector sum is over the first Brillouin zone of the lattice, which may be chosen as $0 \le k_\mu < 2\pi/a$, where a is the lattice spacing and $\mu = 1, \ldots, D$ labels the components. Imposing for convenience periodic boundary conditions, the

---

[4] It turns out that the quartic truncation of the expansion of Eq. (2.40) can be formally justified close to four dimensions. In the physically most interesting dimension $D = 3$ the term involving six powers of the φ_i cannot be neglected. However the truncated φ^4-theory is in the same universality class as the original theory and the effect of all couplings neglected can be absorbed by a redefinition of the remaining couplings.

wave vectors are quantized as $k_\mu = 2\pi n_\mu/L$, with $n_\mu = 0, 1, \ldots, N_\mu - 1$, where N_μ is the number of lattice sites in direction μ such that $\prod_{\mu=1}^{D} N_\mu = N$ is the total number of lattice sites. Using the identity

$$\frac{1}{N} \sum_i e^{i(k-k')\cdot r_i} = \delta_{k,k'} \,, \tag{2.46}$$

we obtain for the Fourier transform of the terms on the right-hand side of Eq. (2.44),

$$\frac{\beta}{2} \sum_{ij} J_{ij}\varphi_i\varphi_j = \frac{\beta}{2} \sum_k J_k\varphi_{-k}\varphi_k \,, \tag{2.47}$$

$$\frac{\beta^2}{2} \sum_i \left[\sum_j J_{ij}\varphi_j\right]^2 = \frac{\beta^2}{2} \sum_k J_{-k}J_k\varphi_{-k}\varphi_k \,, \tag{2.48}$$

$$\frac{\beta^4}{12} \sum_i \left[\sum_j J_{ij}\varphi_j\right]^4 = \frac{\beta^4}{12N} \sum_{k_1,k_2,k_3,k_4} \delta_{k_1+k_2+k_3+k_4,0}$$
$$\times J_{k_1} J_{k_2} J_{k_3} J_{k_4} \varphi_{k_1} \varphi_{k_2} \varphi_{k_3} \varphi_{k_4} \,, \tag{2.49}$$

where J_k is the Fourier transform of the exchange couplings $J_{ij} \equiv J(r_i - r_j)$,

$$J_k = \sum_i e^{-ik\cdot r_i} J(r_i) \,. \tag{2.50}$$

Because φ_i and J_{ij} are real and $J(-r) = J(r)$, we have $\varphi_{-k} = \varphi_k^*$ and $J_{-k} = J_k$. In Fourier space Eq. (2.44) thus reduces to

$$S[\varphi] = -N \ln 2 - \beta^2 J_{k=0} h\sqrt{N}\varphi_{k=0} + \frac{\beta}{2} \sum_k J_k(1 - \beta J_k)\varphi_{-k}\varphi_k$$
$$+ \frac{\beta^4}{12N} \sum_{k_1,k_2,k_3,k_4} \delta_{k_1+k_2+k_3+k_4,0} J_{k_1} J_{k_2} J_{k_3} J_{k_4} \varphi_{k_1} \varphi_{k_2} \varphi_{k_3} \varphi_{k_4}$$
$$+ \mathcal{O}(\varphi_i^6, h^2, h\varphi_i^3) \,. \tag{2.51}$$

Finally, we anticipate that sufficiently close to the critical point only long-wavelength fluctuations (corresponding to small wave vectors) are important. We therefore expand the function J_k appearing in Eq. (2.51) in powers of k. From Eq. (2.50) it is easy to show that for nearest-neighbor interactions on a D-dimensional hypercubic lattice with coordination number $z = 2D$,

$$J_k = J[z - k^2 a^2] + \mathcal{O}(k^4) = T_c \left[1 - \frac{k^2 a^2}{z}\right] + \mathcal{O}(k^4) \,, \tag{2.52}$$

where $T_c = zJ$ is the mean-field result for the critical temperature, see Eq. (2.13). The coefficient of the quadratic term in Eq. (2.51) can then be written as

$$\beta J_k(1 - \beta J_k) = a^2(r_0 + c_0 k^2) + \mathcal{O}(k^4) \,, \tag{2.53}$$

where we have assumed that $|T - T_c| \ll T_c$ and the constants r_0 and c_0 are defined by

$$r_0 = \frac{T - T_c}{a^2 T_c} \,, \tag{2.54}$$

$$c_0 = \frac{1}{z} = \frac{1}{2D} \,. \tag{2.55}$$

In the limit of infinite volume $V = Na^D \to \infty$, the discrete set of allowed wave vectors merges into a continuum, so that the momentum sums can be replaced by integrations according to the following prescription,

$$\frac{1}{V} \sum_k \to \int \frac{d^D k}{(2\pi)^D} \equiv \int_k \,. \tag{2.56}$$

It is then convenient to normalize the fields differently, introducing a new (continuum) field $\varphi(\mathbf{k})$ via[5]

$$\varphi(\mathbf{k}) = a\sqrt{V}\,\varphi_k \,. \tag{2.57}$$

Defining

$$f_0 = -\frac{N}{V}\ln 2 = -a^{-D}\ln 2 \tag{2.58}$$

and the (dimensionful) coupling constants

$$u_0 = 2a^{D-4}(\beta J_{k=0})^4 \approx 2a^{D-4} \,, \tag{2.59}$$

$$h_0 = \frac{\beta^2 J_{k=0} h}{a^{1+D/2}} \approx \frac{\beta h}{a^{1+D/2}} \,, \tag{2.60}$$

where the approximations are again valid for $|T - T_c| \ll T_c$, we find that our lattice action $S[\boldsymbol{\varphi}]$ in Eq. (2.51) reduces to

---

[5] The different normalizations of lattice and continuum fields are represented by different positions of the momentum labels: while in the lattice normalization the momentum label $\mathbf{k}$ is attached to the dimensionless field φ_k as a subscript, the label of the continuum field $\varphi(\mathbf{k})$ is written in brackets after the field symbol.

$$S_{\Lambda_0}[\varphi] = V f_0 - h_0 \varphi(\boldsymbol{k}=0) + \frac{1}{2} \int_{\boldsymbol{k}} \left[r_0 + c_0 k^2\right] \varphi(-\boldsymbol{k})\varphi(\boldsymbol{k})$$
$$+ \frac{u_0}{4!} \int_{\boldsymbol{k}_1} \int_{\boldsymbol{k}_2} \int_{\boldsymbol{k}_3} \int_{\boldsymbol{k}_4} (2\pi)^D \delta(\boldsymbol{k}_1 + \boldsymbol{k}_2 + \boldsymbol{k}_3 + \boldsymbol{k}_4)\varphi(\boldsymbol{k}_1)\varphi(\boldsymbol{k}_2)\varphi(\boldsymbol{k}_3)\varphi(\boldsymbol{k}_4) ,$$

$$(2.61)$$

where it is understood that the momentum integrations in Eq. (2.61) have an implicit *ultraviolet cutoff* $|\boldsymbol{k}| < \Lambda_0 \ll a^{-1}$ which takes into account that in deriving Eq. (2.61) we have expanded $J_{\boldsymbol{k}}$ for small wave vectors. The functional $S_{\Lambda_0}[\varphi]$ is called the *Ginzburg–Landau–Wilson action* and describes the long-wavelength order-parameter fluctuations of the D-dimensional Ising model, in the sense that the integration over the fields $\varphi(\boldsymbol{k})$ yields the contribution of the associated long-wavelength fluctuations to the partition function. The contribution of the neglected short-wavelength fluctuations to the partition function can be taken into account implicitly by simply redefining the field-independent part f_0 and the coupling constants r_0, c_0, and u_0 of our effective action (2.61). Moreover, also the prefactor $\sqrt{\det \tilde{\boldsymbol{J}}} = e^{\frac{1}{2}\mathrm{Tr}\ln\tilde{\boldsymbol{J}}}$ of Eq. (2.39) can be absorbed into a redefinition of the field-independent constant, $f_0 - \frac{1}{2V}\mathrm{Tr}\ln\tilde{\boldsymbol{J}} \to f_0$. Actually, for periodic boundary conditions the eigenvalues of the matrix $\tilde{\boldsymbol{J}}$ are simply given by $J_{\boldsymbol{k}}/T \approx J_{\boldsymbol{k}=0}/T_c = 1$, so that at long wavelengths and to leading order in $T - T_c$ we may approximate the factor $\sqrt{\det \tilde{\boldsymbol{J}}}$ by unity. The functional integral (2.39) representing the partition function of the Ising model can thus be written as

$$\mathcal{Z} = \int \mathcal{D}[\varphi] e^{-S_{\Lambda_0}[\varphi]} . \qquad (2.62)$$

It is sometimes more convenient to consider the effective action $S_{\Lambda_0}[\varphi]$ in real space. Defining the real-space Fourier transforms of the continuum fields $\varphi(\boldsymbol{k})$ via

$$\varphi(\boldsymbol{r}) = \int_{\boldsymbol{k}} e^{i\boldsymbol{k}\cdot\boldsymbol{r}} \varphi(\boldsymbol{k}) , \qquad (2.63)$$

and using the identity

$$\int_{\boldsymbol{k}} e^{i\boldsymbol{k}\cdot\boldsymbol{r}} = \delta(\boldsymbol{r}) , \qquad (2.64)$$

we obtain from Eq. (2.61),

$$S_{\Lambda_0}[\varphi] = \int d^D r \left[f_0 + \frac{r_0}{2}\varphi^2(\boldsymbol{r}) + \frac{c_0}{2}\left[\nabla\varphi(\boldsymbol{r})\right]^2 + \frac{u_0}{4!}\varphi^4(\boldsymbol{r}) - h_0\varphi(\boldsymbol{r}) \right] . \qquad (2.65)$$

For obvious reasons the classical field theory defined by this expression is called φ^4-theory. In the first part of this book, we shall use this field theory to illustrate the main concepts of the RG. Strictly speaking, the identity (2.64) is not quite correct if we keep in mind that we should impose an implicit ultraviolet cutoff Λ_0 on the k-integration. Roughly, the continuum field $\varphi(\boldsymbol{r})$ corresponds to the coarse-grained magnetization, which is averaged over spatial regions with volume of order Λ_0^{-D} containing $(\Lambda_0 a)^{-D}$ spins.

The quantity $e^{-S_{\Lambda_0}[\varphi]}$ is proportional to the probability density for observing an order-parameter distribution specified by the field configuration $\varphi(\boldsymbol{r})$. According to Eq. (2.62), the partition function is given by the average of $e^{-S_{\Lambda_0}[\varphi]}$ over all possible order-parameter configurations. With this probabilistic interpretation of the effective action (2.65), we can now give a more satisfactory interpretation of the mean-field Landau function $\mathcal{L}_{\mathrm{MF}}(T, h; m)$ defined in Eqs. (2.8) and (2.9). Let us therefore evaluate the functional integral (2.62) in saddle point approximation, where the entire integral is estimated by the integrand at a single constant value $\bar{\varphi}_0$ of the field which minimizes the action $S_{\Lambda_0}[\varphi]$. Setting $\varphi(\boldsymbol{r}) \to \bar{\varphi}$ in Eq. (2.62) we obtain

$$\mathcal{Z} \approx \int_{-\infty}^{\infty} \frac{d\bar{\varphi}}{\sqrt{2\pi}} e^{-S_{\Lambda_0}[\bar{\varphi}]} , \tag{2.66}$$

with

$$S_{\Lambda_0}[\bar{\varphi}] = V\left[f_0 + \frac{r_0}{2}\bar{\varphi}^2 + \frac{u_0}{4!}\bar{\varphi}^4 - h_0\bar{\varphi} \right] . \tag{2.67}$$

Note that in Eq. (2.66) we still integrate over all configurations of the homogeneous components $\bar{\varphi}$ of the order-parameter field; the quantity $e^{-S_{\Lambda_0}[\bar{\varphi}]}$ is therefore proportional to the probability of observing for the homogeneous component of the order parameter the value $\bar{\varphi}$. Requiring consistency of Eq. (2.67) with our previous definitions (2.8) and (2.15) of the Landau function, we should identify

$$S_{\Lambda_0}[\bar{\varphi}] = \beta N \mathcal{L}_{\mathrm{MF}}(T, h; m) = \beta N \left[f + \frac{r}{2}m^2 + \frac{u}{4!}m^4 - hm \right] , \tag{2.68}$$

which is indeed the case if

$$\bar{\varphi} = a^{1-D/2}m , \tag{2.69}$$

keeping in mind the definitions of f, r, and u in Eqs. (2.16a), (2.16b), and (2.16c) on the one hand, and of f_0, r_0, and u_0 in Eqs. (2.58), (2.54), and (2.59) on the other hand. For $V \to \infty$ the value of the one-dimensional integral (2.66) is essentially determined by the saddle point of the integrand, corresponding to the most probable value of $\bar{\varphi}$. The physical value $\bar{\varphi}_0$ of the order-parameter field is therefore fixed by the saddle point condition,

$$\left. \frac{\partial S_{\Lambda_0}[\bar{\varphi}]}{\partial \bar{\varphi}} \right|_{\bar{\varphi}_0} = r_0\bar{\varphi}_0 + \frac{u_0}{6}\bar{\varphi}_0^3 - h_0 = 0 , \tag{2.70}$$

which is equivalent to the mean-field self-consistency equation (2.17).

In summary, the mean-field Landau function defined in Eq. (2.8) can be recovered from the functional integral representation (2.62) of the partition function by ignoring spatial fluctuations of the order parameter; the corresponding saddle point equation is equivalent to the mean-field self-consistency equation (2.11). Because of its probabilistic interpretation the Landau function cannot be defined within the framework of thermodynamics, which only makes statements about averages. The concept of a Landau function is further illustrated in Exercise 2.2, where the Landau functions of noninteracting bosons are defined and calculated.

2.3 The Gaussian Approximation

While the mean-field approximation amounts to evaluating the functional integral (2.62) in saddle point approximation, the Gaussian approximation retains quadratic fluctuations around the saddle point and thus includes the lowest-order correction to the mean-field approximation in an expansion in fluctuations around the saddle point. In the field-theory language, the Gaussian approximation corresponds to describing fluctuations in terms of a free field theory, where the fluctuations with different momenta and frequencies are independent. In condensed matter physics, the Gaussian approximation is closely related to the so-called random phase approximation. It turns out that quite generally the Gaussian approximation is only valid if the dimensionality D of the system is larger than a certain upper critical dimension D_{up}, which we have already encountered in the context of the scaling hypothesis in Sect. 1.3, see the discussion after Eq. (1.32). Because for the Ising universality class $D_{\mathrm{up}} = 4$, the Gaussian approximation is not sufficient to describe the critical behavior of Ising magnets in experimentally accessible dimensions. Nevertheless, in order to motivate the RG it is instructive to see how and why the Gaussian approximation breaks down for $D < D_{\mathrm{up}}$. For simplicity, we set $h = 0$ in this section.

2.3.1 Gaussian Effective Action

To derive the Gaussian approximation, we go back to our Ginzburg–Landau–Wilson action $S_{\Lambda_0}[\varphi]$ defined in Eq. (2.65). Let us decompose the field $\varphi(r)$ describing the coarse-grained fluctuating magnetization as follows,

$$\varphi(r) = \bar{\varphi}_0 + \delta\varphi(r) \,, \tag{2.71}$$

or in wave vector space

$$\varphi(k) = (2\pi)^D \delta(k)\bar{\varphi}_0 + \delta\varphi(k) \,. \tag{2.72}$$

Here, $\bar{\varphi}_0$ is the mean-field value of the order parameter satisfying the saddle point equation (2.70) and $\delta\varphi(r)$ describes inhomogeneous fluctuations around the saddle point. Substituting Eq. (2.71) into the action (2.65) and retaining all terms up to quadratic order in the fluctuations, we obtain in momentum space

$$
\begin{aligned}
S_{\Lambda_0}[\bar{\varphi}_0 + \delta\varphi] \approx\; & V\left[f_0 + \frac{r_0}{2}\bar{\varphi}_0^2 + \frac{u_0}{4!}\bar{\varphi}_0^4\right] \\
& + \left[r_0\bar{\varphi}_0 + \frac{u_0}{6}\bar{\varphi}_0^3\right]\delta\varphi(\boldsymbol{k}=0) \\
& + \frac{1}{2}\int_{\boldsymbol{k}}\left[r_0 + \frac{u_0}{2}\bar{\varphi}_0^2 + c_0 k^2\right]\delta\varphi(-\boldsymbol{k})\delta\varphi(\boldsymbol{k}) .
\end{aligned}
\tag{2.73}
$$

This is the *Gaussian approximation* for the Ginzburg–Landau–Wilson action. To further simplify Eq. (2.73), we note that the second line on the right-hand side of Eq. (2.73) vanishes because the coefficient of $\delta\varphi(\boldsymbol{k}=0)$ satisfies the saddle point condition (2.70). The first line on the right-hand side of Eq. (2.73) can be identified with the mean-field free energy. Explicitly substituting for $\bar{\varphi}_0$ in Eq. (2.73) the saddle point value,

$$
\bar{\varphi}_0 = \begin{cases} 0 & \text{for}\quad r_0 > 0 , \\ \sqrt{-6r_0/u_0} & \text{for}\quad r_0 < 0 , \end{cases}
\tag{2.74}
$$

we obtain for the effective action in Gaussian approximation for $T > T_c$, where $\bar{\varphi}_0 = 0$ and $\delta\varphi = \varphi$,

$$
\boxed{\,S_{\Lambda_0}[\varphi] = V f_0 + \frac{1}{2}\int_{\boldsymbol{k}}\left[r_0 + c_0 k^2\right]\varphi(-\boldsymbol{k})\varphi(\boldsymbol{k}) , \quad T > T_c .\,}
\tag{2.75}
$$

On the other hand, for $T < T_c$, where $r_0 < 0$, we have

$$
\frac{r_0}{2}\bar{\varphi}_0^2 + \frac{u_0}{4!}\bar{\varphi}_0^4 = -\frac{3}{2}\frac{r_0^2}{u_0} ,
\tag{2.76}
$$

$$
r_0 + \frac{u_0}{2}\bar{\varphi}_0^2 = -2r_0 ,
\tag{2.77}
$$

and hence

$$
\boxed{\,S_{\Lambda_0}[\varphi] = V\left[f_0 - \frac{3}{2}\frac{r_0^2}{u_0}\right] + \frac{1}{2}\int_{\boldsymbol{k}}\left[-2r_0 + c_0 k^2\right]\delta\varphi(-\boldsymbol{k})\delta\varphi(\boldsymbol{k}) , \quad T < T_c .\,}
$$

$$
\tag{2.78}
$$

With the help of the Gaussian effective action given in Eqs. (2.75) and (2.78), we may now estimate the effect of order-parameter fluctuations on the mean-field results for the thermodynamic critical exponents. Moreover, because the Gaussian approximation takes the spatial inhomogeneity of the order-parameter fluctuations into account, we may also calculate the critical exponents ν and η associated with the order-parameter correlation function.

2.3.2 Gaussian Corrections to the Specific Heat Exponent

Because in the thermodynamic limit both the Gaussian approximation and the mean-field approximation predict the same contribution from the homogeneous fluctuations represented by $\bar{\varphi}$ to the free energy[6], Gaussian fluctuations do not modify the mean-field predictions for the exponents β, γ, and δ, which are related to the homogeneous order-parameter fluctuations. Within the Gaussian approximation, we therefore still obtain $\beta = 1/2$, $\gamma = 1$, and $\delta = 3$. On the other hand, the Gaussian approximation for the specific heat exponent α is different from the mean-field prediction $\alpha = 0$, because the fluctuations with finite wave vectors give a nontrivial contribution Δf to the free energy per lattice site, which according to Eqs. (2.75) and (2.78) can be written as

$$e^{-\beta N \Delta f} = \int \mathcal{D}[\varphi] \exp\left[-\frac{c_0}{2} \int_k (\xi^{-2} + k^2)\delta\varphi(-k)\delta\varphi(k)\right]. \qquad (2.79)$$

Here, we have introduced the length ξ via

$$\frac{c_0}{\xi^2} = \begin{cases} r_0 & \text{for } T > T_c\,, \\ -2r_0 & \text{for } T < T_c\,. \end{cases} \qquad (2.80)$$

In Sect. 2.3.3 we shall identify the length ξ with the order-parameter correlation length introduced in Sect. 1.2. Recall that for $T > T_c$ we may write $\delta\varphi(k) = \varphi(k)$ because in this case $\bar{\varphi}_0 = 0$. To evaluate the Gaussian integral in Eq. (2.79), it is convenient to discretize the integral $\int_k$ in the exponent with the help of the underlying lattice and use the associated lattice normalization of the fields, which according to Eq. (2.57) amounts to the substitution $\delta\varphi(k) = a\sqrt{V}\delta\varphi_k$. Then we obtain from Eq. (2.79),

$$e^{-\beta N \Delta f} = \int_\infty^\infty \frac{d\delta\varphi_0}{\sqrt{2\pi}} \left[\prod_{k,\,k\cdot\hat{n}>0} \int_\infty^\infty \frac{d\mathrm{Re}\delta\varphi_k}{\sqrt{2\pi}} \int_\infty^\infty \frac{d\mathrm{Im}\delta\varphi_k}{\sqrt{2\pi}}\right]$$

$$\times \exp\left[-\frac{c_0 a^2}{2}\sum_k (\xi^{-2} + k^2)\,|\delta\varphi_k|^2\right], \qquad (2.81)$$

where the product in the square braces is restricted to those wave vectors k whose projection $k \cdot \hat{n}$ onto an arbitrary direction $\hat{n}$ is positive. This restriction is necessary in order to avoid double counting of the finite k-fluctuations, which are represented by a real field φ_i whose Fourier components satisfy $\varphi_{-k} = \varphi_k^*$. Since the integrations

---

[6] Recall that the mean-field approximation for the free energy is $f_{\mathrm{MF}}(t, 0) = f_0$ for $T > T_c$, and $f_{\mathrm{MF}}(t, 0) = f_0 - \frac{3}{2}\frac{r^2}{u}$ for $T < T_c$, see Eq. (2.22). Taking the different normalization of $S_{A_0}[\varphi]$ into account, these expressions correspond to the field-independent terms on the right-hand sides of Eqs. (2.75) and (2.78).

in Eq. (2.81) treat the real and the imaginary parts of $\delta\varphi_k$ as independent variables, we have to correct for this double counting by integrating only over half of the possible values of k. The functional integral is now reduced to a product of one-dimensional Gaussian integrals, which are of course a special case of our general Gaussian integration formula (2.25) for $N = 1$,

$$\int_{-\infty}^{\infty} \frac{dx}{\sqrt{2\pi}}\, e^{-\frac{A}{2}x^2} = A^{-1/2} = e^{-\frac{1}{2}\ln A}\,. \tag{2.82}$$

We thus obtain from Eq. (2.81) for the correction from Gaussian fluctuations to the free energy per lattice site,

$$\Delta f = \frac{T}{2N}\sum_{k}\ln\left[c_0 a^2(\xi^{-2} + k^2)\right]\,, \tag{2.83}$$

where it is understood that the sum is regularized via an ultraviolet cutoff $|k| < \Lambda_0$. The corresponding correction to the specific heat per lattice site is

$$\begin{aligned}
\Delta C &= -T\frac{\partial^2 \Delta f}{\partial T^2} = -\frac{T}{2N}\frac{\partial^2}{\partial T^2}\left\{T\sum_{k}\ln\left[c_0 a^2(\xi^{-2} + k^2)\right]\right\}\\
&= -\frac{T}{2N}\frac{\partial}{\partial T}\left\{\sum_{k}\ln\left[c_0 a^2(\xi^{-2} + k^2)\right] + T\left[\frac{\partial}{\partial T}\frac{1}{\xi^2}\right]\sum_{k}\frac{1}{\xi^{-2} + k^2}\right\}\\
&= \frac{T^2}{2}\left[\frac{\partial}{\partial T}\frac{1}{\xi^2}\right]^2\frac{1}{N}\sum_{k}\frac{1}{[\xi^{-2} + k^2]^2} - T\left[\frac{\partial}{\partial T}\frac{1}{\xi^2}\right]\frac{1}{N}\sum_{k}\frac{1}{\xi^{-2} + k^2}\,. \tag{2.84}
\end{aligned}$$

In the thermodynamic limit, the lattice sums can be converted into integrals using Eq. (2.56) so that for $n = 1, 2$

$$\begin{aligned}
\lim_{N\to\infty}\frac{1}{N}\sum_{k}\frac{1}{[\xi^{-2} + k^2]^n} &= a^D\int_{k}\frac{1}{[\xi^{-2} + k^2]^n}\\
&= K_D a^D\int_{0}^{\Lambda_0} dk\,\frac{k^{D-1}}{[\xi^{-2} + k^2]^n} = K_D a^D \xi^{2n-D}\int_{0}^{\Lambda_0\xi} dx\,\frac{x^{D-1}}{[1 + x^2]^n}\,. \tag{2.85}
\end{aligned}$$

Here, the numerical constant K_D is defined by

$$K_D = \frac{\Omega_D}{(2\pi)^D} = \frac{1}{2^{D-1}\pi^{D/2}\Gamma(D/2)}\,, \tag{2.86}$$

where Ω_D is the surface area of the D-dimensional unit sphere. Using the fact that according to Eqs. (2.54) and (2.80) the derivative of the square of the inverse correlation length with respect to temperature is given by

$$\frac{\partial}{\partial T}\frac{1}{\xi^2} = \left\{ \begin{array}{ll} c_0^{-1}\frac{\partial r_0}{\partial T}\,, & \text{for } T > T_c \\[2ex] -2c_0^{-1}\frac{\partial r_0}{\partial T}\,, & \text{for } T < T_c \end{array} \right\} = \frac{A_t}{c_0 a^2 T_c}\,, \tag{2.87}$$

where

$$A_t = \left\{ \begin{array}{ll} 1\,, & \text{for } T > T_c\,, \\[2ex] -2\,, & \text{for } T < T_c\,, \end{array} \right. \tag{2.88}$$

we obtain from Eq. (2.84) for $|t| \equiv |T - T_c|/T_c \ll 1$,

$$\begin{aligned} \Delta C = \; & \frac{K_D}{2}\frac{A_t^2}{c_0^2}\left(\frac{\xi}{a}\right)^{4-D}\int_0^{\Lambda_0\xi} dx\,\frac{x^{D-1}}{[1+x^2]^2} \\ & - K_D\frac{A_t}{c_0}\left(\frac{a}{\xi}\right)^{D-2}\int_0^{\Lambda_0\xi} dx\,\frac{x^{D-1}}{[1+x^2]}\,. \end{aligned} \tag{2.89}$$

Keeping in mind that close to the critical point the dimensionless parameter $\Lambda_0\xi$ is large compared with unity, it is easy to see that the second term on the right-hand side of Eq. (2.89) can be neglected in comparison with the first one, because the dimensionless integral is for $D > 2$ proportional to $(\Lambda_0\xi)^{D-2}$, and has for $D < 2$ a finite limit for $\Lambda_0\xi \to \infty$. On the other hand, the behavior of the first term on the right-hand side of Eq. (2.89) depends on whether D is larger or smaller than the upper critical dimension $D_{\mathrm{up}} = 4$. For $D > D_{\mathrm{up}}$ the integral depends on the ultraviolet cutoff and is proportional to $(\Lambda_0\xi)^{D-4}$, so that $\Delta C \propto (\xi/a)^{4-D}(\Lambda_0\xi)^{D-4} = (\Lambda_0 a)^{D-4}$ which gives rise to a finite correction to the specific heat. Hence, for $D > 4$ long-wavelength Gaussian fluctuations merely modify the size of the jump discontinuity in the specific heat, but do not qualitatively modify the mean-field result $\alpha = 0$. On the other hand, for $D < 4$ the first integral on the right-hand side of Eq. (2.89) remains finite for $\Lambda_0\xi \to \infty$,

$$I_D = \int_0^\infty dx\,\frac{x^{D-1}}{[1+x^2]^2} = \frac{(D-2)\pi}{4\sin(\frac{(D-2)\pi}{2})}\,, \tag{2.90}$$

so that to leading order for large $\xi \propto |r_0|^{-1/2} \propto |t|^{-1/2}$ the contribution from Gaussian fluctuations to the specific heat is

$$\Delta C = \frac{K_D}{2}I_D\frac{A_t^2}{c_0^2}\left(\frac{\xi}{a}\right)^{4-D} \propto |t|^{-(2-D/2)}\,. \tag{2.91}$$

This is more singular than the jump discontinuity predicted by mean-field theory and implies $\alpha = 2 - D/2$. We thus conclude that in Gaussian approximation the specific heat exponent is given by

$$\alpha = \begin{cases} 2 - \dfrac{D}{2} & \text{for } D < D_{\text{up}} = 4 \,, \\ 0 & \text{for } D \geq D_{\text{up}} \,. \end{cases} \qquad (2.92)$$

It should be noted that within the Gaussian model considered here the scaling relations $2 - \alpha = 2\beta + \gamma = \beta(\delta + 1)$ [see Eqs. (1.23) and (1.24)] are violated for $D < 4$. This failure of the Gaussian approximation will be further discussed in Sect. 2.3.4.

2.3.3 Correlation Function

Next, we calculate the order-parameter correlation function $G(r_i - r_j)$ of the Ising model and the associated critical exponents ν and η in Gaussian approximation. We define $G(r_i - r_j)$ via the following thermal average,

$$G(r_i - r_j) = a^{2-D} \langle \delta s_i \delta s_j \rangle \equiv a^{2-D} \frac{\sum_{\{s_i\}} e^{-\beta H} \delta s_i \delta s_j}{\sum_{\{s_i\}} e^{-\beta H}} \,, \qquad (2.93)$$

where $\delta s_i = s_i - \langle s_i \rangle = s_i - m$ is the deviation of the microscopic spin from its average. Using the identity (2.43) we can express $G(r_i - r_j)$ for distances large compared with the range of the J_{ij} as a functional average over all configurations of the fluctuating order-parameter field,

$$G(r_i - r_j) = a^{2-D} \langle \delta\varphi_i \delta\varphi_j \rangle_S \equiv a^{2-D} \frac{\int \mathcal{D}[\varphi] e^{-S[\varphi]} \delta\varphi_i \delta\varphi_j}{\int \mathcal{D}[\varphi] e^{-S[\varphi]}} \,. \qquad (2.94)$$

The prefactor a^{2-D} is introduced such that $G(r_i - r_j)$ can also be written as

$$G(r_i - r_j) = \langle \delta\varphi(r_i) \delta\varphi(r_j) \rangle_S \,, \qquad (2.95)$$

where the continuum field $\varphi(r)$ is related to the real-space lattice field φ_i via

$$\varphi(r_i) = a^{1-D/2} \varphi_i \,, \qquad (2.96)$$

which follows from the definitions (2.63), (2.45), and (2.57), see also Eq. (2.69). Expanding $\delta\varphi_i$ and $\delta\varphi_j$ in terms of their Fourier components $\delta\varphi_k$ defined analogously to Eq. (2.45), we obtain

$$G(r_i - r_j) = \frac{a^2}{V} \sum_{k_1, k_2} e^{i(k_1 \cdot r_i + k_2 \cdot r_j)} \langle \delta\varphi_{k_1} \delta\varphi_{k_2} \rangle_S$$

$$= \frac{1}{V} \sum_{k_1} e^{ik_1 \cdot (r_i - r_j)} G(k_1) \,, \qquad (2.97)$$

where

$$G(\boldsymbol{k}_1) = a^2 \langle \delta\varphi_{\boldsymbol{k}_1} \delta\varphi_{-\boldsymbol{k}_1} \rangle_S = a^2 \langle |\delta\varphi_{\boldsymbol{k}_1}|^2 \rangle_S \tag{2.98}$$

is the Fourier transform of $G(\boldsymbol{r}_i - \boldsymbol{r}_j)$. In the second line of Eq. (2.97) we have used the fact that by translational invariance the functional average is only nonzero if the total wave vector $\boldsymbol{k}_1 + \boldsymbol{k}_2$ vanishes,

$$\langle \delta\varphi_{\boldsymbol{k}_1} \delta\varphi_{\boldsymbol{k}_2} \rangle_S = \delta_{\boldsymbol{k}_1, -\boldsymbol{k}_2} \langle \delta\varphi_{\boldsymbol{k}_1} \delta\varphi_{-\boldsymbol{k}_1} \rangle_S \; . \tag{2.99}$$

In the infinite volume limit, it is more convenient to work with the continuum fields $\varphi(\boldsymbol{k}) = a\sqrt{V}\,\varphi_{\boldsymbol{k}}$ introduced in Eq. (2.57); then the relations (2.98) and (2.99) can be written in the compact form,[7]

$$\langle \delta\varphi(\boldsymbol{k}_1) \delta\varphi(\boldsymbol{k}_2) \rangle_S = (2\pi)^D \delta(\boldsymbol{k}_1 + \boldsymbol{k}_2) G(\boldsymbol{k}_1) \; . \tag{2.100}$$

Let us now evaluate the functional average in Eq. (2.98) within the Gaussian approximation where the long-wavelength effective action $S_{\Lambda_0}[\varphi]$ is given by Eqs. (2.75) and (2.78). Similar to Eq. (2.81), we write the Gaussian effective action for finite V as

$$S_{\Lambda_0}[\varphi] \approx S_{\Lambda_0}[\bar{\varphi}_0] + \frac{c_0 a^2}{2} \sum_{\boldsymbol{k}} (\xi^{-2} + \boldsymbol{k}^2) |\delta\varphi_{\boldsymbol{k}}|^2$$

$$= c_0 a^2 \left(\xi^{-2} + \boldsymbol{k}_1^2 \right) \left| \delta\varphi_{\boldsymbol{k}_1} \right|^2 + \text{terms independent of } \delta\varphi_{\boldsymbol{k}_1} \; , \tag{2.101}$$

where in the second line we have used the fact that the terms $\boldsymbol{k} = \boldsymbol{k}_1$ and $\boldsymbol{k} = -\boldsymbol{k}_1$ give the same contribution to the sum. Using the fact that within the Gaussian approximation fluctuations with different wave vectors are decoupled, the evaluation of the functional average in Eq. (2.98) is now reduced to simple one-dimensional Gaussian integrations. Renaming again $\boldsymbol{k}_1 \to \boldsymbol{k}$, we see that the Gaussian approximation $G_0(\boldsymbol{k})$ for the Fourier transform of the correlation function $G(\boldsymbol{k})$ is for $\boldsymbol{k} \neq 0$,

$$
\begin{aligned}
G_0(\boldsymbol{k}) &= a^2 \frac{\int_{-\infty}^{\infty} \frac{d\mathrm{Re}\delta\varphi_{\boldsymbol{k}}}{\sqrt{2\pi}} \int_{-\infty}^{\infty} \frac{d\mathrm{Im}\delta\varphi_{\boldsymbol{k}}}{\sqrt{2\pi}} |\delta\varphi_{\boldsymbol{k}}|^2 \exp\left[-c_0 a^2 (\xi^{-2} + \boldsymbol{k}^2) |\delta\varphi_{\boldsymbol{k}}|^2 \right]}{\int_{-\infty}^{\infty} \frac{d\mathrm{Re}\delta\varphi_{\boldsymbol{k}}}{\sqrt{2\pi}} \int_{-\infty}^{\infty} \frac{d\mathrm{Im}\delta\varphi_{\boldsymbol{k}}}{\sqrt{2\pi}} \exp\left[-c_0 a^2 (\xi^{-2} + \boldsymbol{k}^2) |\delta\varphi_{\boldsymbol{k}}|^2 \right]} \\[2mm]
&= \frac{\int_0^{\infty} d\rho \, \rho^3 \, e^{-c_0(\xi^{-2} + \boldsymbol{k}^2)\rho^2}}{\int_0^{\infty} d\rho \, \rho \, e^{-c_0(\xi^{-2} + \boldsymbol{k}^2)\rho^2}} \\[2mm]
&= \frac{1}{c_0(\xi^{-2} + \boldsymbol{k}^2)} = \frac{1}{A_t r_0 + c_0 \boldsymbol{k}^2} \; ,
\end{aligned}
\tag{2.102}
$$

---

[7] In a finite volume V the singular expression $(2\pi)^D \delta(\boldsymbol{k} = 0)$ should be regularized with the volume of the system, $(2\pi)^D \delta(\boldsymbol{k} = 0) \to V$.

where A_t is defined in Eq. (2.88) and in the second line we have introduced $\rho = a\,|\delta\varphi_k|$ as a new integration variable. In particular, at the critical point, where $r_0 = 0$, we obtain

$$G_0(k) = \frac{1}{c_0 k^2}\,, \quad T = T_c\,. \tag{2.103}$$

Comparing this with the definition of the correlation function exponent η in Eq. (1.14), we conclude that in Gaussian approximation

$$\boxed{\eta = 0\,.} \tag{2.104}$$

Next, let us calculate the correlation length exponent ν and justify our identification of the length ξ defined in Eq. (2.80) with the correlation length. Therefore, we calculate the correlation function $G(r)$ in real space and compare the result with Eq. (1.11). For $V \to \infty$ the Fourier transformation of the correlation function in Gaussian approximation can be written as

$$G_0(r) = \int_k e^{ik\cdot r} G_0(k) = \frac{1}{c_0} \int_k \frac{e^{ik\cdot r}}{\xi^{-2} + k^2}\,. \tag{2.105}$$

Although there is an implicit ultraviolet cutoff Λ_0 hidden in our notation, for $\min\{|r|, \xi\} \gg \Lambda_0^{-1}$ the integral is dominated by wave vectors much smaller than Λ_0, so that we may formally move the cutoff to infinity. To evaluate the D-dimensional Fourier transform in Eq. (2.105), let us denote by $k_\parallel$ the component of k parallel to the direction of r, and collect the other components of k into a $(D-1)$-dimensional vector $k_\perp$. Then the integration over $k_\parallel$ can be performed using the theorem of residues. Introducing $(D-1)$-dimensional spherical coordinates in $k_\perp$-space we obtain

$$G_0(r) = \frac{K_{D-1}}{2c_0} \int_0^\infty dk_\perp k_\perp^{D-2} \frac{\exp\left[-\sqrt{\xi^{-2} + k_\perp^2}\,|r|\right]}{\sqrt{\xi^{-2} + k_\perp^2}}\,. \tag{2.106}$$

Although for general D the integration cannot be performed analytically, the leading asymptotic behavior for $|r| \ll \xi$ and for $|r| \gg \xi$ is easily extracted. In the regime $|r| \ll \xi$, the exponential factor cuts off the $k_\perp$-integration at $k_\perp \approx 1/|r|$. Since for $|r| \ll \xi$ this cutoff is large compared with $1/\xi$, we may neglect ξ^{-2} as compared with $k_\perp^2$ in the integrand.[8] The integral can then be expressed in terms of the Gamma-function $\Gamma(D-2)$ and we obtain

------

[8] For $D \le 2$ this approximation is not justified because it generates an artificial infrared divergence. In this case a more careful evaluation of the integral (2.106) is necessary. Because for $D \le 2$ our simple Gaussian approximation is not justified anyway, we focus here on $D > 2$.

$$G_0(\boldsymbol{r}) \sim \frac{K_{D-1}}{2c_0} \frac{\Gamma(D-2)}{|\boldsymbol{r}|^{D-2}} \,, \quad \text{for } |\boldsymbol{r}| \ll \xi \,. \tag{2.107}$$

In particular, at the critical point where $\xi = \infty$, this expression is valid for all $\boldsymbol{r}$. A comparison with Eq. (1.13) confirms again $\eta = 0$.

In the opposite limit $|\boldsymbol{r}| \gg \xi$ the integration in Eq. (2.105) is dominated by the regime $k_\perp \lesssim (\xi|\boldsymbol{r}|)^{-1/2} \ll \xi^{-1}$. Then we may ignore the $k_\perp^2$-term as compared with ξ^{-2} in the denominator of Eq. (2.106) and approximate in the exponent,

$$\sqrt{\xi^{-2} + k_\perp^2}\,|\boldsymbol{r}| \approx \frac{|\boldsymbol{r}|}{\xi} + \frac{\xi|\boldsymbol{r}|k_\perp^2}{2} \,, \tag{2.108}$$

so that

$$G_0(\boldsymbol{r}) \sim \frac{K_{D-1}}{2c_0}\xi e^{-|\boldsymbol{r}|/\xi} \int_0^\infty dk_\perp k_\perp^{D-2} e^{-\frac{\xi|\boldsymbol{r}|k_\perp^2}{2}} \,. \tag{2.109}$$

The Gaussian integration is easily carried out and we finally obtain

$$G_0(\boldsymbol{r}) \sim \frac{K_{D-1}}{2c_0} 2^{\frac{D-3}{2}} \Gamma\!\left(\frac{D-1}{2}\right) \frac{e^{-|\boldsymbol{r}|/\xi}}{\sqrt{\xi^{D-3}|\boldsymbol{r}|^{D-1}}} \,, \quad \text{for } |\boldsymbol{r}| \gg \xi \,. \tag{2.110}$$

This has the same form as postulated in Eq. (1.11), justifying the identification of ξ in Eq. (2.80) with the order-parameter correlation length. Hence, within Gaussian approximation we obtain

$$\xi = \sqrt{\frac{c_0}{|A_t r_0|}} \propto |t|^{-1/2} \,, \tag{2.111}$$

implying for the correlation length exponent in Gaussian approximation

$$\boxed{\nu = 1/2 \,.} \tag{2.112}$$

Note that for $D < D_{\mathrm{up}} = 4$, where the Gaussian approximation yields $\alpha = 2 - D/2$ for the specific heat exponent [see Eq. (2.92)], the Gaussian results $\eta = 0$ and $\nu = 1/2$ are also consistent with the hyper-scaling relations (1.29) and (1.32) connecting ν and η with the thermodynamic critical exponents, $\alpha = 2 - D\nu = 2 - D/2$ and $\gamma = (2 - \eta)\nu = (2 - 0) \times \frac{1}{2} = 1$.

2.3.4 Failure of the Gaussian Approximation in $D < 4$

The Gaussian approximation amounts to the quadratic truncation (2.73) of the Ginzburg–Landau–Wilson action $S[\varphi]$ defined in Eq. (2.40). The important question is now whether the Gaussian approximation is sufficient to calculate the critical exponents or not. As already mentioned, the answer depends crucially on the dimensionality of the system: only for $D > D_{up} = 4$ the critical exponents obtained within the Gaussian approximation are correct, while for $D < D_{up}$ the Gaussian approximation is not sufficient. To understand the special role of the upper critical dimension $D_{up} = 4$ for the Ising model, let us attempt to go beyond the Gaussian approximation by retaining the quartic term in the expansion of $S[\varphi]$ in powers of the field. As shown in Sect. 2.2.2, for small wave vectors our effective action then reduces to the φ^4-theory $S_{\Lambda_0}[\varphi]$ defined via Eq. (2.61) or (2.65). At the first sight it seems that if we assume that the coupling constant u_0 associated with the quartic term in Eqs. (2.61) and (2.65) is arbitrarily small, then we may calculate the corrections to the Gaussian approximation perturbatively in powers of u_0. However, this strategy fails in $D < 4$ in the vicinity of the critical point, because the effective dimensionless parameter which is relevant for the perturbative expansion is

$$\bar{u}_0 = \frac{u_0 \xi^{4-D}}{c_0^2} = 2(2D)^2 \left(\frac{\xi}{a}\right)^{4-D} , \tag{2.113}$$

where we have used Eq. (2.59) to approximate $u_0 \approx 2a^{D-4}$. The fact that the perturbative expansion of the interaction in the effective action (2.61) is controlled by the dimensionless coupling $\bar{u}_0$ can be made manifest by introducing dimensionless wave vectors

$$\bar{\boldsymbol{k}} = \boldsymbol{k}\xi , \tag{2.114}$$

and dimensionless fields

$$\bar{\varphi}(\bar{\boldsymbol{k}}) = \sqrt{\frac{c_0}{\xi^{2+D}}} \varphi(\bar{\boldsymbol{k}}/\xi) . \tag{2.115}$$

Assuming for simplicity $T > T_c$ so that $r_0 = c_0/\xi^2$, it is easy to show that in terms of these dimensionless variables our effective Ginzburg–Landau–Wilson action (2.61) takes the form

$$S_{\Lambda_0}[\bar{\varphi}] = V f_0 + \frac{1}{2} \int_{\bar{\boldsymbol{k}}} [1 + \bar{\boldsymbol{k}}^2] \bar{\varphi}(-\bar{\boldsymbol{k}}) \bar{\varphi}(\bar{\boldsymbol{k}})$$
$$+ \frac{\bar{u}_0}{4!} \int_{\bar{\boldsymbol{k}}_1} \int_{\bar{\boldsymbol{k}}_2} \int_{\bar{\boldsymbol{k}}_3} \int_{\bar{\boldsymbol{k}}_4} (2\pi)^D \delta(\bar{\boldsymbol{k}}_1 + \bar{\boldsymbol{k}}_2 + \bar{\boldsymbol{k}}_3 + \bar{\boldsymbol{k}}_4) \bar{\varphi}(\bar{\boldsymbol{k}}_1) \bar{\varphi}(\bar{\boldsymbol{k}}_2) \bar{\varphi}(\bar{\boldsymbol{k}}_3) \bar{\varphi}(\bar{\boldsymbol{k}}_4) . \tag{2.116}$$

Because the coefficient of the quadratic fluctuations in this expression is fixed to unity for small wave vectors, the coefficient $\bar{u}_0$ in front of the quartic part directly gives the relative strength of the quartic interaction between the fluctuations as compared with the free field theory described by the Gaussian part. The crucial point is now that close to the critical point the correlation length $\xi \propto |t|^{-\nu}$ diverges, so that below four dimensions the effective dimensionless coupling $\bar{u}_0$ defined in Eq. (2.113) becomes arbitrarily large for $T \to T_c$, no matter how small the bare coupling constant u_0 is chosen. As a consequence, for $D < 4$ any attempt to calculate perturbatively corrections to the Gaussian approximation is bound to fail for temperatures sufficiently close to T_c. Because the critical exponents are defined in terms of the asymptotic behavior of physical observables for $T \to T_c$, they cannot be obtained perturbatively for $D < D_{\mathrm{up}} = 4$. Obviously, this problem can only be solved with the help of nonperturbative methods, where all orders in the relevant dimensionless coupling $\bar{u}_0$ are taken into account. The most powerful and general method available to deal with this problem is the RG.

In the ordered phase where $\langle \varphi(\boldsymbol{r}) \rangle = \bar{\varphi}_0$ is finite the dimensionless interaction $\bar{u}_0$ in Eq. (2.113) has a nice interpretation in terms of the relative importance of order-parameter fluctuations, which is measured by the following dimensionless ratio,

$$
Q = \frac{\int^{\xi} d^D r \, \langle \delta\varphi(\boldsymbol{r})\delta\varphi(\boldsymbol{r}=0)\rangle_S}{\int^{\xi} d^D r \, \bar{\varphi}_0^2} = \frac{\int^{\xi} d^D r \, G_0(\boldsymbol{r})}{\xi^D \bar{\varphi}_0^2} \, . \tag{2.117}
$$

Here, the integrals $\int^{\xi} d^D r$ extend over a cube with volume ξ^D enclosing the origin, the mean-field value $\bar{\varphi}_0$ of the order parameter is given in Eq. (2.74), and the correlation function in the numerator is evaluated in Gaussian approximation. Intuitively, the quantity Q sets the strength of order-parameter fluctuations in relation to the average order parameter, taking into account that only within a volume of linear extension ξ the fluctuations are correlated. If Q is small compared with unity, then fluctuations are relatively weak and it is reasonable to expect that a mean-field description of the thermodynamics and the Gaussian approximation for the correlation function are sufficient. Taking into account that for $|\boldsymbol{r}| \gtrsim \xi$ the function $G_0(\boldsymbol{r})$ decays exponentially, the value of the integral in the numerator of Eq. (2.117) is not changed if we extend the integration regime over the entire volume of the system, so that we arrive at the estimate,

$$
\int^{\xi} d^D r \, G_0(\boldsymbol{r}) \approx \int d^D r \, G_0(\boldsymbol{r}) = G_0(\boldsymbol{k}=0) = \frac{\xi^2}{c_0} \, . \tag{2.118}
$$

But according to Eqs. (2.74) and (2.80) for $T < T_c$ we may write

$$
\bar{\varphi}_0^2 = \frac{-6r_0}{u_0} = \frac{3c_0}{\xi^2 u_0} \, , \tag{2.119}
$$

so that we finally obtain

$$
Q = \frac{\xi^{4-D} u_0}{3c_0^2} = \frac{\bar{u}_0}{3} \, . \tag{2.120}
$$

Hence, our dimensionless coupling constant $\bar{u}_0$ defined in Eq. (2.113) is proportional to the ratio Q which measures the relative importance of fluctuations. Obviously, for $Q \ll 1$ fluctuations are weak and it is allowed to treat the interaction term in Eq. (2.116) perturbatively. On the other hand, in the regime $Q \gg 1$ we are dealing with a strongly interacting system, so that mean-field theory and the Gaussian approximation are not sufficient. The condition $Q \ll 1$ is called the *Ginzburg criterion* (Ginzburg 1960, Amit 1974).

Exercises

2.1 Mean-field Analysis of the Ising Model in a Transverse Field

Below a few Kelvin, the magnetic properties of the rare-earth insulator lithium holmium fluoride ($LiHoF_4$) are well described by the Ising model in a transverse field (also known as the quantum Ising model). Its Hamiltonian reads

$$\hat{H} = -J \sum_{\langle i,j \rangle} \sigma_i^z \sigma_j^z - \Gamma \sum_i \sigma_i^x \ .$$

Here, $J > 0$ is the exchange coupling for nearest-neighbor spins, $\Gamma \geq 0$ denotes the strength of the transverse field and σ_i^z and σ_i^x are Pauli matrices which measure the x- and z-components of the spins residing on the lattice sites of a hypercubic lattice. Let us denote the eigenstates of the σ_i^z by $| \uparrow \rangle_i$ and $| \downarrow \rangle_i$ (with eigenvalues $s_i = \pm 1$).

(a) Show that for $\Gamma = 0$ and in the above basis spanned by the eigenstates of the σ_i^z, the Hamiltonian $\hat{H}$ reduces to the Hamiltonian of the familiar classical Ising model.

(b) Introduce $g \equiv \Gamma / zJ$ (where $z = 2D$ is the coordination number) to tune a quantum phase transition at $T = 0$. Show that for $g = 0$ ($\Gamma = 0$) and for $g \to \infty$ ($J \to 0$) you get two qualitatively different ground states. These states are called the ferromagnetic and the paramagnetic ground states. Why?

(c) To derive the mean-field Hamiltonian $\hat{H}_{MF}$ of the quantum Ising model, write $\sigma_i^z = m^z + \delta \sigma_i^z$ with the fluctuation matrix $\delta \sigma_i^z = \sigma_i^z - m^z$ and

$$m^z = \langle \sigma_i^z \rangle = \frac{\mathrm{Tr}\,[\sigma_i^z e^{-\beta \hat{H}}]}{\mathrm{Tr}\,[e^{-\beta \hat{H}}]} \ .$$

Expand the term $\sigma_i^z \sigma_j^z$ in $\hat{H}$ to linear order in $\delta \sigma_i^z$ and neglect terms quadratic in the fluctuation matrices.

(d) Calculate the eigenvalues of the single site mean-field Hamiltonian and use these eigenvalues to evaluate the partition function $\mathcal{Z}_{MF}$ and the Landau function $\mathcal{L}_{\mathrm{MF}}$. Minimize the Landau function with respect to the magnetization m^z to obtain a self-consistency condition for $m_0^z = m_0^z(T, \Gamma)$.

(e) Following the procedure outlined in Sect. 2.1.1, solve the mean-field consistency equation graphically. Analyze in particular regions in parameter space where $m^z \neq 0$ and draw the phase boundary (in the T-g-plane) separating this ferromagnetic phase from the paramagnetic phase with $m^z = 0$. What kind of phase transitions can you identify?

(f) For $T = 0$ calculate m^z explicitly and sketch m^z as a function of g. Determine the mean-field value for the critical exponent β_g defined by $m^z \propto (g_c - g)^{\beta_g}$.

(g) Calculate also $m^x = \langle \sigma_i^x \rangle$ and for $T = 0$ sketch m^x as a function of g.

2.2 Landau Functions for the Free Bose Gas

Consider N noninteracting bosons in a D-dimensional harmonic trap with Hamiltonian $\hat{H} = \sum_{\mathbf{m}} E_{\mathbf{m}} b_{\mathbf{m}}^+ b_{\mathbf{m}}$, where $E_{\mathbf{m}} = \hbar\omega(m_1 + \cdots + m_D)$, and $m_i = 0, 1, 2, \ldots$. The partition function in the canonical ensemble is given by $Z_N = \mathrm{Tr}_N e^{-\beta \hat{H}}$, where the trace runs over the Hilbert space with fixed particle number N. It can be expressed as $Z_N = \sum_{n=0}^{\infty} Z_N(n)$, where $Z_N(n) = \mathrm{Tr}_N(\delta_{n, b_0^+ b_0} e^{-\beta \hat{H}})$ has a fixed occupation of the single particle ground state. We define the dimensionless Landau function for Bose–Einstein condensation via $Z_N(n) = e^{-N \mathcal{L}_N^{\mathrm{BEC}}(n/N)}$. Provided $\mathcal{L}_N^{\mathrm{BEC}}(q)$ is smooth and has an absolute minimum at $q \neq 0$ in the thermodynamic limit, the single particle ground state will be macroscopically occupied.

(a) Using $\mathrm{Tr}_N \hat{A} = \int_0^{2\pi} \frac{d\theta}{2\pi} \mathrm{Tr}\left[e^{i(N - \hat{N})\theta} \hat{A} \right]$, where $\hat{N} = \sum_{\mathbf{m}} b_{\mathbf{m}}^\dagger b_{\mathbf{m}}$ and the trace on the right-hand side runs over the entire Hilbert space containing any number of particles (this is the so-called Fock space), show that the dimensionless Landau function $\mathcal{L}_N^{\mathrm{BEC}}(q)$ is given by

$$\mathcal{L}_N^{\mathrm{BEC}}(q) = -\frac{1}{N} \ln \left[\int_0^{2\pi} \frac{d\theta}{2\pi} \, e^{i\theta N(1-q) - \sum_{\mathbf{m} \neq 0} \ln\left(1 - e^{-(\epsilon_{\mathbf{m}} + i\theta)}\right)} \right] .$$

Here, $\epsilon_{\mathbf{m}} \equiv E_{\mathbf{m}}/T$. What is the physical interpretation of $\mathcal{L}_N^{\mathrm{BEC}}(q)$?

(b) Show that alternatively, we can write $Z_N = \int d^2\phi \, e^{-N \mathcal{L}_N^{\mathrm{SSB}}(\phi)}$, where $d^2\phi = d[\mathrm{Re}\phi]\, d[\mathrm{Im}\phi]$ and the Landau function for spontaneous symmetry breaking is given by

$$\mathcal{L}_N^{\mathrm{SSB}}(\phi) = |\phi|^2 - \frac{1}{N} \ln \left[\frac{N}{\pi} \sum_{n=0}^{N} \frac{(N|\phi|^2)^n}{n!} e^{-N \mathcal{L}_N^{\mathrm{BEC}}(n/N)} \right] .$$

Hint: Carry out the partial trace over $\mathbf{m} = 0$ by using coherent states $|z\rangle = e^{z b_0^+} |0\rangle$, i.e., eigenstates of b_0, $b_0|z\rangle = z|z\rangle$. Recall that $\mathrm{Tr}_{\mathbf{m}=0} \hat{A} = \int \frac{d^2 z}{\pi} e^{-|z|^2}$

$\langle z|\hat{A}|z\rangle$ (see e.g., Shankar 1994, Chap. 21). You will also need the overlap $\langle n|z\rangle = \frac{z^n}{\sqrt{n!}}$ with a state of fixed boson number.

(c) Show the recursion relation $Z_N = \frac{1}{N}\sum_{k=1}^{N} Z_{N-k}Z_1(k)$, where $Z_1(k) = \sum_{\mathbf{m}} e^{-k\epsilon_{\mathbf{m}}} = [1 - e^{-k/\tau}]^{-D}$, and the dimensionless temperature is $\tau = T/(\hbar\omega)$. *Hint:* Show for the generating function $\mathscr{Z}(u) \equiv \sum_N Z_N u^N = \prod_{\mathbf{m}}(1 - ue^{-\epsilon_m})^{-1}$, by expressing it as a trace over Fock space. Take derivatives on both sides of this equality and express the right-hand side again in terms of $\mathscr{Z}(u)$.

(d) Show that $Z_N(n) = Z_{N-n} - Z_{N-n-1}$.
Hint: Derive a relation between the generating functions $\mathscr{Z}(u)$ and $h(u) := \sum_N Z_N(n)u^N$.

(e) Plot $\mathcal{L}_N^{\mathrm{BEC}}(q)$ and $\mathcal{L}_N^{\mathrm{SSB}}(\phi)$ for $N = 10$ and $D = 3$ as a function of $\sqrt{q}$ and ϕ, respectively, evaluating the expressions derived above numerically for different temperatures τ. What is the temperature at which the Landau functions start to develop nontrivial minima? More details and a discussion of $N \to \infty$ can be found in Sinner et al. (2006).

References

Amit, D. J. (1974), *The Ginzburg criterion-rationalized*, J. Phys. C: Solid State Phys. **7**, 3369.

Ginzburg, V. L. (1960), *Some remarks on second order phase transitions and microscopic theory of ferroelectrics*, Fiz. Tverd. Tela **2**, 2031.

Ginzburg, V. L. and L. D. Landau (1950), *Concerning the theory of superconductivity*, Zh. Eksp. Teor. Fiz. **20**, 1064.

Huang, K. (1987), *Statistical Mechanics*, Wiley, New York, 2nd ed.

Hubbard, J. (1959), *Calculation of partition functions*, Phys. Rev. Lett. **3**, 77.

Kopietz, P. (1997), *Bosonization of Interacting Fermions in Arbitrary Dimensions*, Springer, Berlin.

Mattis, D. C. (2006), *The Theory of Magnetism Made Simple*, World Scientific, New Jersey.

Onsager, L. (1944), *A Two-Dimensional model with an order-disorder transition*, Phys. Rev. **65**, 117.

Shankar, R. (1994), *Principles of Quantum Mechanics*, Plenum Press, New York, 2nd ed.

Sinner, A., F. Schütz, and P. Kopietz (2006), *Landau functions for noninteracting bosons*, Phys. Rev. A **74**, 023608.

Stratonovich, R. L. (1957), *On a method of calculating quantum distribution functions*, Sov. Phys. Dokl. **2**, 416.

Wannier, G. H. (1966), *Statistical Physics*, Dover Publications, New York.

Chapter 3
Wilsonian Renormalization Group

In this central chapter of Part I we introduce the basic concepts of the RG method invented by Wilson and coauthors in a series of pioneering articles (Wilson 1969, 1971b,c, 1972, Wilson and Fisher 1972, Wilson and Kogut 1974, Wilson 1975). A very nice introduction to the basic concepts of the Wilsonian RG can also be found in the book by Goldenfeld (1992), which has influenced our presentation.

For a microscopic derivation of the critical behavior of thermodynamic observables, we have to calculate the partition function $\mathcal{Z}$ of the system. For Ising models, this amounts to performing the nested spin summations in Eq. (2.3). More generally, we shall consider many-body systems whose partition function can be expressed in terms of some suitably defined functional integral over a field Φ representing the relevant degrees of freedom,

$$\mathcal{Z}(\boldsymbol{g}) = \int \mathcal{D}[\Phi] e^{-S[\Phi;\boldsymbol{g}]} \, , \tag{3.1}$$

where $S[\Phi;\boldsymbol{g}]$ is an effective action depending on a set of coupling constants $g_1, g_2, g_3, \ldots$, which we collect into a vector $\boldsymbol{g} = (g_1, g_2, g_3, \ldots)$. We have shown in Sect. 2.2 how the partition function of the Ising model can be written in such a form. In that case the field Φ can be identified with the fluctuating order-parameter field φ defined in Eq. (2.37). However, Eq. (3.1) is more general: the partition function of any quantum mechanical many-body system consisting of bosons, fermions, or mixtures thereof can also be written in this form. The fields depend not only on the position $\boldsymbol{r}$, but also carry an (imaginary) time label τ, see Eq. (1.37). We shall explain this in detail in Part II, where we develop the general functional RG method for these systems. Of course, for boson–fermion mixtures the field Φ has several components representing the different types of degrees of freedom. Unfortunately, in almost all cases of interest the integration in Eq. (3.1) cannot be performed exactly, so we have to rely on approximations. Moreover, simple approximations such as the mean-field theory developed in Sect. 2.1 or the Gaussian approximation introduced in Sect. 2.3 are not reliable if the dimensionality of the system is smaller than its upper critical dimension D_{up}. The Wilsonian RG method provides us with a general strategy to attack this difficult problem.

Kopietz, P. et al.: *Wilsonian Renormalization Group*. Lect. Notes Phys. **798**, 53–89 (2010)
DOI 10.1007/978-3-642-05094-7_3   © Springer-Verlag Berlin Heidelberg 2010

3.1 The Basic Idea

The basic idea underlying the Wilsonian RG is conceptually very simple, although from the technical point of view it is usually rather difficult to carry out the RG procedure in practice – after all, we would like to solve a strongly interacting many-body problem close to the critical point! The strategy is to perform the integration over the degrees of freedom represented by the field Φ in Eq. (3.1) *iteratively* in small steps by integrating over suitably chosen subsets of fields and calculating the resulting change of the effective action. One iteration of the RG procedure consists of the following two steps:

Step 1: Mode Elimination (Decimation)

The first RG step consists of the elimination of the degrees of freedom (also called *modes*) representing short-distance fluctuations involving a certain interval of small wavelengths. Sometimes, this is also called the *decimation step*. If we represent the partition function via a functional integral of the type (3.1) and work in momentum space, this means that we integrate over all fields $\Phi(\boldsymbol{k})$ with wave vectors $\boldsymbol{k}$ belonging to a certain high-momentum regime.[1] There is considerable freedom in the choice of the high-momentum regime, and the most convenient choice depends on the model of interest and on the requirements on the accuracy of the calculation at hand. For example, for the φ^4-theory defined by $S_{\Lambda_0}[\varphi]$ in Eq. (2.61) it is often convenient to integrate over fields $\varphi(\boldsymbol{k})$ whose wave vectors lie in the momentum shell $\Lambda < |\boldsymbol{k}| < \Lambda_0$. Formally, the separation into a small wave vector and a large wave vector regime amounts to writing the field Φ as a sum of two terms,

$$\Phi = \Phi^< + \Phi^> \, , \tag{3.2}$$

where the "smaller part" $\Phi^<$ (slow modes) contains fluctuations with wave vectors smaller than a certain scale Λ, while the "greater part" $\Phi^>$ (fast modes) contains the complementary fluctuations involving large wave vectors. A simple way to implement the above decomposition is by multiplying the Fourier components $\Phi(\boldsymbol{k})$ in momentum space by $1 = \Theta(\Lambda - |\boldsymbol{k}|) + \Theta(|\boldsymbol{k}| - \Lambda)$ and writing

$$\Phi(\boldsymbol{k}) = \Theta(\Lambda - |\boldsymbol{k}|)\Phi(\boldsymbol{k}) + \Theta(|\boldsymbol{k}| - \Lambda)\Phi(\boldsymbol{k}) \, , \tag{3.3}$$

so that

$$\Phi^<(\boldsymbol{k}) = \Theta(\Lambda - |\boldsymbol{k}|)\Phi(\boldsymbol{k}) \, , \tag{3.4a}$$

$$\Phi^>(\boldsymbol{k}) = \Theta(|\boldsymbol{k}| - \Lambda)\Phi(\boldsymbol{k}) \, . \tag{3.4b}$$

---

[1] For quantum systems the fields Φ are not only labeled by the position (or wave vector), but also by imaginary time (or frequency). Then the degrees of freedom involving large frequencies should also be eliminated iteratively. This gives additional freedom for the implementation of the mode-elimination procedure, see the seminal work by Hertz (1976) on quantum critical phenomena.

However, such a sharp cutoff in momentum space sometimes leads to technical complications, which can be circumvented by smoothing out the boundary between integrated and unintegrated momenta (Wilson and Kogut 1974), or by implementing the partial integration over short-wavelength fluctuations differently. For example, for low-dimensional spin systems it can be advantageous to carry out the mode elimination in real space, see Sect. 3.2. Formally, the mode-elimination step corresponds to the integration over the "larger field" $\Phi^>$ in the functional integral,

$$\mathcal{Z} = \int \mathcal{D}[\Phi^<] \int \mathcal{D}[\Phi^>] e^{-S[\Phi^< + \Phi^>;g]}$$

$$= \int \mathcal{D}[\Phi^<] e^{-S_\Lambda^<[\Phi^<;g^<]} , \tag{3.5}$$

where $S_\Lambda^<$ is simply defined by

$$\boxed{e^{-S_\Lambda^<[\Phi^<;g^<]} = \int \mathcal{D}[\Phi^>] e^{-S[\Phi^< + \Phi^>;g]} .} \tag{3.6}$$

The coupling constants $g^< = \left(g_1^<, g_2^<, g_3^<, \ldots\right)$ in $S^<[\Phi^<; g^<]$ will in general be different from the original couplings $g = (g_1, g_2, g_3, \ldots)$, except in the case where $S[\Phi]$ is Gaussian so that the modes with different wave vectors (or frequencies) are not coupled. In practice the functional integration in Eq. (3.6) cannot be carried out exactly, so that approximations are necessary. Moreover, the set of nonzero coupling constants $g^<$ in $S_\Lambda^<[\Phi^<; g^<]$ will in general be different from the couplings g in the initial action. For example, the effective φ^4-theory defined by the action $S_{\Lambda_0}[\varphi]$ given in Eqs. (2.61) and (2.65) depends on the four coupling constants $g = (f_0, r_0, c_0, u_0)$. We identify in this case $\Phi \to \varphi$. Due to the presence of the quartic interaction, the integration in Eq. (3.6) cannot be performed exactly. However, even without explicitly doing the integration, we know from the structure of the perturbative expansion of $S_\Lambda^<[\Phi^<; g^<]$ in powers of the fields $\varphi^<$ that after the mode-elimination step the exact effective action must be of the form

$$S_\Lambda^<[\varphi^<; g^<] = V f^< + \frac{1}{2} \int_k^\Lambda \left[r^< + c^< k^2 + c_4^< k^4 + c_6^< k^6 + \ldots \right] \varphi^<(-k) \varphi^<(k)$$

$$+ \frac{u^<}{4!} \int_{k_1}^\Lambda \int_{k_2}^\Lambda \int_{k_3}^\Lambda \int_{k_4}^\Lambda (2\pi)^D \delta(k_1 + k_2 + k_3 + k_4) \varphi^<(k_1) \varphi^<(k_2) \varphi^<(k_3) \varphi^<(k_4)$$

$$+ \frac{u_6^<}{6!} \left(\prod_{i=1}^6 \int_{k_i}^\Lambda \right) (2\pi)^D \delta\left(\sum_{i=1}^6 k_i \right) \varphi^<(k_1) \varphi^<(k_2) \varphi^<(k_3) \varphi^<(k_4) \varphi^<(k_5) \varphi^<(k_6)$$

$$+ \text{ terms involving } \varphi^8, \text{ or } \varphi^4 \text{ and higher powers of } k^2, \tag{3.7}$$

where the notation $\int_k^\Lambda$ means that the upper limit for the integration is given by the reduced ultraviolet cutoff $\Lambda = \Lambda_0/b$. In contrast to our original action

$S_{\Lambda_0}[\varphi; f_0, r_0, c_0, u_0]$, which depends only on four coupling constants, the new effective action depends on infinitely many coupling constants $g^< = (f^<, r^<, c^<, c_4^<, c_6^<, u^<, u_6^<, \ldots)$, which are generated via the integration over the short-wavelength fluctuations $\varphi^>$. Obviously, the new couplings $g^<$ are functions of the initial couplings f_0, r_0, c_0, u_0 defining the original model. In other words, the mode-elimination procedure thus defines a noninvertible mapping of the original coupling space spanned by (f_0, r_0, c_0, u_0) onto the infinite-dimensional space spanned by the couplings $g^< = (f^<, r^<, c^<, c_4^<, c_6^<, u^<, u_6^<, \ldots)$ which is necessary to completely specify the new effective action $S^<[\Phi^<; g^<]$. Fortunately, a truncation of the coupling space retaining only the four coupling constants $f^<, r^<, c^<$ and $u^<$ turns out to be sufficient to obtain the leading corrections to the mean-field results for the critical exponents to first order in $\epsilon = 4 - D$. For more accurate calculations it is necessary to retain a larger number of coupling constants which can be technically quite demanding. In Part III of this book we shall show that the functional renormalization group offers an efficient method to keep track of the RG flow of infinitely many coupling constants.

Step 2: Rescaling

In the second step of the iterative RG procedure, we rescale wave vectors and fields and express the functional $S^<[\Phi^<; g^<]$ in terms of rescaled quantities such that it has the same form as before the mode elimination. The combined effect of mode elimination and rescaling should then be taken into account via a modification of the coupling constants. Therefore, we define rescaled wave vectors k' via

$$k' = bk \,, \tag{3.8}$$

where the dimensionless parameter $b = \Lambda_0/\Lambda$ defines the "step size" of the RG transformation.[2] In Fourier space, the rescaled field $\Phi'(k')$ is related to the original field $\Phi^<(k')$ via

$$\Phi'(k') = \zeta_b^{-1} \Phi^<(k'/b) \,, \tag{3.9}$$

---

[2] For quantum systems the field depends not only on wave vectors but also on frequencies ω. Then we should also rescale $\omega' = b^z \omega$, where the value of the dynamic exponent z depends on the low-energy dynamics of the underlying quantum system and on the critical point under consideration. The fact that we choose here for b and z the same notations as in the formulation of the scaling hypothesis in Sect. 1.3 (see Eqs. (1.16) and (1.38)) is not accidental, because we shall show later that the parameters b and z in the scaling hypothesis are indeed identical with b and z in the rescaling step of the RG transformation.

where the so-called field rescaling factor ζ_b is in general a product of two b-dependent factors,[3]

$$\zeta_b = b^{D_\Phi} \sqrt{Z_b} \, . \tag{3.10}$$

The first factor b^{D_Φ} is determined by the so-called canonical dimension D_Φ of the field $\Phi(k)$ which measures how many powers of inverse length are needed to make $\Phi(k)$ dimensionless. The value of D_Φ follows from the bare action by simple dimensional analysis. For example, the continuum field $\varphi(k)$ in the Ginzburg–Landau–Wilson action (2.61) has units of (length)$^{1+D/2}$, so that in this case $D_\varphi = 1 + D/2$.

The second factor $\sqrt{Z_b}$ in Eq. (3.10) cannot be deduced by dimensional analysis and is closely related to the correlation function exponent η defined in Eq. (1.13) and the so-called wave function renormalization factor Z in quantum mechanical many-body systems. We shall explain this in more detail in Sect. 4.2.3. While there is some freedom in the choice of Z_b, for our simple φ^4-theory it is convenient to choose

$$Z_b = \frac{c_0}{c^<}, \tag{3.11}$$

so that after the rescaling step

$$\boxed{\varphi^<(k) = \zeta_b \varphi'(k') = b^{1+D/2} \sqrt{\frac{c_0}{c^<}} \varphi'(k') \, .} \tag{3.12}$$

The Gaussian part of the effective action $S^<[\Phi^<; g^<]$ in Eq. (3.7) then reads

$$\frac{1}{2} \int_k^\Lambda \left[r^< + c^< k^2 + c_4^< k^4 + c_6^< k^6 + \ldots \right] \varphi^<(-k)\varphi^<(k)$$
$$= \frac{1}{2} \int_{k'}^{\Lambda_0} \left[r' + c_0 k'^2 + c_4' k'^4 + c_6' k'^6 + \ldots \right] \varphi'(-k')\varphi'(k') \, , \tag{3.13}$$

with the new coupling constants

$$r' = b^2 Z_b r^< \, , \tag{3.14a}$$
$$c_4' = b^{-2} Z_b c_4^< \, , \tag{3.14b}$$
$$c_6' = b^{-4} Z_b c_6^< \, . \tag{3.14c}$$

Note that the factor Z_b is chosen such that after one iteration of the two RG steps the coefficient c_0 of the quadratic term in the expansion for small wave vectors does not

---

[3] For multicomponent fields ζ_b is in general a matrix acting on field space. For simplicity we assume here that our field Φ has only a single component, or that for multicomponent fields the matrix is proportional to the unit matrix.

change under the combined operation of decimation and rescaling. If we ignore the couplings c_4', c_6', ..., then the second line in Eq. (3.13) indeed has precisely the same form as the Gaussian part of the bare action $S_{\Lambda_0}[\varphi]$ given in Eq. (2.61), but with the bare coupling r_0 replaced by the renormalized coupling r'. The mode-elimination step defines $r^<$ as a function of the bare couplings.

The combination of the mode-elimination step with the rescaling step defines now a mapping between the initial couplings $g = (g_1, g_2, g_3, \ldots)$ of a given model system with action $S[\Phi; g]$ and a modified set of couplings $g' = \left(g_1', g_2', g_3', \ldots\right)$ appearing in the new effective action $S'[\Phi'; g']$. Let us write this mapping as

$$g_i' = \mathcal{R}_i(b; g_1, g_2, g_3, \ldots), \quad i = 1, 2, 3, \ldots , \tag{3.15a}$$

or in compact vector notation,

$$\boxed{g' = \mathcal{R}(b; g) .} \tag{3.15b}$$

The function $\mathcal{R}(b; g)$ represents an *RG transformation*, acting on the (in general infinite-dimensional) space of coupling constants that specifies a specific model system. In general $\mathcal{R}(b; g)$ is a very complicated nonlinear function of the couplings g, whose precise form depends also on the length rescaling factor b. Mathematically, the set of transformations $\mathcal{R}(b; g)$ labeled by the continuous parameter b is a *semigroup*, which is characterized by the same composition law as a group, but does not require that each transformation has an inverse. The group composition law follows from the fact that two successive transformations with scale factors b and b' are equivalent to a single transformation with scale factor $b'' = b'b$, so that the two iterated transformations

$$g' = \mathcal{R}(b; g) \text{ and } g'' = \mathcal{R}(b'; g') , \tag{3.16}$$

are by construction equivalent with

$$g'' = \mathcal{R}(b'; \mathcal{R}(b; g)) = \mathcal{R}(b'b; g) . \tag{3.17}$$

The nonexistence of an inverse transformation follows from the fact that many different microscopic models can have the same long-wavelength properties. Formally, the elimination of the short-wavelength fluctuations generates a projection of the coupling space defining the microscopic model onto a reduced coupling space associated with a new effective model with the same long-wavelength properties.

The complete RG procedure is now defined by iterating the above two-step procedure, defining for given initial couplings $g^{(0)} = g$ a chain of renormalized couplings $g^{(n)}$ via

$$g^{(n)} = \mathcal{R}(b; g^{(n-1)}) = \mathcal{R}(b^n; g) . \tag{3.18}$$

For $n \to \infty$ the total rescaling factor b^n diverges, which means that we have integrated out all degrees of freedom, so that in principle we can identify the field-independent part $f^{(n)}$ of the resulting effective action for $n \to \infty$ with the exact free energy density of the system. Of course, in almost all physically interesting cases the RG transformation (3.15b) can only be carried out approximately, so that we cannot calculate the exact free energy in this way. The crucial point is, however, that the iteration of rather simple approximate RG transformations $\mathcal{R}(b; g)$ with an RG step size $b > 1$ of the order of unity can generate nonperturbative expressions for various physical quantities. Under certain conditions (which will be discussed in Chap. 4), results based on simple approximate RG transformations are quantitatively accurate even in the critical regime, so that the controlled calculation of the critical exponents becomes possible.

In practice, there are many different ways of implementing the RG procedure: for example, the mode-elimination step can be performed in real space, in wave vector space, or (for quantum systems) in frequency space. Moreover, there is some freedom in the choice of the RG step size b and there are many possibilities of introducing a cutoff Λ separating long-wavelength from short-wavelength fluctuations into a given model. Unfortunately, there is no unique implementation of the RG procedure which can be used as a black box in all situations. Instead, special versions of the RG procedure have proven to be useful to study certain classes of problems. Let us briefly list some important cases:

(a) *Migdal–Kadanoff real space RG*: For spin systems where the spins are localized on lattice sites it can be useful to perform the mode-elimination step directly in real space by performing partial traces over the Hilbert spaces associated with certain blocks of spins. After mode elimination, one then obtains an effective *coarse-grained* spin Hamiltonian where the spin blocks are replaced by new effective spins associated with a diluted lattice with a larger lattice spacing. In the second RG step, length scales are then reduced such that the lattice spacing has again the original size. This procedure has been suggested by Migdal (1975) and by Kadanoff (1976), and is therefore called the Migdal–Kadanoff RG procedure. Unfortunately, in more than one dimension the Migdal–Kadanoff procedure requires approximations whose accuracy is difficult to estimate (see Exercise 3.2). Nevertheless, for certain types of systems (such as some quantum spin models which cannot be described in terms of a local field theory with a well-defined Gaussian part) the Migdal–Kadanoff RG is still the most convenient implementation of the RG. In order to explain the basic ideas of the Migdal–Kadanoff RG, we shall use it in Sect. 3.2 to calculate the partition function of the one-dimensional Ising model, whose exact solution has been obtained in Exercise 1.1 via the transfer matrix method.

(b) *Momentum space RG:* For translationally invariant systems whose critical properties can be described by an effective field theory with a well-defined Gaussian part, perhaps the most popular implementation of the RG is based on the mode elimination in momentum space, as explained above. In the simplest case, one introduces a sharp boundary between the integrated and the unintegrated momenta as in Eqs. (3.4a) and (3.4b). Such a sharp cutoff is very convenient to evaluate the lowest-order fluctuation correction to the Gaussian approximation (the so-called one-loop approximation). Since this approximation is sufficient in many problems in condensed matter physics and statistical mechanics, the momentum shell RG with a sharp cutoff has been very popular and successful in this field. On the other hand, beyond the one-loop approximation, the sharp cutoff in momentum space leads to very cumbersome integrations and unphysical nonanalyticities, so that it is better to smooth out the boundary between the integrated and unintegrated regimes in momentum space (Wilson and Kogut 1974). Unfortunately, for a smooth cutoff the integrations appearing in the Wilsonian RG can only be performed numerically, so that for two-loop calculations even condensed matter physicists often abandon the Wilsonian RG and go back to the field-theoretical RG if the model under consideration is renormalizable.

(c) *Functional renormalization group (FRG):* In a seminal paper, Wegner and Houghton (1973) realized that if the mode-elimination step in the momentum space RG is carried out in infinitesimal momentum shells $\Lambda - d\Lambda < |\boldsymbol{k}| < \Lambda$, then the resulting change of the effective action can be expressed in terms of a formally exact functional differential equation describing the evolution of the effective action as a function of the cutoff Λ. This so-called Wegner–Houghton equation is equivalent to an infinite hierarchy of coupled integro-differential equations for the correlation functions of the theory, which are obtained by expanding the effective action in powers of the fields. Of course, usually this hierarchy of flow equations cannot be solved exactly, so that approximations are necessary in order to calculate the critical properties of the system. Still, it is very useful to have formally exact equations describing the mode-elimination step in the Wilsonian RG, because it opens the way for new approximation strategies and in some cases simplifies conceptual problems such as renormalizability proofs of quantum field theories (Polchinski 1984, Keller and Kopper 1991, 1996).

Due to some technical complications related to the sharp cutoff, the Wegner–Houghton equation has not been widely used to solve problems of physical interest. However, in the past 20 years alternative exact FRG flow equations describing the mode-elimination step of the Wilsonian RG have been derived which circumvent some of the technical complications inherent in the Wegner–Houghton equation (Wetterich 1993, Morris 1994). One of the purposes of this book is to give a self-contained introduction to these FRG methods.

(d) *Numerical implementations of the RG:* In condensed matter physics two important numerical implementations of the RG have been developed. The first is the numerical RG approach (usually abbreviated NRG) for quantum impurity systems, which has been developed by Wilson (1975) to study the Kondo problem describing a single localized spin $S = 1/2$ coupled to a bath of non-interacting conduction electrons. Due to the coupling to the bath, the quantum dynamics of the spin is nontrivial and can only be obtained by means of nonperturbative methods. The mode-elimination step is then carried out numerically by eliminating states in the Hilbert space of the system corresponding to certain energy windows. For a comprehensive introduction to this method see the book by Hewson (1993) or the more recent review article by Bulla et al. (2008).

The second mainly numerical RG approach is the *density matrix renormalization group method* (White 1992, Peschel et al. 1999, Schollwöck 2005), which is particularly useful to study one-dimensional quantum systems. Roughly, this DMRG is a sort of real-space RG approach where the density matrix of the system plays an essential role for the selection of states to be eliminated by the RG iteration. In this book we shall not further discuss these numerical RG methods.

(e) *Continuous unitary transformations:* Another implementation of the RG idea which directly manipulates the Hamiltonian of the system is the method of *continuous unitary transformations*, also called *flow equation method*, which was proposed independently by Wegner (1994), and by Głazek and Wilson (1993, 1994). In this method one defines a family of unitary transformations depending on a certain continuous flow parameter which iteratively transform the Hamiltonian into diagonal form or at least block-diagonal form. For a survey of this method, see the recent book by Kehrein (2006).

3.2 Real-Space RG for the One-Dimensional Ising Model

The Migdal–Kadanoff real-space RG is perhaps the best starting point for gaining an intuitive understanding of the basic concepts underlying the RG. For other pedagogical introductions to this method see (Maris and Kadanoff 1978, Chaikin and Lubensky 1995, McComb 2004). Here we shall explain this method by using it to solve the one-dimensional Ising model with nearest-neighbor interactions. It turns out that for this model the mode-elimination step can be performed exactly, which should not be surprising because we already know from Exercise 1.1 that the model can be easily solved using the transfer matrix method.

3.2.1 Exact Decimation

The Hamiltonian of the one-dimensional Ising model with ferromagnetic nearest-neighbor coupling J in a magnetic field h can be written as

$$H = -J \sum_{i=1}^{N} s_i s_{i+1} - h \sum_{i=1}^{N} s_i - E_0 \, , \qquad (3.19)$$

where $s_i = \pm 1$ and we assume periodic boundary conditions, so that $s_{i+N} = s_i$. For later convenience, we have subtracted an arbitrary constant E_0 which defines the zero of energy. Introducing the dimensionless couplings

$$g = \beta J \, , \quad \tilde{h} = \beta h \, , \quad \tilde{f} = \frac{\beta E_0}{N} \, , \qquad (3.20)$$

the partition function of our model can be written as follows,

$$
\begin{aligned}
\mathcal{Z} &= \sum_{s_1=\pm1} \cdots \sum_{s_N=\pm1} \exp\left[g \sum_{i=1}^{N} s_i s_{i+1} + \tilde{h} \sum_{i=1}^{N} s_i + N\tilde{f} \right] \\
&= \sum_{s_1=\pm1} \cdots \sum_{s_N=\pm1} \exp\left[\sum_{i=1}^{N} \left(g s_i s_{i+1} + \frac{\tilde{h}}{2}(s_i + s_{i+1}) + \tilde{f} \right) \right] \\
&= \sum_{\{s_i\}} \left(e^{g s_1 s_2 + \frac{\tilde{h}}{2}(s_1+s_2)+\tilde{f}} \right) \left(e^{g s_2 s_3 + \frac{\tilde{h}}{2}(s_2+s_3)+\tilde{f}} \right) \cdots \left(e^{g s_N s_1 + \frac{\tilde{h}}{2}(s_N+s_1)+\tilde{f}} \right) \\
&= \sum_{s_1=\pm1} \cdots \sum_{s_N=\pm1} T_{s_1 s_2} T_{s_2 s_3} \cdots T_{s_N s_1} = \mathrm{Tr}\left[\mathbf{T}^N \right] ,
\end{aligned}
\tag{3.21}
$$

where the four numbers

$$
T_{ss'} = \exp\left[gss' + \frac{\tilde{h}}{2}(s+s') + \tilde{f} \right]
\tag{3.22}
$$

are arranged to form the 2×2 *transfer matrix*,

$$
\mathbf{T} = \begin{pmatrix} T_{1,1} & T_{1,-1} \\ T_{-1,1} & T_{-1,-1} \end{pmatrix} = e^{\tilde{f}} \begin{pmatrix} e^{g+\tilde{h}} & e^{-g} \\ e^{-g} & e^{g-\tilde{h}} \end{pmatrix} .
\tag{3.23}
$$

Assuming that N is even, a possible implementation of the RG mode-elimination step is now to carry out a partial summation over all sites with even labels in Eq. (3.21),

$$
\begin{aligned}
\mathcal{Z} &= \sum_{s_1 s_3 \ldots s_{N-1}} \left(\sum_{s_2} T_{s_1 s_2} T_{s_2 s_3} \right) \left(\sum_{s_4} T_{s_3 s_4} T_{s_4 s_5} \right) \cdots \left(\sum_{s_N} T_{s_{N-1} s_N} T_{s_N s_1} \right) \\
&= \sum_{s_1 s_3 \ldots s_{N-1}} [\mathbf{T}^2]_{s_1 s_3} [\mathbf{T}^2]_{s_3 s_5} \cdots [\mathbf{T}^2]_{s_{N-1} s_1} = \mathrm{Tr}\left[(\mathbf{T}')^{N'} \right] ,
\end{aligned}
\tag{3.24}
$$

where $N' = N/2$ is the number of sites after mode elimination, and the new transfer matrix is simply

$$
\mathbf{T}' = \mathbf{T}^2 = e^{2\tilde{f}} \begin{pmatrix} e^{2g+2\tilde{h}} + e^{-2g} & e^{\tilde{h}} + e^{-\tilde{h}} \\ e^{\tilde{h}} + e^{-\tilde{h}} & e^{2g-2\tilde{h}} + e^{-2g} \end{pmatrix} .
\tag{3.25}
$$

After the mode-elimination step, the partition function of our model has the same form as the partition function of an effective Ising model with a new transfer matrix $\mathbf{T}'$ and a new lattice with half as many lattice sites as our original model and twice the original lattice spacing $a' = 2a$. This mode-elimination procedure is illustrated in Fig. 3.1. We now define new coupling constants g', $\tilde{h}'$, and $\tilde{f}'$ by demanding that

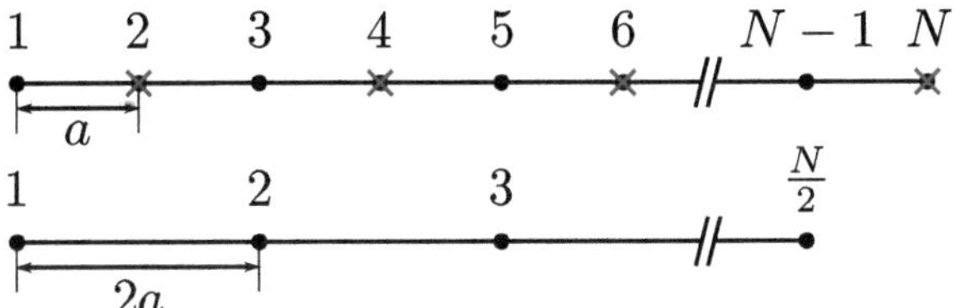

Fig. 3.1 Pictorial representation of the mode-elimination step in the Ising chain corresponding to Eq. (3.24) where all even sites are eliminated

our new transfer matrix $\mathbf{T}'$ should be of the same form as the original one given in Eq. (3.23), i.e.,

$$\mathbf{T}' = e^{\tilde{f}'}\begin{pmatrix} e^{g'+\tilde{h}'} & e^{-g'} \\ e^{-g'} & e^{g'-\tilde{h}'} \end{pmatrix} = e^{2\tilde{f}}\begin{pmatrix} e^{2g+2\tilde{h}}+e^{-2g} & e^{\tilde{h}}+e^{-\tilde{h}} \\ e^{\tilde{h}}+e^{-\tilde{h}} & e^{2g-2\tilde{h}}+e^{-2g} \end{pmatrix} . \qquad (3.26)$$

A priori it is not clear whether after mode elimination the transfer matrix can be parameterized in terms of the same number of parameters as the original transfer matrix. In fact, in two and higher dimensions the mode-elimination step in the Migdal–Kadanoff RG usually generates interactions beyond the nearest neighbors so that the new transfer matrix does not have the same form as the transfer matrix of the original nearest-neighbor model, see Exercise 3.2. Fortunately, for our one-dimensional Ising model the mode-elimination procedure does not generate new couplings. Mathematically, this implies that the matrix equation (3.26) can uniquely be solved for the new parameters g', $\tilde{h}'$ and $\tilde{f}'$ as functions of the original couplings $g, \tilde{h}$, and $\tilde{f}$. To obtain the explicit solution, we note that the matrices in Eq. (3.26) are symmetric, so that this matrix equation is equivalent to the following three non-linear equations for our three unknowns g', $\tilde{h}'$, and $\tilde{f}'$,

$$e^{\tilde{f}'+g'+\tilde{h}'} = e^{2\tilde{f}}\left(e^{2g+2\tilde{h}}+e^{-2g}\right) = 2e^{2\tilde{f}+\tilde{h}}\cosh(2g+\tilde{h}) , \qquad (3.27a)$$

$$e^{\tilde{f}'+g'-\tilde{h}'} = e^{2\tilde{f}}\left(e^{2g-2\tilde{h}}+e^{-2g}\right) = 2e^{2\tilde{f}-\tilde{h}}\cosh(2g-\tilde{h}) , \qquad (3.27b)$$

$$e^{\tilde{f}'-g'} = e^{2\tilde{f}}\left(e^{\tilde{h}}+e^{-\tilde{h}}\right) = 2e^{2\tilde{f}}\cosh\tilde{h} . \qquad (3.27c)$$

Let us now explicitly solve these equations for $\tilde{f}'$, g', and $\tilde{h}'$. Therefore, we take the logarithm of both sides of Eqs. (3.27a), (3.27b), and (3.27c) to obtain

$$\tilde{f}' + g' + \tilde{h}' = 2\tilde{f} + \tilde{h} + \ln[2\cosh(2g+\tilde{h})] , \qquad (3.28a)$$
$$\tilde{f}' + g' - \tilde{h}' = 2\tilde{f} - \tilde{h} + \ln[2\cosh(2g-\tilde{h})] , \qquad (3.28b)$$
$$\tilde{f}' - g' = 2\tilde{f} + \ln[2\cosh\tilde{h}] . \qquad (3.28c)$$

Subtracting Eq. (3.28b) from Eq. (3.28a) yields for the new magnetic field,

$$\tilde{h}' = \tilde{h} + \frac{1}{2}\ln\left[\frac{\cosh(2g+\tilde{h})}{\cosh(2g-\tilde{h})}\right] . \tag{3.29}$$

Adding Eq. (3.28c) to half of the sum of Eqs. (3.28a) and (3.28b) we obtain for the renormalized free energy,

$$\tilde{f}' = 2\tilde{f} + \frac{1}{4}\ln\left[16\cosh^2\tilde{h}\,\cosh(2g+\tilde{h})\cosh(2g-\tilde{h})\right] . \tag{3.30}$$

Finally, we subtract Eq. (3.28c) from half of the sum of Eqs. (3.28a) and (3.28b) and obtain for the new dimensionless nearest-neighbor interaction,

$$g' = \frac{1}{4}\ln\left[\frac{\cosh(2g+\tilde{h})\cosh(2g-\tilde{h})}{\cosh^2\tilde{h}}\right] . \tag{3.31}$$

For simplicity, we now set the magnetic field equal to zero, $\tilde{h} = 0$. Eq. (3.29) then implies that the new magnetic field $\tilde{h}'$ also vanishes, while the RG equation (3.31) for the renormalized coupling constant becomes

$$g' = \ln\left(\sqrt{\cosh 2g}\right) . \tag{3.32}$$

To rewrite this relation in a more convenient form, we exponentiate both sides, so that $e^{g'} = \sqrt{\cosh 2g}$ or $e^{-g'} = 1/\sqrt{\cosh 2g}$, and hence

$$e^{g'} + e^{-g'} = \sqrt{\cosh 2g}\left(1 + \frac{1}{\cosh 2g}\right) , \tag{3.33a}$$

$$e^{g'} - e^{g'} = \sqrt{\cosh 2g}\left(1 - \frac{1}{\cosh 2g}\right) . \tag{3.33b}$$

Dividing Eq. (3.33b) by (3.33a) and using the identity $\cosh 2g = 2\cosh^2 g - 1 = 1 + 2\sinh^2 g$ we finally obtain

$$\boxed{\tanh g' = \tanh^2 g ,} \tag{3.34}$$

while the RG equation (3.30) for the rescaled free energy simplifies for $\tilde{h} = 0$ to

$$\boxed{\tilde{f}' = 2\tilde{f} + \ln\left(2\sqrt{\cosh 2g}\right) .} \tag{3.35}$$

Using Eq. (3.34) and the identities $\cosh(2g) = \cosh^4 g - \sinh^4 g$ and $\cosh g' = 1/\sqrt{1 - \tanh^2 g'}$, we may rewrite the argument of the logarithm in Eq. (3.35) as follows,

$$\sqrt{\cosh 2g} = = \sqrt{\cosh^4 g - \sinh^4 g} = \cosh^2 g \sqrt{1 - \tanh^4 g}$$
$$= \cosh^2 g \sqrt{1 - \tanh^2 g'} = \frac{\cosh^2 g}{\cosh g'} \,, \tag{3.36}$$

so that Eq. (3.35) can also be written as

$$\tilde{f}' = 2\tilde{f} + \ln\left(2 \cosh g\right) + \ln\left(\frac{\cosh g}{\cosh g'}\right) \,. \tag{3.37}$$

All three terms on the right-hand side of Eq. (3.37) have a simple interpretation: the first term $2\tilde{f}$ describes the change of the free energy per site due to the reduction of the number of sites. Since after the mode-elimination step the number of sites $N' = N/2$ has been reduced by a factor of two, the free energy per site increases by the same factor. The term $2\tilde{f}$ can thus be interpreted as a contribution from the rescaling of the number of sites. The second term $\ln\left(2 \cosh g\right)$ on the right-hand side of Eq. (3.37) can be interpreted as the contribution from $N/2$ noninteracting spins to the free energy per site. However, the spins which are eliminated are not free, which is taken into account by the last term on the right-hand side of Eq. (3.37) via the change of the coupling constant. It is important to realize that the RG procedure does not modify the partition function, it just calculates it iteratively, so that with $N' = N/2$,

$$\mathcal{Z}_N(g, \tilde{f}) = \mathcal{Z}_{N'}(g', \tilde{f}') \,. \tag{3.38}$$

On the other hand, since our initial value $\tilde{f} = \beta E_0 / N$ is just an additive constant, the partition function for a system of N spins satisfies

$$\mathcal{Z}_N(g, \tilde{f}) = e^{N\tilde{f}} \mathcal{Z}_N(g, \tilde{f} = 0) \,. \tag{3.39}$$

Combining Eqs. (3.38) and (3.39), we obtain the RG equation for the partition function

$$\mathcal{Z}_N(g, 0) = e^{-N\tilde{f} + \frac{N}{2}\tilde{f}'} \mathcal{Z}_{\frac{N}{2}}(g', 0) = e^{\frac{N}{2}(\tilde{f}' - 2\tilde{f})} \mathcal{Z}_{\frac{N}{2}}(g', 0)$$
$$= \left(2\sqrt{\cosh 2g}\right)^{\frac{N}{2}} \mathcal{Z}_{\frac{N}{2}}(g', 0) = \left(\frac{2 \cosh^2 g}{\cosh g'}\right)^{\frac{N}{2}} \mathcal{Z}_{\frac{N}{2}}(g', 0) \,, \tag{3.40}$$

where in the second line we have used Eqs. (3.35) and (3.37).

   The above recursion relations are exact: we have been able to map the problem of calculating the partition function of the original one-dimensional Ising model with Hamiltonian $H(g, \tilde{h}, \tilde{f})$ on an N-site lattice with lattice spacing a onto the problem of solving an effective Ising model on a lattice with only $\frac{N}{2}$ sites and twice the lattice spacing, but with renormalized Hamiltonian $H'(g', \tilde{h}', \tilde{f}')$. The fact that H' has precisely the same form as H is a special property of the one-dimensional Ising model.

In higher dimensions, the mode-elimination step in the Migdal–Kadanoff RG generates new couplings describing longer range interactions between the spins. As a result, one cannot write down exact RG equations describing the mode-elimination step and has to rely on approximations, see Exercise 3.2. This is certainly a weak point of the Migdal–Kadanoff RG procedure. On the other hand, as first pointed out by Wegner and Houghton (1973), for a quite general class of problems which can be formulated in terms of functional integrals exact RG equations describing the mode-elimination step can be written down if the mode elimination step is performed in infinitesimal steps. This functional renormalization group (FRG) approach will be explained in detail in Part II of this book. Although the resulting functional differential equations typically cannot be solved exactly (see, however, Sect. 11.5 for a nontrivial exception), the FRG approach opens new possibilities for approximations, so that rather simple truncations of the exact FRG equations usually yield nonperturbative results.

3.2.2 Iteration and Fixed Points of the RG

At the first sight the rescaling step of our general RG procedure is still missing. However, this is only partially true, because by expressing all dimensionful quantities after the mode elimination in terms of the new lattice constant $a' = 2a$, we implicitly take the canonical dimension of the coupling constant into account. For example, the first term $2\tilde{f}$ on the right-hand side of the flow equation (3.37) for the rescaled free energy $\tilde{f}$ is due to the rescaling of the number of sites $N' = N/2 = L/a'$, where L is the length of the system. Because the spin variables in real space are not rescaled in the above RG transformation, the corresponding correlation function $G(r)$ also does not scale at the critical point, which according to Eq. (1.13) implies $D - 2 + \eta = 0$, or $\eta = 2 - D = 1$ in $D = 1$. As pointed out by Fisher (1983), the fact that even without spin rescaling (corresponding to $\eta = 1$ in $D = 1$) one obtains a sensible zero temperature fixed point for the one-dimensional Ising model should be considered as a lucky accident; in $D > 1$ one usually needs a nontrivial spin-rescaling factor in order to obtain an RG fixed point.

Let us now iterate the RG equation (3.34) for the coupling constant $g = J/T$ in the absence of a magnetic field. Suppose that the physical temperature is small but finite ($0 < T \ll J$), so that the quantity

$$x_0 = \tanh\left(\frac{J}{T}\right) = \tanh g \tag{3.41}$$

is close to unity. After one step in our RG procedure we obtain from Eq. (3.34) that the coupling g_1 of the new effective Ising model is given by

$$\tanh g_1 = x_1 = x_0^2 \, . \tag{3.42}$$

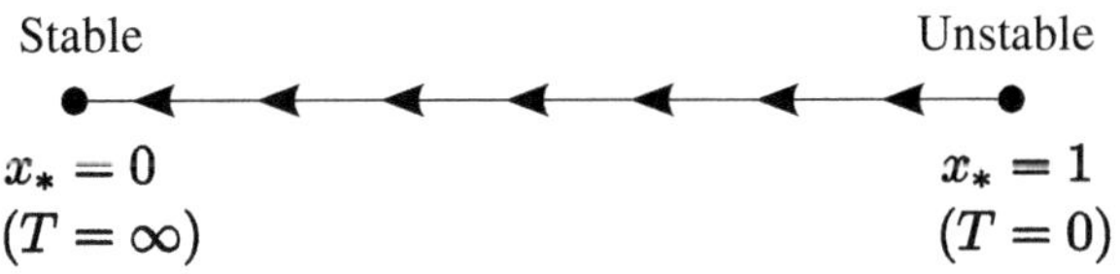

$$\text{Stable} \qquad\qquad\qquad\qquad\qquad\qquad\qquad\qquad \text{Unstable}$$

$$x_* = 0 \qquad\qquad\qquad\qquad\qquad\qquad\qquad\qquad x_* = 1$$
$$(T = \infty) \qquad\qquad\qquad\qquad\qquad\qquad\qquad\qquad (T = 0)$$

Fig. 3.2 Graphical representation of the RG flow described by the recursion relation (3.43). The *arrows* indicate the direction of change of the variables x_n as the RG is iterated

The new parameter g_1 is therefore smaller than the initial value g. Since g_1 is the ratio of the exchange coupling and the temperature, we can interpret this in two possible ways: either the new exchange coupling becomes smaller at constant temperature, or the temperature becomes larger while J is fixed. In the latter picture the RG maps the initial low-temperature Ising model onto a new high-temperature Ising model. Iterating the RG n times, we obtain a sequence of temperatures $T, T_1, T_2, T_3 \ldots$ which are determined by

$$x_{n+1} = x_n^2, \qquad x_n = \tanh\left(\frac{J}{T_n}\right). \tag{3.43}$$

The flow of successive x_n as a function of the initial x_0 is conveniently represented graphically as the RG flow diagram shown in Fig. 3.2. Obviously, the recursion relation (3.43) has two *fixed points* which are invariant under the RG transformation: a stable fixed point at $x_* = 0$ describing the Ising model at infinite temperature, and an unstable fixed point at $x_* = 1$, which determines the critical behavior of the Ising model for $T \to 0$. At any finite initial temperature ($0 < x_0 < 1$), the RG maps the Ising model to an effective model at infinite temperature. But at sufficiently high temperatures the spins are only weakly correlated and there is certainly no spontaneous magnetization. We therefore expect that this is also the case at any finite $T > 0$. To see this explicitly, let us consider the correlation length ξ. Recall that ξ is defined in terms of the asymptotic behavior of the correlation function $G(r_i - r_j)$ for $|r_i - r_j| \to \infty$, or, equivalently, in terms of the behavior of its Fourier transform $G(k)$ for small wave vectors. Since r_i and r_j are well separated, we can always choose sites i and j which are not eliminated in the RG mode-elimination step. Now, eliminating modes does not effect the correlations between any modes which are not integrated out.[4] As a consequence, only the rescaling step (which changes the unit of length) modifies the value of ξ. Since ξ has units of length and all length scales shrink by a factor of $1/b$ under the rescaling transformation, we conclude that after one complete iteration of the RG the correlation length transforms as

$$\xi' \equiv \xi(x') = \frac{\xi(x)}{b} . \tag{3.44}$$

---

[4] The invariance of the correlation length under the mode-elimination step can be shown mathematically via the following chain of identities,

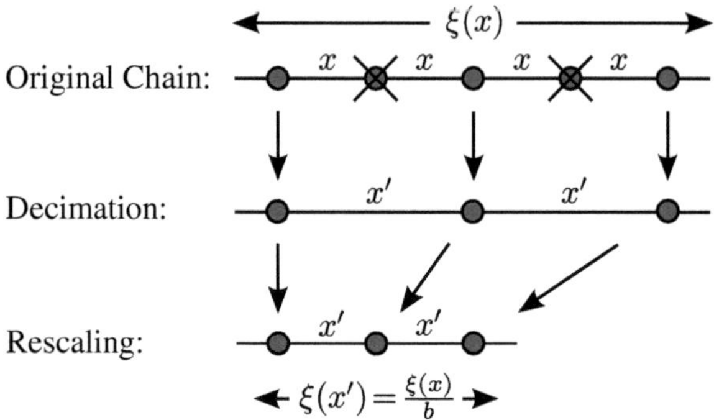

Fig. 3.3 Pictorial representation of the change of the correlation length under the RG transformation. After decimation the new lattice constant is a factor of b larger than the old one, while the correlation length is constant. After rescaling, all length scales are scaled down by a factor of $1/b$

The transformation behavior of the correlation length is illustrated graphically for $b = 2$ in Fig. 3.3. According to our recursion relation (3.43), we have $x' = x^2$, so that Eq. (3.44) implies

$$\xi(x^2) = \frac{\xi(x)}{2} \, . \tag{3.45}$$

This functional equation has the solution

$$\xi(x) = -\frac{a_0}{\ln x} \, , \tag{3.46}$$

where a_0 is an arbitrary length scale. At low temperatures, where $g = J/T \gg 1$, we may approximate $x = \tanh g \approx 1 - 2e^{-2g}$. Identifying $a_0 = a$ with the lattice

$$
\begin{aligned}
G(\boldsymbol{r}_i - \boldsymbol{r}_j) &\equiv a^{2-D} \langle \delta s_i \delta s_j \rangle \equiv a^{2-D} \frac{\sum_{\{s_i\}} \delta s_i \delta s_j \, e^{-\tilde{H}(s_1,s_2,\dots s_N)}}{\sum_{\{s_i\}} e^{-\tilde{H}(s_1,s_2,\dots s_N)}} \\
&= a^{2-D} \frac{\sum_{s_1,s_3,\dots,s_i,\dots,s_j,\dots,s_{N-1}} \delta s_i \delta s_j \sum_{s_2,s_4,\dots,s_N} e^{-\tilde{H}(s_1,s_2,\dots s_N)}}{\sum_{s_1,s_3,\dots,s_i,\dots,s_j,\dots,s_{N-1}} \sum_{s_2,s_4,\dots,s_N} e^{-\tilde{H}(s_1,s_2,\dots s_N)}} \\
&= a^{2-D} \frac{\sum_{s_1,s_3,\dots,s_i,\dots,s_j,\dots,s_{N-1}} \delta s_i \delta s_j \, e^{-\tilde{H}'(s_1,s_3,\dots s_{N-1})}}{\sum_{s_1,s_3,\dots,s_i,\dots,s_j,\dots,s_{N-1}} e^{-\tilde{H}'(s_1,s_3,\dots s_{N-1})}} \, ,
\end{aligned}
$$

where $\tilde{H} = \beta H$ and the reduced Hamiltonian $\tilde{H}'$ is defined by

$$e^{-\tilde{H}'(s_1,s_3,\dots s_{N-1})} \equiv \sum_{s_2,s_4,\dots,s_N} e^{-\tilde{H}(s_1,s_2,\dots s_N)} \, .$$

Apparently the mode-elimination step eliminates modes which by the definition of the correlation function have to be eliminated anyway. Consequently, the correlation function and the correlation length ξ do not change under the explicit mode-elimination process.

constant, we obtain for the low-temperature behavior of the correlation length of the one-dimensional Ising model,

$$\xi \sim \frac{a}{2} e^{2J/T} ,$$

(3.47)

which agrees with the exact low-temperature result for the correlation length obtained in Exercise 1.1. Obviously, for any finite temperature the correlation length is finite, so that the system is disordered. In the zero temperature limit the correlation length diverges. The point $T = 0$ is therefore a critical point of the one-dimensional Ising model. The corresponding critical exponents and the scaling functions have been calculated explicitly in Exercise 1.1. Note that the critical point corresponds to the unstable RG fixed point $x_* = 1$ in Fig. 3.2. We shall show in Sect. 3.3.2 that there is a general connection between critical points of continuous phase transitions and unstable RG fixed points.

3.2.3 Infinitesimal Form of RG Recursion Relations

So far, we have eliminated every second spin, so that after one iteration of the RG the system has $N' = \frac{N}{2}$ sites and lattice spacing $a' = 2a$. But there is no unique way of implementing the decimation procedure; we may just as well eliminate blocks of two neighboring spins as shown in Fig. 3.4. In this case $N' = N/3$ and $a' = 3a$. More generally, we may eliminate blocks of b spins, so that after one mode-elimination step the lattice has $N' = N/b$ sites and lattice spacing $a' = ba$. The RG recursion relations for the parameters of the new Ising model after mode elimination can be obtained in analogy with Eq. (3.24) by expressing the partition function in terms of the renormalized T-Matrix as

$$\mathcal{Z} = \mathrm{Tr}\big[\mathbf{T}^N\big] = \mathrm{Tr}\big[(\mathbf{T}^b)^{N/b}\big] \equiv \mathrm{Tr}\big[(\mathbf{T}')^{N'}\big] .$$

(3.48)

The recursion relations for any integer b can be obtained again by demanding that $\mathbf{T}' = \mathbf{T}^b$ is again of the same form as the original T-matrix,

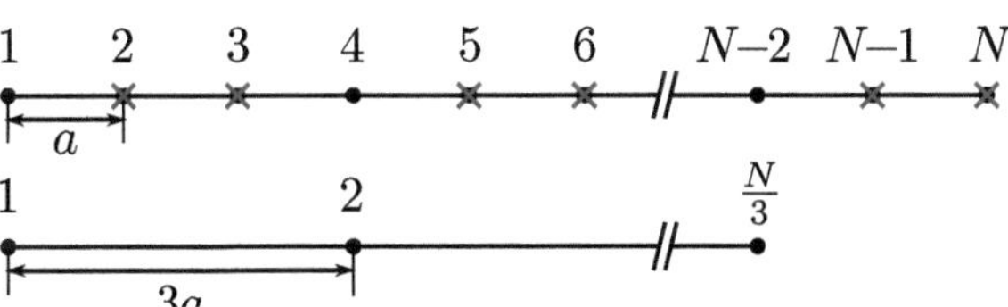

Fig. 3.4 Pictorial representation of a mode-elimination step in the Ising chain where blocks of two neighboring spins are eliminated

$$\mathbf{T}' = e^{\tilde{f}'} \begin{pmatrix} e^{g'+\tilde{h}'} & e^{-g'} \\ e^{-g'} & e^{g'-\tilde{h}'} \end{pmatrix} = \left[e^{\tilde{f}} \begin{pmatrix} e^{g+\tilde{h}} & e^{-g} \\ e^{-g} & e^{g-\tilde{h}} \end{pmatrix} \right]^{b} . \tag{3.49}$$

Setting for simplicity $\tilde{h} = 0$, this implies for the renormalized coupling constant (see Exercise 3.1),

$$\tanh g' = (\tanh g)^{b} , \tag{3.50}$$

and for the renormalized free energy per site,

$$\tilde{f}' = b\tilde{f} + \ln\left[\frac{2^{b-1}(\cosh g)^{b}}{\cosh g'}\right]$$

$$= b\tilde{f} + (b-1)\ln(2\cosh g) + \ln\left(\frac{\cosh g}{\cosh g'}\right) . \tag{3.51}$$

Eqs. (3.50) and (3.51) generalize Eqs. (3.34) and (3.37) for arbitrary b. Although these equations have been derived for integer b, they can be analytically continued for any real b. Using this freedom, we may set

$$b = e^{l} = 1 + l + \mathcal{O}(l^{2}) \tag{3.52}$$

with infinitesimal l, and consider the limit $l \to 0$, where the discrete recursion relations become differential equations. Setting $g' \equiv g_{l}$ and $x_{l} = \tanh g_{l}$ with initial value $g_{l=0} = g$, and expanding for small l,

$$x_{l} = x_{0}^{b} = x_{0}^{1+l+\mathcal{O}(l^{2})} \approx x_{0}[1 + l \ln x_{0} + \mathcal{O}(l^{2})] , \tag{3.53}$$

we obtain from Eq. (3.50)

$$\frac{x_{l} - x_{0}}{l} = x_{0} \ln x_{0} + \mathcal{O}(l) . \tag{3.54}$$

Iterating this equation and taking the limit $l \to 0$ this reduces to the differential equation

$$\boxed{\frac{dx_{l}}{dl} = x_{l} \ln x_{l} .} \tag{3.55}$$

Integrating this differential equation with initial condition $x_{0} = \tanh g$ up to a finite l, we obtain the coupling constant $x_{l} = \tanh g_{l}$ of the effective Ising model which results from the elimination of $b - 1 = e^{l} - 1$ spins. Eq. (3.55) can be explicitly integrated by separation of variables; the result is

$$x_{l} = x_{0}^{e^{l}} , \tag{3.56}$$

which agrees of course with Eq. (3.53). The right-hand side of the differential RG flow equation (3.55) is called RG β-function, which is usually denoted by

$$\beta(x) = x \ln x \ . \tag{3.57}$$

Note that the RG fixed points where $dx_l/dl = 0$ correspond to the zeros of the β-function, in our case $x_* = 0$ and $x_* = 1$.

If we use the variable g_l instead of $x_l = \tanh g_l$, then Eq. (3.55) looks more complicated: abbreviating $\partial_l = \partial/\partial l$ and using $\partial_l \tanh g_l = \partial_l g_l / \cosh^2 g_l$ we obtain

$$\partial_l g_l = \cosh g_l \sinh g_l \ln(\tanh g_l) \ . \tag{3.58}$$

At low temperatures, where $g_l \gg 1$, this simplifies to

$$\partial_l g_l \approx \frac{1}{4} e^{2g_l} \ln\left(1 - 2e^{-2g_l}\right) \approx -\frac{1}{2} \ , \tag{3.59}$$

which yields for the dimensionless temperatures $t_l = 1/g_l = T_l/J$,

$$\partial_l t_l \approx \frac{t_l^2}{2} \ , \tag{3.60}$$

with solution

$$t_l = \frac{t_0}{1 - \frac{t_0}{2}l} \ . \tag{3.61}$$

Since Eq. (3.60) has been derived under the assumption $t_l \ll 1$, the solution of Eq. (3.61) can only be trusted up to $l \lesssim l_*$ where $t_* = t_{l_*}$ is of the order of unity. Roughly, this is the scale where the denominator in Eq. (3.61) vanishes, $l_* = 2/t_0$. But at high temperatures of order J, the correlation length should be of the order of the lattice spacing, so that according to Eq. (3.44) the correlation length in units of the lattice spacing can be estimated to be proportional to $b_* = e^{l_*} = e^{2/t_0}$ at low temperatures, in agreement with our previous result (3.47).

The infinitesimal form of the RG equation for the free energy per site can be obtained analogously from Eq. (3.51),

$$\begin{aligned}
\partial_l \tilde{f}_l &= \tilde{f}_l + \ln(2\cosh g_l) + \partial_l \left[\ln\left(\frac{\cosh g_0}{\cosh g_l}\right)\right]_{l=0} \\
&= \tilde{f}_l + \ln(2\cosh g_l) - (\tanh g_l)\partial_l g_l \\
&= \tilde{f}_l + \ln(2\cosh g_l) - \sinh^2 g_l \ln(\tanh g_l) \ . \tag{3.62}
\end{aligned}$$

The first term on the right-hand side in the last line is due to the fact that at RG step l the number of sites $N_l = N/b = Ne^{-l}$ has decreased by a factor of e^{-l}. In D dimensions, the corresponding flow equation for the free energy per unit volume would start with a term $\partial_l \tilde{f} = D\tilde{f} + \dots$. More generally, the RG flow equation of any coupling g_l which can be made dimensionless by multiplication with a factor $(\text{length})^{D_g}$ contains a term of the form

$$\partial_l g = D_g g + \dots , \tag{3.63}$$

where D_g is called the *canonical dimension* (also called *engineering dimension*) of the coupling g_l. The value of D_g follows from simple dimensional analysis; the first term on the right-hand side of the differential RG flow equation (3.63) simply expresses the fact that the RG rescaling step changes the length scale. As already pointed out at the beginning of Sect. 3.2.2, in the Migdal–Kadanoff RG the length rescaling step is implicitly performed by introducing dimensionless couplings by multiplying the dimensionful couplings with a factor depending on the reduced lattice spacing, as is obvious from Eq. (3.62).

3.3 General Properties of RG Flows

In this section we shall discuss some basic properties of RG flows which are independent of any specific model. In particular, we explain the deep relation between fixed points of the RG transformation and critical points of the underlying physical system, which is the key to understand the microscopic origin of the scaling hypothesis formulated in Sect. 1.3. We also show how the linearized RG flow around RG fixed points determines the critical exponents, and discuss some global properties of RG flows.

3.3.1 RG Fixed Points and the Critical Surface

Given a general RG transformation $\mathcal{R}(b; g)$ acting on the (possibly infinite) set of couplings $g = (g_1, g_2, g_3, \dots)$, a *fixed point* of the RG is a special point $g^* = (g_1^*, g_2^*, g_3^*, \dots)$ in coupling space which is invariant under the transformation $\mathcal{R}(b; g)$, i.e.,

$$\boxed{g^* = \mathcal{R}(b; g^*) \,.} \tag{3.64}$$

To understand the physical significance of RG fixed points, recall that in Sect. 3.2 we have found that the RG transformation for the one-dimensional Ising model has two different fixed points illustrated in Fig. 3.2: an unstable fixed point associated with the critical point of the Ising model at zero temperature, and a stable fixed point describing completely decoupled spins at infinite temperature. From Eq. (3.47) it is obvious that the correlation length at the zero temperature fixed point is infinite; on the other hand, at infinite temperature the spins are uncorrelated, so that the correlation length vanishes. The fact that at a fixed point of the RG the correlation length is either infinite or zero is not a special property of the Ising chain: any RG fixed point describes either a critical system with infinite correlation length, or a completely decoupled system with vanishing correlation length. To see this, note that according to Eq. (3.44) the correlation length ξ transforms under one iteration of the RG as

$$\xi(\boldsymbol{g}') = \frac{\xi(\boldsymbol{g})}{b} \,, \tag{3.65}$$

with arbitrary $b > 1$. Let us emphasize that Eq. (3.65) is an exact and quite general scaling relation which holds by definition for any mode-elimination process in the Wilsonian RG procedure. To see this, note that the correlation length can be defined via the correlation function $G(\boldsymbol{k})$ for small wave vectors.[5] In particular, $|\boldsymbol{k}|$ should be chosen much smaller than the cutoff Λ, such that we may formally integrate out degrees of freedom between Λ and Λ_0, leaving the correlation function unchanged. More precisely, using Eq. (2.98) we have

$$G(\boldsymbol{k}) = a^2 \frac{\int \mathcal{D}[\varphi]\, |\varphi_{\boldsymbol{k}}|^2 e^{-S[\varphi;\boldsymbol{g}]}}{\int \mathcal{D}[\varphi]\, e^{-S[\varphi;\boldsymbol{g}]}} = a^2 \frac{\int \mathcal{D}[\varphi^<]\, |\varphi_{\boldsymbol{k}}|^2 \int \mathcal{D}[\varphi^>]\, e^{-S[\varphi^<+\varphi^>;\boldsymbol{g}]}}{\int \mathcal{D}[\varphi^<]\, \int \mathcal{D}[\varphi^>]\, e^{-S[\varphi^<+\varphi^>;\boldsymbol{g}]}}$$

$$= a^2 \frac{\int \mathcal{D}[\varphi^<]\, |\varphi_{\boldsymbol{k}}|^2 e^{S^<[\varphi^<;\boldsymbol{g}^<]}}{\int \mathcal{D}[\varphi^<]\, e^{S^<[\varphi^<;\boldsymbol{g}^<]}} \,, \tag{3.66}$$

which does not depend on whether we have already eliminated modes explicitly (as in the last line) or not. Using now the fact that at the fixed point $\boldsymbol{g} = \boldsymbol{g}' = \boldsymbol{g}^*$, we conclude from Eq. (3.65) that

$$\xi(\boldsymbol{g}^*) = \frac{\xi(\boldsymbol{g}^*)}{b} \,, \tag{3.67}$$

which can only be satisfied if $\xi(\boldsymbol{g}^*)$ is either infinite or zero. Fixed points with $\xi = \infty$ describe critical points associated with continuous phase transitions and are therefore called *critical fixed points*, while fixed points with $\xi = 0$ are called *trivial fixed points*. In general, a given RG transformation will have several fixed

---

[5] If we write the exact correlation function in Fourier space as $G(\boldsymbol{k}) = [c_0 \boldsymbol{k}^2 + \Sigma(\boldsymbol{k})]^{-1}$, and expand the so-called self-energy function $\Sigma(\boldsymbol{k})$ in powers of $\boldsymbol{k}$, then a simple calculation analogous to the one in Sect. 2.3.3 implies that the correlation length is given by $c_0/\xi^2 = Z\Sigma(0)$ with $Z^{-1} = 1 + \partial\Sigma(\boldsymbol{k})/\partial(c_0\boldsymbol{k}^2)|_{\boldsymbol{k}=0}$. We shall derive these relations in Sect. 4.2.3, see Eqs. (4.97) and (4.98). In Gaussian approximation $\Sigma(\boldsymbol{k}) = A_t r_0$ (see Eq. (2.102)), so that $c_0/\xi^2 = Z\Sigma(0) \approx A_t r_0$ reduces to our previous result (2.80). For a system where all fluctuations with wave vectors with $|\boldsymbol{k}| > \Lambda = \Lambda_0/b$ have been eliminated via the RG the self-energy $\Sigma_\Lambda(\boldsymbol{k})$ and Z_Λ depend on the cutoff Λ, so that we may define a cutoff-dependent length ξ_Λ by setting $c_0/\xi_\Lambda^2 = Z_\Lambda \Sigma_\Lambda(0)$. By definition, in the limit $\Lambda \to 0$ the length ξ_Λ can be identified with the physical correlation length ξ.

points or even a continuum of fixed points forming a certain manifold in coupling space. For example, the RG flow of coupling constants for the two-dimensional Ising model on the square lattice is characterized by the three fixed points shown in Fig. 3.5. Obviously, each fixed point has its own *basin of attraction*, which is the set of points in coupling constant space which flows into the fixed point when the RG is iterated. The basin of attraction of the critical fixed point C where the correlation length is infinite is called its *critical surface* or *critical manifold*. It turns out that the correlation length is not only infinite precisely for the special values g^* of the coupling constants associated with a critical fixed point, but also on the entire critical surface which is mapped into g^* by iterating the RG transformation. Therefore all couplings on the critical surface describe a physical system precisely at the critical point. To see this, suppose we adjust the bare couplings g of some system such that they lie on the critical surface of a certain critical fixed point. For the Ising model with nearest and next-nearest-neighbor interactions we may fine-tune the temperature until we hit the critical surface of the critical fixed point C shown in Fig. 3.5. Suppose now that we iterate the elementary RG step n times and keep track of the correlation length $\xi(g^{(n)})$, where $g^{(n)}$ denotes the value of the renormalized coupling constants after n iterations. From Eq. (3.65) we obtain

$$\xi(g) = b\xi(g^{(1)}) = b^2\xi(g^{(2)}) = \ldots = b^n\xi(g^{(n)}) . \tag{3.68}$$

Assuming now that the initial values g of the couplings lie on the critical surface, then by definition $\lim_{n\to\infty} g^{(n)} = g^*$. But for a critical fixed point $\xi(g^*) = \infty$, so that Eq. (3.68) implies that also the original correlation length $\xi(g)$ must be infinite.

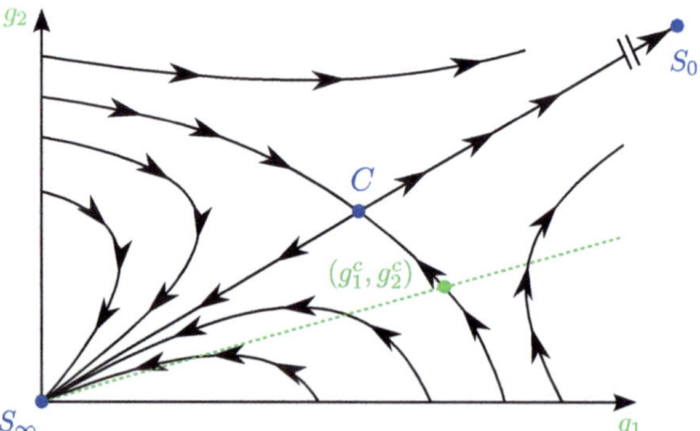

Fig. 3.5 Projected RG flow of the square lattice Ising model in the plane spanned by the dimensionless couplings $g_1 = \beta J$ and $g_2 = \beta J'$ representing nearest (J) and next-nearest-neighbor (J') interactions. There are two trivial fixed points (the zero-temperature fixed point S_0 and the infinite-temperature fixed point S_∞) and one critical fixed point (C). All points flowing into the critical fixed point C constitute its critical manifold and are characterized by an infinite correlation length. Monte Carlo calculations supporting the qualitative features of this schematic flow diagram can be found in Swendsen (1984)

3.3.2 Local RG Flow Close to a Fixed Point: Classification of Couplings and Justification of the Scaling Hypothesis

One of the most fundamental insights provided by the RG is the connection between critical exponents and the local behavior of the RG flow in the vicinity of a given RG fixed point. Consider a general RG transformation of the form (3.15b) and a certain fixed point g^* satisfying the fixed point condition (3.64). Subtracting Eq. (3.64) from Eq. (3.15b) we obtain

$$\delta g' \equiv g' - g^* = \mathcal{R}(b; g) - \mathcal{R}(b; g^*) . \tag{3.69}$$

Assuming now that the deviation $\delta g = g - g^*$ from the fixed point in coupling space is sufficiently small, we may expand the right-hand side of Eq. (3.69) to linear order in δg and obtain the linearized RG flow in the vicinity of the RG fixed point $g^* = \left(g_1^*, g_2^*, g_3^*, \ldots\right)$,

$$\delta g' = \mathbf{R}(b; g^*)\delta g , \tag{3.70}$$

where $\mathbf{R}(b; g^*)$ is a matrix in coupling space with matrix elements

$$R_{ij}(b; g^*) = \left.\frac{\partial \mathcal{R}_i(b; g)}{\partial g_j}\right|_{g^*} . \tag{3.71}$$

Note that each fixed point of a general RG transformation $\mathcal{R}(b; g)$ is characterized by a different matrix $\mathbf{R}(b; g^*)$. To classify the linearized RG flow in the vicinity of a given fixed point, it is useful to determine the eigenvalues of the matrix $\mathbf{R}(b; g^*)$. However, there is no reason why the matrix $\mathbf{R}(b; g^*)$ should be symmetric, so that in general it cannot be diagonalized and we should distinguish between left and right eigenvectors. For our purpose we need only the left eigenvectors v_α^T and the corresponding eigenvalues $\lambda_\alpha(b)$, which we label by an index $\alpha = 1, 2, 3, \ldots$. By definition, the left eigenvectors satisfy the eigenvalue equation

$$v_\alpha^T \mathbf{R}(b; g^*) = v_\alpha^T \lambda_\alpha(b) . \tag{3.72}$$

To discuss the RG flow in the vicinity of the fixed point g^*, it is useful to project the coupling vector δg onto the left eigenvectors v_α^T of the matrix $\mathbf{R}(b; g^*)$,

$$\boxed{u_\alpha = v_\alpha^T \delta g = \sum_i v_{\alpha,i} \delta g_i .} \tag{3.73}$$

These special linear combinations of the coupling constants are called *scaling variables*, because they do not mix under the linearized RG transformation (3.70), which follows simply from Eqs. (3.72) and (3.73),

$$u'_\alpha = v^T_\alpha \delta g' = v^T_\alpha \mathbf{R}(b; g^*)\delta g = \lambda_\alpha(b)v^T_\alpha \delta g = \lambda_\alpha(b)u_\alpha \ . \qquad (3.74)$$

The crucial point is now that the semigroup property imposes a strong constraint on the b-dependence of the eigenvalues $\lambda_\alpha(b)$. According to Eq. (3.17) the composition of two RG transformations satisfies

$$\mathcal{R}(b'; \mathcal{R}(b; g)) = \mathcal{R}(b'b; g) = \mathcal{R}(bb'; g) = \mathcal{R}(b; \mathcal{R}(b'; g)) \ , \qquad (3.75)$$

implying for the transformation matrices associated with the linearized RG flow around a given fixed point,

$$\mathbf{R}(b; g^*)\mathbf{R}(b'; g^*) = \mathbf{R}(bb'; g^*) = \mathbf{R}(b'; g^*)\mathbf{R}(b; g^*) \ . \qquad (3.76)$$

For different values of b all members of the continuous family of matrices $\mathbf{R}(b; g^*)$ therefore commute, so that their eigenvectors v^T_α are independent of b. Moreover, Eq. (3.76) implies that the eigenvalues satisfy

$$\lambda_\alpha(b)\lambda_\alpha(b') = \lambda_\alpha(bb') \ . \qquad (3.77)$$

This functional equation has the solution

$$\boxed{\lambda_\alpha(b) = b^{y_\alpha} \ ,} \qquad (3.78)$$

with some b-independent exponent y_α. To linear order in the deviations from the fixed point, the RG equation for the scaling variables u_α defined in Eq. (3.73) is therefore

$$u'_\alpha = b^{y_\alpha} u_\alpha \ , \qquad (3.79)$$

or with $b = e^l$ in differential form

$$\boxed{\frac{du_\alpha}{dl} = y_\alpha u_\alpha \ ,} \qquad (3.80)$$

where one should keep in mind that corrections of order u^2_α are neglected on the right-hand side. The y_α are called the renormalization group eigenvalues associated with the scaling variables u_α. By construction, the fixed point now corresponds to $u^*_\alpha = 0$. Eq. (3.80) describes the growth or decay of particular linear combinations $u_\alpha = \sum_i v_{\alpha,i} \left(g_i - g^*_i\right)$ of couplings in the vicinity of a given fixed point g^*. Obviously, for $y_\alpha > 0$ a small initial deviation $u_{\alpha,0} \neq 0$ from the fixed point grows exponentially when the RG is iterated so that the RG flow is repelled from the fixed point; scaling variables u_α with positive y_α are therefore called *relevant*. On the other hand, for $y_\alpha < 0$ small initial deviations from the fixed point decrease as we iterate the RG, so that the corresponding scaling variables are called *irrelevant*. Finally, there are *marginal* scaling variables with $y_\alpha = 0$; in this case one has to retain the corrections of quadratic order in the coupling constants in order to decide whether the RG flow in coupling space approaches a given fixed point or flows away from it. In the former case the coupling is called *marginally irrelevant*, while in the latter case it is called *marginally relevant*.[6]

The exponents y_α associated with relevant couplings are closely related to the critical exponents. The precise relation depends on the relevant scaling variables of the system and on the nature of the fixed point. For simplicity, consider here the nearest-neighbor Ising model in a magnetic field h, which can be described by the effective field theory $S[\varphi]$ in Eq. (2.40). To obtain the critical behavior of the system below four dimensions, we may use the simpler φ^4-theory $S_{\Lambda_0}[\varphi]$ defined in Eq. (2.65). We shall show in Chap. 4 that in $D < 4$ the critical RG fixed point which determines the critical behavior of the system is characterized by two relevant scaling variables: a thermal variable t_l, which is proportional to the deviation of the temperature from the critical temperature, and a suitably defined magnetic field variable h_l. The RG equations describing the linearized flow in the vicinity of the critical fixed point are then of the form

$$\partial_l t_l = y_t t_l \, , \quad \partial_l h_l = y_h h_l \, , \tag{3.81}$$

with positive exponents y_t and y_h. The solutions are simply

$$t_l = t_0 e^{y_t l} = t_0 b^{y_t}, \quad h_l = h_0 e^{y_h l} = h_0 b^{y_h} \, . \tag{3.82}$$

Taking into account that after the rescaling step the volume $V' = V/b^D$ shrinks by a factor of $1/b^D$, we conclude that the singular part of the free energy density transforms under the RG as

$$f_{\text{sing}}(t_0, h_0) = b^{-D} f_{\text{sing}}(t_l, h_l) = b^{-D} f_{\text{sing}}(b^{y_t} t_0, b^{y_h} h_0) \, , \tag{3.83}$$

which is precisely the scaling hypothesis (1.16) for the free energy. Due to the linearization of the RG flow equations (3.81) and the neglect of marginal and irrelevant couplings, Eq. (3.83) applies only to the singular part of the free energy density, which is dominated by the scale dependence of the relevant scaling variables in the vicinity of the RG fixed point.

Next, let us justify the scaling hypothesis for the correlation function given in Eq. (1.25). Therefore, we should calculate how the correlation function transforms

---

[6] Since the matrix $\mathbf{R}(b; \mathbf{g}^*)$ is in general not symmetric, one cannot exclude the possibility that some of its eigenvalues are complex, so that the associated exponents y_α are also complex. In this case more complicated RG flows are possible, such as limit cycles or even chaotic RG flows. Although this possibility was discussed by Wilson (1971a), for a long time no models exhibiting this type of behavior could be found. However, recently some models whose RG flows exhibit limit cycles have been constructed (Glazek and Wilson 2002, LeClair et al. 2003) and a physically relevant model exhibiting a cyclic RG flow has been found (Moroz et al. 2009), see also the discussion at the end of Sect. 3.3.3.

under the RG. For simplicity, consider the Fourier transform $G(k)$ of the correlation function of the Ising model, which according to Eq. (2.100) can be written as a functional average,

$$(2\pi)^D \delta(k_1 + k_2) G(k_1; g) = \langle \delta\varphi(k_1)\delta\varphi(k_2)\rangle_S \equiv \frac{\int \mathcal{D}[\varphi] e^{-S[\varphi]} \delta\varphi(k_1)\delta\varphi(k_2)}{\int \mathcal{D}[\varphi] e^{-S[\varphi]}} ,$$

$$(3.84)$$

where the effective action $S[\varphi]$ is defined in Eq. (2.40) and g is the infinite set of coupling constants defining the lattice action $S[\varphi]$. But according to Eq. (3.9), after one iteration of the RG the fields with momenta smaller than the cutoff should be rescaled as $\delta\varphi(k) = \zeta_b \delta\varphi'(k')$ with $k' = bk$, so that

$$\langle \delta\varphi(k_1)\delta\varphi(k_2)\rangle_S = \zeta_b^2 \langle \delta\varphi'(k_1')\delta\varphi'(k_2')\rangle_{S'} = \zeta_b^2 (2\pi)^D \delta(k_1' + k_2') G(k_1'; g') . \quad (3.85)$$

Keeping in mind that $\delta(k') = \delta(bk) = b^{-D}\delta(k)$, we obtain from Eqs. (2.100) and (3.85)

$$\boxed{G(k; g) = \frac{\zeta_b^2}{b^D} G(k'; g') .}$$

$$(3.86)$$

To determine the field rescaling factor ζ_b, we note that the field term $h_0\varphi(k = 0)$ in our effective action (2.61) is invariant under the RG transformation, because it involves only the $k = 0$ component of the field which is not affected by the RG. Hence,

$$h_0\varphi(k = 0) = h_l\varphi'(k' = 0) = \frac{h_l}{\zeta_b}\varphi(k = 0) . \qquad (3.87)$$

This implies $h_l = \zeta_b h_0$ and hence

$$\zeta_b = b^{y_h} . \qquad (3.88)$$

Finally, we use again the fact that the singular part of the correlation function is dominated by the scaling of the relevant couplings t_l and h_l and obtain

$$G_{\text{sing}}(k; t_0, h_0) = b^{2y_h - D} G_{\text{sing}}(bk; b^{y_t} t_0, b^{y_h} h_0) , \qquad (3.89)$$

which is the Fourier transform of the scaling hypothesis (1.25) for the correlation function. Recall that in Sect. 1.3 we have shown that the scaling relations (3.83) and (3.89) imply that the correlation length exponent $\nu = 1/y_t$ can be identified with the inverse of the eigenvalue of the thermal scaling variable, see Eq. (1.28). In this way the numerical values of the critical exponents are related to the linearized RG flow in the vicinity of critical fixed points.

The scaling relation (3.89) does not depend on the irrelevant scaling variables u_α with $y_\alpha < 0$. On the other hand, microscopic lattice models are usually characterized by infinitely many coupling constants with negative scaling dimensions, implying the existence of infinitely many irrelevant scaling variables. For example, the expansion of the exact order-parameter field theory $S[\varphi]$ for the Ising model defined in Eq. (2.40) in powers of the fields contains infinitely many terms with a negative scaling dimension. From Eq. (2.51) it is obvious that the terms in this expansion depend on the Fourier transform J_k of the exchange coupling. Each term in the expansion of J_k in powers of k defines a different coupling constant, such as the couplings c_4' and c_6' in Eqs. (3.14b) and (3.14c).

As the RG is iterated there is an enormous reduction of the dimensionality of the coupling space and all irrelevant terms drop out. The leading singular behavior of physical observables is controlled by a small number of relevant (and marginally relevant) couplings. But what is then the effect of the irrelevant couplings? It turns out that *the effect of irrelevant couplings can be implicitly taken into account by redefining the numerical initial values of the relevant couplings.* Let us illustrate this with a simple example, following Polchinski (1984). Consider a *toy model* whose RG flow can be described by one relevant coupling $u = u_l$ and one irrelevant coupling $v = v_l$, with RG flow equations,

$$\partial_l u = \epsilon u + A(u, v) , \tag{3.90a}$$

$$\partial_l v = -\lambda v + B(u, v) , \tag{3.90b}$$

where for small u, v the functions $A(u, v)$ and $B(u, v)$ have an expansion

$$A(u, v) = a_1 u^2 + a_2 u v + a_3 v^2 + \dots , \tag{3.91a}$$

$$B(u, v) = b_1 u^2 + b_2 u v + b_3 v^2 + \dots . \tag{3.91b}$$

Assuming $y_u = \epsilon > 0$ and $y_v = -\lambda < 0$, the coupling u is relevant, while v is irrelevant. The linearization of the RG flow around the fixed point $(u^*, v^*) = (0, 0)$, where $du/dl = dv/dl = 0$, corresponds to setting $A = B = 0$. In this case the solution of Eqs. (3.90a) and (3.90b) with initial condition $u_{l=0} = u_0$ and $v_{l=0} = v_0$ is simply $u_l = u_0 e^{+\epsilon l}$ and $v_l = v_0 e^{-\lambda l}$, such that u_l grows exponentially while v_l vanishes exponentially for $l \to \infty$. Even if we include the effect of the functions A and B the flow is still attracted by a one-dimensional submanifold which is determined by integrating the flow equations from $(u_0, v_0) = (\pm\delta, 0)$ with small $\delta > 0$. However, before the deviation of the flow from this submanifold is damped away, it will cause the relevant coupling u_l to run a little faster or slower than it would run for $v_0 = 0$. This is shown graphically in Fig. 3.6 where we have integrated the flow equations (3.90a) and (3.90b) for parameters given in the figure caption. The net effect of the irrelevant coupling is that the value of $u_l(u_0, v_0)$ is changed by a finite constant $\Delta u_l = |u_l(u_0, v_0) - u_l(u_0, v_0 = 0)|$. We can therefore absorb the effect of the irrelevant coupling v_0 by redefining the initial value u_0 of the relevant coupling u.

In some cases, however, the effect of the irrelevant couplings cannot be implicitly taken into account via a finite redefinition of the relevant ones. This happens if the scaling functions exhibit a singular dependence on some special irrelevant coupling,

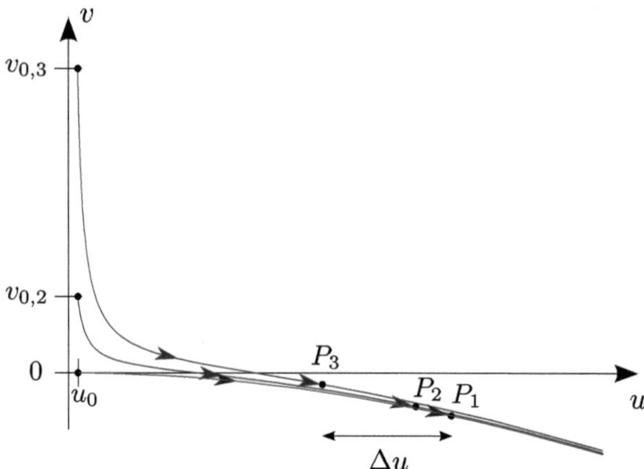

Fig. 3.6 RG flow of the toy model with RG flow equations (3.90a) and (3.90b), $\epsilon = \lambda = 1$, $A(u, v) = -uv$, and $B(u, v) = -u^2$ in the u–v-plane. While the initial relevant coupling u_0 is chosen the same for all three trajectories, we have chosen different initial values $v_{0,i}$ (with $i = 1$, 2, 3) for the irrelevant coupling v_l. Clearly, independent of the irrelevant initial value v_0 the flow approaches the critical one-dimensional submanifold as the flow progresses. For the parameters chosen here, a positive initial value v_0 of the irrelevant coupling v_l causes the relevant coupling u_l to run a little slower and thus changes its value at a given RG time l by a finite amount. This can be seen explicitly by considering the three points P_i, which all correspond to the same l. The shift $\Delta u_l = |u_l(u_0, v_0) - u_l(u_0, v_0 = 0)|$, however, can be compensated by a redefinition of the initial value u_0

so that we cannot simply set it equal to zero. The corresponding irrelevant coupling is then called *dangerously irrelevant*. An example is the quartic interaction u in the φ^4-theory (2.65) above four dimensions. In Chap. 4 we shall derive the RG flow equation for this coupling within the so-called one-loop approximation. The Ginzburg criterion discussed in Sect. 2.3.4 suggests that for $D > 4$ the quartic coupling u is indeed irrelevant, which means that $y_u < 0$. Nevertheless, we cannot ignore u in the scaling of the free energy, so that we should replace Eq. (1.16) by

$$f_{\text{sing}}(t, h, u) = b^{-D} f_{\text{sing}}(b^{y_t} t, b^{y_h} h, b^{y_u} u) . \tag{3.92}$$

The crucial point is now that for small u the singular part of the free energy exhibits a singular dependence on u, e.g. for $h = 0$ we have

$$f_{\text{sing}}(-1, h = 0, u_0) \propto u_0^{-1} , \tag{3.93}$$

such that

$$f_{\text{sing}}(t, h = 0, u) \propto |t|^{(d-y_u)/y_t}/u = |t|^2/u . \tag{3.94}$$

This implies $\alpha = 0$ which (for $D > D_{\text{up}}$) is the exact result. The critical exponents β and γ will be discussed Exercise 3.3, see also (Goldenfeld 1992, pp. 359–361).

3.3.3 Global Properties of RG Flows and Classification of Fixed Points

In the previous section we have shown that the *local* properties of the RG flow near a given fixed point determines the critical exponents. More precisely, the critical exponents can be expressed in terms of the eigenvalues of the matrix associated with the linearized RG flow in the vicinity of the fixed point. The fact that these eigenvalues are independent of the initial values of the couplings is the origin of universality. In this section we shall discuss the *global* behavior of RG flows. Usually, any point in coupling space flows under the iterated RG to some fixed point. The state of the system described by this fixed point represents the phase at the original point in the phase diagram. Since the critical exponents are determined by the RG flow in the vicinity of special types of critical fixed points (see below), all Hamiltonians whose couplings lie within the basin of attraction of a given critical fixed point have the same critical exponents; this is the microscopic reason for the existence of universality classes. The phases of the system and possible transitions between them are determined by the global topology of the RG flows connecting the fixed points.

It is convenient to classify the different types of fixed points according to the number of independent relevant couplings, corresponding to the RG flow away from the fixed point. Recall that relevance/irrelevance is not a global property, but is always defined with respect to a given fixed point. Another criterion for classifying RG fixed points is the value of the correlation length ξ, which according to our general considerations in Sect. 3.3.1 must be either zero or infinite at a fixed point. Let us briefly list the most important types of fixed points, following mainly Goldenfeld (1992):

(a) *Critical fixed points:* These are characterized by two independent relevant couplings and an infinite correlation length. For example, at the Gaussian fixed point describing the critical behavior of the Ising model in dimensions $D > 4$ the couplings $r \propto t$ and h are relevant. The RG flow close to this fixed point in the infinite-dimensional space of all couplings is shown schematically in Fig. 3.7. The two relevant couplings r and h drive the RG flow away from the fixed point, so that the manifold in coupling space covered by the RG trajectories that flow away from the fixed point is two-dimensional. The dimensionality of this manifold is called the *codimension* of the fixed point. Critical fixed points therefore have codimension 2. In order to flow precisely into the fixed point, one has to fine-tune the initial conditions: for given initial values g_0 of the irrelevant couplings, there is a special point $(r_0(g_0), h_0(g_0))$ of the relevant couplings which is mapped into the critical fixed point $\mathcal{C}$ by the RG. By definition, the critical manifold of $\mathcal{C}$ is the set of all points in coupling space that flows into $\mathcal{C}$ under the RG, see Sect. 3.3.1. Trajectories which start slightly off the critical manifold initially flow towards $\mathcal{C}$, but ultimately are repelled from it. However, in a large range of logarithmic flow parameters $l = \ln b$ these slightly *off-critical* trajectories remain close to the fixed point, so that the physical properties of a

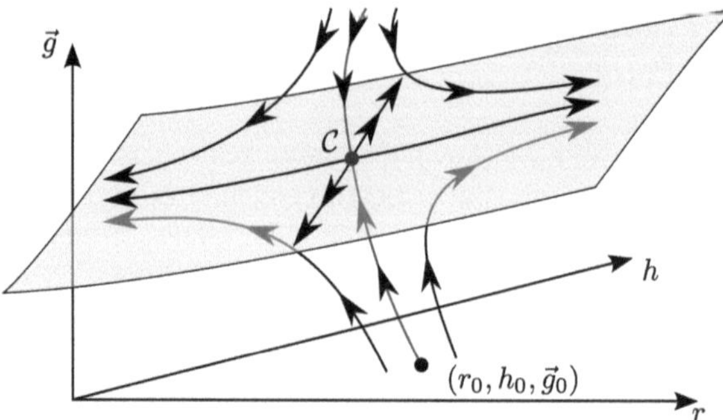

Fig. 3.7 Typical RG flow in the vicinity of a critical fixed point which is characterized by two relevant couplings r and h. The axis labeled **g** represents the (possibly infinite) set of marginal and irrelevant couplings of a given model. The *dot* marked by a $\mathcal{C}$ is the critical fixed point

system corresponding to a slightly off-critical initial condition in coupling space is dominated by the local properties of the associated critical fixed point.

(b) *Multicritical fixed points:* These are critical fixed points ($\xi = \infty$) with more than two relevant couplings. In the special case of three relevant couplings, one calls such a fixed point *tricritical*. To reach a multicritical fixed point, the initial condition of all relevant couplings have to be fine-tuned to move the initial parameters on the critical surface of the fixed point. Experimentally, this is achieved by adjusting the experimentally controllable external parameters, such as the magnetic field, the temperature, or the pressure. However, the number of external parameters which can be controlled experimentally is finite, so that in practice multicritical points of higher order than three are hard to study experimentally.

An exception seems to be the Fermi surface characterizing the normal state of interacting electrons at zero temperature, which forms a $(D - 1)$-dimensional continuum in $D > 1$ dimensions. To calculate the Fermi surface within the framework of the RG, one has to introduce a continuum of relevant coupling constants $r_l(\boldsymbol{k}_F)$ for each point $\boldsymbol{k}_F$ on the Fermi surface, which all have to be fine-tuned to flow into a fixed point associated with the true Fermi surface of the system (Kopietz and Busche 2001, Ledowski and Kopietz 2003, Ledowski et al. 2005, Ledowski and Kopietz 2007).[7] The Fermi surface in $D > 1$ can therefore be viewed as a property of a multicritical point of infinite order. In fact, as recently pointed out by Shiwa (2006), there is a formal similarity between the RG theory for the Fermi surface and the so-called Brazovskii universality class (Brazovskii 1975, Hohenberg and Swift 1995) describing classical systems where the fluctuation spectrum has a minimum at some nonzero wave vector (such as cholesteric liquid crystals), so that infinite-order multicritical points appear also in classical systems.

---

[7] In Sect. 10.4 we shall discuss the problem of calculating the Fermi surface within the framework of the FRG in more detail.

(c) *Sinks:* These are fixed points without relevant directions, so that RG trajectories can only flow into them. Sinks correspond to stable bulk phases of matter and describe a state with vanishing correlation length. As an example, the RG flow diagram of the three-dimensional Ising ferromagnet in a magnetic field shown in Fig. 3.8 has two sinks S_+ and S_- corresponding to infinite temperature and $h = \pm\infty$. These stable fixed points describe the simple fact that an infinite positive or negative magnetic field aligns all spins along its direction. If we start the RG procedure at any point in the coupling space with $h \neq 0$, successive RG iterations will drive the system to one of the sinks with the appropriate sign of h. Because the sinks are associated with infinite temperature, the corresponding correlation length vanishes.

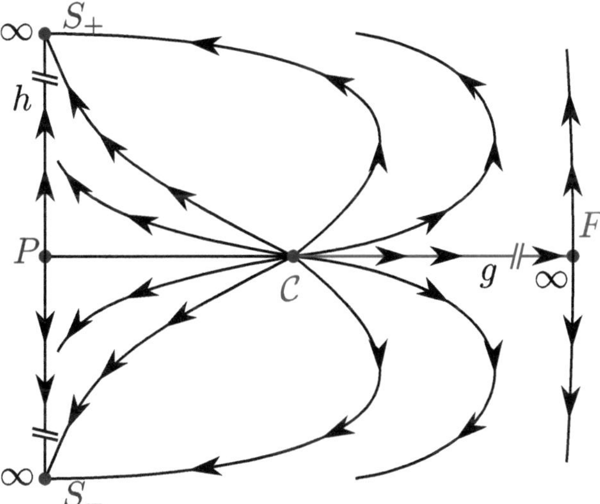

Fig. 3.8 RG flow of the three-dimensional ferromagnetic Ising model with dimensionless nearest-neighbor coupling $g = \beta J$ in a magnetic field. The critical fixed point is denoted by C, while the two fixed points S_+ and S_- at $h = \pm\infty$ and $T = \infty$ are sinks. The paramagnetic fixed point P is an example for a continuity fixed point, while the ferromagnetic fixed point F is a discontinuity fixed point

(d) *Fixed points with one relevant coupling:* The fixed points P and F in Fig. 3.8 have only one relevant direction in coupling space, the magnetic field. The paramagnetic fixed point P is an example for a *continuity fixed point*: it represents a phase of the system with $\xi = 0$, which becomes unstable when the field corresponding to the relevant coupling is switched on. The ferromagnetic fixed point F is called a *discontinuity fixed point*: for all points in its basin of attraction the magnetization jumps discontinuously as the $h = 0$ line is crossed, corresponding to the phase boundary where the two phases with up and down magnetization can coexist. At the fixed point $\xi = 0$.

(e) *Multiple coexistence fixed points:* These are fixed points with more than one relevant coupling describing the coexistence of more than two phases with

$\xi = 0$. In the case of two relevant couplings, such a fixed point is called a *triple point*, which is the $\xi = 0$ version of a critical fixed point.

(f) *Fixed lines and higher-dimensional fixed-point manifolds:* Although most RG flows describing physical systems are controlled by isolated fixed points as described above, in some cases one obtains a continuum of fixed points in coupling space. A famous example is the RG flow diagram of the two-dimensional classical XY-model with nearest-neighbor exchange coupling J, which exhibits a line of fixed points in the two-dimensional space of marginal couplings formed by the dimensionless thermal coupling constant $x = \pi\beta J - 2$ and the fugacity variable $y = 4\pi e^{-E_c/T}$, where E_c is an energy scale related to the creation of vortex-like spin configurations in the system (see e.g., Chaikin and Lubensky 1995, Chap. 9). For small values of x and y the RG flow is

$$\partial_l x = -y^2 , \quad \partial_l y = -xy . \tag{3.95}$$

A graph of this RG flow is shown in Fig. 3.9. Obviously, the entire line $y = 0$ is a fixed line. A similar flow diagram is also found in the RG theory for the one-dimensional electron gas (Sólyom 1979) for a half-filled band,[8] as will be discussed in more detail in Sect. 10.5.3.

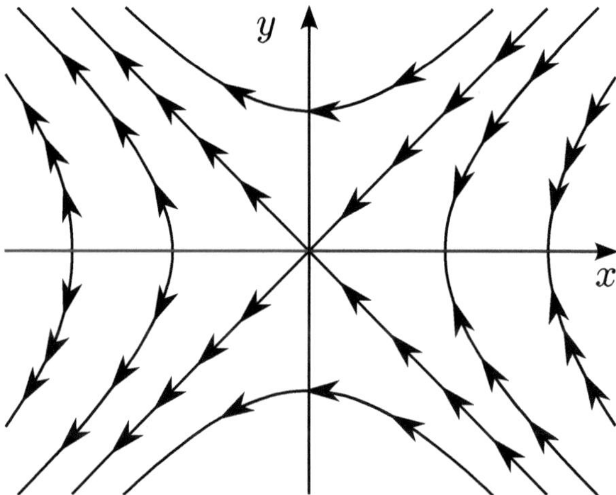

Fig. 3.9 Graph of the RG flow defined via Eq. (3.95), which describes the flow of the marginal couplings of the two-dimensional XY-model. The variable $x = \pi\beta J - 2$ is the suitably shifted inverse temperature, while $y = 4\pi e^{-E_c/T}$ is the fugacity associated with the creation of vortices

---

[8] By comparing Fig. 3.9 with Fig. 10.9 in Chap. 10 we see that in the one-dimensional electron gas the role of the coupling x is played by the momentum-conserving interaction constant g_c in the charge channel, while y corresponds to the so-called Umklapp component g_3 of the interaction involving momentum transfers of twice the Fermi momentum, see Eq. (10.136). The stability of the fixed line for $x > 0$ corresponds in the one-dimensional electron gas to a stable metallic phase called Luttinger liquid (Haldane 1981).

(g) *Exotic RG flows:* As already mentioned in the footnote of Sect. 3.3.2, in principle RG flows can also exhibit more exotic behavior such as limit cycles (where at least some couplings satisfy $g_{l+l_0} = g_l$) or exhibit even chaotic behavior without fixed points (Glazek and Wilson 2002, LeClair et al. 2003). A physically relevant system which exhibits a cyclic RG flow has recently been discussed by Moroz et al. (2009), who used functional RG methods to study the problem of three interacting bosons or fermions in vacuum. They showed that for bosons with $U(1)$-symmetry and for fermions with $SU(3)$ symmetry the existence of three-body bound states (so-called Efimov states) manifests itself in cyclic RG flows of a suitably defined interaction vertex.

Exercises

3.1 Migdal–Kadanoff RG for the Ising Chain

One can exactly solve the Ising model in one dimension (the so-called Ising chain) using the real-space RG, keeping only every b-th spin and tracing over intermediate spins. Fig. 3.4 illustrates this for $b = 3$. In the transfer matrix formalism, the partition function can be written as $\mathcal{Z} = \text{Tr}[\mathbf{T}]^N = \text{Tr}[\mathbf{T}^b]^{\frac{N}{b}}$.

(a) For $h = 0$, $g = \beta J$, show that $\mathbf{T}^b = \text{const} \times \mathbf{T}'$ with

$$
\mathbf{T}' = \begin{pmatrix} e^{g'} & e^{-g'} \\ e^{-g'} & e^{g'} \end{pmatrix} ,
$$

and the recursion relation $g'(g) = \text{Artanh}(\tanh^b g)$.
Hint: You might find it advantageous to diagonalize the matrix $\mathbf{T}$.

(b) Rewrite the recursion relation in terms of $y \equiv e^{-2g}$ and $y' \equiv e^{-2g'}$ as $y'(y)$. Determine the RG β-function and the fixed points and sketch the flow of y under repeated transformations.

(c) Linearize $y'(y)$ around the unstable fixed point $y = 0$ to show that $y' \approx by$. Argue that the correlation length fulfills $\xi(y) = b\xi(y') \approx b\xi(by)$. By an appropriate choice of b show that $\xi \propto y^{-1} = e^{2g} = e^{2\beta J}$. Compare this with the exact result obtained in Exercise 1.1.

3.2 Migdal–Kadanoff RG for the Two-Dimensional Ising Model

Consider the ferromagnetic Ising model on a square lattice with $h = 0$ and $g = \beta J$. To solve this problem, construct an approximate RG transformation according to the following procedure:

- Shift bonds on alternate rows and columns by one lattice spacing to obtain the modified interaction $\tilde{g} = 2g$ (with $b = 2$, see figure).

- Trace over the spins on the sites marked by the crosses to obtain a rescaled system with a new interaction g'.

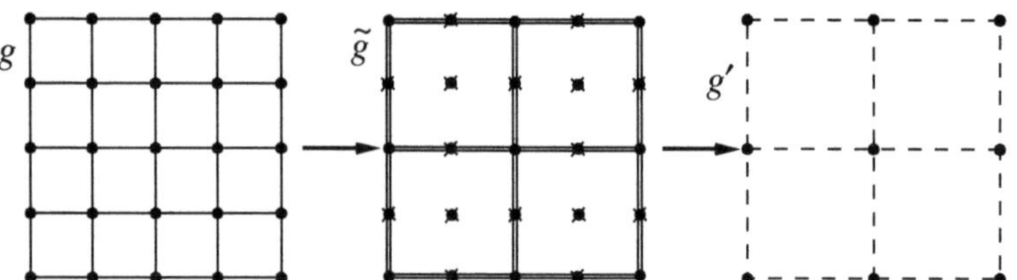

(a) What is the recursion relation $g'(g)$?

(b) Find the nontrivial fixed point g^* of the transformation. What is the critical temperature T_c? Sketch the flow of g under repeated transformations. Linearize $g'(g)$ around g^* and determine the exponent y_g in $g' - g^* = b^{y_g}(g - g^*)$. Argue that the correlation length should fulfill $\xi(g) = b\xi(g') \approx b\xi(g^* + b^{y_g}(g - g^*))$. Use this to obtain the critical exponent ν.

(c) Generalize the Migdal–Kadanoff bond-moving procedure to a hypercubic lattice in D dimensions. What is the recursion relation in this case?

(d) Sketch the fixed point $g^*(D)$ and the RG flow in the vicinity of $D = 1$ (i.e., for $D = 1 + \epsilon$) and deduce the *lower critical dimension* D_{lo} for the Ising model, which is defined as the largest dimension where the critical temperature vanishes; a finite temperature phase transition is therefore only possible for $D > D_{\mathrm{lo}}$. Compute the exponent y_g to first order in ϵ.

3.3 Dangerously Irrelevant Coupling in φ^4-Theory for $D > 4$

(a) Consider the Gaussian field theory in the presence of a magnetic field h,

$$S_{\Lambda_0}[\varphi] = \frac{1}{2} \int^{\Lambda_0} \frac{d^D k}{(2\pi)^D} [r_0 + c_0 k^2]\varphi(-k)\varphi(k) - h_0\varphi(k = 0) ,$$

where $\varphi(k = 0) = V\bar{\varphi}$ and $h_0 = \beta h/a^{1+D/2}$, see Eqs. (2.60) and (2.61). Integrate out the fields with momenta $\Lambda_0/b < |k| < \Lambda_0$ (trivial) and rescale momenta and fields keeping c_0 fixed to derive the RG recursion relations

$$r' = b^{y_t} r_0 , \qquad h' = b^{y_h} h_0 ,$$

with $y_t = 2$ and $y_h = 1 + D/2$.

(b) Now add an interaction term of the form

$$S_n[\varphi] = \frac{u_n}{n!} \int d^D r \, [\varphi(r)]^n ,$$

where $n = 2, 4, 6, \ldots$ is an even integer and

$$\varphi(r) = \int^{\Lambda_0} \frac{d^D k}{(2\pi)^D} e^{ik\cdot r} \varphi(k) \, .$$

Using $\int d^D r \, e^{ik\cdot r} = (2\pi)^D \delta(k)$, write the interaction term in terms of the fields $\varphi(k)$ and carry out the rescaling step to derive the scaling relation $u'_n = b^{y_{u_n}} u_n + \mathcal{O}\left(u_n^2\right)$, with $y_{u_n} = D(1 - n/2) + n$. Determine the dimension $D_{\mathrm{up}}(n)$ for which y_{u_n} becomes marginal, i.e., equal to zero. This dimension is called the upper critical dimension, see also the discussion in Sect. 2.3.

(c) For $D > D_{\mathrm{up}}(4) = 4$, the coupling $u \equiv u_4$ is irrelevant and the critical behavior of the Ising model is expected to be determined by the Gaussian fixed point. Use the scaling relations and the above result for y_t and y_h to calculate all critical exponents. Compare your result with the values obtained in Sect. 2.3 directly from the Gaussian approximation without using RG arguments. Which exponents do and which do not agree?

(d) To track down the failure of the scaling argument for some exponents for $D > 4$, argue that the magnetization has the scaling form

$$m(t, h, u) = b^{y_h - D} m(b^{y_t} t, b^{y_h} h, b^{y_u} u) \, .$$

For $h = 0$ and $b = |t|^{-1/y_t}$, this yields the scaling relation

$$m(t, 0, u) = |t|^{-(y_h - D)/y_t} m(\pm 1, 0, |t|^{-y_u/y_t} u) \, .$$

Close to the critical point, the last argument becomes small and one might be tempted to set $u = 0$ to obtain the scaling result $\beta = (D - y_h)/y_t$. However, according to Landau theory a finite u is needed to get a spontaneous magnetization. Show explicitly that within Landau theory $m(-1, 0, \tilde{u}) \propto \tilde{u}^{-1/2}$. Use this and the scaling form of m to obtain the correct (mean-field) value $\beta = \frac{1}{2}$. Can you construct a similar argument for the exponent δ? Note that the existence of a dangerously irrelevant variable for $D > 4$ implies the violation of the hyperscaling relation (1.29). You may consult Appendix D of the review article by Fisher (1983) if you need some help.

References

Brazovskii, S. A. (1975), *Phase transition of an isotropic system to the heterogeneous state*, Zh. Eksp. Teor. Fiz **68**, 175.

Bulla, R., T. A. Costi, and T. Pruschke (2008), *Numerical renormalization group method for quantum impurity systems*, Rev. Mod. Phys. **80**, 395.

Chaikin, P. M. and T. C. Lubensky (1995), *Principles of Condensed Matter Physics*, Cambridge University Press, Cambridge.

Fisher, M. E. (1983), *Scaling, Universality and Renormalization Group Theory*, in F. J. W. Hahne, editor, *Lecture Notes in Physics*, volume 186, Springer, Berlin.

Głazek, S. D. and K. G. Wilson (1993), *Renormalization of overlapping transverse divergences in a model light-front Hamiltonian*, Phys. Rev. D **47**, 4657.

Głazek, S. D. and K. G. Wilson (1994), *Perturbative renormalization group for Hamiltonians*, Phys. Rev. D **49**, 4214.

Głazek, S. D. and K. G. Wilson (2002), *Limit cycles in quantum theories*, Phys. Rev. Lett. **89**, 230401.

Goldenfeld, N. (1992), *Lectures on Phase Transitions and the Renormalization Group*, Addison-Wesley, Reading.

Haldane, F. D. M. (1981), *'Luttinger liquid theory' of one-dimensional quantum fluids. I. Properties of the Luttinger model and their extension to the general 1D interacting spinless Fermi gas*, J. Phys. C: Solid State Phys. **14**, 2585.

Hertz, J. A. (1976), *Quantum critical phenomena*, Phys. Rev. B **14**, 1165.

Hewson, A. C. (1993), *The Kondo Problem to Heavy Fermions*, Cambridge University Press, New York.

Hohenberg, P. C. and J. B. Swift (1995), *Metastability in fluctuation-driven first-order transitions: Nucleation of lamellar phases*, Phys. Rev. E **52**, 1828.

Kadanoff, L. P. (1976), *Notes on Migdal's recursion formulas*, Ann. Phys. **100**, 359.

Kehrein, S. (2006), *The Flow Equation Approach to Many-Particle Systems*, Springer, Berlin.

Keller, G. and C. Kopper (1991), *Perturbative renormalization of QED via flow equations*, Phys. Lett. B **273**, 323.

Keller, G. and C. Kopper (1996), *Renormalizability proof for QED based on flow equations*, Commun. Math. Phys. **176**, 193.

Kopietz, P. and T. Busche (2001), *Exact renormalization group flow equations for nonrelativistic fermions: Scaling toward the Fermi surface*, Phys. Rev. B **64**, 155101.

LeClair, A., J. M. Rom´an, and G. Sierra (2003), *Russian doll renormalization group and Kosterlitz–Thouless flows*, Nucl. Phys. B **675**, 584.

Ledowski, S. and P. Kopietz (2003), *An exact integral equation for the renormalized Fermi surface*, J. Phys.: Condens. Matter **15**, 4779.

Ledowski, S., P. Kopietz, and A. Ferraz (2005), *Self-consistent Fermi surface renormalization of two coupled Luttinger liquids*, Phys. Rev. B **71**, 235106.

Ledowski, S. and P. Kopietz (2007), *Fermi surface renormalization and confinement in two coupled metallic chains*, Phys. Rev. B **75**, 045134.

Maris, H. J. and L. P. Kadanoff (1978), *Teaching the renormalization group*, Am. J. Phys. **46**, 652.

McComb, W. (2004), *Renormalization Methods: A Guide for Beginners*, Oxford University Press, Oxford.

Migdal, A. A. (1975), *Phase transitions in gauge and spin-lattice systems*, Sov. Phys. JETP **42**, 743.

Moroz, S., S. Floerchinger, R. Schmidt, and C.Wetterich (2009), *Efimov effect from functional renormalization*, Phys. Rev. A **79**, 013603.

Morris, T. R. (1994), *The exact renormalisation group and approximate solutions*, Int. J. Mod. Phys. A **9**, 2411.

Peschel, I., X. Want, M. Kaulke, and K. Hallberg (1999), *Density-Matrix Renormalization, a New Numerical Method in Physics*, Springer, Berlin.

Polchinski, J. (1984), *Renormalization and effective Lagrangians*, Nucl. Phys. B **231**, 269.

Schollwöck, U. (2005), *The density-matrix renormalization group*, Rev. Mod. Phys. **77**, 259.

Shiwa, Y. (2006), *Exact renormalization group for the Brazovskii model of striped patterns*, J. Stat. Phys. **124**, 1207.

Sólyom, J. (1979), *The Fermi gas model of one-dimensional conductors*, Adv. Phys. **28**, 201.

Swendsen, R. H. (1984), *Monte Carlo calculation of renormalized coupling parameters. I. $d = 2$ Ising model*, Phys. Rev. B **30**, 3866.

Wegner, F. J. (1994), *Flow equations for hamiltonians*, Ann. Phys. (Leipzig) **3**, 77.

Wegner, F. J. and A. Houghton (1973), *Renormalization group equation for critical phenomena*, Phys. Rev. A **8**, 401.

Wetterich, C. (1993), *Exact evolution equation for the effective potential*, Phys. Lett. B **301**, 90.

White, S. R. (1992), *Density matrix formulation for quantum renormalization groups*, Phys. Rev. Lett. **69**, 2863.

Wilson, K. G. (1969), *Non-Lagrangian models of current algebra*, Phys. Rev. **179**, 1499.

Wilson, K. G. (1971a), *Renormalization group and strong interactions*, Phys. Rev. D **3**, 1818.

Wilson, K. G. (1971b), *The renormalization group and critical phenomena I: Renormalization group and the Kadanoff scaling picture*, Phys. Rev. B **4**, 3174.

Wilson, K. G. (1971c), *The renormalization group and critical phenomena II: Phase-space cell analysis of critical behavior*, Phys. Rev. B **4**, 3184.

Wilson, K. G. (1972), *Feynman-graph expansion for critical exponents*, Phys. Rev. Lett. **28**, 548.

Wilson, K. G. (1975), *The renormalization group: Critical phenomena and the Kondo problem*, Rev. Mod. Phys. **47**, 773.

Wilson, K. G. and J. Kogut (1974), *The renormalization group and the ϵ expansion*, Phys. Rep. **12**, 75.

Wilson, K. G. and M. E. Fisher (1972), *Critical exponents in 3.99 dimensions*, Phys. Rev. Lett. **28**, 240.

Chapter 4
Critical Behavior of the Ising Model Close to Four Dimensions

In this chapter, we shall use the momentum space RG with a sharp cutoff to study the critical behavior of the Ising model close to four dimensions, where the parameter $\epsilon = 4 - D$ is small. In this case the critical exponents and other physical quantities can be expanded in a series in powers of ϵ (Wilson et al. 1972). For sufficiently small ϵ, a truncation of this expansion retaining only a small number of terms yields quantitatively accurate results. Of course, in order to calculate the critical exponents in the physically relevant dimension $D = 3$, one has to calculate as many terms in the ϵ-expansion as possible and then extrapolate the series to $\epsilon = 1$ (Guida and Zinn-Justin 1998, Zinn-Justin 2002, Pelissetto and Vicari 2002). In order to explain this method, we shall focus here on the calculation of the correlation length exponent ν to linear order in ϵ. To this end, it is convenient to carry out the mode-elimination step in momentum space, using a sharp momentum-shell cutoff. The relevant integrals will be calculated perturbatively in powers of the interaction. In order to keep track of the various terms generated in perturbation theory, it is useful to represent them in terms of Feynman diagrams. We therefore begin this chapter with a brief but self-contained introduction to diagrammatic perturbation theory.

4.1 Diagrammatic Perturbation Theory

In Chap. 2 we have shown that the partition function of the D-dimensional Ising model can be written as a functional integral over a real field φ representing the fluctuating magnetization,

$$\mathcal{Z} = \int \mathcal{D}[\varphi] e^{-S[\varphi]} , \tag{4.1}$$

where according to Eq. (2.51) the effective action is of the following form,

$$S[\varphi] = S_0[\varphi] + S_{\text{int}}[\varphi] . \tag{4.2}$$

The Gaussian part $S_0[\varphi]$ is quadratic in the fields, and the interaction part $S_{\text{int}}[\varphi]$ involves four and in general also higher powers of the fields. Throughout this chapter we shall set the magnetic field equal to zero, so that only even powers of the fields

Kopietz, P. et al.: *Critical Behavior of the Ising Model Close to Four Dimensions*. Lect. Notes Phys. **798**, 91–121 (2010)
DOI 10.1007/978-3-642-05094-7_4

appear in the effective action. For convenience, we have absorbed the factor $\sqrt{\det \tilde{\mathbf{J}}}$ appearing in Eq. (2.39) and the constant f_0 in Eq. (2.61) into the definition of the integration measure $\mathcal{D}[\varphi]$ in Eq. (4.1). From Eq. (2.61), we see that the Gaussian part $S_0[\varphi]$ of our effective action can be written as

$$S_0[\varphi] = \frac{\beta}{2} \sum_k E_k \varphi_{-k} \varphi_k \,, \tag{4.3}$$

with the energy E_k given by

$$E_k = J_k(1 - \beta J_k) \,. \tag{4.4}$$

Note that for small wave vectors $\mathbf{k}$ and for temperatures close to the mean-field critical temperature $T_c^{\mathrm{MF}} = J_{k=0} = zJ$, we may approximate

$$E_k \approx T_c^{\mathrm{MF}} a^2 (r_0 + c_0 k^2) = T - T_c^{\mathrm{MF}} + T_c^{\mathrm{MF}} c_0 (ka)^2 \,. \tag{4.5}$$

Retaining only the term involving four powers of the fields, the interaction part $S_{\mathrm{int}}[\varphi]$ can be written as

$$S_{\mathrm{int}}[\varphi] \approx \frac{\beta}{4!N} \sum_{k_1,k_2,k_3,k_4} \delta_{k_1+k_2+k_3+k_4,0} U(\mathbf{k}_1, \mathbf{k}_2, \mathbf{k}_3, \mathbf{k}_4) \varphi_{k_1} \varphi_{k_2} \varphi_{k_3} \varphi_{k_4} \,, \tag{4.6}$$

with the momentum-dependent interaction energy

$$U(\mathbf{k}_1, \mathbf{k}_2, \mathbf{k}_3, \mathbf{k}_4) = 2\beta^3 J_{k_1} J_{k_2} J_{k_3} J_{k_4} \,. \tag{4.7}$$

Although for the RG transformation it is most convenient to work with continuum fields $\varphi(\mathbf{k}) = a\sqrt{V} \varphi_k$ (see Eq. (2.57)), we shall here develop the perturbation theory using the lattice normalization of the Fourier transformation with dimensionless fields $\varphi_k = \frac{1}{\sqrt{N}} \sum_i e^{-i\mathbf{k}\cdot\mathbf{r}_i} \varphi_i$ (see Eq. (2.45)), which facilitates the counting of the degrees of freedom.

To calculate the partition function $\mathcal{Z}$ perturbatively, we expand the exponential in Eq. (4.1) in powers of the interaction part $S_{\mathrm{int}}[\varphi]$ of our effective action,

$$\mathcal{Z} = \int \mathcal{D}[\varphi] e^{-S_0[\varphi]-S_{\mathrm{int}}[\varphi]} = \sum_{\nu=0}^{\infty} \frac{(-1)^\nu}{\nu!} \int \mathcal{D}[\varphi] e^{-S_0[\varphi]} (S_{\mathrm{int}}[\varphi])^\nu \,. \tag{4.8}$$

Dividing both sides of this expression by the partition function in Gaussian approximation,

$$\mathcal{Z}_0 = \int \mathcal{D}[\varphi] e^{-S_0[\varphi]} \,, \tag{4.9}$$

we may rewrite Eq. (4.8) as

$$\frac{\mathcal{Z}}{\mathcal{Z}_0} = \sum_{\nu=0}^{\infty} \frac{(-1)^\nu}{\nu!} \langle (S_{\text{int}}[\varphi])^\nu \rangle_0 \ , \tag{4.10}$$

where the Gaussian average of any functional $F[\varphi]$ of the fields is defined by

$$\langle F[\varphi] \rangle \rangle_0 = \frac{\int \mathcal{D}[\varphi] e^{-S_0[\varphi]} F[\varphi]}{\int \mathcal{D}[\varphi] e^{-S_0[\varphi]}} \ . \tag{4.11}$$

Similarly, the correlation function of a product of n fields can be expanded as

$$G_n(\boldsymbol{k}_1, \ldots, \boldsymbol{k}_n) \equiv \frac{\int \mathcal{D}[\varphi] e^{-S_0[\varphi]-S_{\text{int}}[\varphi]} \varphi_{k_1} \cdots \varphi_{k_n}}{\int D[\varphi] e^{-S_0[\varphi]-S_{\text{int}}[\varphi]}}$$

$$= \frac{\sum_{\nu=0}^{\infty} \frac{(-1)^\nu}{\nu!} \langle (S_{\text{int}}[\varphi])^\nu \varphi_{k_1} \cdots \varphi_{k_n} \rangle_0}{\sum_{\nu=0}^{\infty} \frac{(-1)^\nu}{\nu!} \langle (S_{\text{int}}[\varphi])^\nu \rangle_0} \ . \tag{4.12}$$

4.1.1 Wick Theorem

For the evaluation of Eqs. (4.10) and (4.12) we need to calculate the Gaussian averages of products of an arbitrary number of fields with different labels. These are conveniently evaluated using the Wick theorem, which is the following property of multidimensional Gaussian integrals,

$$\boxed{\begin{aligned} \langle x_{i_1} x_{i_2} \ldots x_{i_n} \rangle_0 &\equiv \frac{\left(\prod_{i=1}^{N} \int_{-\infty}^{\infty} dx_i \right) e^{-\frac{1}{2} \boldsymbol{x}^T \mathbf{A} \boldsymbol{x}} x_{i_1} x_{i_2} \ldots x_{i_n}}{\left(\prod_{i=1}^{N} \int_{-\infty}^{\infty} dx_i \right) e^{-\frac{1}{2} \boldsymbol{x}^T \mathbf{A} \boldsymbol{x}}} \\ &= \sum_{\text{all pairs}} (\mathbf{A}^{-1})_{i_{p_1} i_{p_2}} (\mathbf{A}^{-1})_{i_{p_3} i_{p_4}} \ldots (\mathbf{A}^{-1})_{i_{p_{n-1}} i_{p_n}} \ . \end{aligned}} \tag{4.13}$$

Here, $\mathbf{A}$ is a real symmetric $N \times N$ matrix, $\boldsymbol{x} = (x_1, \ldots, x_N)^T$ is an N-component column vector, and each of the external indices $i_1, \ldots, i_n$ can take any value in the set of integers $\{1, \ldots, N\}$. The sum in the second line of Eq. (4.13) extends over all possible ways of partitioning the n integers $i_1, \ldots, i_n$ into pairs, (i_{p_1}, i_{p_2}), $(i_{p_3}, i_{p_4}), \ldots, (i_{p_{n-1}}, i_{p_n})$. For even n the sum has $(n-1)!! = (n-1)(n-3) \cdots 3 \cdot 1$ terms, while for odd n the Gaussian integral vanishes by symmetry. In the special case $n = 2$, there is only one pair and the Wick theorem reduces to

$$\langle x_{i_1} x_{i_2} \rangle_0 = (\mathbf{A}^{-1})_{i_1 i_2} \ , \tag{4.14}$$

while for $n = 4$ there are 3 possible pairings,

$$\langle x_{i_1} x_{i_2} x_{i_3} x_{i_4} \rangle_0 = \langle x_{i_1} x_{i_2} \rangle_0 \langle x_{i_3} x_{i_4} \rangle_0 + \langle x_{i_1} x_{i_3} \rangle_0 \langle x_{i_2} x_{i_4} \rangle_0 + \langle x_{i_1} x_{i_4} \rangle_0 \langle x_{i_2} x_{i_3} \rangle_0 . \tag{4.15}$$

The Wick theorem for quantum mechanical many-particle systems can be formulated in a similar way in terms of Gaussian integrals involving either complex fields for bosons, or anticommuting Grassmann fields for fermions (Negele and Orland 1988). To prove Eq. (4.13) we use again the identity (2.41) to write

$$\langle x_{i_1} x_{i_2} \ldots x_{i_n} \rangle_0 = \lim_{y \to 0} \frac{\partial}{\partial y_{i_1}} \frac{\partial}{\partial y_{i_2}} \cdots \frac{\partial}{\partial y_{i_n}} \left\langle e^{x^T y} \right\rangle_0 . \tag{4.16}$$

With the help of our Gaussian integration formula (2.25), we obtain for the Gaussian average,

$$\left\langle e^{x^T y} \right\rangle_0 = e^{\frac{1}{2} y^T A^{-1} y} . \tag{4.17}$$

Taking in Eq. (4.16) the derivative with respect to y_{i_1}, we obtain

$$\langle x_{i_1} x_{i_2} \ldots x_{i_n} \rangle_0 = \lim_{y \to 0} \frac{\partial}{\partial y_{i_2}} \cdots \frac{\partial}{\partial y_{i_n}} \left[\sum_{i=1}^{N} (A^{-1})_{i_1 i} \, y_i \, e^{\frac{1}{2} y^T A^{-1} y} \right] . \tag{4.18}$$

In order to obtain a finite limit $y \to 0$, one of the remaining $n - 1$ derivatives must act on the prefactor, so that

$$\begin{aligned}
\langle x_{i_1} x_{i_2} \ldots x_{i_n} \rangle_0 = \lim_{y \to 0} \bigg[& (A^{-1})_{i_1 i_2} \frac{\partial}{\partial y_{i_3}} \frac{\partial}{\partial y_{i_4}} \cdots \frac{\partial}{\partial y_{i_n}} e^{\frac{1}{2} y^T A^{-1} y} \\
& + (A^{-1})_{i_1 i_3} \frac{\partial}{\partial y_{i_2}} \frac{\partial}{\partial y_{i_4}} \cdots \frac{\partial}{\partial y_{i_n}} e^{\frac{1}{2} y^T A^{-1} y} \\
& + \ldots \\
& + (A^{-1})_{i_1 i_n} \frac{\partial}{\partial y_{i_2}} \frac{\partial}{\partial y_{i_3}} \cdots \frac{\partial}{\partial y_{i_{n-1}}} e^{\frac{1}{2} y^T A^{-1} y} \bigg] ,
\end{aligned} \tag{4.19}$$

where the square bracket contains $n-1$ terms involving $n-2$ derivatives with respect to the auxiliary variables y_i. Taking two more y-derivatives, each of the $n - 1$ terms can again be written as a sum of $n - 3$ new terms involving only $n - 4$ y-derivatives. Iterating this procedure until no y-derivatives are left and taking the limit $y \to 0$ we obtain in total $(n - 1)!! = (n - 1)(n - 3) \cdots 3 \cdot 1$ terms consisting of products of $n/2$ matrix elements of A^{-1} with all possible combinations of labels. This is precisely what is meant by the sum over all pairs in Eq. (4.13). This completes the proof of the Wick theorem.

4.1.2 Feynman Diagrams and Linked Cluster Theorem

The perturbative evaluation of the Gaussian averages $\langle (S_{\text{int}}[\varphi])^\nu \rangle_0$ appearing in the perturbation series (4.8) for the partition function (or more complicated averages of the form $\langle (S_{\text{int}}[\varphi])^\nu \varphi_{k_1} \dots \varphi_{k_n} \rangle_0$ in the perturbation series (4.12) for the correlation function) with the help of the Wick theorem leads even for rather small values of ν to quite a few terms. To keep track of these terms, to classify them, and to develop some intuition for their physical meaning, it is useful to represent them graphically in terms of *Feynman diagrams*. For our simple classical field theory, these involve the following graphical elements:

$$\text{line}: \quad \boldsymbol{k} \underline{\qquad\qquad} -\boldsymbol{k} \;=\; \langle \varphi_k \varphi_{-k} \rangle_0 = \frac{T}{E_k} = \frac{1}{\beta E_k}\,, \tag{4.20}$$

$$\text{vertex}: \quad \begin{array}{c} \boldsymbol{k}_1 \qquad \boldsymbol{k}_4 \\[-2pt] \diagdown\!\!\!\bullet\!\!\!\diagup \\[-2pt] \boldsymbol{k}_2 \qquad \boldsymbol{k}_3 \end{array} \;=\; \frac{\beta}{4!N}\delta_{k_1+k_2+k_3+k_4,0}\,U(\boldsymbol{k}_1,\boldsymbol{k}_2,\boldsymbol{k}_3,\boldsymbol{k}_4)\,. \tag{4.21}$$

Note that the sum of the four momenta attached to the vertex vanishes. Since the interaction function $U(\boldsymbol{k}_1,\boldsymbol{k}_2,\boldsymbol{k}_3,\boldsymbol{k}_4)$ defined in Eq. (4.7) is symmetric with respect to the exchange of all labels, the ordering of the external lines attached to the black circle in Eq. (4.21) is irrelevant.[1] With the above *dictionary*, each term in the perturbation expansion is represented by a *labeled Feynman diagram*. The rules for translating any diagram into a mathematical expression are given by the dictionary in Eqs. (4.20) and (4.21) together with the convention that each closed loop corresponds to an independent momentum summation and that the sum of the four momenta associated with the four legs attached to each interaction vertex vanishes. For example, from Eq. (4.10) we obtain for the first-order interaction correction to the partition function

---

[1] The Feynman diagrams representing the perturbation series of quantum mechanical many-body systems of nonrelativistic bosons or fermions are constructed from different graphical elements (Negele and Orland 1988). In this case the lines representing the Gaussian average of a pair of fields carry an arrow indicating the direction of the particle-number flow. Furthermore, the vertex representing the two-body interaction has two incoming and two outgoing arrows, corresponding to the annihilation and the creation of two particles. Formally, this is due to the fact that the functional integral representation of the partition function of bosons involves a complex field, while the fermionic functional integral involves pairs of anticommuting Grassmann fields.

$$\frac{\mathcal{Z}}{\mathcal{Z}_0} \approx 1 - \langle S_{\text{int}}[\varphi] \rangle_0 = 1 - 3 \times$$

$$= 1 - \frac{3T}{4!N} \sum_{k,k'} \frac{U(k,-k,k',-k')}{E_k E_{k'}} . \tag{4.22}$$

Note that the diagrams representing the interaction corrections to the partition function have no external legs, so that all solid lines must always end at a vertex.[2] Such diagrams without external legs are called *vacuum diagrams*. Symbolically, the perturbation series (4.10) for the partition function can be written as

$$\frac{\mathcal{Z}}{\mathcal{Z}_0} = \sum_{\nu=0}^{\infty} \frac{(-1)^\nu}{\nu!} \langle (S_{\text{int}}[\varphi])^\nu \rangle_0 = 1 + \sum_{\substack{\text{All vacuum} \\ \text{diagrams}}} \bigcirc , \tag{4.23}$$

where the empty circle represents any vacuum diagram including the correct combinatorial factor and sign. All diagrams can be further classified into two types: so-called *connected diagrams*, where all graphical elements are connected via lines and vertices, and *disconnected diagrams* consisting of two or more connected parts. Examples are shown in Fig. 4.1. Fortunately, all disconnected diagrams cancel if we take the logarithm of $\mathcal{Z}/\mathcal{Z}_0$, so that in the diagrammatic expansion of the (dimensionless) free energy $F = -\ln \mathcal{Z}$, only connected diagrams need to be taken into account,

$$F - F_0 = -\ln\left(\frac{\mathcal{Z}}{\mathcal{Z}_0}\right) = - \sum_{\substack{\text{All connected} \\ \text{vacuum diagr.}}} \bullet . \tag{4.24}$$

We represent connected vacuum diagrams by a shaded circle to emphasize that in a sense the density of graphical elements of connected vacuum diagrams is higher than in the vacuum diagrams of Eq. (4.23). The relation (4.24) is called the *linked cluster theorem*, which saves a lot of work in calculations beyond the leading order because only connected diagrams have to be calculated. The simplest way to prove Eq. (4.24) is based on the so-called replica trick (Negele and Orland 1988). Using the fact that

---

[2] Diagrams with external legs represent correlation functions and will be discussed in Sect. 4.1.3.

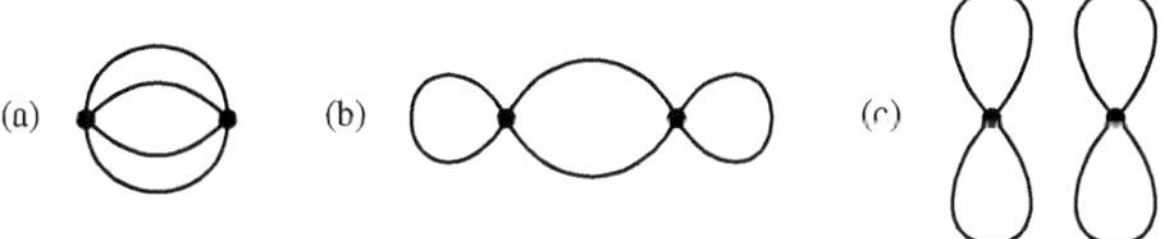

Fig. 4.1 Diagrams representing interaction corrections to the partition function in second-order perturbation theory. Diagrams (**a**) and (**b**) are connected, while the disconnected diagram (**c**) is simply the square of the first-order correction in Eq. (4.22)

$$\left(\frac{\mathcal{Z}}{\mathcal{Z}_0}\right)^n = e^{n\,\ln(\mathcal{Z}/\mathcal{Z}_0)} = 1 + n\,\ln\left(\frac{\mathcal{Z}}{\mathcal{Z}_0}\right) + \mathcal{O}(n^2)\,, \tag{4.25}$$

we may write

$$F - F_0 = -\lim_{n\to 0}\frac{d}{dn}\left(\frac{\mathcal{Z}}{\mathcal{Z}_0}\right)^n\,. \tag{4.26}$$

The replica trick is based on the assumption that the ratio $(\mathcal{Z}/\mathcal{Z}_0)^n$ can be obtained for infinitesimal n by analytic continuation of the same ratio for integer n. The crucial point is that for integer n the perturbative expansion of $(\mathcal{Z}/\mathcal{Z}_0)^n$ can be generated by introducing n copies φ^r, $r = 1, \ldots, n$, of the field and writing

$$\left(\frac{\mathcal{Z}}{\mathcal{Z}_0}\right)^n = \frac{\int \mathcal{D}[\varphi^r]e^{-S_0^{(n)}[\varphi^r]-S_{\mathrm{int}}^{(n)}[\varphi^r]}}{\int \mathcal{D}[\varphi^r]e^{-S_0^{(n)}[\varphi^r]}}\,, \tag{4.27}$$

where the Gaussian part of the replicated action is

$$S_0^{(n)}[\varphi^r] = \frac{\beta}{2}\sum_{r=1}^{n}\sum_{k} E_k \varphi^r_{-k}\varphi^r_k\,, \tag{4.28}$$

and the interaction part is

$$S_{\mathrm{int}}^{(n)}[\varphi^r] = \frac{\beta}{4!N}\sum_{r=1}^{n}\sum_{k_1,k_2,k_3,k_4} \delta_{k_1+k_2+k_3+k_4,0}\,U(k_1,k_2,k_3,k_4)\varphi^r_{k_1}\varphi^r_{k_2}\varphi^r_{k_3}\varphi^r_{k_4}\,. \tag{4.29}$$

If we now expand Eq. (4.27) in powers of $S_{\mathrm{int}}^{(n)}[\varphi^r]$ and evaluate the resulting terms using the Wick theorem, each solid line carries an extra replica index, $\underset{k\ \ r\ \ -k}{\longleftarrow}$, and all lines attached to a given vertex carry the same replica index. As a consequence, all lines in connected diagrams carry the same replica index, so that after summation over $\sum_{r=1}^{n}$, the contribution from any connected diagram to $(\mathcal{Z}/\mathcal{Z}_0)^n$ is proportional to n. On the other hand, the contribution of a disconnected diagram consisting of $k_c > 1$ connected pieces is proportional to n^{k_c} and hence does not contribute to the limit in Eq. (4.26). As a result, only connected diagrams can contribute to the logarithm of the partition function, which proves the linked cluster theorem (4.24).

4.1.3 Diagrams for Correlation Functions and the Irreducible Self-Energy

In contrast to the vacuum diagrams representing interaction corrections to the partition function, the terms arising in Gaussian averages of the type $\langle (S_{\text{int}})^\nu \varphi_{k_1} \dots \varphi_{k_n} \rangle_0$ which appear in the calculation of the correlation function $G_n(k_1, \dots, k_n)$ defined in Eq. (4.12) are represented by diagrams with n external legs associated with the n fields $\varphi_{k_1}, \dots, \varphi_{k_n}$. These diagrams are called *open* and we represent a general diagram of this type by an empty circle with n external legs. The symbolic representation analogous to Eq. (4.23) for the numerator in Eq. (4.12) is then

$$\sum_{\nu=0}^{\infty} \frac{(-1)^\nu}{\nu!} \langle (S_{\text{int}})^\nu \varphi_{k_1} \dots \varphi_{k_n} \rangle_0 = \sum_{\substack{\text{All open} \\ \text{diagrams}}} \quad . \qquad (4.30)$$

Each open diagram contributing to the right-hand side of this expression can be classified according to whether or not it contains vacuum diagrams. Let us give three examples to illustrate this:

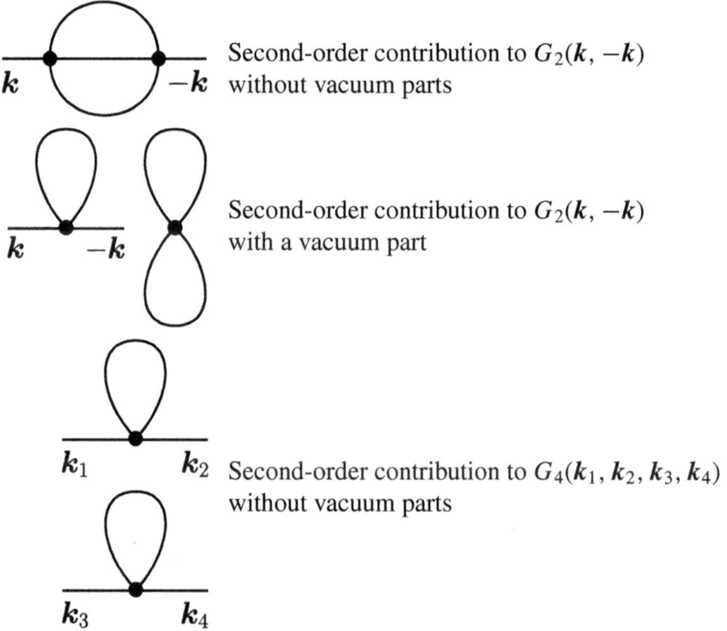

Note that open diagrams with two external legs but no vacuum parts contributing to $G_2(k, -k)$ are automatically connected, but open diagrams with four external legs contributing to $G_4(k_1, k_2, k_3, k_4)$ can be disconnected even if they do not contain any vacuum parts, as shown in the above example. Given the factorization of any open diagram into a product of an open diagram without vacuum parts (connected diagram) and any number of vacuum diagrams, we may rearrange the infinite series over all open diagrams with n externals legs on the right-hand side of Eq. (4.30) as follows:

$$\sum_{\nu=0}^{\infty} \frac{(-1)^\nu}{\nu!} \left\langle (S_{\text{int}}[\varphi])^\nu \varphi_{k_1} \cdots \varphi_{k_n} \right\rangle_0$$

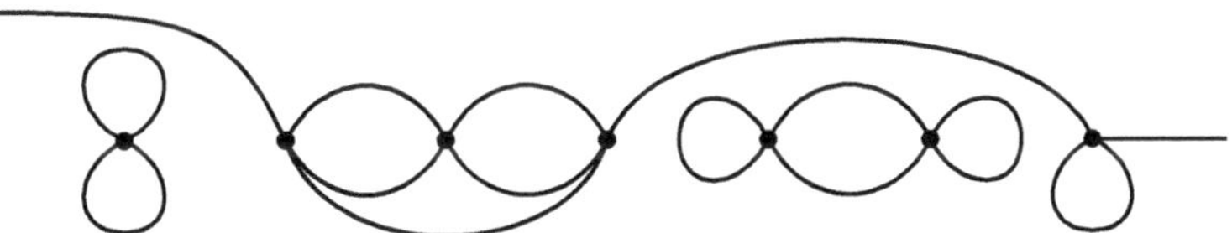

$$\times \left[1 + \sum_{\substack{\text{All vacuum} \\ \text{diagrams}}} \bigcirc \right]. \tag{4.31}$$

To see that all combinatorial factors come out right, let us classify all open diagrams according to the number of vertices ν' which are linked to the external legs. Here is an example for the case of $n = 2$ external legs with $\nu = 7$ and $\nu' = 4$:

Obviously there are $\binom{\nu}{\nu'} = \frac{\nu!}{\nu'!(\nu-\nu')!}$ possibilities of choosing the ν' vertices which are linked to at least one of the external legs. We can therefore write

$$\sum_{\nu=0}^{\infty} \frac{(-1)^\nu}{\nu!} \left\langle (S_{\text{int}}[\varphi])^\nu \varphi_{k_1} \cdots \varphi_{k_n} \right\rangle_0$$

$$= \sum_{\nu=0}^{\infty} \sum_{\nu'=0}^{\nu} \frac{(-1)^{\nu'}}{\nu'!} \frac{(-1)^{(\nu-\nu')}}{(\nu-\nu')!} \left\langle (S_{\text{int}}[\varphi])^{\nu'} \varphi_{k_1} \cdots \varphi_{k_n} \right\rangle_0^{\text{conn}} \left\langle (S_{\text{int}}[\varphi])^{\nu-\nu'} \right\rangle_0$$

$$= \sum_{\nu'=0}^{\infty} \frac{(-1)^{\nu'}}{\nu'!} \left\langle (S_{\text{int}}[\varphi])^{\nu'} \varphi_{k_1} \cdots \varphi_{k_n} \right\rangle_0^{\text{conn}} \sum_{\nu''=0}^{\infty} \frac{(-1)^{\nu''}}{\nu''!} \left\langle (S_{\text{int}}[\varphi])^{\nu''} \right\rangle_0 , \tag{4.32}$$

where in the last step we have rearranged the sum by introducing $v'' = v - v'$. Comparing with the graphical representation given in Eqs. (4.23) and (4.30) completes our proof of Eq. (4.31).

According to Eq. (4.23) the second factor on the right-hand side of Eq. (4.31) is nothing but the denominator in our perturbative expression (4.12) for the correlation function. The sum over all vacuum diagrams therefore cancels in Eq. (4.12) and we obtain

$$G_n(k_1, \ldots, k_n) = \sum_{\substack{\text{All open diagrams} \\ \text{without vacuum parts}}} \quad . \tag{4.33}$$

Hence, in the perturbative calculation of the correlation function, all vacuum diagrams can be omitted.

Of particular interest is the correlation function with two external legs, which by translational invariance is of the form,

$$G_2(k, k') = \langle \varphi_k \varphi_{k'} \rangle = \delta_{k,-k'} T G(k) = \delta_{k,-k'} T \underline{\quad k \qquad -k \quad} . \tag{4.34}$$

Graphically, we represent $G(k)$ by a thick line. The analogous quantity for quantum mechanical many-body systems is called *single-particle Green function* or *propagator* (Negele and Orland 1988). The factor of T in Eq. (4.34) is factored out for convenience. With this normalization we obtain from Eq. (4.20) within Gaussian approximation $G(k) \approx G_0(k) = 1/E_k$. The inverse Green function within Gaussian approximation is therefore simply given by the energy E_k defined in Eq. (4.4). It is useful to identify the exact inverse Green function $G^{-1}(k)$ with a modified energy, which differs from the energy E_k by some self-energy Σ_k. We therefore write

$$G^{-1}(k) = G_0^{-1}(k) + \Sigma_k = E_k + \Sigma_k . \tag{4.35}$$

This relation between the exact Green function and the self-energy is called *Dyson equation*. By expanding Σ_k instead of $G(k)$ in powers of the interaction, we effectively expand the inverse Green function; if we truncate the expansion of Σ_k at some finite order in the interaction and then consider $G(k) = [E_k + \Sigma_k]^{-1}$, we resum infinite orders in the interaction, which is obvious if we expand

$$G(k) = \frac{1}{E_k + \Sigma_k} = \frac{1}{E_k} - \frac{1}{E_k}\Sigma_k\frac{1}{E_k} + \frac{1}{E_k}\Sigma_k\frac{1}{E_k}\Sigma_k\frac{1}{E_k} - \ldots$$
$$= G_0(k) - G_0(k)\Sigma_k G(k) . \tag{4.36}$$

A graphical representation of this geometric series is shown in Fig. 4.2.

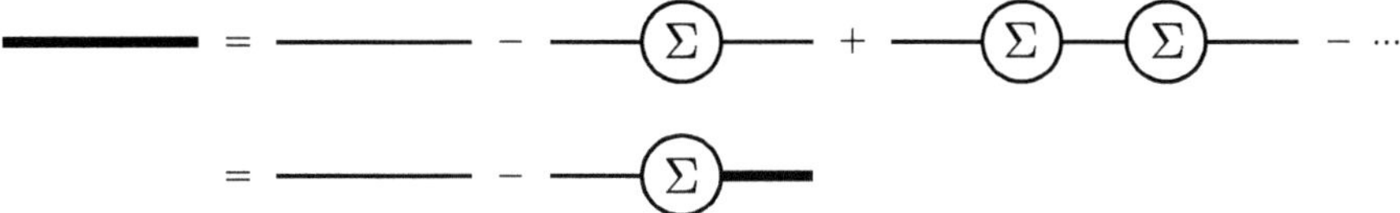

Fig. 4.2 Diagrammatic representation of the expansion (4.36) of the exact Green function in powers of the self-energy Σ_k. The *thick line* represents the exact $G(k)$. Note that in the *second line* the exact $G(k)$ appears also on the right-hand side, so that iteration generates the geometric series in the *first line* of Eq. (4.36)

The part of the diagrams in the first line of Fig. 4.2 contributing to the self-energy Σ_k cannot be separated into two parts by cutting a single propagator line. The self-energy Σ_k is therefore called *one-line irreducible*. Examples for second-order contributions to the irreducible self-energy are shown in Fig. 4.3 (a) and (b), while the second-order diagram (c) is not one-line irreducible and hence does not contribute to the perturbative expansion of Σ_k.

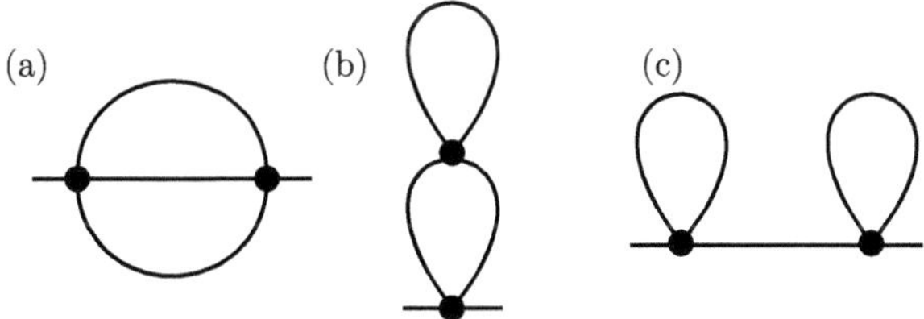

Fig. 4.3 The diagrams (**a**) and (**b**) are second-order contributions to the irreducible self-energy Σ_k. The external legs indicate the attached wave vectors and should not be counted as part of the diagrams. The diagram (**c**) is one-line reducible (it can be separated into two parts by cutting the middle propagator line), so that it does not contribute to the self-energy

4.2 One-Loop Momentum Shell RG

We are now ready to apply our general RG procedure to the effective classical field theory describing the critical properties of the Ising model derived in Sect. 2.2. At this point it is more convenient to use again the continuum normalization of the fields, which according to Eq. (2.57) is related to the lattice normalization via $\varphi(k) = a\sqrt{V}\varphi_k$. Consider the truncated Ginzburg–Landau–Wilson action $S_{\Lambda_0}[\varphi]$ in Eq. (2.61), which describes the long-wavelength properties of the Ising model. Setting for simplicity the magnetic field equal to zero, the partition function can be written as

$$\mathcal{Z} = \int \mathcal{D}[\varphi]e^{-S_{\Lambda_0}[\varphi]}, \tag{4.37}$$

with the effective long-wavelength action given by

$$S_{\Lambda_0}[\varphi] = V f_0 + \frac{1}{2} \int_k^{\Lambda_0} [r_0 + c_0 k^2] \varphi(-k)\varphi(k)$$

$$+ \frac{u_0}{4!} \int_{k_1}^{\Lambda_0} \int_{k_2}^{\Lambda_0} \int_{k_3}^{\Lambda_0} \int_{k_4}^{\Lambda_0} (2\pi)^D \delta(k_1 + k_2 + k_3 + k_4)\varphi(k_1)\varphi(k_2)\varphi(k_3)\varphi(k_4) ,$$

$$(4.38)$$

where we have explicitly written out the ultraviolet cutoff Λ_0. This truncated effective action depends on the four bare coupling constants f_0, r_0, c_0, and u_0, which are related to the parameters of the underlying lattice model as given in Eqs. (2.54), (2.55), (2.58), and (2.59). For simplicity, we shall focus on the disordered regime $T > T_c$ in this section.

4.2.1 Derivation of the RG Flow Equations

We first perform the mode-elimination step of the RG by integrating over all fields whose wave vectors lie in the shell $\Lambda < |\mathbf{k}| < \Lambda_0$, as illustrated in Fig. 4.4. Therefore, we adopt the procedure outlined in Sect. 3.1 and decompose

$$\varphi(k) = \varphi^<(k) + \varphi^>(k) = \Theta(\Lambda - |k|)\varphi(k) + \Theta(|k| - \Lambda)\varphi(k) . \qquad (4.39)$$

Substituting this into Eq. (4.38) we find that our Ginzburg–Landau–Wilson action can we written as

$$S_{\Lambda_0}[\varphi^< + \varphi^>] = S_\Lambda^<[\varphi^<; f_0, r_0, c_0, u_0] + S_{\Lambda,\Lambda_0}^>[\varphi^>] + S_{\mathrm{mix}}[\varphi^<, \varphi^>] . \qquad (4.40)$$

The "smaller part" $S_\Lambda^<[\varphi^<]$ of the effective action has the same form as the original action (4.38), but with reduced ultraviolet cutoff $\Lambda < \Lambda_0$ and with the original fields $\varphi(k)$ replaced by the "smaller component" $\varphi^<(k)$ of the fields,

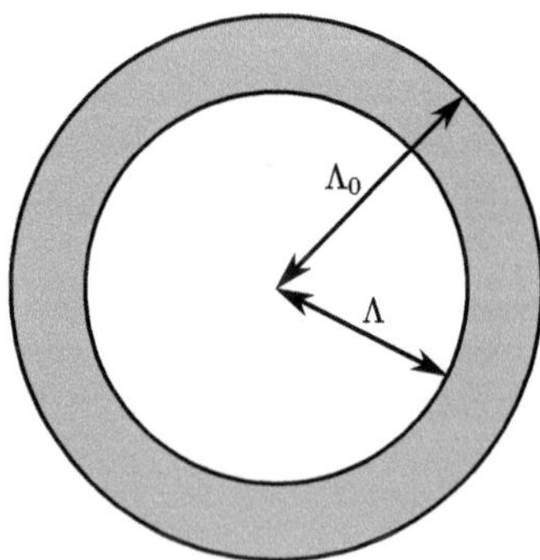

Fig. 4.4 In one iteration of the mode-elimination step of the momentum shell RG all fields with wave vectors in the shaded regime $\Lambda < |\mathbf{k}| < \Lambda_0$ are integrated out

$$S_\Lambda^<[\varphi^<; f_0, r_0, c_0, u_0] = V f_0 + \frac{1}{2} \int_k^\Lambda [r_0 + c_0 k^2] \varphi^<(-k)\varphi^<(k)$$

$$+ \frac{u_0}{4!} \left(\prod_{i=1}^4 \int_{k_i}^\Lambda \right) (2\pi)^D \delta(\sum_{i=1}^4 k_i) \varphi^<(k_1)\varphi^<(k_2)\varphi^<(k_3)\varphi^<(k_4) . \qquad (4.41)$$

The "larger part" $S_{\Lambda,\Lambda_0}^>[\varphi^>]$ of the action depends only on the "larger component" $\varphi^>(k)$ of the field; moreover, the momentum integrations are restricted to the shell $\Lambda < |k| < \Lambda_0$,

$$S_{\Lambda,\Lambda_0}^>[\varphi^>] = \frac{1}{2} \int_{|k|>\Lambda}^{\Lambda_0} [r_0 + c_0 k^2] \varphi^>(-k)\varphi^>(k)$$

$$+ \frac{u_0}{4!} \left(\prod_{i=1}^4 \int_{|k_i|>\Lambda}^{\Lambda_0} \right) (2\pi)^D \delta(\sum_{i=1}^4 k_i) \varphi^>(k_1)\varphi^>(k_2)\varphi^>(k_3)\varphi^>(k_4) . \qquad (4.42)$$

Finally, the interaction part of Eq. (4.38) gives rise to a mixing term involving both components of the field,

$$S_{\text{mix}}[\varphi^<, \varphi^>] = 6 \frac{u_0}{4!} \int_{k_1}^\Lambda \int_{k_2}^\Lambda \int_{|k_3|>\Lambda}^{\Lambda_0} \int_{|k_4|>\Lambda}^{\Lambda_0} (2\pi)^D \delta(\sum_{i=1}^4 k_i)$$

$$\times \varphi^<(k_1)\varphi^<(k_2)\varphi^>(k_3)\varphi^>(k_4)$$

$$+ \text{ terms involving } \varphi^<(\varphi^>)^3 \text{ or } (\varphi^<)^3\varphi^> . \qquad (4.43)$$

The combinatorial factor of 6 is due to the $\binom{4}{2}$ possibilities of picking two $\varphi^<$-fields out of the four available fields.

We now integrate over the "larger part" $\varphi^>$ of the field. Using the same notation as in Eqs. (3.5) and (3.6), the partition function can then formally be written as

$$\mathcal{Z} = \int \mathcal{D}[\varphi^<] e^{-S_\Lambda^<[\varphi^<; f^<, r^<, c^<, u^<]} , \qquad (4.44)$$

with the new effective action defined by

$$e^{-S_\Lambda^<[\varphi^<; f^<, r^<, c^<, u^<]} = e^{-S_\Lambda^<[\varphi^<; f_0, r_0, c_0, u_0]} \int \mathcal{D}[\varphi^>] e^{-S_{\Lambda,\Lambda_0}^>[\varphi^>] - S_{\text{mix}}[\varphi^<, \varphi^>]} , \qquad (4.45)$$

or equivalently,

$$S_\Lambda^<[\varphi^<; f^<, r^<, c^<, u^<] = S_\Lambda^<[\varphi^<; f_0, r_0, c_0, u_0]$$

$$- \ln \left(\int \mathcal{D}[\varphi^>] e^{-S_{\Lambda,\Lambda_0}^>[\varphi^>] - S_{\text{mix}}[\varphi^<, \varphi^>]} \right) . \qquad (4.46)$$

Since the functional integration over $\varphi^>$ in the second line cannot be performed exactly, we evaluate the integral approximately using diagrammatic perturbation theory. In the simplest approximation, we retain in the perturbative evaluation of the second term on the right-hand side of Eq. (4.46) only terms with one internal wave vector integration. This is called the *one-loop approximation* because diagrammatically each wave vector integration can be associated with a closed propagator loop. For the free energy per unit volume, the one-loop correction is due to the quadratic part of $S^>_{\Lambda,\Lambda_0}[\varphi^>]$ given in the first line of Eq. (4.42). The required integration in Eq. (4.46) is therefore Gaussian and has already been carried out in Sect. 2.3.2. We obtain

$$\boxed{f^< = f_0 + \frac{1}{2} \int_\Lambda^{\Lambda_0} \frac{d^D k}{(2\pi)^D} \ln\!\big[a^2(r_0 + c_0 k^2)\big]} \,. \tag{4.47}$$

Note that for $\Lambda \to 0$ this expression would be equivalent to Eq. (2.83), taking into account that in this chapter we consider the free energy per volume in units of temperature, whereas Δf in Eq. (2.83) is the free energy per lattice site. However, in contrast to Eq. (2.83), the wave vector integration in Eq. (4.47) is restricted to the momentum shell $\Lambda < |\mathbf{k}| < \Lambda_0$.

To obtain the new parameters $r^<$ and $c^<$ associated with the Gaussian part of the action after mode elimination, we need to calculate the irreducible self-energy due to $S_{\mathrm{mix}}[\varphi^<, \varphi^>]$ to first order in u_0. To derive this, we expand the integral in the logarithm of Eq. (4.46) in powers of the mixed part $S_{\mathrm{mix}}[\varphi^<, \varphi^>]$ of the effective action,

$$\int \mathcal{D}[\varphi^>] e^{-S^>_{\Lambda,\Lambda_0}[\varphi^>] - S_{\mathrm{mix}}[\varphi^<,\varphi^>]} =$$
$$\int \mathcal{D}[\varphi^>] e^{-S^>_{\Lambda,\Lambda_0}[\varphi^>]} \left[1 - S_{\mathrm{mix}}[\varphi^<, \varphi^>] + \frac{1}{2} S^2_{\mathrm{mix}}[\varphi^<, \varphi^>] + \dots \right]. \tag{4.48}$$

To leading order, we may calculate the functional average with the Gaussian part of $S^>_{\Lambda,\Lambda_0}[\varphi^>]$ given in the first line of Eq. (4.42). Denoting the corresponding Gaussian average by $\langle \dots \rangle_{0,>}$, we obtain for the logarithm of Eq. (4.48) to second order in the interaction,

$$-\ln\left(\int \mathcal{D}[\varphi^>] e^{-S^>_{\Lambda,\Lambda_0}[\varphi^>] - S_{\mathrm{mix}}[\varphi^<,\varphi^>]} \right)$$

$$= \underbrace{-\ln\left(\int \mathcal{D}[\varphi^>] e^{-S^>_{\Lambda,\Lambda_0}[\varphi^>]} \right)}_{f^< - f_0} - \ln\left[\frac{\int \mathcal{D}[\varphi^>] e^{-S^>_{\Lambda,\Lambda_0}[\varphi^>] - S_{\mathrm{mix}}[\varphi^<,\varphi^>]}}{\int \mathcal{D}[\varphi^>] e^{-S^>_{\Lambda,\Lambda_0}[\varphi^>]}} \right]$$

$$\approx f^< - f_0 - \ln\left[1 - \langle S_{\mathrm{mix}}[\varphi^<, \varphi^>]\rangle_{0,>} + \frac{1}{2}\langle S^2_{\mathrm{mix}}[\varphi^<, \varphi^>]\rangle_{0,>} \right]$$

$$\approx f^< - f_0 + \langle S_{\mathrm{mix}}[\varphi^<, \varphi^>]\rangle_{0,>}$$
$$- \frac{1}{2}\left[\langle S_{\mathrm{mix}}^2[\varphi^<, \varphi^>]\rangle_{0,>} - \langle S_{\mathrm{mix}}[\varphi^<, \varphi^>]\rangle_{0,>}^2\right]. \tag{4.49}$$

The term of first order in $S_{\mathrm{mix}}[\varphi^<, \varphi^>]$ in the last equality gives the leading renormalization of the parameters $r^<$ and $c^<$ associated with the quadratic part of the action,

$$\frac{1}{2}(r^< + c^< k^2) = \frac{1}{2}(r_0 + c_0 k^2) + \langle S_{\mathrm{mix}}[\varphi^<, \varphi^>]\rangle_{0,>}^{\varphi^<\text{-fields amputated}}$$

$$= \frac{1}{2}(r_0 + c_0 k^2) + 6 \ \underset{k \qquad -k}{\bigcirc}{}^{\varphi^>}$$

$$= \frac{1}{2}(r_0 + c_0 k^2) + 6 \ \frac{u_0}{4!} \int_\Lambda^{\Lambda_0} \frac{d^D k'}{(2\pi)^D} \frac{1}{r_0 + c_0 k'^2}, \tag{4.50}$$

where the subscript of the bracket in the first line denotes averaging with respect to the "larger fields" in Gaussian approximation, and the superscript means that the two "smaller fields" $\varphi^<(k_1)$ and $\varphi^<(k_2)$ in Eq. (4.43) should be dropped (or "amputated"). Comparing the wave vector independent terms on both sides of Eq. (4.50), we obtain

$$\boxed{\ r^< = r_0 + \frac{u_0}{2} \int_\Lambda^{\Lambda_0} \frac{d^D k}{(2\pi)^D} \frac{1}{r_0 + c_0 k^2}\ .} \tag{4.51}$$

Moreover, because the loop-integral on the right-hand side of Eq. (4.50) is independent of k we conclude that to this order

$$\boxed{\ c^< = c_0\ .} \tag{4.52}$$

Finally, to obtain the new interaction constant $u^<$ after mode elimination, we need to calculate the renormalization of the connected part of the four-field correlation function $G_4(k_1, k_2, k_3, k_4)$ which is generated by the second-order term in the last line of the expansion (4.49). This term is simply the Gaussian average of $(S_{\mathrm{mix}} - \langle S_{\mathrm{mix}}\rangle)^2$. With S_{mix} given by Eq. (4.43), the averaging of this term contains (apart from other terms of order u_0^2 which are not interesting for our purpose) the following term involving four powers of the "smaller field",

$$-\frac{1}{2}\left[\left\langle S_{\text{mix}}^2[\varphi^<,\varphi^>]\right\rangle_{0,>}-\left\langle S_{\text{mix}}[\varphi^<,\varphi^>]\right\rangle_{0,>}^2\right]^{\text{four }\varphi^<\text{ fields}}$$

$$=-\frac{1}{2}\times 2\times 6^2$$

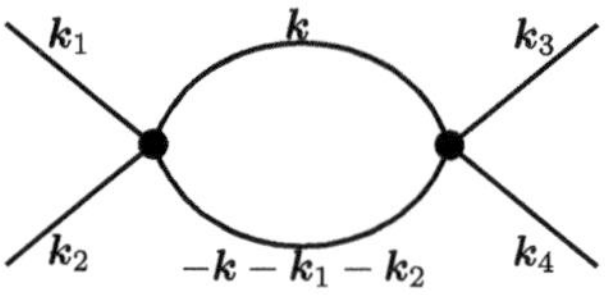

$$=-\frac{1}{2}\times 2\times\left(6\frac{u_0}{4!}\right)^2\left(\prod_{i=1}^{4}\int_{k_i}^{\Lambda}\right)(2\pi)^D\delta(\sum_{i=1}^{4}k_i)\varphi^<(k_1)\varphi^<(k_2)\varphi^<(k_3)\varphi^<(k_4)$$

$$\times\int_{\Lambda}^{\Lambda_0}\frac{d^Dk}{(2\pi)^D}\frac{1}{[r_0+c_0k^2][r_0+c_0(k+k_1+k_2)^2]}\,,\tag{4.53}$$

where the second factor of 2 in the second line is due to the two possibilities of connecting the internal loop to the external legs. Note that the factor in the last line of Eq. (4.53) depends on the external wave vectors k_1 and k_2 carried by the fields. However, keeping in mind that the wave vector k in the loop integration is restricted to the shell $\Lambda<|k|<\Lambda_0$ while the magnitude of the external wave vectors $k_1,\ldots,k_4$ are smaller than Λ, it is plausible to set $k_1=k_2=0$ in the last factor of Eq. (4.53). Formally, this procedure can be justified by noting that the wave vector dependence of the interaction can be described by an infinite set of irrelevant couplings (with respect to the Gaussian fixed point) which therefore do not affect the critical exponents, see the discussion in the last paragraph of Sect. 3.3.2. Setting $k_1=k_2=0$ in the last line of Eq. (4.53) and comparing the resulting expression with Eq. (4.41), we conclude that the mode elimination leads to the following modification of the interaction,

$$u^< = u_0 - \frac{3u_0^2}{2}\int_{\Lambda}^{\Lambda_0}\frac{d^Dk}{(2\pi)^D}\frac{1}{[r_0+c_0k^2]^2}\,.\tag{4.54}$$

In summary, after eliminating fluctuations with wave vectors in the shell $\Lambda<|k|<\Lambda_0$, the partition function can be written in the form (4.44) involving only the "smaller component" of the field, where the new effective action $S_\Lambda^<[\varphi^<;f^<,r^<,c^<,u^<]$, which is defined via Eq. (4.46), is approximately given by

$$S_\Lambda^<[\varphi^<;f^<,r^<,c^<,u^<]\approx Vf^<+\frac{1}{2}\int_{k}^{\Lambda}[r^<+c^<k^2]\varphi^<(-k)\varphi^<(k)$$

$$+\frac{u^<}{4!}\left(\prod_{i=1}^{4}\int_{k_i}^{\Lambda}\right)(2\pi)^D\delta(\sum_{i=1}^{4}k_i)\varphi^<(k_1)\varphi^<(k_2)\varphi^<(k_3)\varphi^<(k_4)\,,\tag{4.55}$$

with the new coupling constants $f^<,r^<,c^<,u^<$ given in Eqs. (4.47), (4.51), (4.52), and (4.54). Note that Eq. (4.55) is approximate for several reasons: first of all, the values of the new couplings $f^<, r^<, c^<, u^<$ have only been calculated within the one-loop approximation; moreover, the integration over the "larger component" of the field in Eq. (4.46) also generates terms involving six and more powers of the "smaller field," which we have neglected in the derivation of Eq. (4.55); finally, in Eq. (4.53) we have ignored the momentum dependence of the quartic interaction vertex.

Next, we perform the RG rescaling step with scale factor $b = \Lambda_0/\Lambda$ to rewrite our effective action in the same form as the original one. Following the general procedure outlined in Sect. 3.1, we introduce new wave vectors $k' = bk$ and the rescaled field

$$\varphi'(k') = \zeta_b^{-1} \varphi^<(k'/b) , \tag{4.56}$$

with the field rescaling factor given by

$$\zeta_b = b^{1+D/2} \sqrt{Z_b} , \qquad Z_b = \frac{c_0}{c^<} , \tag{4.57}$$

see Eqs. (3.8), (3.9), (3.10), and (3.11). Although within our simple one-loop approximation $c^< = c_0$ and hence $Z_b = 1$, it is instructive to explicitly retain the factor Z_b to emphasize that in general the field rescaling factor must be chosen such that the coefficient in front of the k^2-term in the Gaussian part of the action is not changed after one complete iteration of the RG procedure,

$$\begin{aligned}
&\frac{1}{2} \int^\Lambda \frac{d^D k}{(2\pi)^D} (r^< + c^< k^2) \varphi^<(-k) \varphi^<(k) \\
&= \frac{1}{2} b^{-D} \int^{\Lambda_0} \frac{d^D k'}{(2\pi)^D} (r^< + c^< b^{-2} k'^2) b^{2+D} Z_b \varphi'(-k') \varphi'(k') \\
&= \frac{1}{2} \int^{\Lambda_0} \frac{d^D k'}{(2\pi)^D} (b^2 Z_b r^< + c_0 k'^2) \varphi'(-k') \varphi'(k') ,
\end{aligned} \tag{4.58}$$

see also the discussion after Eq. (3.14) in Sect. 3.1. Hence, after one complete iteration of the RG, $c' = c_0$ and the new momentum-independent term r' in the Gaussian part of the effective action is

$$r' = b^2 Z_b r^< = b^2 Z_b \left[r_0 + \frac{u_0}{2} \int_\Lambda^{\Lambda_0} \frac{d^D k}{(2\pi)^D} \frac{1}{r_0 + c_0 k^2} \right] , \tag{4.59}$$

where we have used Eq. (4.51). Similarly, using Eq. (4.54) and the fact that $b^{-3D} \zeta_b^4 = b^{4-D} Z_b^2$, we obtain for the renormalized quartic coupling constant

$$u' = b^{4-D} Z_b^2 u^{<} = b^{4-D} Z_b^2 \left[u_0 - \frac{3u_0^2}{2} \int_\Lambda^{\Lambda_0} \frac{d^D k}{(2\pi)^D} \frac{1}{[r_0 + c_0 k^2]^2} \right] . \tag{4.60}$$

Finally, keeping in mind that under rescaling the volume changes as $V' = b^{-D} V$, we obtain from Eq. (4.47) for the renormalized free energy density

$$f' = b^D f^{<} = b^D \left[f_0 + \frac{1}{2} \int_\Lambda^{\Lambda_0} \frac{d^D k}{(2\pi)^D} \ln[a^2(r_0 + c_0 k^2)] \right] . \tag{4.61}$$

4.2.2 The Wilson–Fisher Fixed Point

For a quantitative analysis of the above RG flow equations, it is convenient to work with the equivalent differential equations describing the iteration of the RG procedure in infinitesimal steps, as explained in Sect. 3.2.3. To derive the differential RG equations we set $\Lambda = \Lambda_0 e^{-l}$ in Eqs. (4.59), (4.60), and (4.61) and consider all quantities as functions of the logarithmic flow parameter $l = \ln b = \ln(\Lambda_0/\Lambda)$. We then take the derivative of these equations with respect to l. Consider first the flow equation (4.59) describing the renormalization of r_0. Defining $r_l = r'$ and

$$Z_l \equiv Z_{b=e^l} , \tag{4.62}$$

and using the fact that

$$\frac{\partial}{\partial l} \int_{\Lambda_0 e^{-l}}^{\Lambda_0} \frac{d^D k}{(2\pi)^D} \frac{1}{r_0 + c_0 k^2} = K_D \frac{\partial}{\partial l} \int_{\Lambda_0 e^{-l}}^{\Lambda_0} dk \frac{k^{D-1}}{r_0 + c_0 k^2} = \frac{K_D \Lambda^D}{r_0 + c_0 \Lambda^2} , \tag{4.63}$$

we obtain from Eq. (4.59)

$$\partial_l r_l = (2 - \eta_l) r_l + b^2 Z_l \frac{u_0}{2} \frac{K_D \Lambda^D}{r_0 + c_0 \Lambda^2} . \tag{4.64}$$

Here, the numerical constant K_D is the surface area of the D-dimensional unit sphere divided by $(2\pi)^D$ (see Eq. (2.86)), and we have introduced the so-called *flowing anomalous dimension*,

$$\eta_l = -\frac{\partial_l Z_l}{Z_l} = -\partial_l \ln Z_l . \tag{4.65}$$

As will be discussed in more detail in Sect. 4.2.3, the correlation function exponent η defined via Eq. (1.13) can be identified with the limit

$$\eta = \lim_{l \to \infty} \eta_l \,. \tag{4.66}$$

Of course, within our simple one-loop approximation $Z_l = 1$ and hence $\eta_l = 0$, but a more accurate two-loop calculation[3] gives a finite value for η_l. Assuming now that l is infinitesimally small and ignoring terms of order l, we may replace $\Lambda \to \Lambda_0$ and $b^2 Z_l \to 1$ on the right-hand side of Eq. (4.64). To the same accuracy we may set $r_0 \to r_l$ and $u_0 \to u_l$ in Eq. (4.64), so that we obtain

$$\partial_l r_l = (2 - \eta_l)r_l + \frac{u_l}{2} \frac{K_D \Lambda_0^D}{r_l + c_0 \Lambda_0^2} \,. \tag{4.67}$$

Integration of this differential equation for finite l yields the flowing coupling r_l resulting from the repeated iteration of the elementary RG step.

The derivation of the differential RG flow equations for the quartic coupling constant u_l and for the free energy density f_l is analogous to the derivation of Eq. (4.67). From Eq. (4.60) we obtain the infinitesimal version of the RG flow equation for the quartic coupling constant u_l,

$$\partial_l u_l = (4 - D - 2\eta_l)u_l - \frac{3u_l^2}{2} \frac{K_D \Lambda_0^D}{\left(r_l + c_0 \Lambda_0^2\right)^2} \,, \tag{4.68}$$

and Eq. (4.61) yields for the flow of the free energy density,

$$\partial_l f_l = D f_l + \frac{K_D}{2} \Lambda_0^D \ln\left[a^2 \left(r_l + c_0 \Lambda_0^2\right)\right] \,. \tag{4.69}$$

To get rid of numerical factors in the above equations, it is convenient to define dimensionless couplings

$$\bar{r}_l = \frac{r_l}{c_0 \Lambda_0^2}, \qquad \bar{u}_l = K_D \frac{u_l}{c_0^2 \Lambda_0^{4-D}} \,. \tag{4.70}$$

Setting now explicitly $\eta_l = 0$, the coupled system of differential equations (4.67) and (4.68) can be written as

---

[3] As will be shown in Sect. 8.4 [see the discussion after Eq. (8.105)], in the symmetry-broken phase one obtains already a finite result for η_l within the one-loop approximation, so that in this case a one-loop calculation is sufficient to obtain a first estimate for the flowing anomalous dimension (Sinner et al. 2008). It turns out that two-loop calculations within the Wilsonian momentum-shell RG are rather difficult and for calculations beyond the one-loop approximation in renormalizable theories it is more convenient to use the field theoretical RG (see e.g., Zinn-Justin 2002). For a two-loop calculation of the RG β-function of φ^4-theory using the FRG version of the Wilsonian RG see (Kopietz 2001).

$$\partial_l \bar{r}_l = 2\bar{r}_l + \frac{1}{2}\frac{\bar{u}_l}{1+\bar{r}_l} , \tag{4.71}$$

$$\partial_l \bar{u}_l = (4 - D)\bar{u}_l - \frac{3}{2}\frac{\bar{u}_l^2}{(1+\bar{r}_l)^2} . \tag{4.72}$$

The associated RG flow in the $\bar{u}$-$\bar{r}$-plane is shown in Fig. 4.5. Obviously, above four dimensions, the critical behavior of the Ising model is controlled by the Gaussian fixed point G at $(\bar{u}_*, \bar{r}_*) = (0, 0)$. In this case the effective interaction vanishes at the fixed point, so that the critical exponents are given by the Gaussian approximation, in particular $\nu = 1/2$ and $\eta = 0$, see Sect. 2.3.3. On the other hand, for $D < 4$ the Gaussian fixed point becomes unstable and a new critical fixed point emerges, the Wilson–Fisher fixed point WF, which controls the critical behavior of the Ising universality class below four dimensions. This is the reason for the breakdown of the Gaussian approximation below four dimensions discussed in Sect. 2.3.4. If we fine-tune the initial conditions such that the RG trajectory starts very close to the critical surface, then in an intermediate interval $l_c \lesssim l \lesssim l_*$ of the logarithmic flow parameter the RG flow remains almost stationary in the vicinity of the critical fixed point, as shown in Fig. 4.6. For $D < 4$ and $\bar{u}_0 \ll \bar{u}_*$, the scale l_c is approximately given by (see Hasselmann et al. 2004)

$$l_c = \frac{\ln(\bar{u}_*/\bar{u}_0)}{4 - D} . \tag{4.73}$$

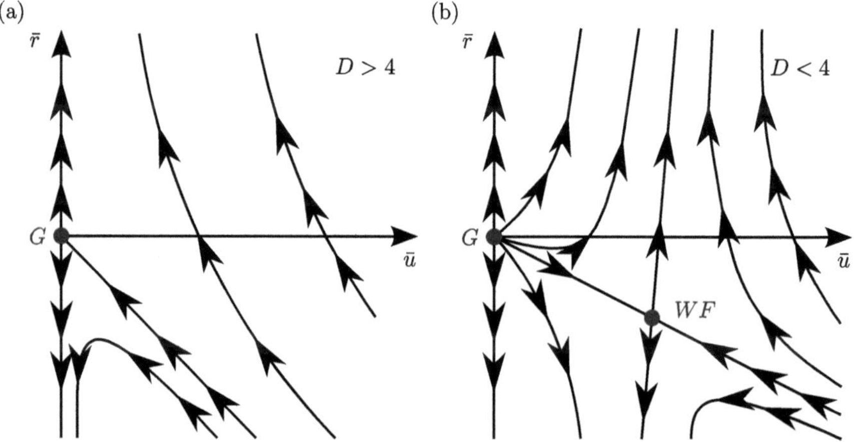

Fig. 4.5 Qualitative behavior of the RG flow in the $\bar{u}$-$\bar{r}$-plane generated by the one-loop flow equations (4.71) and (4.72). The *left figure* (**a**) corresponds to $D > 4$, where the critical behavior of the system is controlled by the Gaussian fixed point G. On the other hand, below four dimensions, shown in (**b**), the critical behavior is determined by the Wilson–Fisher fixed point WF, which is characterized by a finite value of the interaction

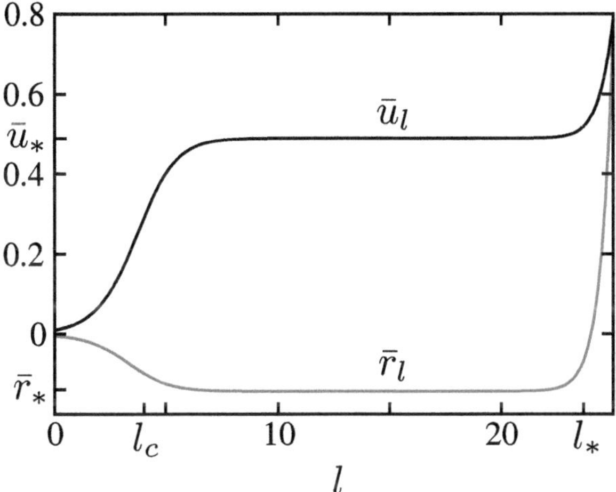

Fig. 4.6 Flow of the couplings $\bar{r}_l$ and $\bar{u}_l$ for a nearly critical system as a function of the logarithmic RG flow parameter l obtained from Eqs. (4.71) and (4.72) for $D = 3$. The two characteristic RG scales l_c and l_* separate three distinct regimes: In the initial regime $0 < l \lesssim l_c$ the couplings $\bar{r}_l$ and $\bar{u}_l$ flow toward the values $\bar{r}_*$ and $\bar{u}_*$ associated with the critical fixed point; in the intermediate interval $l_c \lesssim l \lesssim l_*$ the RG flow is very slow in the vicinity of the fixed point; finally, for $l_* \lesssim l$ the trajectory rapidly flows away from the fixed point

To obtain a physical interpretation of l_c and l_*, it is necessary to calculate the momentum-dependent correlation function $G(\boldsymbol{k})$, which we shall do in Sect. 8.3. Let us anticipate here that the scale $\Lambda_0 e^{-l_*} = \xi^{-1}$ can be identified with the inverse correlation length (which vanishes at the critical point, corresponding to $l_* \to \infty$ on the critical surface), while the interaction dependent scale $\Lambda_0 e^{-l_c} = k_c$ defines the upper cutoff in momentum space where the asymptotic scaling $G(\boldsymbol{k}) \propto |\boldsymbol{k}|^{-2+\eta}$ at the critical point can be observed. The scale k_c is usually called the Ginzburg scale (Ginzburg 1960, Amit 1974, Hasselmann et al. 2007).

The crucial point is now that if we consider the dimensionality D as a continuous parameter and assume that D is only slightly smaller than the upper critical dimension $D_{\mathrm{up}} = 4$, then the smallness of the parameter

$$\epsilon = D_{\mathrm{up}} - D = 4 - D \tag{4.74}$$

justifies an expansion of the critical exponents in powers of ϵ. To see this more clearly, consider the fixed point values $\bar{u}_*$ and $\bar{r}_*$ of our dimensionless couplings, which are obtained by setting the left-hand sides of Eqs. (4.71) and (4.72) equal to zero,

$$0 = 2\bar{r}_* + \frac{1}{2} \frac{\bar{u}_*}{1 + \bar{r}_*} , \tag{4.75}$$

$$0 = \epsilon \bar{u}_* - \frac{3}{2} \frac{\bar{u}_*^2}{(1 + \bar{r}_*)^2} . \tag{4.76}$$

The solutions to first order in ϵ are easily obtained,

$$\bar{u}_* = \frac{2}{3}\epsilon + \mathcal{O}(\epsilon^2) \,, \tag{4.77}$$

$$\bar{r}_* = -\frac{\bar{u}_*}{4} + \mathcal{O}(\epsilon^2) = -\frac{\epsilon}{6} + \mathcal{O}(\epsilon^2) \,. \tag{4.78}$$

Because in the derivation of our one-loop RG equations we have ignored higher orders in u_l, it would be inconsistent to retain the correction of order ϵ^2 in Eqs. (4.77) and (4.78). It is now important that the fixed point values of $\bar{u}_l$ and $\bar{r}_l$ are both of order ϵ. If the initial values of $\bar{u}_l$ and $\bar{r}_l$ are also chosen of order ϵ, then these couplings remain of order ϵ on the entire critical surface, so that the one-loop approximation underlying Eqs. (4.71) and (4.72) is justified. This procedure can be systematically improved by retaining more loops in the RG flow equations, leading to an expansion of critical exponents in powers of ϵ (Pelissetto and Vicari 2002).

Let us now proceed and calculate the scaling variables and the eigenvalues of the linearized RG flow at the Wilson–Fisher fixed point for $D < 4$ to order ϵ. Substituting $\bar{r}_l = \bar{r}_* + \delta\bar{r}_l$ and $\bar{u}_l = \bar{u}_* + \delta\bar{u}_l$ in our one-loop RG equations (4.71) and (4.72), and expanding the right-hand sides to first order in the deviations $\delta\bar{r}_l$ and $\delta\bar{u}_l$ from the fixed point, we obtain the linearized RG equations

$$\partial_l \begin{pmatrix} \delta\bar{r}_l \\ \delta\bar{u}_l \end{pmatrix} = \begin{pmatrix} 2 - \frac{\epsilon}{3} & \frac{1}{2} + \frac{\epsilon}{12} \\ 0 & -\epsilon \end{pmatrix} \begin{pmatrix} \delta\bar{r}_l \\ \delta\bar{u}_l \end{pmatrix} \,. \tag{4.79}$$

Note that the lower off-diagonal element of the coefficient matrix vanishes to order ϵ; in a more accurate calculation to order ϵ^2 the zero in the lower off-diagonal element would be replaced by a finite number of order ϵ^2. Although the 2×2-matrix in Eq. (4.79) cannot be diagonalized, it is easy to find its left-eigenvectors, which according to our general considerations of Sect. 3.3.2 are needed to construct the scaling variables. By inspection, one easily finds one of the left-eigenvectors of the matrix in Eq. (4.79),

$$\boldsymbol{v}_u^T = (0, 1) \,. \tag{4.80}$$

The corresponding eigenvalue

$$y_u = -\epsilon = D - 4 \tag{4.81}$$

is negative for $D < 4$, so that the associated scaling variable

$$\delta\bar{u}_l = \boldsymbol{v}_u^T \begin{pmatrix} \delta\bar{r}_l \\ \delta\bar{u}_l \end{pmatrix} \tag{4.82}$$

is irrelevant at the Wilson–Fisher fixed point. To determine the other left-eigenvector of the matrix in Eq. (4.79), we make the ansatz $\boldsymbol{v}_t^T = (1, x)$, so that the corresponding eigenvector equation reads

$$(1, x) \begin{pmatrix} 2 - \frac{\epsilon}{3} & \frac{1}{2} + \frac{\epsilon}{12} \\ 0 & -\epsilon \end{pmatrix} = \left(2 - \frac{\epsilon}{3} ,\ \frac{1}{2} + \frac{\epsilon}{12} - \epsilon x \right) = y_t(1, x) . \quad (4.83)$$

This implies

$$y_t = 2 - \frac{\epsilon}{3} \tag{4.84}$$

and

$$x = \frac{1 + \frac{\epsilon}{6}}{4 \left[1 + \frac{\epsilon}{3} \right]} = \frac{1}{4} \left[1 - \frac{\epsilon}{6} \right] + O(\epsilon^2). \tag{4.85}$$

Our second left-eigenvector is therefore

$$\boldsymbol{v}_t^T = \left(1, \frac{1 + \frac{\epsilon}{6}}{4 \left[1 + \frac{\epsilon}{3} \right]} \right) = \left(1, \frac{1}{4} \left[1 - \frac{\epsilon}{6} \right] + O(\epsilon^2) \right) , \tag{4.86}$$

and the associated relevant scaling variable is, to first order in ϵ, the following linear combination of $\delta \bar{r}_l$ and $\delta \bar{u}_l$,

$$t_l = \boldsymbol{v}_t^T \begin{pmatrix} \delta \bar{r}_l \\ \delta \bar{u}_l \end{pmatrix} \approx \delta \bar{r}_l + \frac{1}{4} \left[1 - \frac{\epsilon}{6} \right] \delta \bar{u}_l . \tag{4.87}$$

But according to Eq. (1.28), the inverse of the exponent y_t associated with the relevant thermal scaling variable can be identified with the correlation length exponent v, so that to linear order in ϵ we obtain

$$\boxed{ v = \frac{1}{y_t} = \frac{1}{2 - \frac{\epsilon}{3}} = \frac{1}{2} + \frac{\epsilon}{12} + \mathcal{O}(\epsilon^2) . } \tag{4.88}$$

Using the fact that to linear order in ϵ the correlation function exponent η vanishes[4], the thermodynamic exponents slightly below four dimensions can be obtained with the help of the scaling relations (1.33a), (1.33b), (1.33c), and (1.33d).

4.2.3 Wave Function Renormalization and Anomalous Dimension

To conclude this chapter, let us make some remarks about the physical meaning of the flowing field renormalization factor $Z_l \equiv Z_{b=e^l}$ appearing in our definition (4.57) of the field rescaling factor ζ_b and the associated flowing anomalous

---

[4] A more accurate two-loop calculation to order ϵ^2 yields $\eta = \epsilon^2/54 + \mathcal{O}(\epsilon^3)$ (see e.g., Pelissetto and Vicari 2002).

dimension $\eta_l = -\partial_l \ln Z_l$, see Eq. (4.65). Although within our simple one-loop approximation $Z_l = 1$ and hence $\eta_l = 0$, a more accurate calculation would yield a nontrivial result for Z_l. From the definition $Z_b = c_0/c^<$ in Eq. (3.11), it is clear that only if the self-energy exhibits some wave vector dependence the factor Z_l can be different from unity. Given the fact that the one-loop diagram for the self-energy shown in Eq. (4.50) is wave vector independent, it is not surprising that we have obtained $Z_l = 1$ in the one-loop approximation. On the other hand, as already mentioned in the footnote after Eq. (4.66), for $T < T_c$ the one-loop approximation gives rise to a momentum-dependent self-energy, so that in this case one obtains a nontrivial result for Z_l at the one-loop level (Sinner et al. 2008). In Chap. 8 we shall explicitly write down exact and approximate FRG flow equations for the momentum-dependent self-energy in the symmetry-broken phase of φ^4-theory, see Eqs. (8.18) and (8.104).

Let us begin by elaborating on the relation between the wave function renormalization factor Z_l and the flowing anomalous dimension η_l. The defining equation (4.65) of η_l can also be written as a formal differential equation for Z_l,

$$\partial_l Z_l = -\eta_l Z_l \; ,\tag{4.89}$$

which integrates to

$$Z_l = Z_0 e^{-\int_0^l dt \, \eta_t} \; .\tag{4.90}$$

One should keep in mind that in general η_l depends also on Z_l, so that Eq. (4.90) should be understood as an integral equation for Z_l. Suppose now that η_l approaches a finite positive limit for $l \to \infty$,

$$\eta = \lim_{l \to \infty} \eta_l .\tag{4.91}$$

Then Eq. (4.90) implies that $\lim_{l \to \infty} Z_l = 0$. Conversely, if the limit

$$Z = \lim_{l \to \infty} Z_l\tag{4.92}$$

is nonzero, then the integral in the exponent of Eq. (4.90) must approach a finite limit for $l \to \infty$, which is only possible if $\lim_{l \to \infty} \eta_l = 0$. Our notation anticipates that the limit $\eta = \lim_{l \to \infty} \eta_l$ in Eq. (4.91) can be identified with the critical exponent η which we have introduced phenomenologically in Sect. 1.2 via the asymptotic long-distance (or small wave vector) behavior of the correlation function, see Eqs. (1.13) and (1.14). To see this, recall that according to Eqs. (3.86) and (4.57), the correlation function in momentum space transforms under the RG as

$$G(k; g) = b^2 Z_l G(bk; g') \; .\tag{4.93}$$

But if η is finite, then Z_l is proportional to $e^{-\eta l} \propto b^{-\eta}$ for sufficiently large l. Keeping in mind that for a critical system the coupling constants approach their fixed point values g^* for large l, we see that Eq. (4.93) reduces to the homogeneity relation

$$G(k; g^*) \propto b^{2-\eta} G(bk; g^*) \; .\tag{4.94}$$

This implies the critical scaling $G(\boldsymbol{k}) \propto |\boldsymbol{k}|^{-2+\eta}$ postulated in Eq. (1.14), which completes the proof that $\eta = \lim_{l \to \infty} \eta_l$ can indeed be identified with the critical exponent η. Note that η can be viewed as an interaction-dependent contribution to the scaling dimension of the field and is therefore called *anomalous dimension*. The scaling law (4.94) is anomalous in the sense that it does not follow from dimensional analysis. If η is finite, then the scaling of the coupling constants of the system is not only determined by their canonical dimension (which can be found by dimensional analysis), but receives also a contribution from the anomalous dimension η. The canonical and anomalous contributions combine to form the scaling dimension of a given coupling constant.

On the other hand, for noncritical systems the limit $\eta = \lim_{l \to \infty} \eta_l$ necessarily vanishes. However, for finite l the flowing η_l is usually finite and positive such that $Z < 1$. In fact, if we express the exact two-field correlation function $G(\boldsymbol{k})$ in terms of the irreducible self-energy,[5]

$$G(\boldsymbol{k}) = \frac{1}{c_0 \boldsymbol{k}^2 + \Sigma(\boldsymbol{k})} \, , \tag{4.95}$$

and expand $\Sigma(\boldsymbol{k})$ to second order in $\boldsymbol{k}$ we obtain for small wave vectors $\boldsymbol{k}$

$$G(\boldsymbol{k}) = \frac{Z}{c_0(\boldsymbol{k}^2 + \xi^{-2})} \, , \tag{4.96}$$

where

$$Z = \frac{1}{1 + \left. \frac{\partial \Sigma(\boldsymbol{k})}{\partial (c_0 \boldsymbol{k}^2)} \right|_{k=0}} \tag{4.97}$$

and

$$\frac{c_0}{\xi^2} = Z \Sigma(\boldsymbol{k} = 0) \, . \tag{4.98}$$

The identification (4.97) of Z with the limit of Z_l for $l \to \infty$ in Eq. (4.92) follows from the definition of $Z_b = c_0/c^>$ in Eq. (3.11), see also the footnote in Sect. 3.3.1.

The field renormalization factor also plays an important role in quantum mechanical many-body systems and quantum field theories. In this context it is also called *wave function renormalization factor*. In quantum field theories the self-energy $\Sigma(\omega, \boldsymbol{k})$ depends also on frequency, and the wave function renormalization factor Z is usually defined in terms of the frequency derivative of the self-energy. For example, for normal Fermi liquids one defines for each wave vector $\boldsymbol{k}$ (Abrikosov et al. 1963, Nozières 1964, Negele and Orland 1988),

---

[5] Recall that in Sect. 4.1 we introduced diagrammatic perturbation theory using the lattice normalization. The corresponding self-energy Σ_k appearing in the Dyson equation (4.36) has units of energy. In contrast, here we use the continuum normalization where the self-energy $\Sigma(\boldsymbol{k})$ has units of $1/(\text{length})^2$. We indicate the different normalizations of the self-energy by different positions of the wave vector labels.

$$Z_k = \frac{1}{1 - \left.\frac{\partial \Sigma(\omega,k)}{\partial \omega}\right|_{\omega=0}} \; . \tag{4.99}$$

For wave vectors k on the Fermi surface k_F, the function Z_{k_F} is usually called quasiparticle residue, because it can be identified with the weight of the quasiparticle pole in the single-particle Green function of a Fermi liquid. The vanishing of Z_{k_F} in the normal state would imply the breakdown of the quasiparticle picture.[6]

Finally, let us point out that the wave function renormalization factor appears also in second-order Rayleigh–Schrödinger perturbation theory of single-particle quantum mechanics (see e.g., Sakurai 1994). Recall that in quantum mechanical perturbation theory one expands the eigenvalues $E_n(\lambda)$ and eigenstates $|E_n\rangle$ of a Hamiltonian of the form $H = H_0 + \lambda V$ in powers of the small parameter λ, assuming that the eigenvalues ϵ_n and eigenstates $|\epsilon_n\rangle$ of H_0 are known. To generate the perturbation series, it is convenient to work with states $|E_n\rangle$ which are not normalized to unity but whose projection onto the corresponding unperturbed states $|\epsilon_n\rangle$ is unity,

$$\langle \epsilon_n | E_n \rangle = 1 \, , \tag{4.100}$$

see Fig. 4.7. Assuming for simplicity that the states are nondegenerate, Rayleigh–Schrödinger perturbation theory yields the formal expansion for the perturbed states (Sakurai 1994),

$$|E_n\rangle = \sum_{\nu=0}^{\infty} \left[\frac{Q_n}{\epsilon_n - H_0}(\epsilon_n - E_n(\lambda) + \lambda V) \right]^{\nu} |\epsilon_n\rangle \, , \tag{4.101}$$

where $Q_n = 1 - |\epsilon_n\rangle\langle\epsilon_n|$ projects onto the subspace orthogonal to $|\epsilon_n\rangle$. The corresponding energies are

$$E_n(\lambda) = \epsilon_n + \langle \epsilon_n | \lambda V | E_n \rangle$$
$$= \epsilon_n + \sum_{\nu=0}^{\infty} \langle \epsilon_n | \lambda V \left[\frac{Q_n}{\epsilon_n - H_0}(\epsilon_n - E_n(\lambda) + \lambda V) \right]^{\nu} |\epsilon_n\rangle \, . \tag{4.102}$$

Unfortunately, the right-hand sides of Eqs. (4.101) and (4.102) depend again on $E_n(\lambda)$, so that for a systematic expansion in powers of the small parameter λ one has to reshuffle the terms corresponding to different powers of ν in the perturbation expansion. This makes calculations within Rayleigh–Schrödinger perturbation

---

[6] For example, for interacting fermions in one dimension $Z_{k_F} = 0$ for arbitrarily weak interaction. The normal state of interacting fermions in $D = 1$ is therefore not a Fermi liquid and is called Luttinger liquid (Haldane 1981).

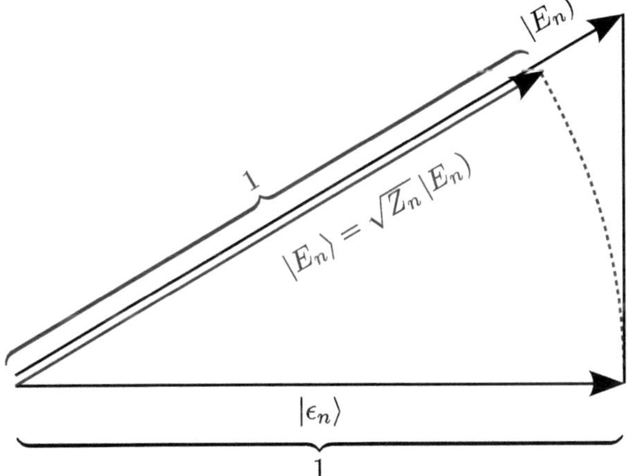

Fig. 4.7 In Rayleigh–Schrödinger perturbation theory the perturbed states $|E_n)$ are normalized such that their projection onto the unperturbed states $|\epsilon_n\rangle$ is unity. To obtain properly normalized states $|E_n\rangle = \sqrt{Z_n}|E_n)$, the length of the states $|E_n)$ has to be reduced by a factor $\sqrt{Z_n} < 1$

theory beyond the second order quite tedious. Nevertheless, in principle we can calculate the perturbed states $|E_n)$ to a given power of λ. Because of the normalization condition (4.100), the states $|E_n)$ are not normalized to unity and should be rescaled at the end of the calculation by a factor $Z_n < 1$ to obtain properly normalized states $|E_n\rangle$,

$$|E_n\rangle = \sqrt{Z_n}|E_n) \,, \tag{4.103}$$

where the wave function renormalization factor is given by

$$Z_n = \frac{1}{(E_n|E_n)} \,. \tag{4.104}$$

It is a simple exercise in quantum mechanics to show that (Sakurai 1994)

$$Z_n = \frac{\partial E_n(\lambda)}{\partial \epsilon_n} \,. \tag{4.105}$$

From Fig. 4.7 it is obvious that Z_n is in general smaller than unity. In fact, if a perturbed state $|E_n)$ is almost orthogonal to the corresponding unperturbed state $|\epsilon_n\rangle$, then the wave function renormalization factor is very small. A vanishing wave function renormalization factor (corresponding to a finite anomalous dimension) therefore means that the true eigenstates of the system cannot be reached perturbatively from the eigenstates of the corresponding noninteracting system.

Exercises

4.1 One-Loop Flow Equations for the $O(N)$-symmetric φ^4-theory

Consider the φ^4-theory with an N-component (vector) field $\boldsymbol{\varphi}(\boldsymbol{k}) \equiv (\varphi_1(\boldsymbol{k}), \ldots, \varphi_N(\boldsymbol{k}))$ which is invariant under $O(N)$-rotations in field component space,

$$S[\varphi] = \frac{1}{2} \int^{\Lambda_0} \frac{d^D k}{(2\pi)^D} [r_0 + c_0 k^2] \sum_{i=1}^{N} |\varphi_i(\boldsymbol{k})|^2 + \frac{u_0}{4!} \int d^D r \sum_{i,j=1}^{N} \varphi_i^2(\boldsymbol{r})\varphi_j^2(\boldsymbol{r}) .$$

Here, Λ_0 is the initial ultraviolet cutoff and

$$\varphi_i(\boldsymbol{r}) = \int^{\Lambda_0} \frac{d^D k}{(2\pi)^D} e^{i\boldsymbol{k}\cdot\boldsymbol{r}} \varphi_i(\boldsymbol{k}) .$$

(a) In analogy to the discussion of the Ising-like φ^4-theory with $N = 1$ eliminate all modes with $\Lambda < k < \Lambda_0$ perturbatively. Rescale momenta and fields appropriately to derive the recursion relations,

$$r' = \frac{\zeta_b^2}{b^D} \left[r_0 + \frac{(N+2)u_0}{6} \int_\Lambda^{\Lambda_0} \frac{d^D k}{(2\pi)^D} \frac{1}{r_0 + c_0 k^2} + \mathcal{O}\left(u_0^2\right) \right] ,$$

$$c' = \frac{\zeta_b^2}{b^{D+2}} [c_0 + \mathcal{O}\left(u_0^2\right)] ,$$

$$u' = \frac{\zeta_b^4}{b^{3D}} \left[u_0 - \frac{(N+8)u_0^2}{6} \int_\Lambda^{\Lambda_0} \frac{d^D k}{(2\pi)^D} \frac{1}{(r_0 + c_0 k^2)^2} + \mathcal{O}\left(u_0^3\right) \right] .$$

Note: Using Feynman diagrams, the interaction term can be represented by a vertex of the form $i \!\!>\!\!\cdots\!\!<\!\! j$ (with $u_0/4!$) to denote the two independent summations over field indices. Internal loops that are not connected to external legs by solid lines then lead to free summations over a field component index and the corresponding diagram is thus proportional to N.

(b) Demand $c' = c_0$ to determine the field rescaling factor ζ_b and derive differential flow equations by setting $b = e^l$ with $l \to 0$,

$$\frac{d\bar{r}}{dl} = 2\bar{r} + \frac{(N+2)\bar{u}}{6(1+\bar{r})} , \qquad \frac{d\bar{u}}{dl} = \epsilon\bar{u} - \frac{(N+8)\bar{u}^2}{6(1+\bar{r})^2} ,$$

where $\epsilon = 4 - D$. For a convenient notation, we have introduced the dimensionless couplings $\bar{r} = r'/c_0\Lambda_0^2$ and $\bar{u} = K_D u'/c_0^2\Lambda_0^\epsilon$. Here, K_D is the surface area of a D-dimensional unit sphere divided by $(2\pi)^D$, see Eq. (2.86).

(c) Determine the fixed points of these flow equations and show that besides the Gaussian fixed point, there is a nontrivial Wilson–Fisher fixed point with

$$\bar{u}_* = \frac{6\epsilon}{N+8} + \mathcal{O}(\epsilon^2)\,, \qquad \bar{r}_* = -\frac{1}{2}\frac{N+2}{N+8}\epsilon + \mathcal{O}(\epsilon^2)\,.$$

(d) Linearize the flow equations around the nontrivial fixed points, sketch the flow in the $\bar{r}$-$\bar{u}$ plane for $D < 4$, and determine all critical exponents to linear order in ϵ. Set $D = 3$ (i.e., $\epsilon = 1$) and compare your results with the known critical exponents of the $3D$-Heisenberg universality class $\alpha = -0.10$, $\beta = 0.36$, $\gamma = 1.39$, $\delta = 5.11$, $\nu = 0.70$, and $\eta = 0.027$.

4.2 Logarithmic Corrections to Scaling

Close to four dimensions the one-loop RG flow equations for φ^4-field theory are given by

$$\partial_l \bar{r}_l = 2\bar{r}_l + \frac{A\bar{u}_l}{1+\bar{r}_l}\,,$$

$$\partial_l \bar{u}_l = \epsilon \bar{u}_l - \frac{B\bar{u}_l^2}{(1+\bar{r}_l)^2}\,.$$

For the Ising universality class, it was shown in Sect. 4.2.2 that $A = 1/2$ and $B = 3/2$, see Eqs. (4.71) and (4.72). This result was generalized to the $\mathcal{O}(N)$-symmetric case in Exercise 4.1. For general N we have $A = (N+2)/6$ and $B = (N+8)/6$. Here, we will discuss implications of the above flow equations for $D = 4$, i.e., $\epsilon = 0$.

(a) Show that $(\bar{r}_*, \bar{u}_*) = 0$ is the only fixed point of the RG flow equations. To obtain the relevant scaling variable use the substitution $t_l \equiv \bar{r}_l + A\bar{u}_l/2$ to eliminate the linear term $A\bar{u}_l$ in the first equation, resulting in

$$\partial_l t_l = y_t t_l - A\bar{u}_l t_l + \dots\,,$$
$$\partial_l \bar{u}_l = -B\bar{u}_l^2 + \dots\,.$$

In fact, up to the factor of A which for the Ising universality class equals $1/2$, the above variable substitution is the same as used in the context of scaling in the vicinity of the Wilson–Fisher fixed point, see Eq. (4.87). The ellipses denote higher-order terms in t_l and $\bar{u}_l$ which can be neglected. Explain this in detail. Why is $\bar{u}_l$ called a marginally irrelevant coupling?

(b) Integrate the last equation to obtain $\bar{u}_l$ and use this result to determine t_l. Starting from a small t_0 and $\bar{u}_0$, we want to iterate the RG until $t_l = \mathcal{O}(1)$. In practice, this final value for t_l should be large enough such that mean field theory is applicable. On the other hand, it should be chosen small enough such that neglecting higher-order terms as done above is still justified. To obtain the l^* where the iteration of the RG should be stopped solve your solution for t_l approximately for l as a function of t. You should obtain an l of the form $l = c_1 \ln[c_2 t] + c_3 \ln[1 + c_4 \ln(c_5 t)]$.

(c) To calculate the specific heat, recall that the singular part of the free energy satisfies

$$f_{\text{sing}}(t_0, \bar{u}_0) = e^{-Dl} f_{\text{sing}}(t_l, \bar{u}_l) \ .$$

Evaluate the right-hand side within mean field theory to derive the leading contribution to the free energy. Use this result to show that the specific heat exhibits the following logarithmic behavior,

$$C(t_0, u_0) \propto (\ln(t_{l*}/t_0))^{(4-N)/(N+8)} \ .$$

(d) Logarithmic corrections to scaling are quite ubiquitous at the upper critical dimension. (This is especially important in systems where the upper critical dimension is $D = 3$.) Verify this explicitly by evaluating the scaling behavior of the correlation length ξ.

References

Abrikosov, A. A., L. P. Gorkov, and I. E. Dzyaloshinski (1963), *Methods of Quantum Field Theory in Statistical Mechanics*, Prentice-Hall, Inc., Englewood Cliffs, New Jersey.

Amit, D. J. (1974), *The Ginzburg criterion-rationalized*, J. Phys. C: Solid State Phys. **7**, 3369.

Ginzburg, V. L. (1960), *Some remarks on second order phase transitions and microscopic theory of ferroelectrics*, Fiz. Tverd. Tela **2**, 2031.

Guida, R. and J. Zinn-Justin (1998), *Critical exponents of the N-vector model*, J. Phys. A: Math. Gen. **31**, 8103.

Haldane, F. D.M. (1981), *'Luttinger liquid theory' of one-dimensional quantum fluids. I. Properties of the Luttinger model and their extension to the general 1D interacting spinless Fermi gas*, J. Phys. C: Solid State Phys. **14**, 2585.

Hasselmann, N., A. Sinner, and P. Kopietz (2007), *Two-parameter scaling of correlation functions near continuous phase transitions*, Phys. Rev. E **76**, 040101.

Hasselmann, N., S. Ledowski, and P. Kopietz (2004), *Critical behavior of weakly interacting bosons: A functional renormalization group approach*, Phys. Rev. A **70**, 063621.

Kopietz, P. (2001), *Two-loop beta-function from the exact renormalization group*, Nucl. Phys. B **595**, 493.

Negele, J. W. and H. Orland (1988), *Quantum Many-Particle Systems*, Addison-Wesley, Redwood City.

Nozières, P. (1964), *Theory of Interacting Fermi Systems*, Benjamin, New York.

Pelissetto, A. and E. Vicari (2002), *Critical phenomena and renormalization group theory*, Phys. Rep. **368**, 549.

Sakurai, J. J. (1994), *Modern Quantum Mechanics*, Addison-Wesley, Reading.

Sinner, A., N. Hasselmann, and P. Kopietz (2008), *Functional renormalization group in the broken symmetry phase: Momentum dependence and two-parameter scaling of the self-energy*, J. Phys.: Condens. Matter **20**, 075208.

Wilson, K. G. and M. E. Fisher (1972), *Critical exponents in 3.99 dimensions*, Phys. Rev. Lett. **28**, 240.

Zinn-Justin, J. (2002), *Quantum Field Theory and Critical Phenomena*, Clarendon Press, Oxford, 4th ed.

Chapter 5
Field-Theoretical Renormalization Group

The concept of the RG was originally introduced in the context of quantum field theory by Stueckelberg and Petermann (1953) and by Gell-Mann and Low (1954), long before Wilson's seminal works. Although the field-theoretical RG is less general than the Wilsonian RG and can only be applied to so-called renormalizable theories (we shall explain in this chapter what this means), calculations beyond the leading order in the number of loop integrations are usually simpler within the field theoretical RG if it is applicable. Here, we shall briefly introduce the main ideas of the field-theoretical RG method.[1] More detailed introductions can be found in the authoritative book by Zinn-Justin (2002) or in many other textbooks on quantum field theory such as (Sterman 1993, Peskin and Schroeder 1995, Zee 2003).

5.1 Divergencies and Their Regularization in Field Theory

By definition, in field theories, space-time is treated as a continuum, so that in momentum-frequency space there is no ultraviolet cutoff. As a consequence, correlation functions can be ultraviolet divergent if the dimensionality of space is high enough because there is no upper bound for the momenta and frequencies circulating around the closed loops encountered in perturbation theory. In the lattice models of condensed matter physics this problem does not arise, because the finite lattice spacing a always provides a physical ultraviolet cutoff $\Lambda_0 \approx 1/a$ for all momentum integrations. In order to make sense out of a (quantum) field theory, one has to regularize the theory at least at intermediate stages of the calculation by introducing an ultraviolet cutoff Λ_0. Of course, eventually we are interested in the limit $\Lambda_0 \to \infty$. In renormalizable field theories the infinities associated with this limit can be absorbed through a redefinition of a *finite* number of coupling constants, which can be fixed experimentally by making a *finite* number of measurements.

In order to compare the field-theoretical RG with the Wilsonian momentum shell RG introduced in Sect. 4.2, we shall explain in this chapter the main ideas of the

---

[1] The presentation in this chapter is influenced by a series of lectures on perturbative renormalization given by G. Sterman in Fall 1987 at the State University of New York at Stony Brook.

Kopietz, P. et al.: *Field-Theoretical Renormalization Group.* Lect. Notes Phys. **798**, 123–139 (2010)
DOI 10.1007/978-3-642-05094-7_5

field-theoretical RG within the context of classical φ^4-theory close to four dimensions. Formally, this theory can be obtained from the Ginzburg–Landau–Wilson model given in Eq. (2.65) by removing the ultraviolet cutoff Λ_0. With a slight change of notation (which is adopted to the field theory context of this chapter), the bare action of our model reads

$$S[\varphi_0; m_0, g_0] = \int d^D r \, \mathcal{L}[\varphi_0; m_0, g_0] \,, \tag{5.1}$$

with Lagrangian density

$$\mathcal{L}[\varphi_0; m_0, g_0] = \frac{1}{2}\left[\nabla \varphi_0(r)\right]^2 + \frac{m_0^2}{2}\varphi_0^2(r) + \frac{g_0}{4!}\varphi_0^4(r) \,. \tag{5.2}$$

In the notation of Eq. (2.65) we have set $\varphi_0(r) = \sqrt{c_0}\varphi(r)$, $m_0^2 = r_0/c_0$, and $g_0 = u_0/c_0^2$. The subscripts on φ_0, m_0 and g_0 indicate that these are bare quantities, which will be renormalized by fluctuation corrections. In wave vector space Eq. (5.1) reads

$$S[\varphi_0; m_0, g_0] = \frac{1}{2}\int_k \left[k^2 + m_0^2\right]\varphi_0(-k)\varphi_0(k)$$
$$+ \frac{g_0}{4!}\left(\prod_{i=1}^{4}\int_{k_i}\right)(2\pi)^D \delta\left(\sum_{i=1}^{4}k_i\right)\varphi_0(k_1)\varphi_0(k_2)\varphi_0(k_3)\varphi_0(k_4) \,. \tag{5.3}$$

In contrast to the Ginzburg–Landau–Wilson action for the Ising model in Eq. (2.61) the above field theory does not have an ultraviolet cutoff so that the momentum integrations are unrestricted, which leads to ultraviolet divergencies in perturbation theory. For example, the first-order correction to the irreducible self-energy is

$$\delta\Sigma = \bigcirc = \frac{g_0}{2}\int_k \frac{1}{k^2 + m_0^2} = \frac{g_0}{2}K_D\int_0^\infty dk\,\frac{k^{D-1}}{k^2 + m_0^2}$$
$$= \frac{g_0}{2}K_D m_0^{D-2}\int_0^\infty dx\,\frac{x^{D-1}}{1 + x^2} = \infty \quad \text{for } D \geq 2 \,. \tag{5.4}$$

Note that in the analogous expression (4.50) appearing in the Wilsonian RG the wave vector integrations are restricted to the shell $\Lambda < |k| < \Lambda_0$, whereas in field theory the integrations are unbounded. Moreover, fluctuation corrections to correlation functions with more than two external legs can also be ultraviolet divergent. For example, the leading interaction correction to the effective interaction is

$$\delta\Gamma = -\frac{1}{2}$$

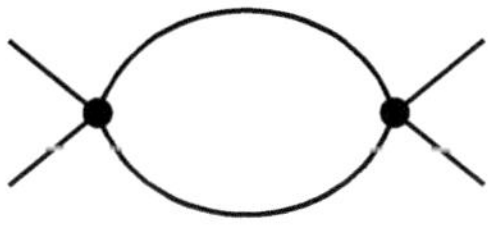

$$= -\frac{3g_0^2}{2}\int_k \frac{1}{\left[k^2+m_0^2\right]^2}$$

$$= -\frac{3g_0^2}{2} K_D m_0^{D-4} \int_0^\infty dx \frac{x^{D-1}}{[1+x^2]^2} = \infty \quad \text{for } D \geq 4 \,. \tag{5.5}$$

The analogous expression in the Wilsonian momentum shell RG given in Eq. (4.53) is again finite due to the restricted wave vector integration.

In order to manipulate the divergent integrals appearing in the perturbation series, we have to make them finite at intermediate stages of the calculation by introducing some kind of ultraviolet cutoff. This is called *regularization*, not to be confused with *renormalization*. At the end of the calculation we should somehow remove artificial cutoffs to obtain a renormalized theory which is independent of the regularization procedure. There are many ways of introducing a regularization. Some useful regularization procedures are:

(a) *Regularization via a momentum cutoff*: The simplest strategy to regularize the ultraviolet divergencies is to cut them off by "brute force" at some scale Λ. In the simplest case, one chooses a sharp momentum cutoff, replacing

$$\int \frac{d^D k}{(2\pi)^D} \rightarrow \int \frac{d^D k}{(2\pi)^D} \Theta(\Lambda - |\boldsymbol{k}|) \,. \tag{5.6}$$

However, for calculations beyond one loop the nonanalyticity of the sharp cutoff gives rise to mathematical complications, so that in practice it is often better to work with a smooth cutoff, for example a Gaussian momentum cutoff,

$$\int \frac{d^D k}{(2\pi)^D} \rightarrow \int \frac{d^D k}{(2\pi)^D} e^{-\boldsymbol{k}^2/\Lambda^2} \,. \tag{5.7}$$

(b) *Dimensional regularization:* Another possibility to regularize the ultraviolet divergencies is to evaluate the Feynman diagrams encountered in perturbation theory for some small enough dimensionality D. Divergencies then enter as poles in $2 - D$ or $D - 4$. Away from the poles, the integrals can be defined through analytic continuation in D.

(c) *Lattice regularization:* In this case one discretizes space (and time for quantum systems) and replaces the spatial continuum by a lattice with finite lattice spacing a. Then all momentum integrations are over a finite domain defined by the first Brillouin zone of the lattice. While in continuum field theories one should eventually take the limit $a \rightarrow 0$, in condensed matter systems this regularization is provided by nature via the physical crystal lattice.

Of course, depending on the regularization scheme, different results for the divergent Feynman diagrams are obtained. For example, if we regularize the loop integration in the first-order self-energy correction (5.4) using the Gaussian momentum cutoff (5.7) we obtain in $D = 4$,

$$
\begin{aligned}
\delta\Sigma(\Lambda) &= \frac{g_0}{2} \int \frac{d^4k}{(2\pi)^4} \frac{1}{k^2 + m_0^2} e^{-k^2/\Lambda^2} \\
&= \frac{g_0}{2} K_4 m_0^2 \int_0^\infty dx \frac{x^3}{1+x^2} e^{-\alpha x^2} \qquad \text{(with } \alpha = m_0^2/\Lambda^2) \\
&= \frac{g_0}{2} K_4 m_0^2 \frac{1}{2} \int_0^\infty dy \frac{1+y-1}{1+y} e^{-\alpha y} \qquad \text{(with } y = x^2) \\
&= \frac{g_0}{4} K_4 m_0^2 \left[\frac{1}{\alpha} - e^\alpha E_1(\alpha) \right],
\end{aligned}
\tag{5.8}
$$

where the exponential integral $E_1(z)$ is defined by (Abramowitz and Stegun 1965)

$$
E_1(z) = \int_z^\infty dt \, \frac{e^{-t}}{t} .
\tag{5.9}
$$

Using the asymptotic expansion of $E_1(z)$ for small z,

$$
E_1(z) \sim -\gamma - \ln z + \mathcal{O}(z) ,
\tag{5.10}
$$

where $\gamma = 0.577\ldots$ is Euler's constant, we finally obtain for $\Lambda^2 \gg m_0^2$ in four dimensions,

$$
\boxed{\delta\Sigma(\Lambda) = \frac{K_4}{2} g_0 \left[\frac{\Lambda^2}{2} - m_0^2 \ln\left(\frac{\Lambda}{m_0}\right) + \mathcal{O}(1) \right], \quad D = 4 .}
\tag{5.11}
$$

Similarly, we obtain for the regularized correction $\delta\Gamma$ to the interaction (5.5) in four dimensions,

$$
\begin{aligned}
\delta\Gamma &= -\frac{3}{2} g_0^2 \int \frac{d^4k}{(2\pi)^4} \frac{1}{\left(k^2 + m_0^2\right)^2} e^{-k^2/\Lambda^2} \\
&= -\frac{3K_4}{4} g_0^2 \int_0^\infty dy \frac{y}{(1+y)^2} e^{-\alpha y} .
\end{aligned}
\tag{5.12}
$$

For small $\alpha = m_0^2/\Lambda^2$ the integral in the second line can be approximated by $\ln(1/\alpha)$ to leading order, so that

$$
\boxed{\delta\Gamma(\Lambda) \sim -\frac{3K_4}{2} g_0^2 \ln\left(\frac{\Lambda}{m_0}\right) + \mathcal{O}(1), \quad D = 4 .}
\tag{5.13}
$$

On the other hand, if we use dimensional regularization then the regularized expressions for $\delta\Sigma$ and $\delta\Gamma$ look different. Assuming that the dimension D is sufficiently small so that the integrals exist, we obtain for the regularized self-energy correction (5.4) using the same substitutions as in Eq. (5.8),

$$\delta\Sigma(\epsilon) = \frac{g_0}{2} \int \frac{d^D k}{(2\pi)^D} \frac{1}{k^2 + m_0^2}$$

$$= \frac{g_0}{4} K_D \, m_0^{D-2} \int_0^\infty dy \frac{y^{\frac{D}{2}-1}}{1+y}$$

$$= \frac{g_0}{4} K_D \, m_0^{D-2} \Gamma\left(\frac{D}{2}\right)\Gamma\left(1 - \frac{D}{2}\right), \tag{5.14}$$

where we have used (Abramowitz and Stegun 1965, p. 256)

$$\int_0^\infty dy \, \frac{y^{z-1}}{1+y} = \pi \csc(\pi z) = \frac{\pi}{\sin(\pi z)} = \Gamma(z)\Gamma(1-z). \tag{5.15}$$

Substituting for K_D the explicit expression given in Eq. (2.86) and using the fact that for small $\epsilon = 4 - D > 0$ we may approximate

$$\Gamma\left(1 - \frac{D}{2}\right) = \Gamma\left(-1 + \frac{\epsilon}{2}\right) = -\frac{2}{\epsilon} + \mathcal{O}(1), \tag{5.16}$$

we obtain slightly below four dimensions

$$\boxed{\delta\Sigma(\epsilon) = \frac{g_0}{2} \frac{m_0^{D-2}}{(4\pi)^{D/2}} \, \Gamma\left(1 - \frac{D}{2}\right) \sim -\frac{g_0}{2} \frac{m_0^2}{8\pi^2\epsilon} + \mathcal{O}(1),} \tag{5.17}$$

which looks rather different from the result (5.11) obtained via regularization with Gaussian momentum cutoff. To evaluate the correction $\delta\Gamma$ to the interaction in Eq. (5.5) in dimensional regularization, we simply use

$$\frac{1}{\left(k^2 + m_0^2\right)^2} = -\frac{\partial}{\partial m_0^2} \frac{1}{k^2 + m_0^2} \tag{5.18}$$

so that with Eqs. (5.14) and (5.17) we can write

$$\boxed{\delta\Gamma(\epsilon) = 3g_0 \frac{\partial}{\partial m_0^2} \delta\Sigma(\epsilon) \sim -\frac{3}{2} g_0^2 \frac{1}{8\pi^2\epsilon} + \mathcal{O}(1).} \tag{5.19}$$

Again, the regularized result for $\delta\Gamma(\epsilon)$ obtained via dimensional regularization looks rather different from the corresponding expression (5.13) obtained with Gaussian momentum cutoff, so that at this point it appears almost like a miracle that by using different regularization schemes we are eventually able to obtain unambiguous results for physical observables from field theories which can be compared with experiments.

5.2 Perturbative Renormalization

5.2.1 The Renormalized Lagrangian

As a first step towards our goal to use field theory to make predictions for experimentally measurable physical quantities we introduce the so-called *renormalized Lagrangian*, which is a function of the renormalized field φ_R, the renormalized mass m_R, and the renormalized coupling constant g_R. These renormalized quantities are defined in terms of the corresponding bare quantities φ_0, m_0 and g_0 via the relations

$$\varphi_0 = \sqrt{Z_\varphi}\varphi_R \,, \tag{5.20a}$$

$$m_0^2 = Z_m m_R^2 \,, \tag{5.20b}$$

$$g_0 = Z_g \mu^\epsilon g_R \,, \tag{5.20c}$$

where the dimensionless multiplicative renormalization constants Z_φ, Z_m and Z_g will be determined iteratively order by order in perturbation theory, and μ in Eq. (5.20c) is an arbitrary mass scale which is introduced to make the renormalized coupling g_R dimensionless. Our strategy is to absorb all infinities encountered in perturbation theory into the relations between the bare quantities and the renormalized ones. The idea is that only the renormalized quantities have a physical meaning and can be related to physical observables. We therefore require that the renormalized quantities have finite values, while the bare quantities are infinite due to the singularities contained in the Z-factors. In order to realize this strategy, let us express our original Lagrangian density $\mathcal{L}[\varphi_0; m_0, g_0]$ defined in Eq. (5.2) in terms of renormalized quantities as follows,

$$
\begin{aligned}
\mathcal{L}[\varphi_0; m_0, g_0] &\equiv \frac{1}{2}\Big[(\nabla\varphi_0)^2 + m_0^2\varphi_0^2\Big] + \frac{g_0}{4!}\varphi_0^4 \\
&= \frac{1}{2}\Big[Z_\varphi(\nabla\varphi_R)^2 + Z_m Z_\varphi m_R^2\varphi_R^2\Big] + Z_g Z_\varphi^2 \frac{\mu^\epsilon g_R}{4!}\varphi_R^4 \\
&= \mathcal{L}_R[\varphi_R; m_R, g_R, Z_\varphi, Z_m, Z_g] \,,
\end{aligned}
\tag{5.21}
$$

where the renormalized Lagrangian $\mathcal{L}_R[\varphi_R; m_R, g_R, Z_\varphi, Z_m, Z_g]$ should be considered to be a function of the renormalized field φ_R, mass m_R, and coupling g_R, and the associated dimensionless Z-factors. Alternatively, the renormalized Lagrangian can be written as

$$\mathcal{L}_R[\varphi_R; m_R, g_R, Z_\varphi, Z_m, Z_g] = \frac{1}{2}\Big[(\nabla\varphi_R)^2 + m_R^2\varphi_R^2\Big] + \frac{\mu^\epsilon g_R}{4!}\varphi_R^4$$

$$+\frac{1}{2}(Z_\varphi - 1)(\nabla\varphi_R)^2 + \frac{1}{2}(Z_m Z_\varphi - 1)m_R^2\varphi_R^2 + \left(Z_g Z_\varphi^2 - 1\right)\frac{\mu^\epsilon g_R}{4!}\varphi_R^4 \,. \qquad (5.22)$$

Note that the first term on the right-hand side of Eq. (5.22) is of the same form as the bare Lagrangian but with the bare quantities replaced by the renormalized ones. The remaining three terms in the second line of Eq. (5.22) resemble the terms appearing in the first line, but with different coefficients which all vanish if the Z-factors are replaced by unity. These terms are called *counterterms* and are considered to be a part of the interaction in perturbative renormalization. Each counterterm therefore gives rise to a new interaction vertex, which we represent by the following graphical elements,

$$\text{Field counterterm:} \quad \frac{1}{2}(Z_\varphi - 1)(\nabla\varphi_R)^2 \;=\; \qquad\qquad , \qquad (5.23\text{a})$$

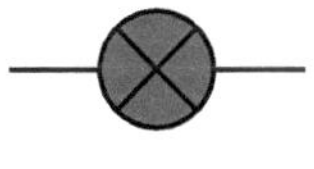

$$\text{Mass counterterm:} \quad \frac{1}{2}(Z_m Z_\varphi - 1)m_R^2\varphi_R^2 \;=\; \qquad\qquad , \qquad (5.23\text{b})$$

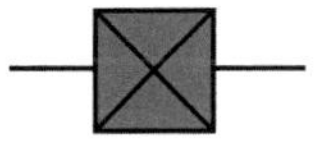

$$\text{Interaction counterterm:} \quad \frac{\mu^\epsilon g_R}{4!}\left(Z_g Z_\varphi^2 - 1\right)\varphi_R^4 \;=\; \qquad\qquad . \qquad (5.23\text{c})$$

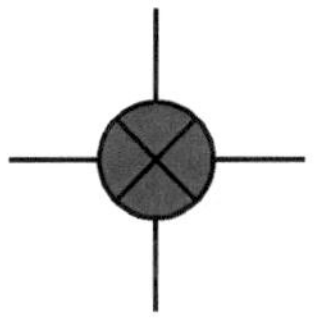

The structure of the above relation, namely

$$\mathcal{L}_R[\varphi_R; m_R, g_R, Z_\varphi, Z_m, Z_g] = \{\text{classical Lagrangian with bare quantities}$$

$$\text{replaced by renormalized ones}\} + \{\text{counterterms}\}\,,$$

$$(5.24)$$

is very general. It is not only the starting point of renormalized perturbation theory in our simple classical φ^4-theory. A similar decomposition is also the starting point of renormalized perturbation theory in quantum electrodynamics and quantum chromodynamics.[2] Of course, in this case one needs different graphical elements to represent the theory and the counterterms.

---

[2] For example, in quantum electrodynamics the classical Lagrangian is (see e.g., Peskin and Schroeder 1995, p. 330)

Let us point out that the strategy underlying field-theoretical perturbative renormalization has also been successfully adopted in condensed matter physics, where it is sometimes called *renormalized perturbation theory* (see e.g., Hewson 1993, Appendix L). For example, in order to calculate the true Fermi surface k_F of an interacting Fermi system, it is convenient to include the exact self-energy $\Sigma(\omega = 0, k_F)$ at vanishing frequency as a counterterm into the quadratic part of the Hamiltonian (Nozières 1964). The a priori unknown self-energy $\Sigma(\omega = 0, k_F)$ is determined self-consistently from the requirement that the self-energy calculated from the residual interaction vanishes for momenta on the Fermi surface and for vanishing frequency. For recent applications of this technique see (Neumayr and Metzner 2003, Ledowski et al. 2005, Ledowski and Kopietz 2007).

5.2.2 Perturbative Calculation of Renormalization Factors

A priori it is not clear whether all divergencies encountered to all orders in perturbation theory can be absorbed in our three dimensionless factors Z_i (where $i = \varphi, m, g$). The fact that this is indeed possible amounts to the statement that our theory is renormalizable. Let us now explain how to calculate the Z_i iteratively order by order in perturbation theory. From the form (5.22) of the renormalized Lagrangian it is obvious that the Z_i are functions of the renormalized mass m_R, the renormalized coupling constant, g_R, and the (arbitrary) mass scale μ. Moreover, because we have to regularize the divergent integrals encountered in perturbation theory, the Z-factors will in general also depend on the regularization procedure: for regularization with momentum cutoff Λ the Z_i will depend on Λ, while in dimensional regularization they will depend on ϵ. In renormalized perturbation theory, we

---

$$\mathcal{L}[\psi, A; m_0, e_0] = \bar{\psi}(i\gamma^\mu \partial_\mu - e_0 \gamma^\mu A_\mu - m_0)\psi - \frac{1}{4} F_{\mu\nu} F^{\mu\nu} \,,$$

where the four-component field ψ represents the bare electrons, the bare photon field is represented by the four-vector potential A_μ with field strength $F_{\mu\nu} = \partial_\mu A_\nu - \partial_\nu A_\mu$, γ^μ, $\mu = 0, 1, 2, 3$ are 4×4-matrices, and m_0 and e_0 are the bare mass and the bare charge of the electron. This is a $U(1)$-gauge theory, so that one needs an extra gauge fixing term, which only affects the photon propagator; the effective bare Lagrangian contains then one additional parameter λ_0 which fixes the gauge, $\mathcal{L}_{\mathrm{eff}}[\psi, A; m_0, e_0, \lambda_0] = \mathcal{L}[\psi, A; m_0, e_0] - \frac{1}{2}\lambda_0(\partial_\mu A^\mu)^2$. The corresponding renormalized Lagrangian is obtained by introducing the renormalized quantities $\psi = \sqrt{Z_\psi}\psi_R$, $A^\mu = \sqrt{Z_A}A_R^\mu$, $m_0 = Z_m m_R$, $e_0 = Z_e e_R$, and has four counterterms,

$$\mathcal{L}_R[\psi_R, A_R; m_R, e_R, \lambda_R, Z_\psi, Z_A, Z_m, Z_e] = \mathcal{L}_{\mathrm{eff}}[\psi_R, A_R; m_R, e_R, \lambda_R]$$

$$+i(Z_\psi - 1)\bar{\psi}_R \gamma^\mu \partial_\mu \psi_R - (Z_m Z_\psi - 1)m_R \bar{\psi}_R \psi_R - (Z_A - 1)\frac{1}{4} F_{R,\mu\nu} F_R^{\mu\nu}$$

$$-\left(Z_\psi \sqrt{Z_A} Z_e - 1\right) e_R \bar{\psi}_R \gamma^\mu A_{R,\mu} \psi_R \,.$$

Gauge invariance implies that there is no counterterm for the gauge-fixing term and that $Z_e = Z_A^{-1/2}$, so that all divergencies can be absorbed into three renormalization constants Z_ψ, Z_m and Z_e. The renormalized Lagrangian for quantum chromodynamics looks even more complicated (see for example (Sterman 1993, Peskin and Schroeder 1995)).

expand the Z_i in powers of the renormalized interaction to obtain a series of the
form

$$Z_i = 1 + \sum_{\nu=1}^{\infty} C_i^{(\nu)}(m_R, \mu, \Lambda \text{ or } \epsilon)g_R^\nu , \quad i = \varphi, m, g, \tag{5.25}$$

with dimensionless expansion coefficients $C_i^{(\nu)}(m_R, \mu, \Lambda \text{ or } \epsilon)$. The notation "$\Lambda$
or ϵ" in the last argument means that for cutoff regularization these coefficients
depend on Λ, while for dimensional regularization they are functions of $\epsilon = 4 - D$.
The crucial step is now to choose the coefficients $C_i^{(\nu)}(m_R, \mu, \Lambda \text{ or } \epsilon)$ such that
the perturbative expansion in powers of the interaction part of the renormalized
Lagrangian $\mathcal{L}_R[\varphi_R; m_R, g_R, Z_\varphi, Z_m, Z_g]$ (which includes the counterterms) is finite
order by order in perturbation theory. In other words, any singularity encountered in
the perturbative expansion can be absorbed into one of the three factors Z_φ, Z_m or
Z_g. The fact that this really works to all orders in perturbation theory is equivalent
with the statement of renormalizability of the theory.

To see how this procedure works in practice, let us here explicitly calculate
the Z-factors to first order in the renormalized coupling. Consider first the correction to the mass, which is determined by the momentum-independent part of
the irreducible self-energy. According to Eqs. (5.11) and (5.17) the interaction part
$(\mu^\epsilon g_R/4!)\varphi_R^4$ of our renormalized Lagrangian (5.22) gives rise to the following self-
energy correction,

$$\delta\Sigma_1 = \frac{K_4}{2}g_R \begin{cases} \frac{\Lambda^2}{2} - m_R^2 \ln\left(\frac{\Lambda}{m_R}\right) + \mathcal{O}(1) , & \text{(cutoff regularization)} \\ -\frac{m_R^2}{\epsilon} + \mathcal{O}(1) . & \text{(dimensional regularization)} \end{cases} \tag{5.26}$$

Note that in dimensional regularization the renormalized coupling is actually $\mu^\epsilon g_R =
e^{\epsilon \ln \mu} g_R = g_R + \mathcal{O}(\epsilon \ln \mu)$, so that the nonsingular correction of order unity
depends on the regularization scheme. However, the mass counterterm $-\!\!\boxtimes\!\!-$
in Eq. (5.22), which should also be considered as part of the interaction, gives an
additional correction $(Z_\varphi Z_m - 1)m_R^2$ to the self-energy, so that in total we obtain

$$\delta\Sigma = \delta\Sigma_1 + (Z_m Z_\varphi - 1)m_R^2 . \tag{5.27}$$

Because to order g_R the self-energy is momentum-independent, the leading correction to the field renormalization factor is of the order g_R^2, so that we may approximate
$Z_\varphi \approx 1$ to first order in g_R. Expanding $Z_m = 1 + C_m^{(1)}g_R + \mathcal{O}(g_R^2)$, we obtain to first
order in g_R,

$$\delta\Sigma = \frac{K_4}{2}g_R \begin{cases} \frac{\Lambda^2}{2} - m_R^2 \ln\left(\frac{\Lambda}{m_R}\right) + \mathcal{O}(1) \\ -\frac{m_R^2}{\epsilon} + \mathcal{O}(1) \end{cases} + C_m^{(1)}g_R m_R^2 + \mathcal{O}\left(g_R^2\right) , \tag{5.28}$$

where it is understood that the upper line in the bracket corresponds to cutoff regularization, whereas the lower line corresponds to dimensional regularization. The
singular part of $C_m^{(1)}$ is now fixed by demanding that the right-hand side of Eq. (5.28)

remains finite for $\Lambda \to \infty$ or $\varepsilon \to 0$. This implies

$$C_m^{(1)} = \frac{K_4}{2} \left\{ \begin{matrix} \ln\left(\frac{\Lambda}{m_R}\right) - \frac{\Lambda^2}{2m_R^2} \\ \frac{1}{\epsilon} \end{matrix} \right\} + c_m^{(1)} , \tag{5.29}$$

where $c_m^{(1)}$ denotes the nonsingular part of $C_m^{(1)}$ which remains finite for $\Lambda \to \infty$ or $\epsilon \to 0$. Obviously, the requirement that perturbation theory remains finite fixes only the singular part of the expansion coefficients $C_i^{(\nu)}$. To first order in the renormalized coupling constant g_R, the mass renormalization factor Z_m is therefore given by

$$\boxed{Z_m = 1 + g_R \left[\frac{K_4}{2} \left\{ \begin{matrix} \ln\left(\frac{\Lambda}{m_R}\right) - \frac{\Lambda^2}{2m_R} \\ \frac{1}{\epsilon} \end{matrix} \right\} + c_m^{(1)} \right] + \mathcal{O}\left(g_R^2\right) ,} \tag{5.30}$$

where $K_4 = 1/(8\pi^2)$ and the finite part $c_m^{(1)}$ depends on the regularization scheme.

Next, let us calculate the singular part of the coupling constant renormalization factor Z_g. To this end we consider the leading correction to the interaction, which according to Eqs. (5.13) and (5.19) is for large Λ or small ϵ given by

$$\delta\Gamma_1 = -\frac{3}{2} K_4 g_R^2 \left\{ \begin{matrix} \ln\left(\frac{\Lambda}{m_R}\right) + \mathcal{O}(1) \\ \frac{1}{\epsilon} + \mathcal{O}(1) \end{matrix} \right\} . \tag{5.31}$$

Again, the interaction counterterm ⊗ in Eq. (5.22) generates an additional contribution to the renormalized coupling constant, so that in total

$$\delta\Gamma = \delta\Gamma_1 + g_R \left(Z_g Z_\varphi^2 - 1\right) . \tag{5.32}$$

But to order g_R^2 we may approximate

$$g_R \left(Z_g Z_\varphi^2 - 1\right) \approx g_R^2 C_g^{(1)} , \tag{5.33}$$

so that the total g_R^2 correction to the coupling constant is

$$\delta\Gamma = -\frac{3}{2} K_4 g_R^2 \left\{ \begin{matrix} \ln\left(\frac{\Lambda}{m_R}\right) + \mathcal{O}(1) \\ \frac{1}{\epsilon} + \mathcal{O}(1) \end{matrix} \right\} + C_g^{(1)} g_R^2 + \mathcal{O}\left(g_R^3\right) . \tag{5.34}$$

Demanding again that this remains finite for $\Lambda \to \infty$ or $\epsilon \to 0$ fixes the singular part of $C_g^{(1)}$,

$$C_g^{(1)} = \frac{3}{2} K_4 \left\{ \begin{matrix} \ln\left(\frac{\Lambda}{m_R}\right) \\ \frac{1}{\epsilon} \end{matrix} \right\} + c_g^{(1)} , \tag{5.35}$$

where the nonsingular part $c_g^{(1)}$ remains undetermined. The coupling constant renormalization factor Z_g to first order in g_R is therefore

$$\boxed{Z_g = 1 + g_R \left[\frac{3K_4}{2} \left\{ \begin{matrix} \ln\left(\frac{\Lambda}{m_R}\right) \\ \frac{1}{\epsilon} \end{matrix} \right\} + c_g^{(1)} \right] + \mathcal{O}\left(g_R^2\right) .} \tag{5.36}$$

As already mentioned (see the discussion after Eq. (5.27)), the field renormalization factor is of the form

$$Z_\varphi = 1 + \mathcal{O}\left(g_R^2\right) , \tag{5.37}$$

so that to first order in g_R we may approximate $Z_\varphi \approx 1$. Note that in dimensional regularization the Z-factors are independent of m_R. This iterative procedure of calculating the renormalization factors can in principle be carried out to arbitrary high orders in g_R. The important point is that for doing the calculation to a given order in g_R we do not need to know in advance the counterterms to this order: they are obtained to this order while doing the calculation. As anticipated, the singularities encountered in perturbation theory can be absorbed in the renormalization factors given by Eqs. (5.30) and (5.36). If we demand that the renormalized quantities are finite, then the bare quantities are actually infinite if we remove the regularization.

5.2.3 Relating Renormalized Perturbation Theory to Experiments

Consider the expansion (5.25) of the renormalization factors Z_i in powers of the renormalized coupling constant g_R. We have shown that the first-order coefficients $C_m^{(1)}(m_R, \mu, \epsilon)$ and $C_g^{(1)}(m_R, \mu, \epsilon)$ have in dimensional regularization the form

$$C_i^{(1)}(m_R, \mu, \epsilon) = \frac{Z_i^{(1)}}{\epsilon} + c_i^{(1)} , \tag{5.38}$$

where the nonsingular parts $c_i^{(1)}$ depend on m_R, μ and ϵ, while the residues $Z_i^{(1)}$ of the singular part are given by $Z_m^{(1)} = K_4/2$ and $Z_g^{(1)} = 3K_4/2$. A similar decomposition is also obtained for the higher order expansion coefficients $C_i^{(v)}(m_R, \mu, \epsilon)$ with $v > 1$. The requirement that renormalized perturbation theory remains finite for $\epsilon \to 0$ fixes only the residues $Z_i^{(v)}$ of the singular parts of the expansion coefficients $C_i^{(v)}(m_R, \mu, \epsilon)$, leaving their nonsingular parts $c_i^{(v)}$ completely arbitrary. As a result, the Z-factors depend on a lot of parameters,

$$Z_i = Z_i(m_R, g_R, \mu, \epsilon, \underbrace{c_m^{(1)}, c_g^{(1)}, c_m^{(2)}, c_g^{(2)}, c_\varphi^{(2)}, \dots}_{\text{Finite parts of counterterms}}) . \tag{5.39}$$

Recall that in dimensional regularization one has to introduce an arbitrary mass scale μ in order to define the dimensionless renormalized coupling constant g_R in arbitrary dimension. But how do we fix the coefficients $c_i^{(\nu)}$ in order to get unique predictions which can be tested experimentally? To answer this, we observe that the bare Lagrangian really depends only on the two parameters m_0, g_0, because according to Eq. (5.21)

$$\mathcal{L}[\varphi_0; m_0, g_0] \equiv \frac{1}{2}\left[(\nabla \varphi_0)^2 + m_0^2 \varphi_0^2\right] + \frac{g_0}{4!} \varphi_0^4$$
$$= \mathcal{L}_R[\varphi_R; m_R, g_R, Z_\varphi, Z_m, Z_g] , \tag{5.40}$$

with $\varphi_0 = \sqrt{Z_\varphi}\,\varphi_R$, $m_0^2 = Z_m m_R^2$, and $g_0 = Z_g \mu^\epsilon g_R$. Hence there can be only two independent parameters in the renormalized Lagrangian that distinguish different physical theories. As a consequence, the finite parts $c_\varphi^{(\nu)}$, $c_m^{(\nu)}$ and $c_g^{(\nu)}$ of the counterterms as well as the mass scale μ can be chosen arbitrarily. In particular, we may fix the finite parts $c_i^{(\nu)}$ of the counterterms in some convenient way which simplifies the calculations. This procedure is usually called *choosing a renormalization scheme*. One possibility is so simply set the finite parts of all counterterms equal to zero, $c_i^{(\nu)} = 0$. This renormalization scheme is called *minimal subtraction*. An alternative scheme is the so-called *momentum subtraction*, where the finite parts of the counterterms are chosen such that the self-energy $\Sigma(k)$ and the (dimensionless) renormalized effective interaction Γ_R satisfy

$$\Sigma(0) = m_R^2 , \tag{5.41a}$$

$$\lim_{k \to 0} \frac{\partial \Sigma(k)}{\partial k^2} = 1 , \tag{5.41b}$$

$$\Gamma_R = m_R^\epsilon g_R . \tag{5.41c}$$

At this stage, the perturbation series generated from $\mathcal{L}_R$ depends only on m_R, g_R, and μ. Now comes the input from experiment: in order to get a theory with predictive powers, we proceed as follows:

(a) Pick two observable quantities, for example the mass m_{phys} of some physical particle and some physical scattering cross-section σ_{phys}. We then calculate these observables perturbatively using renormalized perturbation theory to obtain relations of the form

$$m_{\text{phys}} = f_1(m_R, g_R, \mu) , \tag{5.42}$$

$$\sigma_{\text{phys}} = f_2(m_R, g_R, \mu) , \tag{5.43}$$

where the functions $f_1(m_R, g_R, \mu)$ and $f_2(m_R, g_R, \mu)$ are truncated series expansions to a given order in g_R.

(b) Determine m_{phys} and σ_{phys} from experiment and invert Eqs. (5.42) and (5.43) to obtain m_R and g_R as functions of the physical parameters.

$$m_R = m_R(m_{\text{phys}}, \sigma_{\text{phys.}}, \mu) \,, \tag{5.44}$$

$$g_R = g_R(m_{\text{phys}}, \sigma_{\text{phys}}, \mu) \,. \tag{5.45}$$

(c) With m_R and g_R expressed in terms of physical observables, we can now use our renormalized Lagrangian to calculate other physical observables. In this way our theory has acquired predictive power. If our theory is correct, the results should again agree with experiment.

From this algorithm it is clear why nonrenormalizable field theories do not have any predictive power: In this case the divergencies cannot be absorbed in a finite number of counterterms. As a result, the theory depends on infinitely many parameters, which can only be fixed by measuring infinitely many observables.

The relations (5.44) and (5.45) still depend on the arbitrary mass scale μ defining the renormalized dimensionless coupling constant. If we could calculate the functions $f_1(m_R, g_R, \mu)$ and $f_2(m_R, g_R, \mu)$ in Eqs. (5.42) and (5.43) to all orders in perturbation theory, they would be independent of μ, because we know that our original theory depends only on two independent parameters. But in practice we have to truncate the perturbation series at some finite order in g_R. In this case the functions f_1 and f_2 do not loose their μ-dependence, so that the values of m_R and g_R still depend on the arbitrary mass scale μ. Fortunately, for renormalizable theories the residual μ-dependence is only logarithmic and is therefore very weak (see e.g., (Sterman 1993)).

5.3 Callan–Symanzik Equation

The freedom in the choice of the mass scale μ in the definition of the renormalized Lagrangian $\mathcal{L}_R[\varphi_R; m_R, g_R, \mu]$ implies that correlation functions which are calculated within the renormalized theory for different values of the renormalized parameters m_R, g_R can be related to each other. This relation can be expressed in terms of a partial differential equation, the *Callan–Symanzik equation*, which we now derive. Let us therefore consider the so-called one-line irreducible vertex functions $\Gamma^{(n)}(k_1, \cdots, k_n)$ with n external legs, which we represent graphically by a shaded circle with n external legs,

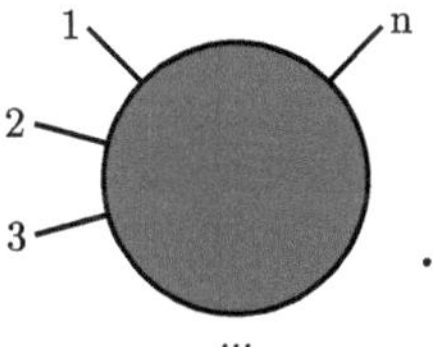

One-line irreducible means that any diagram contributing to the shaded circle cannot be separated into two disconnected parts by cutting a single propagator line. In the special case $n = 2$ the vertex function $\Gamma^{(2)}(k, -k) = \Sigma(k)$ can be identified

with the irreducible self-energy introduced in Eq. (4.95); for $n = 4$ the vertex $\Gamma^{(4)}(k_1, k_2, k_3, k_4)$ is usually called the *effective interaction*. We shall formally define these vertices in Sect. 6.2. Taking into account that the bare and renormalized fields are related by $\varphi_0 = \sqrt{Z_\varphi}\varphi_R$, the relation between bare and renormalized irreducible vertices is

$$\Gamma_0^{(n)}(k_1, \cdots, k_n; m_0, g_0) = Z_\varphi^{-\frac{n}{2}} \Gamma_R^{(n)}(k_1, \cdots, k_n; m_R, g_R, \mu) \,. \tag{5.46}$$

Obviously, the left-hand side of this identity is manifestly independent of the arbitrary mass scale μ, so that the explicit μ-dependence of the right-hand side must also cancel against the implicit μ-dependence contained in Z_φ, m_R and g_R. Taking the derivative of both sides of Eq. (5.46) with respect to μ while keeping the bare parameters m_0 and g_0 constant we obtain

$$\begin{aligned}
0 = \frac{d}{d\mu} \Gamma_0^{(n)}(k_1, \ldots, k_n; m_0, g_0) &= \frac{d}{d\mu}\left[Z_\varphi^{-\frac{n}{2}} \Gamma_R^{(n)}(k_1 \cdots k_n; m_R, g_R, \mu) \right] \\
&= \frac{dZ_\varphi^{-\frac{n}{2}}}{d\mu} \Gamma_R^{(n)}(k_1 \cdots k_n; m_R, g_R, \mu) + Z_\varphi^{-\frac{n}{2}} \frac{d}{d\mu} \Gamma_R^{(n)}(k_1 \cdots k_n; m_R, g_R, \mu) \,.
\end{aligned}$$
$$\tag{5.47}$$

Multiplying this equation by $Z_\varphi^{\frac{n}{2}}\mu$ and writing $\mu\frac{d}{d\mu} = \frac{d}{d\ln\mu}$ we obtain

$$\frac{d}{d\ln\mu} \Gamma_R^{(n)}(k_1 \cdots k_n; m_R, g_R, \mu) + \left[Z_\varphi^{\frac{n}{2}} \frac{dZ_\varphi^{-\frac{n}{2}}}{d\ln\mu} \right] \Gamma_R^{(n)}(k_1 \cdots k_n; m_R, g_R, \mu) = 0 \,. \tag{5.48}$$

The factor in the square braces of Eq. (5.48) can be written as

$$Z_\varphi^{\frac{n}{2}} \frac{dZ_\varphi^{-\frac{n}{2}}}{d\ln\mu} = -\frac{n}{2} Z_\varphi^{-1} \frac{dZ_\varphi}{d\ln\mu} \equiv -\frac{n}{2}\eta(m_R, g_R) \,, \tag{5.49}$$

where the anomalous dimension in the field-theoretical RG approach is given by

$$\eta(m_R, g_R) = \frac{d\ln Z_\varphi}{d\ln\mu}\bigg|_{m_0, g_0} \,. \tag{5.50}$$

The subscripts indicate that the differentiation should be carried out keeping the bare quantities m_0 and g_0 constant. Introducing also the *Gell-Mann–Low β-function*

$$\beta(m_R, g_R) = \frac{\partial g_R}{\partial\ln\mu}\bigg|_{m_0, g_0} \,, \tag{5.51}$$

and the function

$$\gamma(m_R, g_R) = \left.\frac{\partial \ln m_R^2}{\partial \ln \mu}\right|_{m_0, g_0} = \left.\frac{1}{m_R^2}\frac{\partial m_R^2}{\partial \ln \mu}\right|_{m_0, g_0}, \tag{5.52}$$

Eq. (5.48) can be written in the form

$$\left[\mu\frac{\partial}{\partial \mu} + \beta\frac{\partial}{\partial g_R} + \gamma m_R^2\frac{\partial}{\partial m_R^2} - \frac{n}{2}\eta\right]\Gamma_R^{(n)}(k_1, \ldots, k_n; m_R, g_R, \mu) = 0. \tag{5.53}$$

This is the *Callan–Symanzik equation* for the renormalized vertex functions. It tells us how a change of the arbitrary mass scale μ appearing in the definition of the renormalized coupling constant $g_R = Z_g^{-1}\mu^{-\epsilon}g_0$ is compensated by a change of the renormalized parameters φ_R, m_R and g_R such that the renormalized vertex functions are independent of μ if the bare parameters are held constant.

Let us calculate the functions $\beta(m_R, g_R)$ and $\gamma(m_R, g_R)$ to leading order in g_R. Using dimensional regularization with minimal subtraction we obtain from Eqs. (5.30) and (5.36)

$$Z_m = 1 + \frac{1}{2}\frac{K_4}{\varepsilon}g_R + \mathcal{O}\left(g_R^2\right), \tag{5.54}$$

$$Z_g = 1 + \frac{3}{2}\frac{K_4}{\varepsilon}g_R + \mathcal{O}\left(g_R^2\right). \tag{5.55}$$

These expressions are independent of m_R, so that in dimensional regularization the functions $\beta(g_R)$ and $\gamma(g_R)$ depend only on the renormalized coupling constant g_R. From the definition (5.51) we obtain

$$\beta(g_R) = \frac{\partial}{\partial \ln \mu}\left[Z_g^{-1}e^{-\epsilon \ln \mu}g_0\right] = -\epsilon g_R + \frac{\partial Z_g^{-1}}{\partial \ln \mu}\mu^{-\epsilon}g_0$$

$$= -\epsilon g_R + \frac{3}{2}\frac{K_4}{\epsilon}\frac{\partial g_R}{\partial \ln \mu}g_R + \mathcal{O}\left(g_R^3\right). \tag{5.56}$$

In the second term of the last line we may approximate $\frac{\partial g_R}{\partial \ln \mu} = -\epsilon g_R + \mathcal{O}\left(g_R^2\right)$, so that

$$\beta(g_R) \equiv \frac{\partial g_R}{\partial \ln \mu} = -\epsilon g_R + \frac{3K_4}{2}g_R^2 + \mathcal{O}\left(g_R^3\right). \tag{5.57}$$

Similarly, we obtain from Eq. (5.52)

$$\gamma(g_R) = Z_m \frac{\partial Z_m^{-1}}{\partial \ln \mu} = Z_m \frac{\partial}{\partial \ln \mu} \left[1 - \frac{K_4}{2\epsilon} g_R + \mathcal{O}\left(g_R^2\right) \right]$$

$$= -Z_m \frac{K_4}{2\varepsilon} \frac{\partial g_R}{\partial \ln \mu} + \mathcal{O}\left(g_R^2\right) = -[1 + \mathcal{O}(g_R)] \frac{K_4}{2\epsilon} \left[-\epsilon g_R + \mathcal{O}\left(g_R^2\right) \right] . \quad (5.58)$$

To leading order in g_R we thus obtain

$$\boxed{\gamma(g_R) \equiv \frac{1}{m_R^2} \frac{\partial m_R^2}{\partial \ln \mu} = \frac{K_4}{2} g_R + \mathcal{O}\left(g_R^2\right) .} \quad (5.59)$$

Finally, let us compare Eqs. (5.57) and (5.59) with the flow equations (4.71) and (4.72) describing the change of the couplings u_l and r_l due to mode elimination and rescaling in the Wilsonian RG. Setting $\mu = \mu_0 e^{-l}$ so that $\mu \partial_\mu = -\partial_l$ and defining $\bar{g}_l = K_4 g_R$, the field-theoretical RG equation (5.57) for the renormalized coupling constant can be written as

$$\partial_l \bar{g}_l = \epsilon \bar{g}_l - \frac{3}{2} \bar{g}_l^2 , \quad (5.60)$$

which is formally identical with the Wilsonian RG equation (4.72) provided we set $\bar{r}_l \approx 0$ in the second term of Eq. (4.72). This is consistent close to the Wilson–Fisher fixed point where $\bar{r}_l$ is of order $\bar{g}_l$. Note, however the conceptual difference between $\bar{g}_l$ and $\bar{u}_l$: the coupling $\bar{g}_l = K_4 g_R$ is the renormalized field theoretical coupling, while the flowing coupling $\bar{u}_l$ in the Wilsonian approach describes the change of effective coupling as we eliminate degrees of freedom. Although at order $\bar{g}_l^2$ or $\bar{u}_l^2$ the field-theoretical RG equation (5.60) is formally identical to the corresponding Wilsonian RG equation (4.72), at higher orders in g_R the corresponding flow equations disagree. In fact, coefficients of g_R^3 and higher in the perturbative field-theoretical β-function depend on the regularization scheme (Creutz 1983).

To compare the field-theoretical RG equation (5.59) to the Wilsonian flow equation (4.71) for the thermal variable $\bar{r}_l$, let us define the dimensionless coupling $\bar{t}_l = m_R^2/\mu^2$, which satisfies

$$\mu \partial_\mu \bar{t}_l = -2\bar{t}_l + \frac{1}{\mu^2} \frac{\partial m_R^2}{\partial \ln \mu} = [-2 + \gamma(g_R)] \bar{t}_l , \quad (5.61)$$

and to leading order in g_l,

$$\partial_l \bar{t}_l = 2\bar{t}_l - \frac{\bar{g}_l}{2} \bar{t}_l . \quad (5.62)$$

On the other hand, if we linearize the Wilsonian RG equation (4.71) we obtain

$$\partial_l \bar{r}_l = 2\bar{r}_l + \frac{1}{2}\bar{u}_l \tag{5.63}$$

which obviously is not identical with the field-theoretical RG equation (5.62). However, in the vicinity of the Wilson–Fisher fixed point we may set $\bar{g}_l \to \frac{2}{3}\epsilon$, so that Eq. (5.62) reduces to

$$\partial_l \bar{t}_l = y_t \bar{t}_l \quad , \quad y_t = 2 - \frac{\epsilon}{3} \, , \tag{5.64}$$

which agrees with the Wilsonian flow equation for the scaling field $t_l = \delta \bar{r}_l + \frac{1}{4}\left(1 - \frac{\epsilon}{6}\right)\delta \bar{u}_l$ defined in Eq. (4.87). We conclude that both field-theoretical and Wilsonian RG yield the same results for critical exponents, although conceptually and technically the two versions of the RG are rather different.

After this brief excursion into the field-theoretical RG, we will go back to the Wilsonian RG and proceed in the second part of this book with the development of modern functional RG methods.

References

Abramowitz, M. and I. A. Stegun (1965), *Handbook of Mathematical Functions*, Dover, New York.

Creutz, M. (1983), *Quarks, Gluons and Lattices*, Cambridge University Press, New York.

Gell-Mann, M. and F. E. Low (1954), *Quantum electrodynamics at small distances*, Phys. Rev. **95**, 1300.

Hewson, A. C. (1993), *The Kondo Problem to Heavy Fermions*, Cambridge University Press, New York.

Ledowski, S. and P. Kopietz (2007), *Fermi surface renormalization and confinement in two coupled metallic chains*, Phys. Rev. B **75**, 045134.

Ledowski, S., P. Kopietz, and A. Ferraz (2005), *Self-consistent Fermi surface renormalization of two coupled Luttinger liquids*, Phys. Rev. B **71**, 235106.

Neumayr, A. and W. Metzner (2003), *Renormalized perturbation theory for Fermi systems: Fermi surface deformation and superconductivity in the two-dimensional Hubbard model*, Phys. Rev. B **67**, 035112.

Nozières, P. (1964), *Theory of Interacting Fermi Systems*, Benjamin, New York.

Peskin, M. E. and D. V. Schroeder (1995), *An Introduction to Quantum Field Theory*, Addison-Wesley, Reading, MA.

Sterman, G. (1993), *An Introduction to Quantum Field Theory*, Cambridge University Press, Cambridge.

Stueckelberg, E. C. G. and A. Petermann (1953), *La normalisation des constantes dans la théorie des quanta*, Helv. Phys. Acta **26**, 499.

Zee, A. (2003), *Quantum Field Theory in a Nutshell*, Princeton University Press, Princeton, NJ.

Zinn-Justin, J. (2002), *Quantum Field Theory and Critical Phenomena*, Clarendon Press, Oxford, 4th ed.

Part II
Introduction to the Functional Renormalization Group

The mode-elimination step of the Wilsonian RG discussed in Chap. 3 requires integrating over field components describing short-wavelength and high-energy fluctuations. We have seen in Sect. 4.2 that in general this step can only be carried out approximately and usually involves rather tedious calculations. It turns out, however, that if the mode elimination is carried out in infinitesimal steps, its effect can be described mathematically via an exact hierarchy of functional differential equations for certain types of Green functions or vertex functions. This has already been discussed in the early days of the RG by Wegner and Houghton (1973), who derived a formally exact functional differential equation describing the change of the effective action due to an infinitesimal change of the cutoff Λ. This so-called *Wegner–Houghton equation* is equivalent to an infinite hierarchy of coupled functional RG flow equations involving all momentum-dependent vertices of the theory. Unfortunately, the explicit solution of the Wegner–Houghton hierarchy of flow equations leads to some technical complications, so that for practical calculations the Wegner–Houghton equation has not been widely used in the literature. However, there are alternative and in practice more convenient ways of expressing the Wilsonian mode elimination step in terms of formally exact functional differential equations for certain generating functionals, which will be defined in Chap. 6. In the past 20 years, approximate solutions of these formally exact RG flow equations have been extensively studied; the work in this field is now summarized under the name *functional renormalization group* (FRG) or *exact renormalization group*. In this second part of this book, we give a self-contained introduction to the FRG, building on the foundations presented in Part I. For other comprehensive reviews of the FRG, see (Morris 1998, Salmhofer 1999, Bagnuls and Bervillier 2001, Berges et al. 2002, Pawlowski 2007, Delamotte 2007). We shall assume in the rest of this book that the reader is familiar with the functional integral approach to quantum many-body systems, as described, for example, in the textbook by Negele and Orland (1988).

We begin in Chap. 6 with a summary of functional methods in many-body physics; in particular, we introduce several generating functionals for different types of Green functions and vertex functions and interpret them diagrammatically. In Chap. 7, we then derive formally exact FRG flow equations describing the change of the generating functionals due to an infinitesimal Wilsonian mode elimination step.

While there are several equivalent formulations of the FRG involving different types of generating functionals, in many cases it is advantageous to formulate the FRG in terms of the generating functional Γ of the one-line irreducible vertices, which can be obtained from the generating functional of the connected Green functions via a Legendre transformation and is therefore also called the *Legendre effective action*. The advantages of working with the Legendre effective action have already been recognized in the older RG literature by Di Castro et al. (1974) and later by Nicoll, Chang, and coworkers (Nicoll et al. 1976a,b, Nicoll and Chang 1977, Chang et al. 1992). More recently, Wetterich (1993) has written down the exact FRG flow equation for the Legendre effective action in a very concise form given in Eq. (7.54) which is sometimes called the *Wetterich equation*.

Although the exact FRG equation for the generating functional Γ of the one-line irreducible vertices looks deceptively simple, the explicit solution of this equation is rather difficult and usually requires drastic approximations. Two different approximation strategies have been developed: the first is based on the combination of the local potential approximation with the derivative expansion (Bagnuls and Bervillier 2002, Berges et al. 2002, Pawlowski 2007). This approach has been very successful to obtain accurate results for critical exponents (Bagnuls and Bervillier 2002, Berges et al. 2002, Delamotte et al. 2004, Delamotte 2007) and is convenient to describe the phases with broken symmetry (Birse et al. 2005, Krippa et al. 2005, Krippa 2007, 2009, Diehl and Wetterich 2007, Diehl et al. 2007a,b, 2009), because it is based on an expansion of Γ in terms of invariant densities which automatically fulfill all symmetry requirements. In Chap. 9, we shall give a brief introduction to this method.

The other approximation strategy is based on the expansion of Γ in powers of the fields (*vertex expansion*), leading to an infinite hierarchy of coupled integro-differential equations for the one-particle irreducible vertices of the model under consideration. This strategy was pioneered by Morris (1994) and has the advantage of providing information on the momentum and frequency dependence of the vertices at all momentum and energy scales, as will be explained in detail in Chap. 8. For nonrelativistic fermions, the corresponding hierarchy of RG flow equations for the irreducible vertices has been derived by Kopietz and Busche (2001) and independently by Salmhofer and Honerkamp (2001), while for nonrelativistic bosons the corresponding equations are given by Ledowski et al. (2004) and by Hasselmann et al. (2004). Truncations of the exact hierarchy of flow equations for the one-line irreducible vertices of nonrelativistic fermions have been extensively studied in the condensed matter literature for various systems, such as two-dimensional Hubbard models (Honerkamp and Salmhofer 2001b, Honerkamp 2001, 2003, Honerkamp et al. 2004, Honerkamp and Salmhofer 2005, Kampf and Katanin 2003, Katanin and Kampf 2003, 2004, Katanin 2004, 2009, Ossadnik et al. 2008), one-dimensional systems (Busche et al. 2002, Meden et al. 2002, Andergassen et al. 2004, Ledowski et al. 2005, Ledowski and Kopietz 2007), or quantum mechanical impurity models (Hedden et al. 2004, Andergassen et al. 2006, Meden and Marquardt 2006, Karrasch et al. 2007, 2008, Bartosch et al. 2009a). We shall discuss various versions of the vertex expansion for fermions in Part III of this book. In particular, in Chap. 10

we explicitly write down the coupled system of functional RG equations describing the flow of the one-particle irreducible vertices of nonrelativistic fermions. These equations follow as a special case of a more general "master flow equation" derived in Chap. 7, which treats fermions, bosons, or mixtures thereof within a single unifying "superfield-formalism" (Schütz et al. 2005, Schütz 2005, Schütz and Kopietz 2006).

References

Andergassen, S., T. Enss, V. Meden, W. Metzner, U. Schollwöck, and K. Schönhammer (2004), *Functional renormalization group for Luttinger liquids with impurities*, Phys. Rev. B **70**, 075102.

Andergassen, S., T. Enss, and V. Meden (2006), *Kondo physics in transport through a quantum dot with Luttinger-liquid leads*, Phys. Rev. B **73**, 153308.

Bagnuls, C. and C. Bervillier (2001), *Exact renormalization group equations: An introductory review*, Phys. Rep. **348**, 91.

Bagnuls, C. and C. Bervillier (2002), *Classical-to-critical crossovers from field theory*, Phys. Rev. E **65**, 066132.

Bartosch, L., H. Freire, J. J. Ramos Cardenas, and P. Kopietz (2009), *Functional renormalization group approach to the Anderson impurity model*, J. Phys.: Condens. Matter **21**, 305602.

Berges, J., N. Tetradis, and C. Wetterich (2002), *Non-perturbative renormalization flow in quantum field theory and statistical physics*, Phys. Rep. **363**, 223.

Birse, M. C., B. Krippa, J. A. McGovern, and N. R. Walet (2005), *Pairing in many-fermion systems: An exact renormalisation group treatment*, Phys. Lett. B **605**, 287.

Busche, T., L. Bartosch, and P. Kopietz (2002), *Dynamic scaling in the vicinity of the Luttinger liquid fixed point*, J. Phys.: Condens. Matter **14**, 8513.

Chang, T. S., D. D. Vvedensky, and J. F. Nicoll (1992), *Differential renormalization-group generators for static and dynamic critical phenomena*, Phys. Rep. **217**, 279.

Delamotte, B. (2007), *An Introduction to the Nonperturbative Renormalization Group*, arXiv:cond-mat/0702365.

Delamotte, B., D. Mouhanna, and M. Tissier (2004), *Nonperturbative renormalization-group approach to frustrated magnets*, Phys. Rev. B **69**, 134413.

Di Castro, C., G. Jona-Lasinio, and L. Peliti (1974), *Variational principles, renormalization group, and Kadanoff's universality*, Ann. Phys. **87**, 327.

Diehl, S. and C. Wetterich (2007), *Functional integral for ultracold fermionic atoms*, Nucl. Phys. B **770**, 206.

Diehl, S., H. Gies, J. M. Pawlowski, and C. Wetterich (2007a), *Flow equations for the BCS-BEC crossover*, Phys. Rev. A **76**, 021602(R).

Diehl, S., H. Gies, J. M. Pawlowski, and C. Wetterich (2007b), *Renormalization flow and universality for ultracold fermionic atoms*, Phys. Rev. A **76**, 053627.

Diehl, S., S. Floerchinger, H. Gies, J. M. Pawlowski, and C. Wetterich (2009), *Functional renormalization group approach to the BCS-BEC crossover*, arXiv:0907.2193 [cond-mat.quant-gas].

Hasselmann, N., S. Ledowski, and P. Kopietz (2004), *Critical behavior of weakly interacting bosons: A functional renormalization-group approach*, Phys. Rev. A **70**, 063621.

Hedden, R., V. Meden, T. Pruschke, and K. Schönhammer (2004), *A functional renormalization group approach to zero-dimensional interacting systems*, J. Phys.: Condens. Matter **16**, 5279.

Honerkamp, C. (2001), *Electron-doping versus hole-doping in the 2D t-t' Hubbard model*, Eur. Phys. J. B **21**, 81.

Honerkamp, C. (2003), *Instabilities of interacting electrons on the triangular lattice*, Phys. Rev. B **68**, 104510.

Honerkamp, C. and M. Salmhofer (2001), *Temperature-flow renormalization group and the competition between superconductivity and ferromagnetism*, Phys. Rev. B **64**, 184516.

Honerkamp, C. and M. Salmhofer (2005), *Eliashberg Equations derived from the functional renormalization group*, Progr. Theoret. Phys. **113**, 1145.

Honerkamp, C., D. Rohe, S. Andergassen, and T. Enss (2004), *Interaction flow method for many-fermion systems*, Phys. Rev. B **70**, 235115.

Kampf, A. P. and A. A. Katanin (2003), *Competing phases in the extended U-V-J Hubbard model near the Van Hove fillings*, Phys. Rev. B **67**, 125104.

Karrasch, C., T. Hecht, A.Weichselbaum, Y. Oreg, J. von Delft, and V. Meden (2007), *Mesoscopic to universal crossover of the transmission phase of multilevel quantum dots*, Phys. Rev. Lett. **98**, 186802.

Karrasch, C., R. Hedden, R. Peters, T. Pruschke, K. Schönhammer, and V. Meden (2008), *A finite-frequency functional renormalization group approach to the single impurity Anderson model*, J. Phys.: Condens. Matter **20**, 345205.

Katanin, A. A. (2004), *Fulfillment of Ward identities in the functional renormalization group approach*, Phys. Rev. B **70**, 115109.

Katanin, A. A. (2009), *The two-loop functional renormalization group approach to the one- and two-dimensional Hubbard model*, Phys. Rev. B **79**, 235119.

Katanin, A. A. and A. P. Kampf (2003), *Renormalization group analysis of magnetic and superconducting instabilities near van Hove band fillings*, Phys. Rev. B **68**, 195101.

Katanin, A. A. and A. P. Kampf (2004), *Quasiparticle anisotropy and Pseudogap formation from the weak-coupling renormalization group point of view*, Phys. Rev. Lett. **93**, 106406.

Kopietz, P. and T. Busche (2001), *Exact renormalization group flow equations for nonrelativistic fermions: Scaling toward the Fermi surface*, Phys. Rev. B **64**, 155101.

Krippa, B. (2007), *Superfluidity in many fermion systems: Exact renormalization group treatment*, Eur. Phys. J. A **31**, 734.

Krippa, B. (2009), *Exact renormalization group flow for ultracold Fermi gases in the unitary limit*, J. Phys. A: Math. Theor. **42**, 465002.

Krippa, B., M. C. Birse, N. R. Walet, and J. A. McGovern (2005), *Exact renormalisation group and pairing in many-fermion systems*, Nucl. Phys. A **749**, 134.

Ledowski, S., P. Kopietz, and A. Ferraz (2005), *Self-consistent Fermi surface renormalization of two coupled Luttinger liquids*, Phys. Rev. B **71**, 235106.

Ledowski, S. and P. Kopietz (2007), *Fermi-surface renormalization and confinement in two coupled metallic chains*, Phys. Rev. B **75**, 045134.

Ledowski, S., N. Hasselmann, and P. Kopietz (2004), *Self-energy and critical temperature of weakly interacting bosons*, Phys. Rev. A **69**, 061601.

Meden, V. and F. Marquardt (2006), *Correlation-induced resonances in transport through coupled quantum dots*, Phys. Rev. Lett. **96**, 146801.

Meden, V., W. Metzner, U. Schollwöck, and K. Schönhammer (2002), *Scaling behavior of impurities in mesoscopic Luttinger liquids*, Phys. Rev. B **65**, 045318.

Morris, T. R. (1994), *The exact renormalisation group and approximate solutions*, Int. J. Mod. Phys. A **9**, 2411.

Morris, T. R. (1998), *Elements of the continuous renormalization group*, Progr. Theoret. Phys. Suppl. **131**, 395.

Negele, J. W. and H. Orland (1988), *Quantum Many-Particle Systems*, Addison-Wesley, Redwood City.

Nicoll, J. F. and T. S. Chang (1977), *An exact one-particle-irreducible renormalization-group generator for critical phenomena*, Phys. Lett. A **62**, 287.

Nicoll, J. F., T. S. Chang, and H. E. Stanley (1976a), *Exact and approximate differential renormalization-group generators*, Phys. Rev. A **13**, 1251.

Nicoll, J. F., T. S. Chang, and H. E. Stanley (1976b), *A differential generator for the free energy and the magnetization equation of state*, Phys. Lett. A **57**, 7.

Ossadnik, M., C. Honerkamp, T. M. Rice, and M. Sigrist (2008), *Breakdown of Landau theory in overdoped cuprates near the onset of superconductivity*, Phys. Rev. Lett. **101**, 256405.

Pawlowski, J. M. (2007), *Aspects of the functional renormalisation group*, Ann. Phys. **322**, 2831.

Salmhofer, M. (1999), *Renormalization: An Introduction*, Springer, Berlin.

Salmhofer, M. and C. Honerkamp (2001), *Fermionic renormalization group flows*, Progr. Theoret. Phys. **105**, 1.

Schütz, F. (2005), *Aspects of Strong Correlations in Low Dimensions*, Doktorarbeit, Goethe-Universität Frankfurt.

Schütz, F. and P. Kopietz (2006), *Functional renormalization group with vacuum expectation values and spontaneous symmetry breaking*, J. Phys. A: Math. Gen. **39**, 8205.

Schütz, F., L. Bartosch, and P. Kopietz (2005), *Collective fields in the functional renormalization group for fermions, Ward identities, and the exact solution of the Tomonaga-Luttinger model*, Phys. Rev. B **72**, 035107.

Wegner, F. J. and A. Houghton (1973), *Renormalization group equation for critical phenomena*, Phys. Rev. A **8**, 401.

Wetterich, C. (1993), *Exact evolution equation for the effective potential*, Phys. Lett. B **301**, 90.

Chapter 6
Functional Methods

Our derivation of exact FRG equations in Chap. 7 will be based on the concept of generating functionals of several types of Green functions and vertex functions. For a self-contained introduction to the FRG it is therefore necessary to properly define the various generating functionals and develop some intuition for the meaning of the associated correlation functions, which we shall do in this chapter. Moreover, we shall use the invariance properties of generating functionals under shift and symmetry transformations to derive so-called Dyson–Schwinger equations of motion and Ward identities relating different types of correlation functions. Of course, comprehensive introductions to functional methods can be found in several excellent textbooks (Negele and Orland 1988, Zinn-Justin 2002, Rammer 2007). In this context we also mention the more specialized books by Vasiliev (1998) and by Fried (1972, 2002), which emphasize the formulation of field theory in terms of functional derivatives.

To describe fermions, bosons, or even mixtures thereof within a single formalism, it is useful to have a compact notation which allows us to keep track of the combinatorial factors and possible minus signs in an efficient way. Let us therefore collect all different field types of a given theory into a *superfield* Φ labeled by a *superlabel* α, which denotes the different field types and, for a given field, all labels needed to specify the field configuration, such as momentum, frequency, spin or additional flavor labels (Schütz et al. 2005, Schütz 2005, Schütz and Kopietz 2006). A related approach has been developed by Wetterich and coauthors (Baier et al. 2004, 2005, Wetterich 2007). For purely fermionic systems, a similar notation has also been used by Salmhofer and Honerkamp (2001). We assume that the correlation functions of the system can be represented as a functional integral involving the exponential of some general action $S[\Phi]$, with the integration measure normalized such that the exact partition function of the system can be written as

$$\mathcal{Z} = \int \mathcal{D}[\Phi] e^{-S[\Phi]} . \tag{6.1}$$

Furthermore, we assume that the action can be decomposed as

$$S[\Phi] = S_0[\Phi] + S_1[\Phi] , \tag{6.2}$$

with some arbitrary interaction part $S_1[\Phi]$, and a Gaussian part of the general form

Kopietz, P. et al.: *Functional Methods*. Lect. Notes Phys. **798**, 147–180 (2010)
DOI 10.1007/978-3-642-05094-7_6

$$S_0[\Phi] = -\frac{1}{2} \int_\alpha \int_{\alpha'} \Phi_\alpha \left[\mathbf{G}_0^{-1} \right]_{\alpha\alpha'} \Phi_{\alpha'} \equiv -\frac{1}{2} \left(\Phi, \mathbf{G}_0^{-1} \Phi \right) . \tag{6.3}$$

The symbol $\int_\alpha$ denotes integration over the continuous components and summation over the discrete components of the superfield label α, and the Gaussian propagator $\mathbf{G}_0$ is a matrix in superlabel space. The bosonic components of Φ are given by complex fields, while the fermionic components are anticommuting Grassmann fields. It is understood that the matrix $\mathbf{G}_0$ appearing in the Gaussian action (6.3) has been properly symmetrized in the bosonic sectors and antisymmetrized in the fermionic sectors. Assuming that the different sectors do not mix, the Gaussian propagator has the symmetry

$$\mathbf{G}_0^T = \mathbf{Z}\mathbf{G}_0 = \mathbf{G}_0\mathbf{Z} , \tag{6.4}$$

where the statistics matrix $\mathbf{Z}$ is an infinite diagonal matrix in superlabel space with matrix elements

$$\boxed{\mathbf{Z}_{\alpha\alpha'} = \delta_{\alpha\alpha'}\zeta_\alpha, \quad \text{with } \zeta_\alpha = \begin{cases} 1 & \text{if } \alpha \text{ labels a boson,} \\ -1 & \text{if } \alpha \text{ labels a fermion.} \end{cases}} \tag{6.5}$$

For example, to describe a system of nonrelativistic electrons we need a pair of Grassmann fields ψ_σ and $\bar{\psi}_\sigma$ for each spin projection σ. For a one-band model the fields can be labeled by Matsubara frequencies $i\omega$ and momenta $\mathbf{k}$. For simplicity, we shall collect $i\omega$ and $\mathbf{k}$ into a collective label $K = (i\omega, \mathbf{k})$. Then the superlabel α assumes the values $\alpha = (\psi, K, \sigma)$ and $\alpha = (\bar{\psi}, K, \sigma)$, where the first specification denotes the field type, and the other labels give the energy, the momentum, and spin carried by the field. The Gaussian part of the Euclidean action can in this case be written as

$$\begin{aligned}
S_0[\Phi] &= -\sum_\sigma \int_K G_0^{-1}(K)\bar{\psi}_{K\sigma}\psi_{K\sigma} \\
&= -\frac{1}{2}\sum_\sigma \int_K (\psi_{K\sigma}, \bar{\psi}_{K\sigma}) \begin{pmatrix} 0 & -G_0^{-1}(K) \\ G_0^{-1}(K) & 0 \end{pmatrix} \begin{pmatrix} \psi_{K\sigma} \\ \bar{\psi}_{K\sigma} \end{pmatrix} ,
\end{aligned} \tag{6.6}$$

where the minus sign in the upper diagonal has been generated by anticommuting the Grassmann fields, $\bar{\psi}_{K\sigma}\psi_{K\sigma} = -\psi_{K\sigma}\bar{\psi}_{K\sigma}$, and $G_0(K)$ is the imaginary frequency propagator of noninteracting fermions with energy dispersion ϵ_k and chemical potential μ,

$$G_0(K) \equiv G_0(i\omega, \mathbf{k}) = \frac{1}{i\omega - \epsilon_k + \mu} . \tag{6.7}$$

We have normalized the Grassmann fields $\psi_{K\sigma}$ and $\bar{\psi}_{K\sigma}$ such that the integration measure in Eq. (6.6) is

$$\int_K = \frac{1}{\beta V} \sum_{\omega,k} \quad \overset{\beta,V\to\infty}{\longrightarrow} \quad \int \frac{d\omega}{2\pi} \frac{d^D k}{(2\pi)^D} \; . \tag{6.8}$$

Comparing Eq. (6.6) with Eq. (6.3) we conclude that for nonrelativistic fermions the infinite matrix $\mathbf{G}_0^{-1}$ has the following block structure in field space,

$$\mathbf{G}_0^{-1} = \begin{pmatrix} 0 & -\hat{G}_0^{-1} \\ \hat{G}_0^{-1} & 0 \end{pmatrix} , \tag{6.9}$$

where the blocks $\hat{G}_0^{-1}$ are infinite diagonal matrices in frequency, momentum, and spin space, with matrix elements

$$\left[\hat{G}_0^{-1}\right]_{K\sigma,K'\sigma'} = \delta_{K,K'}\delta_{\sigma,\sigma'}G_0^{-1}(K) \; . \tag{6.10}$$

The Kronecker delta symbol

$$\delta_{K,K'} = \beta V \delta_{\omega,\omega'}\delta_{k,k'} \tag{6.11}$$

is normalized such that it reduces to the $(D + 1)$-dimensional Dirac δ-function $\delta_{K,K'} \to (2\pi)^{D+1}\delta(\omega - \omega')\delta^{(D)}(\mathbf{k} - \mathbf{k}')$ in the limit $\beta, V \to \infty$. Obviously, for a theory involving only fermionic fields the matrix $\mathbf{G}_0$ is antisymmetric, $\mathbf{G}_0^T = -\mathbf{G}_0$, in agreement with the general symmetry property (6.4).

In order to obtain a manifestly antisymmetric propagator in Eq. (6.6), we have to pay the price of doubling its number of field components. Such a parameterization has also been introduced by Vasiliev (1998) as well as Salmhofer and Honerkamp (2001); its advantages will become evident when we develop the functional methods in this and the following chapter. In particular, we shall show in Chap. 7 that this construction greatly facilitates the derivation of the proper combinatorial factors in the vertex expansion of the FRG flow equations. Let us also emphasize that the assumed form of the action given in Eqs. (6.2) and (6.3) is very general; it includes interacting fermions or bosons as special cases, but applies also to more general field theories involving fermion–boson mixtures or more general field theories involving real bosonic or fermionic fields.[1] The manipulations in the rest of this chapter are independent of any particular model as long as the partition function can be represented as a functional integral (6.1) with effective action of the form (6.2) and (6.3).

---

[1] Real fermions are also called Majorana fermions, which are useful to construct a representation of the spin algebra in terms of hermitian anticommuting operators, see for example the book by Tsvelik (2003). For a recent application of this representation see (Shnirman and Makhlin 2003).

6.1 Generating Functionals for Green Functions

6.1.1 Disconnected Green Functions

For the class of models whose correlation functions can be represented as functional averages with respect to an action $S[\Phi]$ of the form (6.2) and (6.3), we define the *disconnected Green functions* $G^{(n)}_{\alpha_1...\alpha_n}$ with n external legs via the following functional average,[2]

$$G^{(n)}_{\alpha_1...\alpha_n} = \langle \Phi_{\alpha_n} \ldots \Phi_{\alpha_1} \rangle = \frac{\int \mathcal{D}[\Phi] e^{-S[\Phi]} \Phi_{\alpha_n} \ldots \Phi_{\alpha_1}}{\int \mathcal{D}[\Phi] e^{-S[\Phi]}} \, . \tag{6.12}$$

The functions $G^{(n)}_{\alpha_1...\alpha_n}$ can be obtained as coefficients of the functional Taylor expansion of the generating functional

$$\mathcal{G}[J] = \frac{\int \mathcal{D}[\Phi] e^{-S[\Phi]+(J,\Phi)}}{\int \mathcal{D}[\Phi] e^{-S[\Phi]}} \, , \tag{6.13}$$

where we have introduced the notation

$$(J, \Phi) = \int_\alpha J_\alpha \Phi_\alpha \, . \tag{6.14}$$

The generating functional $\mathcal{G}[J]$ depends on the supersources J whose components J_α are of the same type as the corresponding field components Φ_α, i.e., J_α is bosonic for bosonic field components, and is a Grassmann variable if the corresponding field component Φ_α is fermionic. This assures that all terms $J_\alpha \Phi_\alpha$ in the sum (6.14) commute with all other fields of the theory. Differentiating $\mathcal{G}[\Phi]$ first with respect to J_{α_1}, and then with respect to J_{α_2}, and so on until we reach J_{α_n}, and setting after the differentiation the sources equal to zero, we see that the disconnected Green functions defined in Eq. (6.12) can be represented as derivatives of the corresponding generating functional,

$$G^{(n)}_{\alpha_1...\alpha_n} = \left. \frac{\delta^n \mathcal{G}[J]}{\delta J_{\alpha_n} \ldots \delta J_{\alpha_1}} \right|_{J=0} \, . \tag{6.15}$$

This relation is equivalent with the following functional Taylor expansion of the generating functional (6.13),

---

[2] Note that the ordering of the labels in the functional average $\langle \Phi_{\alpha_n} \ldots \Phi_{\alpha_1} \rangle$ is opposite to the ordering of the labels of $G^{(n)}_{\alpha_1...\alpha_n}$. This convention has the advantage that the ordering of the sources $J_{\alpha_1} \ldots J_{\alpha_n}$ in the functional Taylor expansion (6.16) agrees with the ordering of the labels of $G^{(n)}_{\alpha_1...\alpha_n}$.

$$\mathcal{G}[J] = \sum_{n=0}^{\infty} \frac{1}{n!} \int_{\alpha_1} \cdots \int_{\alpha_n} G^{(n)}_{\alpha_1 \ldots \alpha_n} J_{\alpha_1} \ldots J_{\alpha_n} . \tag{6.16}$$

It turns out that disconnected Green functions are not very convenient to work with, because in general their perturbative expansion contains subdiagrams corresponding to Green functions of lower order. For example, the perturbative calculation of $G^{(4)}_{\alpha_1 \alpha_2 \alpha_3 \alpha_4} = \langle \Phi_{\alpha_4} \Phi_{\alpha_3} \Phi_{\alpha_2} \Phi_{\alpha_1} \rangle$ using the Wick theorem contains also disconnected contributions of the form

$$G^{(2)}_{\alpha_3 \alpha_4} G^{(2)}_{\alpha_1 \alpha_2} = \langle \Phi_{\alpha_4} \Phi_{\alpha_3} \rangle \langle \Phi_{\alpha_2} \Phi_{\alpha_1} \rangle = \quad \tag{6.17}$$

Because these contributions can be expressed in terms of the exact two-field Green function,

$$G^{(2)}_{\alpha_1 \alpha_2} = \langle \Phi_{\alpha_2} \Phi_{\alpha_1} \rangle = \quad , \tag{6.18}$$

they do not exclusively contain information about correlations involving four fields. Note that in quantum systems the function $G^{(2)}_{\alpha_1 \alpha_2}$ corresponds to the single-particle propagator and is represented graphically by an arrow pointing to the *first* index; below we shall use a similar convention for all superfield correlation functions.

6.1.2 Connected Green Functions

By definition, the diagrammatic expansion of connected Green functions does not contain diagrams which can be decomposed into two or more disconnected pieces. Keeping in mind our discussion of the linked cluster theorem in Sect. 4.1.2, it should not be surprising that the generating functional $\mathcal{G}_c[J]$ of the connected Green functions is simply related to the logarithm of the corresponding generating functional $\mathcal{G}[J]$ of the disconnected Green functions. For reasons which will become obvious later on (see Eq. (6.39)) we normalize $\mathcal{G}_c[J]$ as follows

$$\mathcal{G}_c[J] = \ln\left(\frac{Z}{Z_0} \mathcal{G}[J] \right) = \ln\left(\frac{\int \mathcal{D}[\Phi] e^{-S[\Phi]+(J,\Phi)}}{\int \mathcal{D}[\Phi] e^{-S_0[\Phi]}} \right) . \tag{6.19}$$

Note that the functional integral in the denominator involves only the Gaussian part $S_0[\Phi]$ of the action. The connected Green functions $G^{(n)}_{c,\alpha_1 \ldots \alpha_n}$ with n externals legs are then defined similar to Eq. (6.15) in terms of functional derivatives of $\mathcal{G}_c[J]$,

$$G^{(n)}_{c,\alpha_1 \ldots \alpha_n} = \left. \frac{\delta^n \mathcal{G}_c[J]}{\delta J_{\alpha_n} \ldots \delta J_{\alpha_1}} \right|_{J=0} . \tag{6.20}$$

The corresponding functional Taylor expansion analogous to Eq. (6.16) is

$$\mathcal{G}_c[J] = \sum_{n=0}^{\infty} \frac{1}{n!} \int_{\alpha_1} \cdots \int_{\alpha_n} G^{(n)}_{c,\alpha_1\ldots\alpha_n} J_{\alpha_1} \ldots J_{\alpha_n} \; . \qquad (6.21)$$

Graphically we represent the connected Green functions $G^{(n)}_{c,\alpha_1\ldots\alpha_n}$ by oriented vertices (to take into account the ordering for fermions) with n external legs as shown in Fig. 6.1. From Eqs. (6.19) and (6.20) one easily obtains the relations between the connected and disconnected Green functions for $n = 0, 1, 2$,

$$G^{(0)}_c = \ln\left(\mathcal{Z}/\mathcal{Z}_0\right) , \qquad (6.22a)$$

$$G^{(1)}_{c,\alpha_1} = G^{(1)}_{\alpha_1} = \langle \Phi_{\alpha_1} \rangle , \qquad (6.22b)$$

$$G^{(2)}_{c,\alpha_1\alpha_2} = G^{(2)}_{\alpha_1\alpha_2} - G^{(1)}_{\alpha_1} G^{(1)}_{\alpha_2} = \langle \Phi_{\alpha_2} \Phi_{\alpha_1} \rangle - \langle \Phi_{\alpha_2} \rangle \langle \Phi_{\alpha_1} \rangle . \qquad (6.22c)$$

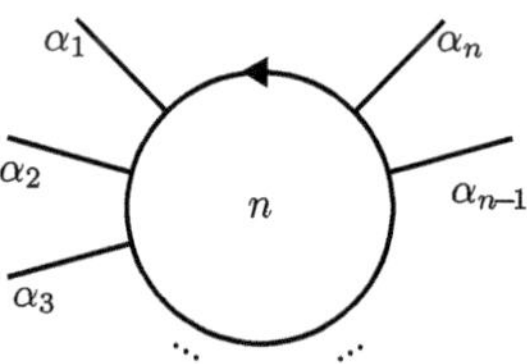

Fig. 6.1 Graphical representation of the connected Green function $G^{(n)}_{c,\alpha_1\ldots\alpha_n}$ with n external legs defined via Eq. (6.20) or (6.21). Because for fermions the order of the external legs matters, we represent $G^{(n)}_{c,\alpha_1\ldots\alpha_n}$ by an *oriented circle* where the *arrow* points to the leg associated with the first label and the subsequent labels follow in the order defined by the arrow

In the language of probability theory (see e.g., Van Kampen 1981), the disconnected Green functions $G^{(n)}_{\alpha_1\ldots\alpha_n}$ are the *moments* of the probability distribution defined by $\mathcal{Z}^{-1} e^{-S[\Phi]}$, while the connected Green functions $G^{(n)}_{c,\alpha_1\ldots\alpha_n}$ are the *cumulants*. Because the zeroth-order expansion coefficient $G^{(0)}_c$ in the Taylor expansion of $\mathcal{G}_c[J]$ is given by the logarithm of the partition function, we know from the linked cluster theorem for vacuum diagrams discussed in Sect. 4.1.2 that diagrammatically only connected vacuum diagrams contribute to $G^{(0)}_c$. To also prove that the higher-order coefficients $G^{(n)}_{c,\alpha_1\ldots\alpha_n}$ with $n \geq 1$ (corresponding to diagrams of the type shown in Fig. 6.1 with n external legs) do not contain any disconnected parts, we proceed as in the proof of the linked cluster theorem in Sect. 4.1.2 and use the replica trick,

$$\mathcal{G}_c[J] = \ln\left(\frac{\mathcal{Z}}{\mathcal{Z}_0}\mathcal{G}[J]\right) = \lim_{n\to 0} \frac{d}{dn}\left(\frac{\mathcal{Z}}{\mathcal{Z}_0}\mathcal{G}[J]\right)^n . \qquad (6.23)$$

Representing $\left(\frac{\mathcal{Z}}{\mathcal{Z}_0}\mathcal{G}[J]\right)^n$ as a functional over n-fold replicated fields $\Phi^r, r = 1, \ldots, n$ as in Eq. (4.27), we see that only those diagrams survive in the limit $n \to 0$

whose contribution is proportional to n. But these are precisely the connected diagrams, which completes our proof.

It is instructive to calculate the generating functional $\mathcal{G}_c[J]$ within the Gaussian approximation, which we denote by $\mathcal{G}_{0c}[J]$. In this case we ignore the interaction part of our action and obtain with the notation (6.3),

$$\mathcal{G}_0[J] = e^{\mathcal{G}_{0c}[J]} = \frac{\int \mathcal{D}[\Phi] e^{\frac{1}{2}(\Phi, \mathbf{G}_0^{-1}\Phi) + (J, \Phi)}}{\int \mathcal{D}[\Phi] e^{\frac{1}{2}(\Phi, \mathbf{G}_0^{-1}\Phi)}} \; . \tag{6.24}$$

In the numerator we shift the integration variable $\Phi = \Phi' - \mathbf{G}_0^T J$, so that the argument of the exponential becomes

$$\begin{aligned}
\frac{1}{2}&\left(\Phi, \mathbf{G}_0^{-1}\Phi\right) + (J, \Phi) \\
&= \frac{1}{2}\left(\Phi' - \mathbf{G}_0^T J, \mathbf{G}_0^{-1}\left(\Phi' - \mathbf{G}_0^T J\right)\right) + \left(J, \Phi' - \mathbf{G}_0^T J\right) \\
&= \frac{1}{2}\left(\Phi', \mathbf{G}_0^{-1}\Phi'\right) - \left(J, \mathbf{G}_0^T J\right) + \frac{1}{2}\left(\mathbf{G}_0^T J, \mathbf{G}_0^{-1}\mathbf{G}_0^T J\right) \\
&\quad + (J, \Phi') \underbrace{- \frac{1}{2}\left(\mathbf{G}_0^T J, \mathbf{G}_0^{-1}\Phi'\right) - \frac{1}{2}\left(\Phi', \mathbf{G}_0^{-1}\mathbf{G}_0^T J\right)}_{\text{cancel}} \\
&= \frac{1}{2}\left(\Phi', \mathbf{G}_0^{-1}\Phi'\right) - \frac{1}{2}\left(J, \mathbf{G}_0^T J\right) \; .
\end{aligned} \tag{6.25}$$

The indicated cancellation follows from the identities

$$\left(\mathbf{G}_0^T J, \mathbf{G}_0^{-1}\Phi'\right) = \left(J, \mathbf{G}_0\mathbf{G}_0^{-1}\Phi'\right) = (J, \Phi') \; , \tag{6.26a}$$

$$\left(\Phi', \mathbf{G}_0^{-1}\mathbf{G}_0^T J\right) = (\Phi', \mathbf{Z} J) = (J, \Phi') \; , \tag{6.26b}$$

where in the second equation we have used the symmetry (6.4) of the superfield matrix $\mathbf{G}_0$. After the shift $\Phi = \Phi' - \mathbf{G}_0^T J$ in the numerator of Eq. (6.24), the Gaussian integral of the field variables cancels and we obtain for the generating functional of the connected Green functions in Gaussian approximation,

$$\boxed{\mathcal{G}_{0c}[J] = -\frac{1}{2}\left(J, \mathbf{G}_0^T J\right) = -\frac{1}{2}\int_\alpha \int_{\alpha'} [\mathbf{G}_0]_{\alpha'\alpha} J_\alpha J_{\alpha'} \; .} \tag{6.27}$$

We conclude that in Gaussian approximation

$$[\mathbf{G}_0]_{\alpha\alpha'} = -\frac{\delta^2 \mathcal{G}_{0c}[J]}{\delta J_\alpha \delta J_{\alpha'}} = -\langle \Phi_\alpha \Phi_{\alpha'} \rangle = -G_{0c,\alpha'\alpha}^{(2)} = -G_{0,\alpha'\alpha}^{(2)} \; . \tag{6.28}$$

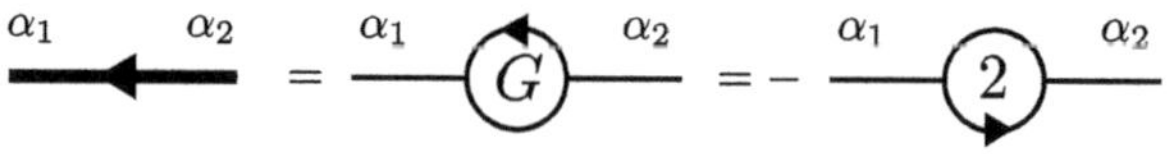

$$\alpha_1 \longleftarrow \alpha_2 \quad = \quad \alpha_1 \, \overbrace{(G)} \, \alpha_2 \quad = - \quad \alpha_1 \, \overbrace{(2)} \, \alpha_2$$

Fig. 6.2 Graphical representation of the relation (6.32) between the exact propagator **G** and the second functional derivative of $\mathcal{G}_c[J]$. The matrix element $[\mathbf{G}]_{\alpha_1\alpha_2}$ is either represented by a *thick arrow* pointing from α_2 to α_1, or by an *oriented circle* enclosing the symbol G. We use the convention that the *arrow* defining the orientation of the circle points to the first index of $[\mathbf{G}]_{\alpha_1\alpha_2}$

For later convenience let us introduce the matrix differential operator $\frac{\delta}{\delta J} \otimes \frac{\delta}{\delta J}$ in superfield space with matrix elements

$$\left[\frac{\delta}{\delta J} \otimes \frac{\delta}{\delta J} \right]_{\alpha\alpha'} = \frac{\delta}{\delta J_\alpha} \frac{\delta}{\delta J_{\alpha'}} \, . \tag{6.29}$$

With this notation Eq. (6.28) can be written as an identity between matrices in superfield space,

$$\mathbf{G}_0 = - \frac{\delta}{\delta J} \otimes \frac{\delta}{\delta J} \, \mathcal{G}_{0c}[J] \, . \tag{6.30}$$

By analogy with the relation (6.28) we identify the second derivative of the exact generating functional $\mathcal{G}_c[J]$ at vanishing external sources with the exact superfield Green function **G**,

$$[\mathbf{G}]_{\alpha\alpha'} = - \left. \frac{\delta^2 \mathcal{G}_c[J]}{\delta J_\alpha \delta J_{\alpha'}} \right|_{J=0} = - G^{(2)}_{c,\alpha'\alpha} \, , \tag{6.31}$$

or in matrix notation

$$\mathbf{G} = - \left[\frac{\delta}{\delta J} \otimes \frac{\delta}{\delta J} \, \mathcal{G}_c[J] \right]_{J=0} \, . \tag{6.32}$$

This relation is shown graphically in Fig. 6.2. In the case of boson–fermion mixtures we assume that the matrix elements $[\mathbf{G}]_{\alpha\alpha'}$ are only nonzero if both α and α' refer either to fermionic or bosonic fields. Then the commuting or anticommuting property of the functional derivatives in Eq. (6.32) implies that the exact **G** satisfies the same symmetry relation (6.4) as $\mathbf{G}_0$,

$$\mathbf{G}^T = \mathbf{Z}\mathbf{G} = \mathbf{G}\mathbf{Z} \, . \tag{6.33}$$

It is convenient to parameterize interaction corrections to $\mathbf{G}_0$ in terms of a superfield self-energy $\mathbf{\Sigma}$, which we define via the matrix Dyson equation

$$\mathbf{G}^{-1} = \mathbf{G}_0^{-1} - \mathbf{\Sigma} \, . \tag{6.34}$$

Note that here the sign convention for the self-energy is opposite to the sign convention for the corresponding self-energy Σ_k introduced in Eq. (4.35) for the classical φ^4-theory. The negative sign in Eq. (6.34) is customary in quantum many-particle physics and has the advantage that all terms in the expansion of $\mathbf{G}$ in powers of $\boldsymbol{\Sigma}$ are positive,

$$\mathbf{G} = \mathbf{G}_0 + \mathbf{G}_0 \boldsymbol{\Sigma} \mathbf{G}_0 + \mathbf{G}_0 \boldsymbol{\Sigma} \mathbf{G}_0 \boldsymbol{\Sigma} \mathbf{G}_0 + \dots \ . \tag{6.35}$$

A graphical representation of Eqs. (6.34) and (6.35) is shown in Fig. 6.3.

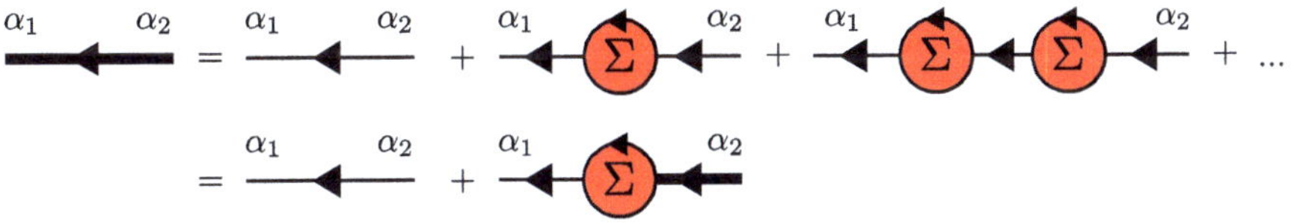

Fig. 6.3 Graphical representation of the Dyson equation (6.34) and its expansion (6.35). *Thick arrows* pointing from α_2 to α_1 represent the exact propagator $[\mathbf{G}]_{\alpha_1\alpha_2}$, while *thin arrows* represent the corresponding propagator $[\mathbf{G}_0]_{\alpha_1\alpha_2}$ in Gaussian approximation. The irreducible self-energy $[\boldsymbol{\Sigma}]_{\alpha_1\alpha_2}$ is represented by a shaded oriented circle with the arrow pointing to the first index α_1

To conclude this section, let us derive a representation of the functional $\frac{\mathcal{Z}}{\mathcal{Z}_0}\mathcal{G}[J] = e^{\mathcal{G}_c[J]}$ defined in Eq. (6.19) where the functional integration is formally replaced by a functional differentiation with respect to the sources J. Using the "source trick"

$$(\Phi_\alpha)^n e^{(J,\Phi)} = \left(\frac{\delta}{\delta J_\alpha}\right)^n e^{(J,\Phi)} \ , \tag{6.36}$$

we may write

$$e^{-S_1[\Phi]+(J,\Phi)} = e^{-S_1\left[\frac{\delta}{\delta J}\right]} e^{(J,\Phi)} \ . \tag{6.37}$$

The term $e^{-S_1\left[\frac{\delta}{\delta J}\right]}$ is now independent of the integration variables Φ_α and may be pulled out of the functional integral in Eq. (6.19),

$$e^{\mathcal{G}_c[J]} = e^{-S_1\left[\frac{\delta}{\delta J}\right]} \underbrace{\frac{1}{\mathcal{Z}_0} \int \mathcal{D}[\Phi] e^{-S_0[\Phi]+(J,\Phi)}}_{e^{\mathcal{G}_{0c}[J]} = e^{-\frac{1}{2}\left(J,\mathbf{G}_0^T J\right)}} \ , \tag{6.38}$$

where we have used Eq. (6.24). We thus arrive at the identity

$$\boxed{e^{\mathcal{G}_c[J]} = \frac{\mathcal{Z}}{\mathcal{Z}_0}\mathcal{G}[J] = \frac{\int \mathcal{D}[\Phi] e^{-S[\Phi]+(J,\Phi)}}{\int \mathcal{D}[\Phi] e^{-S_0[\Phi]}} = e^{-S_1\left[\frac{\delta}{\delta J}\right]} e^{-\frac{1}{2}(J,\mathbf{G}_0^T J)} \ .} \tag{6.39}$$

From this expression it is obvious why it is convenient to normalize the functional integral defining $e^{\mathcal{G}_c[J]}$ with the Gaussian partition function $\mathcal{Z}_0$.

6.1.3 Amputated Connected Green Functions

In Chap. 7 we shall derive formally exact flow equations for generating functionals of Green functions with initial conditions corresponding to the limit $\mathbf{G}_0 \to 0$. From the representation (6.39) of $\mathcal{G}_c[J]$ it is clear that $\mathcal{G}_c \to 0$ in this limit, which is not a sensible starting point for approximations. The exact FRG equation for $\mathcal{G}_c$ is therefore not very useful, so that we should consider other types of Green functions. One possibility is to use so-called *amputated connected Green functions*, which are generated by the functional $\mathcal{G}_{\mathrm{ac}}[\bar{\Phi}]$ defined by

$$\boxed{\; e^{\mathcal{G}_{\mathrm{ac}}[\bar{\Phi}]} = \frac{1}{\mathcal{Z}_0} \int \mathcal{D}[\Phi] e^{-S_0[\Phi]-S_1[\Phi+\bar{\Phi}]} \;,}$$

(6.40)

where $\bar{\Phi}$ is an auxiliary field of the same type as Φ which is not integrated over. The amputated connected Green functions $G^{(n)}_{\mathrm{ac},\alpha_1\ldots\alpha_n}$ with n external legs are then defined via

$$G^{(n)}_{\mathrm{ac},\alpha_1\ldots\alpha_n} = \left. \frac{\delta^n \mathcal{G}_{\mathrm{ac}}[\bar{\Phi}]}{\delta\bar{\Phi}_{\alpha_n} \ldots \delta\bar{\Phi}_{\alpha_1}} \right|_{\bar{\Phi}=0} = \qquad\qquad\qquad ,$$

(6.41)

or equivalently via the functional Taylor expansion,

$$\mathcal{G}_{\mathrm{ac}}[\bar{\Phi}] = \sum_{n=0}^{\infty} \frac{1}{n!} \int_{\alpha_1} \cdots \int_{\alpha_n} G^{(n)}_{\mathrm{ac},\alpha_1\ldots\alpha_n} \bar{\Phi}_{\alpha_1} \ldots \bar{\Phi}_{\alpha_n} \,.$$

(6.42)

To understand why the functions $\mathcal{G}^{(n)}_{\mathrm{ac},\alpha_1\ldots\alpha_n}$ are called *amputated*, let us derive the relation between the generating functionals $\mathcal{G}_{\mathrm{ac}}[\bar{\Phi}]$ and $\mathcal{G}_c[J]$. Therefore, we shift the integration variables Φ in the definition (6.40) via the substitution $\Phi' = \Phi + \bar{\Phi}$, so that

$$e^{\mathcal{G}_{\mathrm{ac}}[\bar{\Phi}]} = \frac{1}{\mathcal{Z}_0} \int \mathcal{D}[\Phi'] e^{-S_0[\Phi'-\bar{\Phi}]-S_1[\Phi']} \,.$$

(6.43)

The Gaussian action with shifted argument can be written as

$$S_0[\Phi' - \bar{\Phi}] = -\frac{1}{2}\big(\Phi' - \bar{\Phi}, \mathbf{G}_0^{-1}(\Phi' - \bar{\Phi})\big)$$

$$= -\frac{1}{2}\big(\Phi', \mathbf{G}_0^{-1}\Phi'\big) - \frac{1}{2}\big(\bar{\Phi}, \mathbf{G}_0^{-1}\bar{\Phi}\big) + \frac{1}{2}\big(\bar{\Phi}, \mathbf{G}_0^{-1}\Phi'\big) + \frac{1}{2}\big(\Phi', \mathbf{G}_0^{-1}\bar{\Phi}\big) \,.$$

(6.44)

The last two terms in the second line are identical because

$$\left(\Phi', \mathbf{G}_0^{-1}\bar{\Phi}\right) = \left(\left[\mathbf{G}_0^{-1}\right]^T \Phi', \bar{\Phi}\right) = \left(\mathbf{Z}\mathbf{G}_0^{-1}\Phi', \bar{\Phi}\right)$$

$$= \int_\alpha \zeta_\alpha \left[\mathbf{G}_0^{-1}\Phi'\right]_\alpha \bar{\Phi}_\alpha = \int_\alpha \bar{\Phi}_\alpha \left[\mathbf{G}_0^{-1}\Phi'\right]_\alpha = \left(\bar{\Phi}, \mathbf{G}_0^{-1}\Phi'\right) . \qquad (6.45)$$

Setting the auxiliary field to the value $\bar{\Phi} = -\mathbf{G}_0^T J$, the shifted Gaussian action (6.44) can be written as

$$S_0\left[\Phi' + \mathbf{G}_0^T J\right] = -\frac{1}{2}\left(\Phi', \mathbf{G}_0^{-1}\Phi'\right) - \frac{1}{2}\left(J, \mathbf{G}_0^T J\right) - (J, \Phi') . \qquad (6.46)$$

The representation of $\mathcal{G}_{\mathrm{ac}}[\bar{\Phi}]$ given in Eq. (6.43) can therefore be transformed into

$$e^{\mathcal{G}_{\mathrm{ac}}[-\mathbf{G}_0^T J]} = e^{\frac{1}{2}(J,\mathbf{G}_0^T J)} \underbrace{\frac{1}{\mathcal{Z}_0}\int \mathcal{D}[\Phi']e^{-S_0[\Phi']-S_1[\Phi']+(J,\Phi')}}_{e^{\mathcal{G}_c[J]}} , \qquad (6.47)$$

which implies

$$\mathcal{G}_c[J] = \mathcal{G}_{\mathrm{ac}}\left[-\mathbf{G}_0^T J\right] - \frac{1}{2}\left(J, \mathbf{G}_0^T J\right) . \qquad (6.48)$$

Substituting again $-\mathbf{G}_0^T J = \bar{\Phi}$ we obtain the following relation between the generating functionals of the amputated connected and the connected Green functions,

$$\boxed{\;\mathcal{G}_{\mathrm{ac}}[\bar{\Phi}] = \mathcal{G}_c\left[-\left(\mathbf{G}_0^T\right)^{-1}\bar{\Phi}\right] + \frac{1}{2}\left(\bar{\Phi}, \mathbf{G}_0^{-1}\bar{\Phi}\right) .\;} \qquad (6.49)$$

The last term simply subtracts the Gaussian part of $\mathcal{G}_c$, so that for a free field theory (where $S_1 = 0$) the generating functional of the amputated connected Green functions vanishes identically. To understand why the functional derivatives of $\mathcal{G}_{\mathrm{ac}}[\bar{\Phi}]$ in Eq. (6.49) can be identified with amputated connected Green functions, let us consider the second functional derivatives $G^{(2)}_{\mathrm{ac},\alpha\alpha'}$, which we collect into a superfield matrix

$$\mathbf{G}^{(2)}_{\mathrm{ac}} = \left(\frac{\delta}{\delta\bar{\Phi}} \otimes \frac{\delta}{\delta\bar{\Phi}}\, \mathcal{G}_{\mathrm{ac}}[\bar{\Phi}]\right)^T_{\bar{\Phi}=0} . \qquad (6.50)$$

Using Eq. (6.49) and the fact that according to Eq. (6.32) the negative second functional derivative of $\mathcal{G}_c[J]$ at vanishing sources can be identified with the exact superfield propagator $\mathbf{G}$, it is easy to show that

$$\begin{aligned}
\mathbf{G}^{(2)}_{\mathrm{ac}} &= \mathbf{G}_0^{-1} - \mathbf{G}_0^{-1}\mathbf{G}\mathbf{G}_0^{-1} \\
&= -[\mathbf{\Sigma} + \mathbf{\Sigma}\mathbf{G}_0\mathbf{\Sigma} + \mathbf{\Sigma}\mathbf{G}_0\mathbf{\Sigma}\mathbf{G}_0\mathbf{\Sigma} + \ldots] ,
\end{aligned} \qquad (6.51)$$

where in the second line we have substituted the Dyson equation (6.34) in the form $\mathbf{G}_0^{-1}\mathbf{G} = 1 + \mathbf{\Sigma}\mathbf{G}$. The relation between the exact propagator $\mathbf{G}$ and $\mathbf{G}_{ac}^{(2)}$ is therefore

$$\mathbf{G} = \mathbf{G}_0 - \mathbf{G}_0\mathbf{G}_{ac}^{(2)}\mathbf{G}_0 \ . \tag{6.52}$$

The relations (6.51) and (6.52) are shown graphically in Fig. 6.4. Obviously, the external Gaussian propagator legs appearing in the diagrams of the perturbation series of the connected Green function $\mathbf{G}$ are amputated in $\mathbf{G}_{ac}^{(2)}$, which explains the name. For the higher-order amputated connected Green functions with more than two external legs the term $\left(\mathbf{G}_0^T\right)^{-1}$ in the argument of $\mathcal{G}_c\left[-\left(\mathbf{G}_0^T\right)^{-1}\bar{\Phi}\right]$ in Eq. (6.49) generates a similar amputation of the Gaussian propagators of all externals legs.

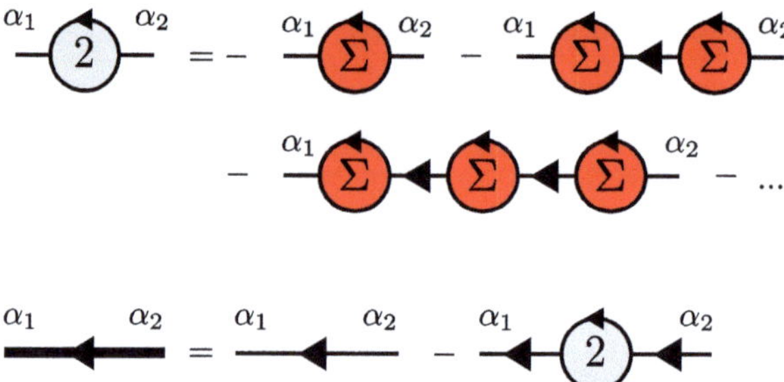

Fig. 6.4 The *upper graph* represents the relation (6.51) between the amputated connected Green function with two external legs $\mathbf{G}_{ac}^{(2)}$ and the irreducible self-energy $\mathbf{\Sigma}$. The *lower graph* represents the relation (6.52) between the connected Green function $\mathbf{G}$ and $\mathbf{G}_{ac}^{(2)}$

Next, let us derive a representation of $\mathcal{G}_{ac}[\bar{\Phi}]$ in terms of a functional differential operator which is analogous to the corresponding identity (6.39) for $\mathcal{G}_c[J]$. With the help of the identity

$$e^{-S_1[\Phi+\bar{\Phi}]} = e^{-S_1\left[\frac{\delta}{\delta J}\right]}e^{(J,\Phi+\bar{\Phi})}\Bigg|_{J=0}, \tag{6.53}$$

we obtain from the definition (6.40),

$$e^{\mathcal{G}_{ac}[\bar{\Phi}]} = e^{-S_1\left[\frac{\delta}{\delta J}\right]}\underbrace{\frac{1}{\mathcal{Z}_0}\int \mathcal{D}[\Phi]e^{-S_0[\Phi]+(J,\Phi+\bar{\Phi})}}_{e^{-\frac{1}{2}\left(J,\mathbf{G}_0^T J\right)+(J,\bar{\Phi})}}\Bigg|_{J=0}. \tag{6.54}$$

Using the identity

$$e^{-\frac{1}{2}(J,\mathbf{G}_0^T J)+(J,\bar{\Phi})} = e^{-\frac{1}{2}\left(\mathbf{Z}\frac{\delta}{\delta\bar{\Phi}},\mathbf{G}_0^T\mathbf{Z}\frac{\delta}{\delta\bar{\Phi}}\right)}e^{(J,\bar{\Phi})} = e^{-\frac{1}{2}\left(\frac{\delta}{\delta\bar{\Phi}},\mathbf{G}_0^T\frac{\delta}{\delta\bar{\Phi}}\right)}e^{(J,\bar{\Phi})} , \tag{6.55}$$

we see that Eq. (6.54) can be written in the form

$$e^{\mathcal{G}_{ac}[\bar{\Phi}]} = e^{-\frac{1}{2}\left(\frac{\delta}{\delta\Phi},\mathbf{G}_0^T\frac{\delta}{\delta\Phi}\right)}\underbrace{e^{-S_1\left[\frac{\delta}{\delta J}\right]}e^{(J,\bar{\Phi})}\Big|_{J=0}}_{e^{-S_1[\bar{\Phi}]}} . \qquad (6.56)$$

We thus conclude that the generating functional of amputated connected Green functions can be written as

$$\boxed{\,e^{\mathcal{G}_{ac}[\bar{\Phi}]} = \frac{1}{\mathcal{Z}_0}\int \mathcal{D}[\Phi]e^{-S_0[\Phi]-S_1[\Phi+\bar{\Phi}]} = e^{-\frac{1}{2}\left(\frac{\delta}{\delta\Phi},\mathbf{G}_0^T\frac{\delta}{\delta\Phi}\right)}e^{-S_1[\bar{\Phi}]} .\,} \qquad (6.57)$$

In Sect. 7.2.3 we shall derive an exact FRG equation for a modified functional $\mathcal{G}_{ac}[\bar{\Phi}]$ where we "switch off" the free propagator, $\mathbf{G}_0 \rightarrow 0$. From Eq. (6.57) it is obvious that in this limit $\mathcal{G}_{ac}[\bar{\Phi}] \rightarrow -S_1[\bar{\Phi}]$ which turns out to be a convenient boundary condition for solving the FRG equations.

Finally, combining Eq. (6.57) with the relation (6.48) between $\mathcal{G}_{ac}[\bar{\Phi}]$ and $\mathcal{G}_c[J]$, we obtain another representation of the generating functional of connected Green functions in terms of a functional differential operator,

$$\boxed{\,e^{\mathcal{G}_c[J]} = e^{-\frac{1}{2}\left(J,\mathbf{G}_0^T J\right)}\left[e^{-\frac{1}{2}\left(\frac{\delta}{\delta\Phi},\mathbf{G}_0^T\frac{\delta}{\delta\Phi}\right)}e^{-S_1[\bar{\Phi}]}\right]_{\bar{\Phi}=-\mathbf{G}_0^T J} .\,} \qquad (6.58)$$

Note that the differential operator in the exponent on the right-hand side of Eq. (6.58) involves only two functional derivatives, in contrast to the differential operator $S_1\left[\frac{\delta}{\delta J}\right]$ in Eq. (6.39).

6.2 One-Line Irreducible Vertices

From Figs. 6.3 and 6.4 it is obvious that the perturbative expansion of the connected Green functions $G_{c,\alpha_1\dots\alpha_n}^{(n)}$ as well as their amputated connected counterparts $G_{ac,\alpha_1\dots\alpha_n}^{(n)}$ contain subclasses of diagrams which can be separated into two parts by cutting a single line associated with a Gaussian propagator. As already explained in Sect. 4.1.3, these diagrams are called *one-line reducible*. It turns out that the isolated propagators connecting different blocks of reducible diagrams can lead to technical complications in FRG flow equations, in particular if one works with a sharp cutoff in momentum space (Morris 1994). These complications can be avoided if one decomposes the connected Green functions $G_{c,\alpha_1\dots\alpha_n}^{(n)}$ into *irreducible vertices* $\Gamma_{\alpha_1\dots\alpha_n}^{(n)}$ and formulates FRG flow equations directly for these vertices. By definition, the diagrams contributing to $\Gamma_{\alpha_1\dots\alpha_n}^{(n)}$ cannot be separated into two parts by cutting a single propagator line. For $n = 2$ this is the graphical definition of the irreducible self-energy, as discussed in Sect. 4.1.3. In order to properly define the higher-order irreducible vertices, it is useful to construct the associated generating functional $\Gamma[\bar{\Phi}]$. In the following we first define this functional and explain its physical meaning. We then derive in Sect. 6.2.2 the so-called *tree expansion* giving the decomposition of connected Green functions into irreducible vertices.

6.2.1 Generating Functional of the Irreducible Vertices

We would like to construct a functional $\Gamma[\bar\Phi]$ of some superfield variables $\bar\Phi_\alpha$ such that the irreducible vertices with n external legs can be expressed as usual via derivatives of this functional,

$$\Gamma^{(n)}_{\alpha_1\ldots\alpha_n} = \left.\frac{\delta^n \Gamma[\bar\Phi]}{\delta\bar\Phi_{\alpha_n}\cdots\delta\bar\Phi_{\alpha_1}}\right|_{\bar\Phi=0} = \ . \tag{6.59}$$

We shall show shortly that the variables $\bar\Phi_\alpha$ can be identified physically with the expectation values $\langle\Phi_\alpha\rangle$ of our superfields Φ_α in the presence of external sources J. We represent $\Gamma^{(n)}_{\alpha_1\ldots\alpha_n}$ graphically by an oriented shaded circle with n short external legs, where the arrow attached to the circle points again to the first index and defines the order of the external legs as in the graph for $G^{(n)}_{c,\alpha_1\ldots\alpha_n}$ shown in Fig. 6.1. The functional Taylor expansion equivalent to Eq. (6.59) is

$$\Gamma[\bar\Phi] = \sum_{n=0}^{\infty} \frac{1}{n!} \int_{\alpha_1}\cdots\int_{\alpha_n} \Gamma^{(n)}_{\alpha_1\ldots\alpha_n}\,\bar\Phi_{\alpha_1}\cdots\bar\Phi_{\alpha_n}\ . \tag{6.60}$$

It turns out that the generating functional $\Gamma[\Phi]$ of the irreducible vertices can be constructed from the generating functional $\mathcal{G}_c[J]$ of the connected Green function via a (functional) Legendre transformation as follows,

$$\boxed{\ \Gamma[\bar\Phi] = \mathcal{L}[\bar\Phi] - S_0[\bar\Phi] = (J[\bar\Phi],\,\bar\Phi) - \mathcal{G}_c[J[\bar\Phi]] + \frac{1}{2}\left(\bar\Phi,\mathbf{G}_0^{-1}\bar\Phi\right)\ .\ } \tag{6.61}$$

The functional Legendre transform $\mathcal{L}[\bar\Phi]$ of $\mathcal{G}_c[J]$ is defined by

$$\mathcal{L}[\bar\Phi] = (J,\,\bar\Phi) - \mathcal{G}_c[J[\bar\Phi]]\ , \tag{6.62}$$

where it is understood that on the right-hand side one should substitute $J = J[\bar\Phi]$ as a functional of the variables $\bar\Phi$ by inverting the relations

$$\bar\Phi_\alpha \equiv \langle\Phi_\alpha\rangle = \frac{\delta\mathcal{G}_c[J]}{\delta J_\alpha}\ . \tag{6.63}$$

The proof that the functional $\Gamma[\bar\Phi]$ defined in Eq. (6.61) indeed generates the irreducible vertices is based on the explicit construction of the relations between the connected Green functions $G^{(n)}_{c,\alpha_1\ldots\alpha_n}$ and the vertices $\Gamma^{(n)}_{\alpha_1\ldots\alpha_n}$ defined via the functional Taylor expansion of $\Gamma[\bar\Phi]$ in Eq. (6.60). In order to derive these relations, let us take the functional derivative of the Legendre transform $\mathcal{L}[\bar\Phi]$ of the generating functional $\mathcal{G}_c[J]$ in Eq. (6.62) with respect to the field components $\bar{\Phi}_\alpha$. Taking into account that the fermionic components of the sources J_α anticommute with $\bar{\Phi}_\alpha$, the chain rule gives

$$\frac{\delta\mathcal{L}[\bar{\Phi}]}{\delta\bar{\Phi}_\alpha} = \zeta_\alpha J_\alpha + \left(\frac{\delta J}{\delta\bar{\Phi}_\alpha},\bar{\Phi}\right) - \int_{\alpha'}\frac{\delta J_{\alpha'}}{\delta\bar{\Phi}_\alpha}\underbrace{\frac{\delta\mathcal{G}_c[J]}{\delta J_{\alpha'}}}_{\bar{\Phi}_{\alpha'}} = \zeta_\alpha J_\alpha\ . \tag{6.64}$$

With the help of the statistics matrix $\mathbf{Z}$ defined in Eq. (6.5) the relation (6.64) can be written in compact form as an identity between supervectors,

$$J = \mathbf{Z}\frac{\delta\mathcal{L}[\bar{\Phi}]}{\delta\bar{\Phi}}\ , \tag{6.65}$$

whereas the complementary relation (6.63) can be written as

$$\bar{\Phi} = \frac{\delta\mathcal{G}_c[J]}{\delta J}\ . \tag{6.66}$$

Using the relation (6.64), the chain rule in superfield space becomes

$$\frac{\delta}{\delta\bar{\Phi}_\alpha} = \int_{\alpha'}\frac{\delta J_{\alpha'}}{\delta\bar{\Phi}_\alpha}\frac{\delta}{\delta J_{\alpha'}} = \int_{\alpha'}\frac{\delta^2\mathcal{L}[\bar{\Phi}]}{\delta\bar{\Phi}_\alpha\delta\bar{\Phi}_{\alpha'}}\zeta_{\alpha'}\frac{\delta}{\delta J_{\alpha'}}\ , \tag{6.67}$$

or in compact vector notation,

$$\frac{\delta}{\delta\bar{\Phi}} = \left[\frac{\delta}{\delta\bar{\Phi}}\otimes\frac{\delta}{\delta\bar{\Phi}}\mathcal{L}[\bar{\Phi}]\right]\mathbf{Z}\frac{\delta}{\delta J}\ , \tag{6.68}$$

where the matrix differential operator $\frac{\delta}{\delta\bar{\Phi}}\otimes\frac{\delta}{\delta\bar{\Phi}}$ is defined as in Eq. (6.29), i.e.,

$$\left[\frac{\delta}{\delta\bar{\Phi}}\otimes\frac{\delta}{\delta\bar{\Phi}}\right]_{\alpha\alpha'} = \frac{\delta}{\delta\bar{\Phi}_\alpha}\frac{\delta}{\delta\bar{\Phi}_{\alpha'}}\ . \tag{6.69}$$

Applying the differential operator (6.68) to both sides of Eq. (6.66), we obtain the following matrix identity in superfield space,

$$\boxed{1 = \left(\frac{\delta}{\delta\bar{\Phi}}\otimes\frac{\delta}{\delta\bar{\Phi}}\mathcal{L}[\bar{\Phi}]\right)\mathbf{Z}\left(\frac{\delta}{\delta J}\otimes\frac{\delta}{\delta J}\mathcal{G}_c[J]\right)\ .} \tag{6.70}$$

The relation (6.70) is the superfield generalization of the well-known fact in magnetic systems that the second derivative of the Helmholtz free energy $F(h)$ with respect to the external magnetic field is the inverse of the second derivative of the

Gibbs potential $G(M)$ with respect to the magnetization M. In fact, all of the above definitions have magnetic analogues.[3]

To prove that the functional $\Gamma[\bar{\Phi}]$ defined in Eq. (6.61) indeed generates the irreducible vertices, let us write Eq. (6.70) in the form

$$\boxed{\frac{\delta}{\delta J} \otimes \frac{\delta}{\delta J} \mathcal{G}_c[J] = \mathbf{Z}\left(\frac{\delta}{\delta\bar{\Phi}} \otimes \frac{\delta}{\delta\bar{\Phi}} \mathcal{L}[\bar{\Phi}]\right)^{-1}.} \qquad (6.71)$$

This relation expresses the second derivative of the generating functional $\mathcal{G}_c[J]$ of the connected Green functions in terms of the second derivative of its Legendre transform $\mathcal{L}[\bar{\Phi}] = \Gamma[\bar{\Phi}] + S_0[\bar{\Phi}]$. Our strategy is to expand both sides of Eq. (6.71) in powers of the fields to obtain the *tree expansion* of the connected Green functions $G^{(n)}_{c,\alpha_1\ldots\alpha_n}$ in terms of the irreducible vertices $\Gamma^{(m)}_{\alpha_1\ldots\alpha_m}$ with $m \leq n$. Before embarking on the general case, consider the special case $n = 2$, which can be obtained directly from Eq. (6.70) by setting the sources and the conjugate fields equal to zero after the differentiation. Using the fact that according to Eq. (6.32) the second functional derivative of $\mathcal{G}_c[J]$ for vanishing sources is simply the negative Green function $-\mathbf{G}$, we obtain from Eq. (6.70),

$$\left[\frac{\delta}{\delta\bar{\Phi}} \otimes \frac{\delta}{\delta\bar{\Phi}} \mathcal{L}[\bar{\Phi}]\right]_{\bar{\Phi}=0} = -\mathbf{Z}\mathbf{G}^{-1} = -[\mathbf{G}^T]^{-1}. \qquad (6.72)$$

Keeping in mind that $\Gamma[\bar{\Phi}] = \mathcal{L}[\bar{\Phi}] + \frac{1}{2}\left(\bar{\Phi}, \mathbf{G}_0^{-1}\bar{\Phi}\right)$ we obtain from Eq. (6.72),

$$\left[\frac{\delta}{\delta\bar{\Phi}} \otimes \frac{\delta}{\delta\bar{\Phi}} \Gamma[\bar{\Phi}]\right]_{\bar{\Phi}=0} = -[\mathbf{G}^T]^{-1} + [\mathbf{G}_0^T]^{-1} = \mathbf{\Sigma}^T, \qquad (6.73)$$

where we have used the Dyson equation (6.34). Taking matrix elements, and using the fact that by definition

$$\left[\frac{\delta}{\delta\bar{\Phi}} \otimes \frac{\delta}{\delta\bar{\Phi}} \Gamma[\bar{\Phi}]\right]_{\alpha_1\alpha_2,\bar{\Phi}=0} = \Gamma^{(2)}_{\alpha_2\alpha_1}, \qquad (6.74)$$

we conclude that the irreducible vertex with two external legs can be identified with the irreducible self-energy in superfield space,

---

[3] For a magnet with Hamiltonian $\hat{H} = \hat{H}_0 - h\hat{M}$ the role of a constant source J is played by the uniform magnetic field h, while the expectation value of the magnetization operator $\langle\hat{M}\rangle = M$ corresponds to $\langle\Phi\rangle = \bar{\Phi}$. The Helmholtz free energy $F(h) = -T\ln\mathrm{Tr}[e^{-\beta\hat{H}_0+\beta h\hat{M}}]$ corresponds to $-\mathcal{G}_c[J]$, while the Gibbs potential $G(M) = h(M)M + F(h(M))$ corresponds to $\mathcal{L}[\bar{\Phi}]$ defined in Eq. (6.62). The magnetic version of the relations (6.65) and (6.66) is $h = \partial G/\partial M$ and $M = -\partial F/\partial h$, while Eq. (6.70) corresponds to $1 = (\partial h/\partial M)(\partial M/\partial h) = (\partial^2 G/\partial M^2)(-\partial^2 F/\partial h^2)$.

$$\Gamma^{(2)}_{\alpha_1\alpha_2} = [\mathbf{\Sigma}]_{\alpha_1\alpha_2} =$$

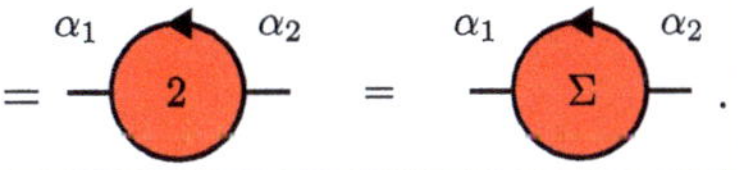

(6.75)

6.2.2 Tree Expansion

The irreducible vertices with more than two external legs characterize the interactions between the particles in the many-body system. For example, the vertex $\Gamma^{(4)}_{\alpha_1\alpha_2\alpha_3\alpha_4}$ with four external legs describes the true interaction between two particles in the many-body system and is called the *effective interaction*. As a special case of our general graphical notation for the irreducible vertices defined in Eq. (6.59) we represent the effective interaction by the following symbol,

$$\Gamma^{(4)}_{\alpha_1\alpha_2\alpha_3\alpha_4} =$$

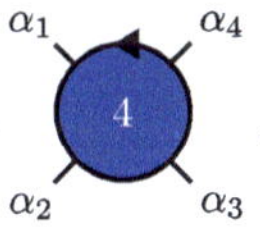

(6.76)

To obtain the general expansion of the connected Green functions in terms of irreducible vertices, it is convenient to rewrite the right-hand side of Eq. (6.71) in terms of the superfield matrix

$$\mathbf{U}[\bar{\Phi}] = \left(\frac{\delta}{\delta\bar{\Phi}} \otimes \frac{\delta}{\delta\bar{\Phi}} \Gamma[\bar{\Phi}]\right)^T - \left(\frac{\delta}{\delta\bar{\Phi}} \otimes \frac{\delta}{\delta\bar{\Phi}} \Gamma[\bar{\Phi}]\right)^T\Bigg|_{\bar{\Phi}=0}$$
$$= \left(\frac{\delta}{\delta\bar{\Phi}} \otimes \frac{\delta}{\delta\bar{\Phi}} \Gamma[\bar{\Phi}]\right)^T - \mathbf{\Sigma} , \tag{6.77}$$

which is a functional of the fields $\bar{\Phi}$ and satisfies $\mathbf{U}[\bar{\Phi} = 0] = 0$. Substituting the functional Taylor expansion (6.60) of $\Gamma[\bar{\Phi}]$ into Eq. (6.77) we obtain explicitly

$$\mathbf{U}[\bar{\Phi}] = \sum_{n=1}^{\infty} \frac{1}{n!} \int_{\alpha_1} \cdots \int_{\alpha_n} \mathbf{\Gamma}^{(n+2)}_{\alpha_1 \ldots \alpha_n} \bar{\Phi}_{\alpha_1} \ldots \bar{\Phi}_{\alpha_n} , \tag{6.78}$$

where we have introduced the supermatrices $\mathbf{\Gamma}^{(n+2)}_{\alpha_1 \ldots \alpha_n}$ with matrix elements

$$\left[\mathbf{\Gamma}^{(n+2)}_{\alpha_1 \ldots \alpha_n}\right]_{\alpha\alpha'} = \Gamma^{(n+2)}_{\alpha\alpha'\alpha_1 \ldots \alpha_n} . \tag{6.79}$$

The second functional derivative of $\mathcal{L}[\bar{\Phi}]$ whose inverse appears on the right-hand side of Eq. (6.71) can then be written as

$$\frac{\delta}{\delta\bar{\Phi}} \otimes \frac{\delta}{\delta\bar{\Phi}} \mathcal{L}[\bar{\Phi}] = \frac{\delta}{\delta\bar{\Phi}} \otimes \frac{\delta}{\delta\bar{\Phi}} \Gamma[\bar{\Phi}] - \left[\mathbf{G}_0^T\right]^{-1}$$
$$= \mathbf{U}^T[\bar{\Phi}] + \mathbf{\Sigma}^T - \left[\mathbf{G}_0^T\right]^{-1} = \mathbf{U}^T[\bar{\Phi}] - [\mathbf{G}^T]^{-1} . \tag{6.80}$$

Next we expand the inverse of this expression in powers of the matrix $\mathbf{U}^T[\bar{\Phi}]$,

$$
\left(\frac{\delta}{\delta\bar{\Phi}}\otimes\frac{\delta}{\delta\bar{\Phi}}\mathcal{L}[\bar{\Phi}]\right)^{-1}=\left(\mathbf{U}^T-[\mathbf{G}^T]^{-1}\right)^{-1}=\left((\mathbf{U}^T\mathbf{G}^T-\mathbf{1})[\mathbf{G}^T]^{-1}\right)^{-1}
$$

$$
=-\mathbf{G}^T\left[\mathbf{1}-\mathbf{U}^T\mathbf{G}^T\right]^{-1}=-\mathbf{G}^T\sum_{\nu=0}^{\infty}\left[\mathbf{U}^T\mathbf{G}^T\right]^{\nu}. \tag{6.81}
$$

Substituting this into Eq. (6.71) and using the fact that $\mathbf{Z}\mathbf{G}^T=\mathbf{G}$ we obtain

$$
\frac{\delta}{\delta J}\otimes\frac{\delta}{\delta J}\mathcal{G}_c[J]=\mathbf{Z}\left(\frac{\delta}{\delta\bar{\Phi}}\otimes\frac{\delta}{\delta\bar{\Phi}}\mathcal{L}[\bar{\Phi}]\right)^{-1}=-\mathbf{G}\sum_{\nu=0}^{\infty}\left[\mathbf{U}^T[\bar{\Phi}]\mathbf{G}^T\right]^{\nu}. \tag{6.82}
$$

Setting on the left-hand side the sources equal to zero after the differentiation we recover from the $\nu=0$ term on the right-hand side the identity (6.32) shown graphically in Fig. 6.2. To obtain the desired tree expansion of connected Green functions in terms of irreducible vertices, we now expand both sides of Eq. (6.82) in powers of the sources. For the left-hand side we obtain from the functional Taylor expansion of $\mathcal{G}_c[J]$ in Eq. (6.21),

$$
\frac{\delta}{\delta J}\otimes\frac{\delta}{\delta J}\mathcal{G}_c[J]=\sum_{n=0}^{\infty}\frac{1}{n!}\int_{\alpha_1}\cdots\int_{\alpha_n}\left(\mathbf{G}_{c,\alpha_1\ldots\alpha_n}^{(n+2)}\right)^T J_{\alpha_1}\ldots J_{\alpha_n}, \tag{6.83}
$$

where the supermatrix $\mathbf{G}_{c,\alpha_1\ldots\alpha_n}^{(n+2)}$ is defined as in Eq. (6.79),

$$
\left[\mathbf{G}_{c,\alpha_1\ldots\alpha_n}^{(n+2)}\right]_{\alpha\alpha'}=G_{c,\alpha\alpha'\alpha_1\ldots\alpha_n}^{(n+2)}. \tag{6.84}
$$

Note that with this notation the relation (6.31) simply reads $\mathbf{G}^T=-\mathbf{G}_c^{(2)}$. On the right-hand side of Eq. (6.82) we substitute the expansion (6.78) of the functional $\mathbf{U}[\bar{\Phi}]$ in powers of the fields $\bar{\Phi}$, so that Eq. (6.82) reduces to the identity

$$
\sum_{n=0}^{\infty}\frac{1}{n!}\int_{\alpha_1}\cdots\int_{\alpha_n}\left(\mathbf{G}_{c,\alpha_1\ldots\alpha_n}^{(n+2)}\right)^T J_{\alpha_1}\ldots J_{\alpha_n}
$$

$$
=-\mathbf{G}\sum_{\nu=0}^{\infty}\left[\sum_{n=1}^{\infty}\frac{1}{n!}\int_{\alpha_1}\cdots\int_{\alpha_n}\left(\mathbf{\Gamma}_{\alpha_1\ldots\alpha_n}^{(n+2)}\right)^T\mathbf{G}^T\bar{\Phi}_{\alpha_1}\ldots\bar{\Phi}_{\alpha_n}\right]^{\nu}
$$

$$
=-\sum_{\nu=0}^{\infty}\sum_{n_1=1}^{\infty}\cdots\sum_{n_\nu=1}^{\infty}\frac{1}{n_1!\ldots n_\nu!}\int_{\alpha_1^1}\cdots\int_{\alpha_{n_1}^1}\cdots\int_{\alpha_1^\nu}\cdots\int_{\alpha_{n_\nu}^\nu}
$$

$$
\times\mathbf{Z}\mathbf{G}^T\left(\mathbf{\Gamma}_{\alpha_1^1\ldots\alpha_{n_1}^1}^{(n_1+2)}\right)^T\mathbf{G}^T\ldots\left(\mathbf{\Gamma}_{\alpha_1^\nu\ldots\alpha_{n_\nu}^\nu}^{(n_\nu+2)}\right)^T\mathbf{G}^T\bar{\Phi}_{\alpha_1^1}\ldots\bar{\Phi}_{\alpha_{n_1}^1}\ldots\bar{\Phi}_{\alpha_1^\nu}\ldots\bar{\Phi}_{\alpha_{n_\nu}^\nu}.
$$

$$
\tag{6.85}
$$

We can now explicitly take the transposed of this matrix equation, resulting in

$$\sum_{n=0}^{\infty} \frac{1}{n!} \int_{\alpha_1} \cdots \int_{\alpha_n} \mathbf{G}_{c,\alpha_1\ldots\alpha_n}^{(n+2)} J_{\alpha_1} \ldots J_{\alpha_n}$$

$$= -\sum_{v=0}^{\infty} \sum_{n_1=1}^{\infty} \cdots \sum_{n_v=1}^{\infty} \frac{1}{n_1!\ldots n_v!} \int_{\alpha_1^1} \cdots \int_{\alpha_{n_1}^1} \cdots \int_{\alpha_1^v} \cdots \int_{\alpha_{n_v}^v}$$

$$\times \mathbf{G}\boldsymbol{\Gamma}_{\alpha_1^v\ldots\alpha_{n_1}^v}^{(n_v+2)} \mathbf{G} \ldots \boldsymbol{\Gamma}_{\alpha_1^1\ldots\alpha_{n_1}^1}^{(n_1+2)} \mathbf{GZ}\, \bar{\Phi}_{\alpha_1^1} \ldots \bar{\Phi}_{\alpha_{n_1}^1} \ldots \bar{\Phi}_{\alpha_1^v} \ldots \bar{\Phi}_{\alpha_{n_v}^v} \, . \tag{6.86}$$

It should be noted that the order of the supermatrices $\boldsymbol{\Gamma}_{\alpha_1\ldots\alpha_n}^{(n+2)}$ has now been reversed. Our strategy is to compare terms involving the same powers of the sources J_α on both sides of Eq. (6.86). Therefore, we still have to express the expectation values $\bar{\Phi}_\alpha = \langle \Phi_\alpha \rangle$ on the right-hand side in terms of the sources using the relation (6.63),

$$\bar{\Phi}_\alpha = \frac{\delta \mathcal{G}_c[J]}{\delta J_\alpha} = \sum_{n=0}^{\infty} \frac{1}{n!} \int_{\alpha_1} \cdots \int_{\alpha_n} G_{c,\alpha\alpha_1\ldots\alpha_n}^{(n+1)} J_{\alpha_1} \ldots J_{\alpha_n} \, . \tag{6.87}$$

Substituting this into Eq. (6.86) we can compare the vertices involving the same powers of the sources on both sides. However, we should be careful to identify terms with the same symmetry under permutation of the field labels. For example, if all fields are bosonic, then all Green functions $G_{c,\alpha_1\ldots\alpha_n}^{(n)}$ and vertices $\Gamma_{\alpha_1\ldots\alpha_n}^{(n)}$ can be chosen to be symmetric under an arbitrary permutation of the labels, while for fermion fields we require that all Green functions and vertices are antisymmetric. For theories involving both bosons and fermions, we require symmetry with respect to permutation of bosonic labels, and antisymmetry for fermionic labels. Before comparing the coefficients of a given power of the sources on both sides of Eq. (6.86), we should therefore properly symmetrize the right-hand side. Therefore it is convenient to introduce a symmetrization operator $\mathcal{S}$ in the following way: Consider any function $F_{\alpha_1\ldots\alpha_n}$ of $n = n_1 + n_2 + \ldots + n_v$ superlabels $\alpha_1, \ldots, \alpha_n$. We assume that the labels can be subdivided into $v \geq 1$ disjunct subsets $s_1 = \{\alpha_1, \ldots, \alpha_{n_1}\}$, $s_2 = \{\alpha_{n_1+1}, \ldots, \alpha_{n_2}\}, \ldots, s_v = \{\alpha_{n-n_v+1}, \ldots, \alpha_n\}$ such that the function $F_{\alpha_1\ldots\alpha_n}$ is already symmetrized with respect to permutations of the n_i labels in the index set $s_i, i = 1, \ldots, v$. The action of the operator $\mathcal{S}$ on the function $F_{\alpha_1\ldots\alpha_n}$ is then

$$\mathcal{S}_{\alpha_1\ldots\alpha_{n_1};\ldots;\alpha_{n-n_v+1}\ldots\alpha_n}\{F_{\alpha_1\ldots\alpha_n}\} = \frac{1}{n_1! \cdots n_v!} \sum_P \text{sgn}_\zeta(P)\, F_{\alpha_{P(1)}\ldots\alpha_{P(n)}} \, , \tag{6.88}$$

where P denotes a permutation of $(1, \ldots, n)$ and sgn_ζ is the sign created by permuting field variables according to the permutation P, i.e.,

$$\bar{\Phi}_{\alpha_1} \cdot \ldots \cdot \bar{\Phi}_{\alpha_n} = \text{sgn}_\zeta(P)\, \bar{\Phi}_{\alpha_{P(1)}} \cdot \ldots \cdot \bar{\Phi}_{\alpha_{P(n)}} \, . \tag{6.89}$$

The effect of $\mathcal{S}$ is rather simple: it acts on an expression already symmetric in the index groups separated by semi-colons to generate an expression symmetric also with respect to the exchange of indices between different groups. Totally the right-hand side of Eq. (6.88) thus contains $n!/(n_1! \cdot \ldots \cdot n_\nu!)$ terms.[4] With this notation, we can write down an analytic expression for the relation between connected Green functions and irreducible vertices resulting from the identification of the coefficients of the same symmetrized products of the sources on both sides of Eq. (6.86),

$$
\begin{aligned}
\mathbf{G}^{(n+2)}_{c,\alpha_1\ldots\alpha_n} = &-\sum_{\nu=0}^{\infty}\sum_{n_1=1}^{\infty}\cdots\sum_{n_\nu=1}^{\infty}\frac{1}{n_1!\ldots n_\nu!}\int_{\beta_1^1}\cdots\int_{\beta_{n_1}^1}\cdots\int_{\beta_1^\nu}\cdots\int_{\beta_{n_\nu}^\nu}\\
&\times\Big(\sum_{m_1^1=1}^{\infty}\cdots\sum_{m_{n_1}^1=1}^{\infty}\Big)\cdots\Big(\sum_{m_1^\nu=1}^{\infty}\cdots\sum_{m_{n_\nu}^\nu=1}^{\infty}\Big)\delta_{n,\sum_{i=1}^{\nu}\sum_{j=1}^{n_i}m_j^i}\\
&\times\mathbf{G}\boldsymbol{\Gamma}^{(n_\nu+2)}_{\beta_1^\nu\ldots\beta_{n_1}^\nu}\mathbf{G}\ldots\boldsymbol{\Gamma}^{(n_1+2)}_{\beta_1^1\ldots\beta_{n_1}^1}\mathbf{GZ}\\
&\times\mathcal{S}_{\alpha_1\ldots\alpha_{m_1^1};\ldots;\alpha_{n-m_{n_\nu}^\nu+1}\ldots\alpha_n}\Big\{\mathbf{G}^{(m_1^1+1)}_{c,\beta_1^1\alpha_1\ldots\alpha_{m_1^1}}\cdots\mathbf{G}^{(m_{n_\nu}^\nu+1)}_{c,\beta_{n_\nu}^\nu\alpha_{n-m_{n_\nu}^\nu+1}\ldots\alpha_n}\Big\}\,.
\end{aligned}
\tag{6.90}
$$

On the right-hand side of this rather cumbersome expression, only connected correlation functions with degree smaller than on the left-hand side appear. We can therefore recursively express all connected correlation functions via their one-line irreducible counterparts. This is the tree expansion. For fixed n, only a finite number of terms contribute on the right-hand side of Eq. (6.90). The simplest case is $n = 0$, where all summations and integrations can be omitted and Eq. (6.90) reduces to $\mathbf{G}^{(2)}_c = -\mathbf{G}^T$. This relation has already been derived earlier in Eq. (6.32) and is shown graphically in Fig. 6.2. For $n = 1$ only the single term with $\nu = 1, n_1 = 1$, $m_1^1 = 1$ contributes on the right-hand side of Eq. (6.90). Explicitly, the tree expansion of the connected Green function with three external legs can be written as

$$
G^{(3)}_{c,\alpha_1\alpha_2\alpha_3} = \int_{\beta_1}\int_{\beta_2}\int_{\beta_3}[\mathbf{G}]_{\alpha_1\beta_1}[\mathbf{G}]_{\alpha_2\beta_2}[\mathbf{G}]_{\alpha_3\beta_3}\Gamma^{(3)}_{\beta_1\beta_2\beta_3}\,,
\tag{6.91}
$$

which is shown graphically in Fig. 6.5. The tree expansion of the connected Green function with four external legs, corresponding to $n = 2$ in Eq. (6.90), is more complicated. In this case the following three terms in the nested sum on the right-hand side of Eq. (6.90) contribute,

term	ν	n_i	m_j^i
1.	1	$n_1 = 1$	$m_1^1 = 2$
2.	1	$n_1 = 2$	$m_1^1 = m_2^1 = 1$
3.	2	$n_1 = n_2 = 1$	$m_1^1 = m_1^2 = 1$.

---

[4] For example, the symmetrized form of a function $F_{\alpha_1\alpha_2}$ which is not symmetric with respect to the exchange $\alpha_1 \leftrightarrow \alpha_2$ is $\mathcal{S}_{\alpha_1;\alpha_2}\{F_{\alpha_1\alpha_2}\} = F_{\alpha_1\alpha_2} + \mathrm{sgn}_\zeta(P_{12})F_{\alpha_2\alpha_1}$, which corresponds to $\nu = n = 2$ and $n_1 = n_2 = 1$ in Eq. (6.88). On the other hand, if the function $F_{\alpha_1\alpha_2}$ is already symmetrized, then $\mathcal{S}_{\alpha_1\alpha_2}\{F_{\alpha_1\alpha_2}\} = F_{\alpha_1\alpha_2}$, corresponding to $\nu = 1$ and $n = n_1 = 2$.

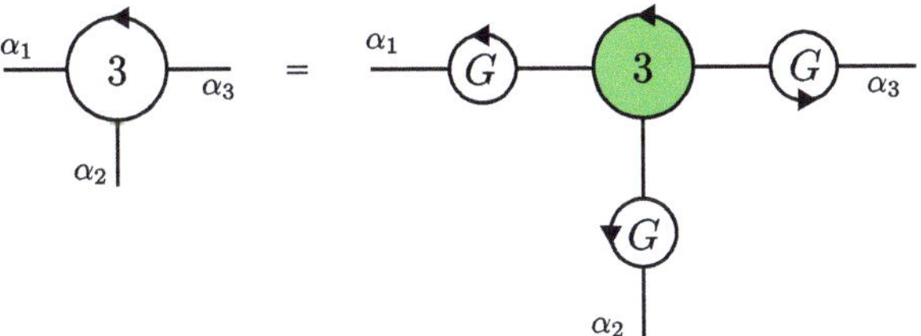

Fig. 6.5 Graphical representation of the tree expansion for the connected Green function with three external legs given in Eq. (6.91). Connected Green functions are drawn as *empty oriented circles* with the indicated number of external legs, as defined in Fig. 6.1. The irreducible vertices are represented by *shaded oriented circles*, see Eq. (6.59)

The resulting tree expansion of the connected four-point function in terms of irreducible vertices is explicitly,

$$
\begin{aligned}
G^{(4)}_{c,\alpha_1\alpha_2\alpha_3\alpha_4} = &- \int_{\beta_1}\cdots\int_{\beta_4} [\mathbf{G}]_{\alpha_1\beta_1}[\mathbf{G}]_{\alpha_2\beta_2}[\mathbf{G}]_{\alpha_3\beta_3}[\mathbf{G}]_{\alpha_4\beta_4}\,\Gamma^{(4)}_{\beta_1\beta_2\beta_3\beta_4} \\
&- \int_{\beta_1}\cdots\int_{\beta_6} [\mathbf{G}]_{\alpha_1\beta_1}[\mathbf{G}]_{\alpha_2\beta_2}[\mathbf{G}]_{\alpha_3\beta_3}[\mathbf{G}]_{\alpha_4\beta_4}\,\Gamma^{(3)}_{\beta_1\beta_2\beta_5}[\mathbf{G}]_{\beta_5\beta_6}\,\Gamma^{(3)}_{\beta_6\beta_3\beta_4} \\
&- \int_{\beta_1}\cdots\int_{\beta_6} \mathcal{S}_{\alpha_1;\alpha_2}\left\{[\mathbf{G}]_{\alpha_1\beta_1}[\mathbf{G}]_{\alpha_2\beta_2}[\mathbf{G}]_{\alpha_3\beta_3}[\mathbf{G}]_{\alpha_4\beta_4}\,\Gamma^{(3)}_{\beta_1\beta_5\beta_4}[\mathbf{G}]_{\beta_5\beta_6}\,\Gamma^{(3)}_{\beta_6\beta_2\beta_3}\right\},
\end{aligned}
$$

$$(6.92)$$

which is shown graphically in Fig. 6.6. From Figs. 6.5 and 6.6 it should now be obvious why the expansion (6.90) is called *tree expansion*: The diagrams representing this expansion consist of vertices $\Gamma^{(n)}$ linked via full propagator lines without loops. Diagrams of this type are called trees. The fact that the vertices $\Gamma^{(n)}$ are one-line irreducible follows from the fact that by construction all one-line irreducible diagrams appear as branches of the tree, so that the vertices where the branches end must be one-line irreducible.

Finally, let us point out that in some textbooks (Negele and Orland 1988, Rammer 2007) the tree expansion is derived graphically by taking higher-order derivatives of the fundamental relation (6.70) between the functionals $\mathcal{G}_c[J]$ and $\mathcal{L}[\bar{\Phi}]$. With the help of our compact notation we have been able to derive an explicit analytic expression for the tree expansion given by Eq. (6.90).

6.3 Symmetries

The vertices $\Gamma^{(n)}_{\alpha_1\ldots\alpha_n}$ defined above are rather abstract objects and depend in a complicated manner on many superlabels. Those labels are in turn composed of multiple variables: a space-time position $X = (\tau, \boldsymbol{x})$ or $K = (i\omega, \boldsymbol{k})$ after Fourier transformation, as well as internal labels, e.g., for field type and spin. For practical calculations,

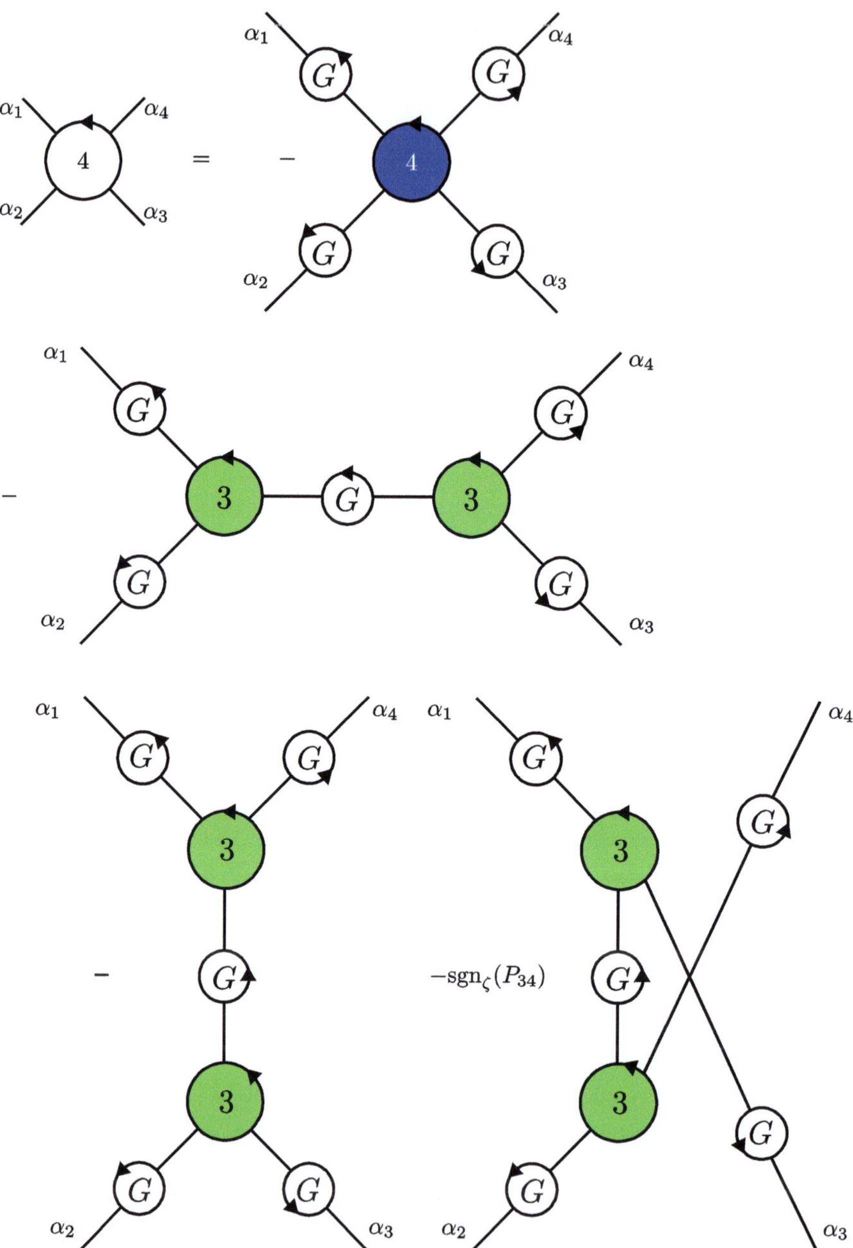

Fig. 6.6 Graph of the tree expansion of the connected Green function $G^{(4)}_{c,\alpha_1\alpha_2\alpha_3\alpha_4}$ with four external legs given in Eq. (6.92). The symbol $\mathrm{sgn}_\zeta(P_{34})$ denotes the parity under the exchange of the labels α_3 and α_4, i.e., -1 for fermions and $+1$ for bosons or one fermion and one boson leg, see Eqs. (6.88) and (6.89)

it is often helpful to use symmetry arguments to simplify the functional form of these
dependencies.

6.3.1 General Properties

We define a symmetry transformation $\mathcal{F}$ as an invertible linear mapping between
the fields,

$$\mathcal{F}: \quad \Phi \mapsto \mathbf{F}\Phi , \qquad [\mathbf{F}\Phi]_\alpha = \int_\beta F_{\alpha\beta}\Phi_\beta , \tag{6.93}$$

that has the additional property of leaving the action and the functional integration
measure invariant, i.e.

$$S[\mathbf{F}\Phi] = S[\Phi] , \tag{6.94}$$

for all configurations of the field Φ, and $\mathcal{D}[\mathbf{F}\Phi] = \mathcal{D}[\Phi]$. We exclude symmetry
breaking in this section, such that the interaction does not contain counterterms that
explicitly break a symmetry. The invariance property in Eq. (6.94) is then valid
separately for the bare interaction S_0 and the interaction part S_1. For a given system,
the set of all symmetry transformations forms the symmetry group G. We can distinguish between space-time and internal symmetries, if $\mathcal{F}$ only acts on space-time or
internal labels, respectively. Furthermore, if G (or a subgroup) only contains a finite
or countably infinite number of elements, the symmetry is discrete. For a continuous
symmetry on the other hand, we can choose a parametrization $\mathbf{F}(l_1, \ldots, l_m) = \mathbf{F}(l)$
with continuous parameters l_i. In all relevant cases, this dependence is differentiable
and we can choose $\mathbf{F}(l) \to \mathbf{1}$ for $l \to 0$.

For a given symmetry transformation $\mathcal{F}$, we can change integration variables
according to Eq. (6.93) in the functional integral for the generating functional to
show that $\mathcal{G}[J]$ is also invariant under the transformation,

$$\mathcal{G}[J] = \frac{1}{\mathcal{Z}} \int \mathcal{D}[\Phi] e^{-S[\Phi]+(J,\Phi)} = \frac{1}{\mathcal{Z}} \int \mathcal{D}[\mathbf{F}\Phi] e^{-S[\mathbf{F}\Phi]+(J,\mathbf{F}\Phi)}$$

$$= \frac{1}{\mathcal{Z}} \int \mathcal{D}[\Phi] e^{-S[\Phi]+(\mathbf{F}^T J,\Phi)} = \mathcal{G}[\mathbf{F}^T J] . \tag{6.95}$$

Consequently,

$$\mathcal{G}_c[\mathbf{F}^T J] = \ln\left(\frac{\mathcal{Z}}{\mathcal{Z}_0}\mathcal{G}[\mathbf{F}^T J]\right) = \ln\left(\frac{\mathcal{Z}}{\mathcal{Z}_0}\mathcal{G}[J]\right) = \mathcal{G}_c[J] . \tag{6.96}$$

Similarly,

$$e^{\mathcal{G}_{\mathrm{ac}}[\bar{\Phi}]} = \frac{1}{Z_0} \int \mathcal{D}[\Phi] e^{-S_0[\Phi]-S_1[\Phi+\bar{\Phi}]} = \frac{1}{Z_0} \int \mathcal{D}[\mathbf{F}\Phi] e^{-S_0[\mathbf{F}\Phi]-S_1[\mathbf{F}\Phi+\bar{\Phi}]}$$

$$= \frac{1}{Z_0} \int \mathcal{D}[\Phi] e^{-S_0[\Phi]-S_1[\Phi+\mathbf{F}^{-1}\bar{\Phi}]} = e^{\mathcal{G}_{\mathrm{ac}}[\mathbf{F}^{-1}\bar{\Phi}]} \, , \qquad (6.97)$$

i.e., $\mathcal{G}_{\mathrm{ac}}[\bar{\Phi}] = \mathcal{G}_{\mathrm{ac}}[\mathbf{F}^{-1}\bar{\Phi}]$. Finally, in order to show a similar relation for the generating functional Γ of the one-line irreducible vertices, consider the symmetry transformation of the relation $\bar{\Phi}[J] = \frac{\delta \mathcal{G}_c[J]}{\delta J}$ which is given by

$$\bar{\Phi}[\mathbf{F}^T J] = \left. \frac{\delta \mathcal{G}_c[\tilde{J}]}{\delta \tilde{J}} \right|_{\tilde{J}=\mathbf{F}^T J} = \mathbf{F}^{-1} \frac{\delta}{\delta J} \left(\mathcal{G}_c[\mathbf{F}^T J] \right) = \mathbf{F}^{-1} \frac{\delta \mathcal{G}_c[J]}{\delta J}$$

$$= \mathbf{F}^{-1} \bar{\Phi}[J] \, . \qquad (6.98)$$

For the inverse relation, this implies $J[\mathbf{F}^{-1}\bar{\Phi}] = \mathbf{F}^T J[\bar{\Phi}]$. Using this in the definition of the effective action, we obtain

$$\mathcal{L}[\mathbf{F}^{-1}\bar{\Phi}] = (J[\mathbf{F}^{-1}\bar{\Phi}], \mathbf{F}^{-1}\bar{\Phi}) - \mathcal{G}_c[J[\mathbf{F}^{-1}\bar{\Phi}]] = (\mathbf{F}^T J[\bar{\Phi}], \mathbf{F}^{-1}\bar{\Phi}) - \mathcal{G}_c[\mathbf{F}^T J[\bar{\Phi}]]$$

$$= (J[\bar{\Phi}], \bar{\Phi}) - \mathcal{G}_c[J[\bar{\Phi}]] = \mathcal{L}[\bar{\Phi}] \, . \qquad (6.99)$$

Finally, this implies,

$$\Gamma[\mathbf{F}^{-1}\bar{\Phi}] = \mathcal{L}[\mathbf{F}^{-1}\bar{\Phi}] - S_0[\mathbf{F}^{-1}\bar{\Phi}] = \mathcal{L}[\bar{\Phi}] - S_0[\bar{\Phi}] = \Gamma[\bar{\Phi}] \, . \qquad (6.100)$$

Summarizing, the generating functionals are invariant under all transformations $\mathcal{F}$ of the symmetry group, i.e.,

$$\boxed{\begin{aligned} \mathcal{G}[\mathbf{F}^T J] &= \mathcal{G}[J] \, , \\ \mathcal{G}_c[\mathbf{F}^T J] &= \mathcal{G}_c[J] \, , \\ \mathcal{G}_{\mathrm{ac}}[\mathbf{F}\bar{\Phi}] &= \mathcal{G}_{\mathrm{ac}}[\bar{\Phi}] \, , \\ \Gamma[\mathbf{F}\bar{\Phi}] &= \Gamma[\bar{\Phi}] \, . \end{aligned}} \qquad (6.101)$$

Expanding both sides of these relations for the generating functionals in powers of the fields, we obtain relations for the vertices, e.g., for the one-line irreducible vertices,

$$\Gamma^{(n)}_{\alpha_1 \ldots \alpha_n} = \int_{\beta_1} \cdots \int_{\beta_n} \Gamma^{(n)}_{\beta_1 \ldots \beta_n} F_{\beta_1 \alpha_1} \ldots F_{\beta_n \alpha_n} \, . \qquad (6.102)$$

For continuous symmetries, we can take derivatives of the symmetry relations in Eq. (6.101) with respect to the parameters l_i to obtain, e.g.,

$$\boxed{\left(\mathbf{T}^{(i)} \bar{\Phi}, \frac{\delta \Gamma}{\delta \bar{\Phi}} \right) = 0 \, ,}$$

(6.103)

where the generators are defined by

$$\mathbf{T}^{(i)} = -i \left. \frac{\partial \mathbf{F}}{\partial l_i} \right|_{l=0} .$$

(6.104)

6.3.2 Nonrelativistic Particles

To illustrate the most important symmetries, we will restrict ourselves in this subsection to a system of nonrelativistic particles with an action of the form given in Eq. (6.6). We now explicitly discuss some important symmetry transformations of this system:

(a) $U(1)$-*gauge transformation:* A global gauge transformation is given by

$$\mathcal{F}(\alpha): \quad \psi_\sigma(X) \mapsto e^{i\alpha} \psi_\sigma(X) \, , \qquad \bar{\psi}_\sigma(X) \mapsto e^{-i\alpha} \bar{\psi}_\sigma(X) \, ,$$

(6.105)

where the phase factor α is independent of the space-time point $X = (\tau, \boldsymbol{x})$. In quantum field theory, this transformation is promoted to a local symmetry with position-dependent gauge fields by including gauge fields into the theory. Nonrelativistic systems are in general not invariant under local gauge transformations. However, in Sects. 11.4 and 12.4 we will see how local gauge transformations can nevertheless be used to derive important Ward identities between vertex functions having different numbers of external legs.

A global gauge transformation is a continuous symmetry that depends on the single parameter α. The corresponding generator $\mathbf{T}^{(\alpha)}$ acts only on the field type indices,

$$\mathbf{T}^{(\alpha)} \begin{pmatrix} \psi_{K\sigma} \\ \bar{\psi}_{K\sigma} \end{pmatrix} = \begin{pmatrix} \psi_{K\sigma} \\ -\bar{\psi}_{K\sigma} \end{pmatrix} .$$

(6.106)

Written out explicitly, Eq. (6.103) then becomes

$$\boxed{\int_{K\sigma} \left(\psi_{K\sigma} \frac{\delta}{\delta \psi_{K\sigma}} - \bar{\psi}_{K\sigma} \frac{\delta}{\delta \bar{\psi}_{K\sigma}} \right) \Gamma = 0 \, .}$$

(6.107)

The functional differential operator in this equation counts the difference between the number of ψ and $\bar{\psi}$ fields when applied to a monomial. Eq. (6.107) then implies that an expansion of Γ in terms of monomials of the fields can only contain terms with an equal number of ψ and $\bar{\psi}$ fields. In other words, the global gauge symmetry implies particle number conservation.

(b) *Translation:* Spatial and time translations transform the fields according to

$$\mathcal{F}(A): \quad \psi_\sigma(X) \mapsto \psi_\sigma(X + A), \qquad \bar{\psi}_\sigma(X) \mapsto \bar{\psi}_\sigma(X + A), \tag{6.108}$$

where $A = (a_\tau, \boldsymbol{a})$. For a homogeneous system, A can take on all possible values, and the symmetry is continuous. For lattice models on the other hand $\boldsymbol{a}$ has to be a translational vector of the Bravais lattice, and the spatial part of the symmetry is discrete. Written in terms of the Fourier transformed fields,

$$\psi_{K\sigma} = \int_X e^{-iK \cdot X} \psi_\sigma(X), \qquad \bar{\psi}_{K\sigma} = \int_X e^{iK \cdot X} \bar{\psi}_\sigma(X), \tag{6.109}$$

with $\int_X = \int d\tau \int d^D x$ and $K \cdot X = \boldsymbol{k} \cdot \boldsymbol{x} - \omega\tau$, the transformation reads

$$\mathcal{F}(A): \quad \psi_{K\sigma} \mapsto e^{iK \cdot A} \psi_{K\sigma}, \qquad \bar{\psi}_{K\sigma} \mapsto e^{-iK \cdot A} \bar{\psi}_{K\sigma}. \tag{6.110}$$

For a homogeneous system, four generators $\mathbf{T}^{(i)}$ with $i = \tau, x, y, z$ generate continuous translations in the four space-time directions. Combining the resulting four Eqs. (6.103) in a vector notation, they read explicitly,

$$\boxed{\int_{K\sigma} K \left(\psi_{K\sigma} \frac{\delta}{\delta \psi_{K\sigma}} - \bar{\psi}_{K\sigma} \frac{\delta}{\delta \bar{\psi}_{K\sigma}} \right) \Gamma = 0.} \tag{6.111}$$

Here, the differential operator yields the sum of all energy-momenta in a monomial, where momenta of incoming and outgoing fields are counted with different signs. Thus, in Γ only monomials can appear that conserve energy–momentum. Hence, translational invariance in time and space implies energy and momentum conservation for the vertices.

For a lattice system, Eqs. (6.102) and (6.110) yield

$$0 = \left[1 - e^{-i\left(K_1' + \cdots + K_n' - K_n - \cdots - K_1\right) \cdot A} \right]$$
$$\times \Gamma^{(2n)}_{(\bar{\psi}, K_1', \sigma_1') \ldots (\bar{\psi}, K_n', \sigma_n')(\psi, K_n, \sigma_n) \ldots (\psi, K_1, \sigma_1)}, \tag{6.112}$$

which is valid for all $A = (\tau_a, \mathbf{a})$, where $\mathbf{a}$ is a translational vector of the Bravais lattice. Thus, the vertex functions are only nonvanishing for momenta such that the total momentum appearing in the exponential is part of the reciprocal lattice. Combining particle number and energy–momentum conservation, we can thus expand the generating functional Γ as

$$\Gamma[\bar{\psi}, \psi] = \sum_{n=0}^{\infty} \frac{1}{(n!)^2} \int_{K_1'\sigma_1'} \cdots \int_{K_n'\sigma_n'} \int_{K_n\sigma_n} \cdots \int_{K_1\sigma_1}$$
$$\times \delta^{(G)}_{K_1'+\cdots+K_n', K_1+\cdots+K_n}$$
$$\times \Gamma^{(2n)}\left(K_1'\sigma_1', \ldots, K_n'\sigma_n'; K_n\sigma_n, \ldots, K_1\sigma_1\right)$$
$$\times \bar{\psi}_{K_1'\sigma_1'} \cdots \bar{\psi}_{K_n'\sigma_n'} \psi_{K_n\sigma_n} \cdots \psi_{K_1\sigma_1} \, . \tag{6.113}$$

Here, the vertices are related to the ones defined in the abstract notation in Eq. (6.60) by

$$\Gamma^{(2n)}_{(\bar{\psi}, K_1', \sigma_1')\ldots(\bar{\psi}, K_n', \sigma_n')(\psi, K_n, \sigma_n)\ldots(\psi, K_1, \sigma_1)}$$
$$= \delta^{(G)}_{K_1'+\cdots+K_n', K_1+\cdots+K_n}$$
$$\times \Gamma^{(2n)}\left(K_1'\sigma_1', \ldots, K_n'\sigma_n'; K_n\sigma_n, \ldots, K_1\sigma_1\right) \, . \tag{6.114}$$

For a lattice system the $\delta^{(G)}$-function appearing in Eq. (6.113) is defined as,

$$\delta^{(G)}_{K,K'} = \sum_{G} \delta_{K-K',G} \, , \tag{6.115}$$

where the sum extends over all vectors G of the reciprocal lattice. For a homogeneous system, the superscript G should be dropped.

If we omit the spin indices, the Taylor expansion (6.113) is also valid for spinless fermions.

(c) *Spin rotation:* Nonrelativistic systems without spin-orbit coupling and in the absence of an external magnetic field are invariant under rotations in spin space, corresponding to the symmetry transformation

$$\mathcal{F}(\mathbf{U}): \quad \begin{pmatrix} \psi_{K\uparrow} \\ \psi_{K\downarrow} \end{pmatrix} \mapsto \mathbf{U} \begin{pmatrix} \psi_{K\uparrow} \\ \psi_{K\downarrow} \end{pmatrix}, \qquad \begin{pmatrix} \bar{\psi}_{K\uparrow} \\ \bar{\psi}_{K\downarrow} \end{pmatrix} \mapsto \mathbf{U}^* \begin{pmatrix} \bar{\psi}_{K\uparrow} \\ \bar{\psi}_{K\downarrow} \end{pmatrix}, \tag{6.116}$$

where $\mathbf{U} \in SU(2)$ is generated by the Pauli matrices, $\mathbf{U}(l) = e^{i\boldsymbol{\sigma}\cdot l}$, with $\boldsymbol{\sigma} = [\sigma^x, \sigma^y, \sigma^z]$. Our general symmetry relation (6.103) involving the generators of continuous symmetries then leads to the equation

$$\boxed{\int_{K\sigma\sigma'} \left([\boldsymbol{\sigma}]_{\sigma\sigma'} \psi_{K\sigma'} \frac{\delta}{\delta\psi_{K\sigma}} - [\boldsymbol{\sigma}^*]_{\sigma\sigma'} \bar{\psi}_{K\sigma'} \frac{\delta}{\delta\bar{\psi}_{K\sigma}} \right) \Gamma = 0 \, .} \tag{6.117}$$

Combining the z component of this equation with Eq. (6.107), we obtain

$$\int_K \left(\psi_{K\sigma} \frac{\delta}{\delta\psi_{K\sigma}} - \bar{\psi}_{K\sigma} \frac{\delta}{\delta\bar{\psi}_{K\sigma}} \right) \Gamma = 0 \, . \tag{6.118}$$

The functional differential operator in this equation counts the difference of the number of incoming and outgoing fields of a fixed spin projection σ. Thus only monomials that conserve the particle number individually for a each spin projection can occur in an expansion of Γ. In other words, the spins of incoming and outgoing fields have to be pairwise identical, and the expansion in Eq. (6.113) can be written as

$$
\begin{aligned}
\Gamma[\bar{\psi}, \psi] = \sum_{n=0}^{\infty} \frac{1}{n!} \sum_{\sigma_1,\ldots,\sigma_n} & \int_{K_1'} \cdots \int_{K_n'} \int_{K_n} \cdots \int_{K_1} \\
& \times \delta_{K_1'+\cdots+K_n',K_1+\cdots+K_n} \\
& \times U^{(2n)}_{\sigma_1\ldots\sigma_n} \left(K_1', \ldots, K_n'; K_n, \ldots, K_1 \right) \\
& \times \bar{\psi}_{K_1'\sigma_1} \cdots \bar{\psi}_{K_n'\sigma_n} \psi_{K_n\sigma_n} \cdots \psi_{K_1\sigma_1} ,
\end{aligned}
\tag{6.119}
$$

where the vertices are symmetric under simultaneous permutations of the in and outgoing indices,

$$
\begin{aligned}
& U^{(2n)}_{\sigma_1\ldots\sigma_n} \left(K_1', \ldots, K_n'; K_n, \ldots, K_1 \right) \\
& = U^{(2n)}_{\sigma_{P(1)}\ldots\sigma_{P(n)}} \left(K_{P(1)}', \ldots, K_{P(n)}'; K_{P(n)}, \ldots, K_{P(1)} \right) .
\end{aligned}
\tag{6.120}
$$

Here, P is a permutation of the integers $1, \ldots, n$. By symmetrizing the expansion in Eq. (6.119) with respect to separate permutations of incoming and outgoing indices, we obtain the relation

$$
\begin{aligned}
& \Gamma^{(2n)} \left(K_1'\sigma_1', \ldots, K_n'\sigma_n'; K_n\sigma_n, \ldots, K_1\sigma_1 \right) \\
& = \sum_{P} \operatorname{sgn}_\zeta(P) \, U^{(2n)}_{\sigma_1'\ldots\sigma_n'} \left(K_1', \ldots, K_n'; K_{P(n)}, \ldots, K_{P(1)} \right) \\
& \quad \times \delta_{\sigma_1',\sigma_{P(1)}} \cdots \delta_{\sigma_n',\sigma_{P(n)}} ,
\end{aligned}
\tag{6.121}
$$

which connects the vertices defined via the expansions in Eqs. (6.119) and (6.113). The form in Eq. (6.119) should for example be used in the presence of a magnetic field when only a spin rotation symmetry around the axis of the magnetic field is present. For full spin rotation invariance, we can further simplify the form of the vertices by using the x-component of Eq. (6.117), which explicitly reads

$$
\int_{K\sigma} \left(\psi_{K\bar{\sigma}} \frac{\delta}{\delta\psi_{K\sigma}} - \bar{\psi}_{K\bar{\sigma}} \frac{\delta}{\delta\bar{\psi}_{K\sigma}} \right) \Gamma = 0 ,
\tag{6.122}
$$

where $\bar{\sigma} = -\sigma$. Applying this to the expansion in Eq. (6.119) yields

$$
\begin{aligned}
0 = {}& \sum_{n=1}^{\infty} \frac{1}{n!} \sum_{\sigma_1,\dots,\sigma_n} \int_{K_1'} \cdots \int_{K_n'} \int_{K_n} \cdots \int_{K_1} \delta_{K_1'+\cdots+K_n',K_1+\cdots+K_n} \\
& \times U^{(2n)}_{\sigma_1\dots\sigma_n}\left(K_1',\dots,K_n';K_n,\dots,K_1\right) \\
& \times \sum_{r=1}^{n} \Big[\bar{\psi}_{K_1'\sigma_1} \cdots \bar{\psi}_{K_n'\sigma_n} \psi_{K_n\sigma_n} \cdots \psi_{K_r\bar{\sigma}_r} \cdots \psi_{K_1\sigma_1} \\
& \qquad\quad - \bar{\psi}_{K_1'\sigma_1} \cdots \bar{\psi}_{K_r'\bar{\sigma}_r} \cdots \bar{\psi}_{K_n'\sigma_n} \psi_{K_n\sigma_n} \cdots \psi_{K_1\sigma_1} \Big] \\
= {}& \sum_{n=1}^{\infty} \frac{1}{(n-1)!} \sum_{\sigma_1,\dots,\sigma_n} \int_{K_1'} \cdots \int_{K_n'} \int_{K_n} \cdots \int_{K_1} \delta_{K_1'+\cdots+K_n',K_1+\cdots+K_n} \\
& \times \Big[U^{(2n)}_{\sigma_1\sigma_2\dots\sigma_n}\left(K_1',\dots,K_n';K_n,\dots,K_1\right) \\
& \qquad - U^{(2n)}_{\bar{\sigma}_1\sigma_2\dots\sigma_n}\left(K_1',\dots,K_n';K_n,\dots,K_1\right) \Big] \\
& \times \bar{\psi}_{K_1'\sigma_1} \cdots \bar{\psi}_{K_n'\sigma_n} \psi_{K_n\sigma_n} \cdots \psi_{K_1\bar{\sigma}_1} \,.
\end{aligned}
\tag{6.123}
$$

The expression in the rectangular brackets vanishes for all momenta. Thus $U^{(2n)}$ is independent of the first spin index σ_1 and by the permutation symmetry in Eq. (6.120) also independent of all other spin indices,

$$
U^{(2n)}_{\sigma_1\dots\sigma_n}\left(K_1',\dots,K_n';K_n,\dots,K_1\right) = U^{(2n)}\left(K_1',\dots,K_n';K_n,\dots,K_1\right) \,.
\tag{6.124}
$$

The generating functional for a system which is invariant under SU(2)-transformations can therefore be written as

$$
\begin{aligned}
\Gamma[\bar{\psi},\psi] = {}& \sum_{n=0}^{\infty} \frac{1}{n!} \sum_{\sigma_1,\dots,\sigma_n} \int_{K_1'} \cdots \int_{K_n'} \int_{K_n} \cdots \int_{K_1} \\
& \times \delta_{K_1'+\cdots+K_n',K_1+\cdots+K_n} \\
& \times U^{(2n)}\left(K_1',\dots,K_n';K_n,\dots,K_1\right) \\
& \times \bar{\psi}_{K_1'\sigma_1} \cdots \bar{\psi}_{K_n'\sigma_n} \psi_{K_n\sigma_n} \cdots \psi_{K_1\sigma_1} \,.
\end{aligned}
\tag{6.125}
$$

(d) *Time and space inversion:* In this case the action in real space is invariant under

$$
\mathcal{F}: \quad \begin{pmatrix} \psi_\sigma(X) \\ \bar{\psi}_\sigma(X) \end{pmatrix} \mapsto \begin{pmatrix} \zeta \bar{\psi}_\sigma(-X) \\ \psi_\sigma(-X) \end{pmatrix} = \begin{pmatrix} 0 & \zeta \\ 1 & 0 \end{pmatrix} \begin{pmatrix} \psi_\sigma(-X) \\ \bar{\psi}_\sigma(-X) \end{pmatrix} \,.
\tag{6.126}
$$

In Fourier space this becomes

$$
\mathcal{F}: \quad \begin{pmatrix} \psi_{K\sigma} \\ \bar{\psi}_{K\sigma} \end{pmatrix} \mapsto \begin{pmatrix} \zeta \bar{\psi}_{K\sigma} \\ \psi_{K\sigma} \end{pmatrix} = \begin{pmatrix} 0 & \zeta \\ 1 & 0 \end{pmatrix} \begin{pmatrix} \psi_{K\sigma} \\ \bar{\psi}_{K\sigma} \end{pmatrix} \,.
\tag{6.127}
$$

For this transformation, the relation (6.102) implies

$$U^{(2n)}\left(K_1',\ldots,K_n';K_n,\ldots,K_1\right)=U^{(2n)}\left(K_1,\ldots,K_n;K_n',\ldots,K_1'\right)\ . \quad (6.128)$$

(e) *Rotation and spatial inversion:* For this spatial symmetry, the transformation reads

$$\mathcal{F}(\mathbf{R}):\quad \psi_\sigma(\tau,\mathbf{x})\mapsto\psi_\sigma(\tau,\mathbf{R}\mathbf{x})\ ,\qquad \bar\psi_\sigma(\tau,\mathbf{x})\mapsto\bar\psi_\sigma(\tau,\mathbf{R}\mathbf{x})\ , \quad (6.129)$$

where $\mathbf{R}\in O(D)$ is either a D-dimensional rotation matrix, or the inversion matrix $-\mathbf{1}$, or a combination thereof. For an isotropic system any such matrix $\mathbf{R}$ leads to a symmetry, whereas on the lattice the allowed matrices form a discrete group called the point group of the lattice. In Fourier space one obtains,

$$\mathcal{F}(\mathbf{R}):\quad \psi_\sigma(i\omega,\mathbf{k})\mapsto\psi_\sigma(i\omega,\mathbf{R}^T\mathbf{k})\ ,\qquad \bar\psi_\sigma(i\omega,\mathbf{k})\mapsto\bar\psi_\sigma(i\omega,\mathbf{R}^T\mathbf{k})\ ,$$
$$(6.130)$$

where $\mathbf{R}^T=\mathbf{R}^{-1}\in O(D)$. For this transformation, the relation (6.102) yields

$$U^{(2n)}\left(\omega_1'\mathbf{k}_1',\ldots,\omega_n'\mathbf{k}_n';\omega_n\mathbf{k}_n,\ldots,\omega_1\mathbf{k}_1\right)$$
$$=U^{(2n)}\left(\omega_1'\mathbf{R}\mathbf{k}_1',\ldots,\omega_n'\mathbf{R}\mathbf{k}_n';\omega_n\mathbf{R}\mathbf{k}_n,\ldots,\omega_1\mathbf{R}\mathbf{k}_1\right)\ . \quad (6.131)$$

6.3.3 Dyson–Schwinger and Skeleton Equations

Techniques similar to the ones presented above to derive symmetry relations can be used to obtain Dyson–Schwinger equations. More precisely, performing a shift $\Phi\mapsto\Phi+\Delta$ in the integration for the generating functional $\mathcal{G}$ and expanding to first order in Δ, we obtain

$$\boxed{\left(\zeta_\alpha J_\alpha-\frac{\delta S}{\delta\Phi_\alpha}\left[\frac{\delta}{\delta J}\right]\right)\mathcal{G}[J]=0\ .} \quad (6.132)$$

By taking further functional derivatives of this relation, one can obtain so-called skeleton equations which provide relations between vertices with a different number of external legs.

As an example, let us consider a system of nonrelativistic spinless particles with the bare action

$$S[\bar\psi,\psi]=-\int_K\bar\psi_K[i\omega-\xi_k]\psi_K+\frac{1}{4}\int_{K_1}\int_{K_2}\int_{K_3}v(K_1,K_2;K_3,K_1+K_2-K_3)$$
$$\times\bar\psi_{K_1}\bar\psi_{K_2}\psi_{K_3}\psi_{K_1+K_2-K_3}\ . \quad (6.133)$$

Note that the coupling v can be chosen to be antisymmetric for fermions and symmetric for bosons under independent exchange of incoming or outgoing momenta. For $\alpha=(\psi,K)$, Eq. (6.132) then reads explicitly

$$0 = \left(\zeta \bar{j}_K + [i\omega - \xi_k]\frac{\delta}{\delta j_K} - \frac{1}{2}\int_{K_1}\int_{K_2} v(K_1, K_2; K, K_1 + K_2 - K) \right.$$

$$\left. \times \frac{\delta^3}{\delta j_{K_1}\delta j_{K_2}\delta \bar{j}_{K_1+K_2-K}} \right)\mathcal{G}[\bar{j}, j] \,. \tag{6.134}$$

We have used here the following notation for the source term

$$(J, \Phi) = (\bar{j}, \psi) + (\bar{\psi}, j) \,, \tag{6.135}$$

to identify $J_{\psi K} = \bar{j}_K$ and $J_{\bar{\psi}K} = \zeta j_K$. Taking one further derivative with respect to $\bar{j}_{K'}$ before setting the external fields to zero and then converting the resulting equation to one for connected Green functions, we obtain

$$\Sigma(K) = i\omega - \xi_k - G^{-1}(K) = -\zeta \int_{K'} v(K, K'; K', K)G(K')$$

$$-\frac{\zeta}{2}\int_{K_1}\int_{K_2} v(K_1, K_2; K, K_1 + K_2 - K)$$

$$\times G_c^{(4)}(K_1 + K_2 - K, K; K_2, K_1)G^{-1}(K) \,, \tag{6.136}$$

where the connected four-point function $G_c^{(4)}(K_1', K_2'; K_2, K_1)$ is defined via the functional Taylor expansion of the corresponding generating functional $\mathcal{G}_c[\bar{j}, j]$ analogously to Eq. (6.113),

$$\mathcal{G}_c[\bar{j}, j] = \sum_n \frac{1}{(n!)^2}\int_{K_1'}\cdots\int_{K_n'}\int_{K_n}\cdots\int_{K_1} \delta^{(G)}_{K_1'+\cdots+K_n',K_n+\cdots+K_1}$$

$$\times G_c^{(2n)}\left(K_1',\ldots, K_n'; K_n,\ldots, K_1\right)\bar{j}_{K_1'}\cdots\bar{j}_{K_n'}j_{K_n}\cdots j_{K_1} \,. \tag{6.137}$$

The tree expansion in Eq. (6.92) expressing the connected four-point function in terms of the one-particle irreducible vertex $\Gamma^{(4)}$ then explicitly reads,

$$G_c^{(4)}\left(K_1', K_2'; K_2, K_1\right) = -G\left(K_1'\right)G\left(K_2'\right)G(K_2)G(K_1)\Gamma^{(4)}\left(K_1', K_2'; K_2, K_1\right) \,. \tag{6.138}$$

Combining this with Eq. (6.136), we obtain

$$\Sigma(K) = i\omega - \xi_k - G(K)^{-1} = -\zeta \int_{K'} v(K, K'; K', K)G(K')$$

$$+\frac{\zeta}{2}\int_{K_1}\int_{K_2} v(K_1, K_2; K, K_1 + K_2 - K)G(K_1)G(K_2)$$

$$\times G(K_1 + K_2 - K)\Gamma^{(4)}(K_1 + K_2 - K, K; K_2, K_1) \,. \tag{6.139}$$

A graphical representation of this relation is shown in Fig. 6.7. Another application of the general Dyson–Schwinger equation (6.132) for a theory involving both bosonic and fermionic fields will be discussed in Chap. 11.3.

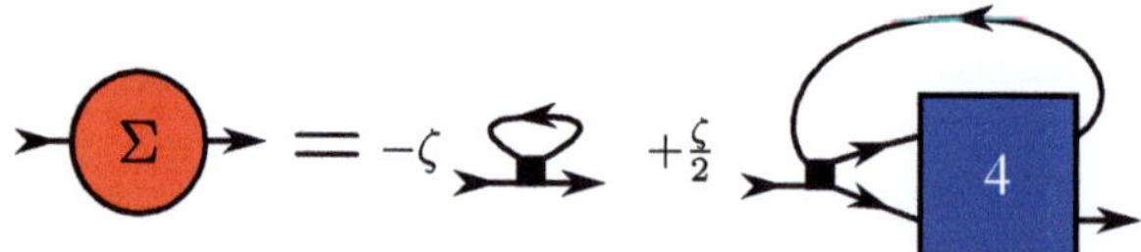

Fig. 6.7 Skeleton equation for the irreducible self-energy of a system of spinless particles. The exact self-energy is given by the sum of the one-loop Hartree-Fock type of diagram with bare vertex (*small black square*) and a two-loop diagram involving the exact effective interaction (*blue square*). All solid arrows represent exact propagators. *Arrows* pointing into a vertex represent incoming fields ψ, while *arrows* pointing out of a vertex represent outgoing fields $\bar{\psi}$. The precise relation between these graphical elements and the elements of the superfield diagrams in Sect. 6.1 will be established in Fig. 10.1

Exercises

6.1 Generating Functions of a Zero-Dimensional Field Theory

To illustrate the machinery of the functional RG in a simple context, consider the generating functions of the zero-dimensional field theory defined by the following integrals over a single real variable φ,

$$e^{g_c(j)} = \frac{Z}{Z_0} g(j) = \frac{1}{Z_0} \int_{-\infty}^{\infty} d\varphi \, e^{-s(\varphi) + j\varphi} \ ,$$

$$e^{g_{ac}(\bar{\varphi})} = \frac{1}{Z_0} \int_{-\infty}^{\infty} d\varphi \, e^{-s_0(\varphi) - s_1(\varphi + \bar{\varphi})} \ .$$

Here, $s(\varphi) = s_0(\varphi) + s_1(\varphi)$, with the bare part $s_0(\varphi) = -\frac{\varphi^2}{2G_0}$, $G_0 < 0$, and the interaction $s_1(\varphi) = \frac{u}{4!}\varphi^4$, $u > 0$. This *toy model* describes a classical one-dimensional anharmonic oscillator.

The full partition function as well as the partition function in the harmonic approximation are given by $Z = \int_{-\infty}^{+\infty} d\varphi \, e^{-s(\varphi)}$ and $Z_0 = \int_{-\infty}^{+\infty} d\varphi \, e^{-s_0(\varphi)} = \sqrt{2\pi(-G_0)}$, respectively. Furthermore, we define the Legendre transform $l(\varphi) = j(\varphi)\varphi - g_c(j(\varphi))$, where $j(\varphi)$ is obtained by inverting the relation $\varphi = \frac{\partial g_c}{\partial j}$, as well as the generating function of one-line irreducible vertices $\gamma(\varphi) = l(\varphi) - s_0(\varphi)$. The vertices $g^{(n)}$, $g_c^{(n)}$, $g_{ac}^{(n)}$, $l^{(n)}$, and $\gamma^{(n)}$ are then the coefficients of the Taylor expansion of the associated generating function, e.g., $g_c(j) = \sum_{n=0}^{\infty} \frac{1}{n!} g_c^{(n)} j^n$.

(a) Show the relations $g_{ac}(\bar{\varphi}) = g_c\left(-G_0^{-1}\bar{\varphi}\right) + \frac{\bar{\varphi}^2}{2G_0}$ and $\frac{\partial^2 l}{\partial \varphi^2} \frac{\partial^2 g_c}{\partial j^2} = 1$.

(b) By definition, the vertices $g^{(n)} = I_n/I_0$ as well as the partition function $Z = I_0$ can be obtained from the integrals

$$I_n = \int_{-\infty}^{+\infty} d\varphi \, \varphi^n e^{-s(\varphi)}$$

which can easily be evaluated numerically. In fact, all integrals I_n can be expressed in terms of the confluent hypergeometric function $U(a, b, z)$. Derive the following explicit relations that yield all other vertices up to $n = 4$, once $g^{(0)}$, $g^{(2)}$ and $g^{(4)}$ are known,

$$\begin{aligned}
g_c^{(0)} &= \ln[Z/Z_0] \,, & g_c^{(2)} &= g^{(2)} \,, & g_c^{(4)} &= g^{(4)} - 3[g^{(2)}]^2 \,, \\
g_{ac}^{(0)} &= g_c^{(0)} \,, & g_{ac}^{(2)} &= G_0^{-2}\left[g_c^{(2)} + G_0\right] \,, & g_{ac}^{(4)} &= G_0^{-4} g_c^{(4)} \,, \\
l^{(0)} &= -g_c^{(0)} \,, & l^{(2)} &= \left[g_c^{(2)}\right]^{-1} \,, & l^{(4)} &= -\left[g_c^{(2)}\right]^{-4} g_c^{(4)} \,, \\
\gamma^{(0)} &= l^{(0)} \,, & \gamma^{(2)} &= l^{(2)} + G_0^{-1} \,, & \gamma^{(4)} &= l^{(4)} \,.
\end{aligned}$$

(c) Calculate the lowest irreducible vertices perturbatively up to order u^2,

$$\gamma^{(0)} = \frac{u}{8}G_0^2 - \frac{u^2}{12}G_0^4 \,, \qquad \gamma^{(2)} = -\frac{u}{2}G_0 + \frac{5u^2}{12}G_0^3 \,, \qquad \gamma^{(4)} = u - \frac{3u^2}{2}G_0^2 \,.$$

References

Baier, T., E. Bick, and C. Wetterich (2004), *Temperature dependence of antiferromagnetic order in the Hubbard model*, Phys. Rev. B **70**, 125111.

Baier, T., E. Bick, and C. Wetterich (2005), *Antiferromagnetic gap in the Hubbard model*, Phys. Lett. B **605**, 144.

Fried, H. M. (1972), *Functional Methods and Models in Quantum Field Theory*, MIT Press, Cambridge.

Fried, H. M. (2002), *Green's Functions and Ordered Exponentials*, Cambridge University Press, Cambridge.

Morris, T. R. (1994), *The exact renormalisation group and approximate solutions*, Int. J. Mod. Phys. A **9**, 2411.

Negele, J. W. and H. Orland (1988), *Quantum Many-Particle Systems*, Addison-Wesley, Redwood City.

Rammer, J. (2007), *Quantum Field Theory of Non-equilibrium States*, Cambridge University Press, Cambridge.

Salmhofer, M. and C. Honerkamp (2001), *Fermionic renormalization group flows*, Progr. Theoret. Phys. **105**, 1.

Schütz, F. (2005), *Aspects of Strong Correlations in Low Dimensions*, Doktorarbeit, Goethe-Universität Frankfurt.

Schütz, F. and P. Kopietz (2006), *Functional renormalization group with vacuum expectation values and spontaneous symmetry breaking*, J. Phys. A: Math. Gen. **39**, 8205.

Schütz, F., L. Bartosch, and P. Kopietz (2005), *Collective fields in the functional renormalization group for fermions, Ward identities, and the exact solution of the Tomonaga-Luttinger model*, Phys. Rev. B **72**, 035107.

Shnirman, A. and Y. Makhlin (2003), *Spin-Spin correlators in the Majorana representation*, Phys. Rev. Lett. **91**, 207204.

Tsvelik, A. M. (2003), *Quantum Field Theory in Condensed Matter Physics*, Cambridge University Press, Cambridge.

Van Kampen, N. G. (1981), *Stochastic Processes in Physics and Chemistry*, North Holland, Amsterdam.

Vasiliev, A. N. (1998), *Functional Methods in Quantum Field Theory and Statistical Physics*, Gordon and Breach, Amsterdam.

Wetterich, C. (2007), *Bosonic effective action for interacting fermions*, Phys. Rev. B **75**, 085102.

Zinn-Justin, J. (2002), *Quantum Field Theory and Critical Phenomena*, Clarendon Press, Oxford, 4th ed.

Chapter 7
Exact FRG Flow Equations

We are now ready to derive FRG flow equations describing the mode-elimination step of the Wilsonian RG exactly. Let us therefore introduce a cutoff Λ into the matrix-propagator $\mathbf{G}_0$ appearing in the Gaussian part $S_0[\Phi]$ of our initial action given in Eq. (6.3). The cutoff scale Λ defines the boundary between long-wavelength (or low energy) fluctuations and short-wavelength (or high energy) fluctuations. The cutoff should be introduced in such a way that fluctuations with wave vectors (or energies) below the cutoff scale are suppressed, while the short-wavelength, high-energy fluctuations are not modified. There is considerable freedom in the implementation of the cutoff procedure, as will be discussed in Sect. 7.1. Ultimately, we shall take the limit $\Lambda \to 0$ where we recover our original theory. The idea is to use the functional representations of the various types of generating functionals introduced in Chap. 6 to obtain exact functional differential equations describing the change of the generating functionals due to an infinitesimal change of the cutoff Λ. We shall derive these surprisingly compact equations in Sect. 7.2 and then show in Sect. 7.3 how the exact FRG flow equation for the generating functional $\Gamma[\bar{\Phi}]$ of the irreducible vertices $\Gamma^{(n)}_{\alpha_1 \ldots \alpha_n}$ can be reduced to an infinite hierarchy of coupled integro-differential equations for these vertices. In the last section, Sect. 7.4, of this chapter we show how to include the possibility of spontaneous symmetry breaking (which is accompanied by a finite vacuum expectation value of some bosonic field) into the exact hierarchy of FRG flow equations for the irreducible vertices (Schütz and Kopietz 2006).

7.1 Cutoffs

Consider the propagator $\mathbf{G}_0$ in the Gaussian part $S_0[\Phi]$ of our general superfield theory describing interacting fermions, bosons, or mixtures thereof, see Eqs. (6.1), (6.2), and (6.3). Let us deform our theory by introducing a cutoff Λ into the Gaussian propagator via the replacement

$$\mathbf{G}_0 \to \mathbf{G}_{0,\Lambda} \,, \tag{7.1}$$

Kopietz, P. et al.: *Exact FRG Flow Equations*. Lect. Notes Phys. **798**, 181–208 (2010)
DOI 10.1007/978-3-642-05094-7_7

where we require that

$$\mathbf{G}_{0,\Lambda} \sim \begin{cases} \mathbf{G}_0 \text{ for } \Lambda \to 0 \,, \\ \mathbf{0} \quad \text{for } \Lambda \to \infty \,. \end{cases} \tag{7.2}$$

This condition implies that when we remove the cutoff Λ, we recover our original physical theory. On the other hand, if the cutoff is larger than all other intrinsic scales of the system, then the Gaussian propagator is switched off such that the particles cannot move. If we introduce the cutoff in momentum space, then we can think of Λ as the boundary between short-wavelength fluctuations ($|\boldsymbol{k}| \gtrsim \Lambda$) which have already been eliminated in the Wilsonian RG and the long-wavelength fluctuations ($|\boldsymbol{k}| \lesssim \Lambda$) which still have to be integrated out. This follows from the observation that Λ acts as an infrared cutoff for the momentum integrations in the Feynman diagrams generated by the theory with propagator $\mathbf{G}_{0,\Lambda}$, which prohibits the propagation of fluctuations with momenta below the cutoff Λ. There are several ways of implementing the condition (7.2). One possibility is to multiply the Gaussian propagator $\mathbf{G}_0$ by a suitable supermatrix $\boldsymbol{\Theta}_\Lambda$,

$$\mathbf{G}_{0,\Lambda} = \boldsymbol{\Theta}_\Lambda \mathbf{G}_0 \,, \tag{7.3}$$

with boundary condition

$$\boldsymbol{\Theta}_\Lambda \sim \begin{cases} \mathbf{1} \text{ for } \Lambda \to 0 \,, \\ \mathbf{0} \text{ for } \Lambda \to \infty \,. \end{cases} \tag{7.4}$$

Alternatively, we may introduce an additive cutoff function $\mathbf{R}_\Lambda$ (regulator) into the inverse Gaussian propagator, defining

$$\mathbf{G}_{0,\Lambda}^{-1} = \mathbf{G}_0^{-1} - \mathbf{R}_\Lambda \,, \tag{7.5}$$

or equivalently

$$\mathbf{G}_{0,\Lambda} = [1 - \mathbf{G}_0 \mathbf{R}_\Lambda]^{-1} \mathbf{G}_0 \,. \tag{7.6}$$

In order to satisfy the condition (7.2), we require

$$|\mathbf{R}_\Lambda| \sim \begin{cases} 0 \ \text{ for } \Lambda \to 0 \,, \\ \infty \text{ for } \Lambda \to \infty \,. \end{cases} \tag{7.7}$$

Of course, we can always express the multiplicative cutoff $\boldsymbol{\Theta}_\Lambda$ in terms of the additive regulator $\mathbf{R}_\Lambda$ by comparing Eqs. (7.3) and (7.6), which yields

$$\boldsymbol{\Theta}_\Lambda = [1 - \mathbf{G}_0 \mathbf{R}_\Lambda]^{-1} \,, \tag{7.8}$$

or

$$\mathbf{R}_\Lambda = -\mathbf{G}_0^{-1}\left(\mathbf{\Theta}_\Lambda^{-1} - 1\right) . \tag{7.9}$$

But in practice one chooses either $\mathbf{\Theta}_\Lambda$ or $\mathbf{R}_\Lambda$ to be some simple function independent of $\mathbf{G}_0$. It is therefore more natural to think of Eqs. (7.3) and (7.6) as two different classes of cutoff procedures.

There is no unique way of choosing the cutoff functions $\mathbf{\Theta}_\Lambda$ or $\mathbf{R}_\Lambda$. In fact, depending on the system under consideration and the required accuracy, different cutoff procedures can be advantageous. Moreover, the cutoff Λ need not be a momentum scale. For example, for quantum systems one can also impose the cutoff in frequency space, or even identify some other parameter with the cutoff, such as the temperature (Honerkamp and Salmhofer 2001), the chemical potential (Sauli and Kopietz 2006), or the strength of the interaction (Honerkamp et al. 2004). However, with the latter three cutoff schemes the intuitive interpretation of the cutoff as the separation between eliminated and not-eliminated degrees of freedom is lost. In fact, cutoff schemes using the temperature, the chemical potential or the interaction strength do not give rise to the standard RG transformations in the sense defined by Fisher (1983), who demanded that an essential part of any RG transformation should be some mode-elimination procedure reducing the number of degrees of freedom.

In order to further specify the matrices $\mathbf{\Theta}_\Lambda$ and $\mathbf{R}_\Lambda$, let us assume that our theory involves several types of fields (including several flavors of fermions and bosons) which we label by a field-type index i. Our superlabel α is then decomposed as $\alpha = (i, K_i)$, where K_i denotes the set of quantum numbers which are necessary to specify the configuration of the field of type i. In general, the matrix $\mathbf{G}_0$ will not be diagonal in field space,[1] but can be diagonalized by means of a suitable rotation in field space. Let us denote the eigenvalues of $\mathbf{G}_0$ by $G_0^{(i)}(K_i)$, which describe the free propagation of a particle associated with the field of type i.

A multiplicative cutoff of the form (7.3) can now be introduced by multiplying the eigenvalues $G_0^{(i)}(K_i)$ of $\mathbf{G}_0$ by suitable cutoff functions $\Theta_\Lambda^{(i)}(K_i)$, so that the eigenvalues of $\mathbf{G}_{0,\Lambda}$ become

$$G_{0,\Lambda}^{(i)}(K_i) = \Theta_\Lambda^{(i)}(K_i)G_0^{(i)}(K_i) . \tag{7.10}$$

Note that in general we may introduce different cutoff functions for each field type. This freedom can be used to construct special cutoff schemes for Bose–Fermi theories which do not violate the symmetries of the system (Schütz et al. 2005). To give an example, suppose that for some classical field component i whose configuration is completely specified by its momentum, we impose a smooth multiplicative cutoff

---

[1] For example, for a theory involving only spin $S = 1/2$ fermions the index i enumerates the four different field types $\psi_\uparrow, \bar\psi_\uparrow, \psi_\downarrow, \bar\psi_\downarrow$. The associated matrix $\mathbf{G}_0$ defined in Eqs. (6.9) and (6.10) is not diagonal, but can be diagonalized by means of a proper rotation in field space. The corresponding eigenvalues are simply the free fermionic propagators $\pm G_0(K)$ in Eq. (6.7). All labels K_i should then be identified with the collective label $K = (i\omega, \mathbf{k})$ denoting fermionic Matsubara frequencies and wave vectors.

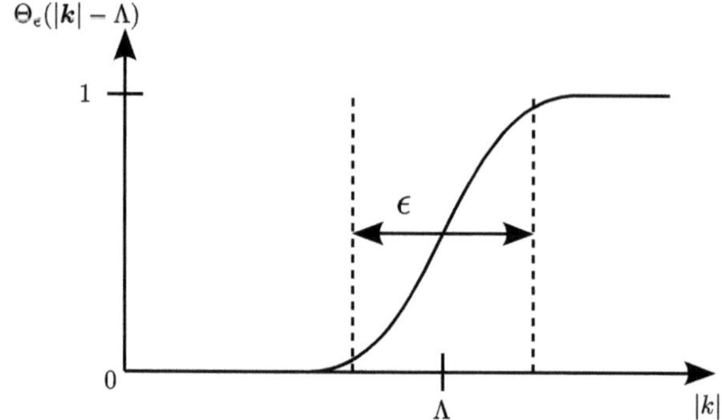

Fig. 7.1 Smoothed unit step function in momentum space with step width of order ϵ. By multiplying the diagonal elements of the superfield matrix propagator in the momentum basis by this function one obtains a cutoff-dependent Gaussian propagator $\mathbf{G}_{0,\Lambda}$ satisfying the boundary conditions (7.2)

in momentum space; in this case we should identify $K_i \to \mathbf{k}$ and substitute for the cutoff function

$$\Theta_\Lambda^{(i)}(K_i) \to \Theta_\epsilon(|\mathbf{k}| - \Lambda) , \tag{7.11}$$

where $\Theta_\epsilon(|\mathbf{k}| - \Lambda)$ is a smoothed step function, where the step is smeared out over an interval of order ϵ, as shown in Fig. 7.1. By construction, the function $\Theta_\epsilon(|\mathbf{k}|-\Lambda)$ is unity for $\Lambda = 0$ and vanishes for $\Lambda \to \infty$, so that the boundary condition (7.4) is satisfied. We recover the sharp cutoff in momentum space used in Sect. 4.2 by letting the step width shrink to zero,

$$\lim_{\epsilon \to 0} \Theta_\epsilon(|\mathbf{k}| - \Lambda) = \Theta(|\mathbf{k}| - \Lambda) . \tag{7.12}$$

It turns out that the exact FRG flow equations involve the derivative of the cutoff function with respect to the cutoff,

$$\delta_\epsilon(|\mathbf{k}| - \Lambda) = -\frac{\partial}{\partial \Lambda} \Theta_\epsilon(|\mathbf{k}| - \Lambda) , \tag{7.13}$$

which reduces to the Dirac delta function $\delta(|\mathbf{k}| - \Lambda)$ for $\epsilon \to 0$. In this limit the integration over $|\mathbf{k}|$ in the loop integrals can be trivially carried out, which makes a sharp multiplicative cutoff for many applications very convenient. However, a sharp momentum cutoff also has some disadvantages. First of all, the nonanalyticity of such a cutoff can give rise to an unphysical nonanalytic momentum dependence of various correlation functions. Although one can in principle take care of this problem by introducing corresponding flowing coupling constants (Hasselmann et al. 2004), in the symmetry-broken phase it is better to avoid a sharp multiplicative cutoff (Sinner et al. 2008). Another technical complication of the sharp cutoff is that in perturbative expansions it can give rise to ambiguous expressions of the type

$$I(x) = \delta(x) f(\Theta(x)) , \tag{7.14}$$

where $f(\Theta(x))$ is some function of the step function $\Theta(x)$. Since at the point $x = 0$ where the δ-function is nonzero the function $\Theta(x)$ is ambiguous, it is at the first sight not clear how to interpret the right-hand side of Eq. (7.14). If the function $f(\Theta(x))$ is known, it is easy to see that the correct interpretation of Eq. (7.14) is given by[2]

$$\boxed{I(x) = \delta(x)f(\Theta(x)) = \delta(x) \int_0^1 dt\, f(t)\,.} \tag{7.15}$$

In the context of the FRG this prescription has first been derived by Morris (1994) and we shall therefore refer to Eq. (7.15) as the *Morris-Lemma*. To prove this, let us regularize $\delta(x)$ and $\Theta(x)$ by their smoothed counter-parts $\delta_\epsilon(x)$ and $\Theta_\epsilon(x)$ defined above and write Eq. (7.14) as

$$I(x) = \lim_{\epsilon \to 0} \frac{\partial}{\partial x} \int_0^{\Theta_\epsilon(x)} dt\, f(t)\,. \tag{7.16}$$

Exchanging the limiting procedure with the differentiation and assuming that the function $f(t)$ is continuous, we obtain

$$I(x) = \frac{\partial}{\partial x} \lim_{\epsilon \to 0} \int_0^{\Theta_\epsilon(x)} dt\, f(t) = \frac{\partial}{\partial x} \Theta(x) \int_0^1 dt\, f(t)\,. \tag{7.17}$$

With $\frac{\partial}{\partial x}\Theta(x) = \delta(x)$ this reduces to Eq. (7.15). Unfortunately, in the context of the FRG the function $f(\Theta(x))$ depends on the cutoff Λ and is in general not known a priori, but has to be determined while solving the RG flow equations. For this reason we cannot directly apply (7.15) unless we have further information about the function $f(\Theta(x))$. This limits the practical usefulness of a sharp cutoff in momentum space (or frequency space for quantum systems).

Finally, let us discuss possible implementations of additive cutoffs $\mathbf{R}_\Lambda$, which are introduced into the inverse Gaussian propagator as in Eq. (7.5). Denoting again the eigenvalues of the supermatrix $\mathbf{G}_0$ by $G_0^{(i)}(K_i)$, Eq. (7.5) is equivalent with the following relation between the corresponding cutoff-dependent eigenvalues $G_{0,\Lambda}^{(i)}(K_i)$ and the eigenvalues $R_\Lambda^{(i)}(K_i)$ of the cutoff matrix $\mathbf{R}_\Lambda$,

$$\left[G_{0,\Lambda}^{(i)}(K_i) \right]^{-1} = \left[G_0^{(i)}(K_i) \right]^{-1} - R_\Lambda^{(i)}(K_i)\,. \tag{7.18}$$

For simplicity, let us assume again that the configuration of a certain field component i is specified by its momentum. A possible choice of the regulator $R_\Lambda^{(i)}(K_i)$ is in this case

---

[2] A special case of Eq. (7.15) is $\delta(x)\Theta(x) = \frac{1}{2}\delta(x)$, which amounts to setting $\Theta(0) = 1/2$. Note, however, that $\delta(x)\Theta^2(x) = \frac{1}{3}\delta(x)$.

$$R_{\Lambda}^{(i)}(K_i) \to C_{\Lambda} \Lambda^z R(|\mathbf{k}|^z/\Lambda^z) \,, \tag{7.19}$$

where z is some positive exponent and $C_{\Lambda} > 0$ is some dimensionful constant (in general cutoff-dependent) which has a finite limit for $\Lambda \to \infty$. In order to satisfy the boundary condition (7.7), the dimensionless function $R(x)$ should be chosen such that

$$R(x) \sim \begin{cases} 0 & \text{for } x \to \infty \quad (\text{corresponding to } \Lambda \to 0) \\ 1 & \text{for } x \to 0 \quad (\text{corresponding to } \Lambda \to \infty) \end{cases} . \tag{7.20}$$

The proper choice of the exponent z, the prefactor C_{Λ} and the cutoff function $R(x)$ depend on the problem under consideration. Usually it is advantageous to choose the exponent z such that the cutoff in Eq. (7.19) matches the scaling of the energy dispersion in the corresponding inverse free propagator.[3] Several regulators $R(x)$ have been proposed in the literature (Berges et al. 2002). A particularly convenient choice is the cutoff function proposed by Litim (2001),

$$R(x) = (1 - x)\Theta(1 - x) \,, \tag{7.21}$$

which has the advantage that often the resulting loop integrations are still elementary. Another possibility is the analytic function (Berges et al. 2002)

$$R(x) = \frac{x}{e^x - 1} \,, \tag{7.22}$$

which is convenient for numerical solutions of the FRG flow equations.

7.2 Exact FRG Flow Equations for Generating Functionals

With the replacement $\mathbf{G}_0 \to \mathbf{G}_{0,\Lambda}$ the Gaussian part of our general action $S = S_0 + S_1$ given in Eqs. (6.2) and (6.3) becomes cutoff-dependent,

$$S_{0,\Lambda}[\Phi] = -\frac{1}{2} \int_{\alpha} \int_{\alpha'} \Phi_{\alpha} \big[\mathbf{G}_{0,\Lambda}^{-1}\big]_{\alpha\alpha'} \Phi_{\alpha'} \equiv -\frac{1}{2}\big(\Phi, \mathbf{G}_{0,\Lambda}^{-1}\Phi\big) \,, \tag{7.23}$$

while the interaction part $S_1[\Phi]$ remains independent of Λ. The cutoff dependence of all generating functionals defined in Chap. 6 therefore arises exclusively from the cutoff dependence of $\mathbf{G}_{0,\Lambda}$. This enables us to derive formally exact functional differential equations for the generating functionals by simply differentiating their

---

[3] For example, for classical φ^4-theory the dispersion of the inverse Gaussian propagator is $c_0 k^2$, so that it is natural to choose $z = 2$. Moreover, it is convenient to choose in this case $C_{\Lambda} = c_0/Z_l$, where Z_l is the flowing wave function renormalization factor as a function of $l = \ln(\Lambda_0/\Lambda)$, see Eqs. (4.57) and (4.62) and the discussion in Sect. 4.2.3.

representations as functional integrals or functional differential operators given in Chap. 6 with respect to Λ. This section is devoted to the derivation of these equations, which form the basis of the FRG.

7.2.1 Disconnected Green Functions

Let us start from the functional integral representation (6.13) of the generating functional of the disconnected Green functions, which for a given cutoff-dependent Gaussian action $S_{0,\Lambda}[\Phi]$ can be written as

$$\mathcal{G}_\Lambda[J] = \frac{1}{\mathcal{Z}_\Lambda} \int \mathcal{D}[\Phi] e^{-S_{0,\Lambda}[\Phi] - S_1[\Phi] + (J,\Phi)} , \tag{7.24}$$

where

$$\mathcal{Z}_\Lambda = \int \mathcal{D}[\Phi] e^{-S_{0,\Lambda}[\Phi] - S_1[\Phi]} \tag{7.25}$$

reduces for $\Lambda \to 0$ to the exact partition function of the system. Differentiating Eq. (7.24) with respect to Λ we obtain

$$\partial_\Lambda \mathcal{G}_\Lambda = -\left(\frac{\partial_\Lambda \mathcal{Z}_\Lambda}{\mathcal{Z}_\Lambda}\right) \mathcal{G}_\Lambda + \frac{1}{\mathcal{Z}_\Lambda} \int \mathcal{D}[\Phi] \frac{1}{2}\left(\Phi, \left[\partial_\Lambda \mathbf{G}_{0,\Lambda}^{-1}\right] \Phi\right) e^{-S_{0,\Lambda}[\Phi] - S_1[\Phi] + (J,\Phi)} .$$

$$\tag{7.26}$$

Using the "source trick" (6.36) we may replace the quadratic form in the integrand by the functional differential operator

$$\left(\Phi, \left[\partial_\Lambda \mathbf{G}_{0,\Lambda}^{-1}\right] \Phi\right) \to \left(\frac{\delta}{\delta J}, \left[\partial_\Lambda \mathbf{G}_{0,\Lambda}^{-1}\right] \frac{\delta}{\delta J}\right) \equiv \int_\alpha \int_{\alpha'} \left[\partial_\Lambda \mathbf{G}_{0,\Lambda}^{-1}\right]_{\alpha\alpha'} \frac{\delta}{\delta J_\alpha} \frac{\delta}{\delta J_{\alpha'}} .$$
$$\tag{7.27}$$

The differentiation can then be pulled out of the integral and we obtain the exact FRG flow equation for the generating functional of the disconnected Green functions,

$$\boxed{\partial_\Lambda \mathcal{G}_\Lambda[J] = \left[\frac{1}{2}\left(\frac{\delta}{\delta J}, \left[\partial_\Lambda \mathbf{G}_{0,\Lambda}^{-1}\right] \frac{\delta}{\delta J}\right) - \partial_\Lambda \ln \mathcal{Z}_\Lambda\right] \mathcal{G}_\Lambda[J] .} \tag{7.28}$$

With the help of the matrix differential operator $\frac{\delta}{\delta J} \otimes \frac{\delta}{\delta J}$ defined in Eq. (6.29), the first term on the right-hand side in Eq. (7.28) can alternatively be written as

$$\left(\frac{\delta}{\delta J}, \left[\partial_\Lambda \mathbf{G}_{0,\Lambda}^{-1}\right] \frac{\delta}{\delta J}\right) \mathcal{G}_\Lambda = \mathrm{Tr}\left[\left[\partial_\Lambda \mathbf{G}_{0,\Lambda}^{-1}\right] \left(\frac{\delta}{\delta J} \otimes \frac{\delta}{\delta J} \mathcal{G}_\Lambda\right)^T\right] , \tag{7.29}$$

where the trace is over all components of the superfield label α.

7.2.2 Connected Green Functions

To obtain the FRG flow equation for the generating functional $\mathcal{G}_c[J]$ of the connected Green functions we simply substitute the relation (6.19) between $\mathcal{G}[J]$ and $\mathcal{G}_c[J]$ into Eq. (7.28). With our normalization of the generating functionals this relation is

$$\mathcal{G}_\Lambda[J] = \frac{\mathcal{Z}_{0,\Lambda}}{\mathcal{Z}_\Lambda} e^{\mathcal{G}_{c,\Lambda}[J]} \,. \tag{7.30}$$

Here $\mathcal{Z}_{0,\Lambda}$ is the cutoff-dependent partition function in Gaussian approximation, which can be obtained by setting $S_1[\Phi] = 0$ in Eq. (7.25) and can be written as

$$\mathcal{Z}_{0,\Lambda} = \int \mathcal{D}[\Phi] e^{-S_{0,\Lambda}[\Phi]} \propto e^{-\frac{1}{2}\mathrm{Tr}[\mathbf{Z}\ln(-\mathbf{G}_{0,\Lambda}^{-1})]} \,. \tag{7.31}$$

With the substitution (7.30) the left-hand side of the flow equation (7.28) becomes

$$\partial_\Lambda \mathcal{G}_\Lambda[J] = \partial_\Lambda \left(\frac{\mathcal{Z}_{0,\Lambda}}{\mathcal{Z}_\Lambda} e^{\mathcal{G}_{c,\Lambda}[J]} \right) = \frac{\mathcal{Z}_{0,\Lambda}}{\mathcal{Z}_\Lambda} e^{\mathcal{G}_{c,\Lambda}[J]} \left[\partial_\Lambda \mathcal{G}_{c,\Lambda} + \partial_\Lambda \ln \left(\frac{\mathcal{Z}_{0,\Lambda}}{\mathcal{Z}_\Lambda} \right) \right] \,. \tag{7.32}$$

To calculate the second functional derivatives of the functional $e^{\mathcal{G}_{c,\Lambda}[J]}$ arising from the right-hand side of Eq. (7.28), one should keep in mind that the second derivatives generate two terms,

$$\left(\frac{\delta}{\delta J}, [\partial_\Lambda \mathbf{G}_{0,\Lambda}^{-1}] \frac{\delta}{\delta J} \right) e^{\mathcal{G}_{c,\Lambda}} = \left(\frac{\delta}{\delta J}, [\partial_\Lambda \mathbf{G}_{0,\Lambda}^{-1}] \frac{\delta \mathcal{G}_{c,\Lambda}}{\delta J} \right) e^{\mathcal{G}_{c,\Lambda}}$$

$$= e^{\mathcal{G}_{c,\Lambda}} \left\{ \left(\frac{\delta \mathcal{G}_{c,\Lambda}}{\delta J}, [\partial_\Lambda \mathbf{G}_{0,\Lambda}^{-1}] \frac{\delta \mathcal{G}_{c,\Lambda}}{\delta J} \right) + \mathrm{Tr}\left[[\partial_\Lambda \mathbf{G}_{0,\Lambda}^{-1}] \left(\frac{\delta}{\delta J} \otimes \frac{\delta}{\delta J} \mathcal{G}_{c,\Lambda} \right)^T \right] \right\} \,, \tag{7.33}$$

where in the second line we have used Eq. (7.29). Substituting Eqs. (7.32) and (7.33) into Eq. (7.28), we obtain the following exact FRG flow equation for the generating functional of the connected Green functions,

$$\boxed{\begin{aligned} \partial_\Lambda \mathcal{G}_{c,\Lambda}[J] = {} & \frac{1}{2}\left(\frac{\delta \mathcal{G}_{c,\Lambda}}{\delta J}, [\partial_\Lambda \mathbf{G}_{0,\Lambda}^{-1}] \frac{\delta \mathcal{G}_{c,\Lambda}}{\delta J} \right) \\ & + \frac{1}{2}\mathrm{Tr}\left[[\partial_\Lambda \mathbf{G}_{0,\Lambda}^{-1}] \left(\frac{\delta}{\delta J} \otimes \frac{\delta}{\delta J} \mathcal{G}_{c,\Lambda} \right)^T \right] - \partial_\Lambda \ln \mathcal{Z}_{0,\Lambda} \,. \end{aligned}} \tag{7.34}$$

If we expand both sides of this equation in powers of the sources J and compare the coefficients, we obtain an infinite hierarchy of coupled FRG flow equations for the cutoff-dependent connected Green functions $G_{c,\Lambda,\alpha_1\ldots\alpha_n}^{(n)}$. Unfortunately, the condition (7.2) that the Gaussian propagator $\mathbf{G}_{0,\Lambda}$ should vanish for $\Lambda \to \infty$ implies the vanishing of all higher-order connected Green functions in this limit, because then all connections between the different parts of any Feynman diagram are switched off. The generating functional of connected Green functions therefore satisfies the boundary condition

$$\mathcal{G}_{c,\Lambda}[J] \to 0 \,, \quad \text{for } \Lambda \to \infty \,. \tag{7.35}$$

Formally, this result can also be deduced from the functional derivative representation (6.39) of the generating functional $\mathcal{G}_{c,\Lambda}[J]$. In practice this boundary condition is not useful, because it contains no information about the system, so that all physical properties of the system have to be generated in the process of integrating the FRG flow equations. It is better to have a boundary condition where the correlation functions for $\Lambda \to \infty$ reduce to some simple solvable limit, such as the noninteracting limit or the result of the mean-field approximation.

7.2.3 Amputated Connected Green Functions

It turns out that the generating functional $\mathcal{G}_{\mathrm{ac},\Lambda}[\bar{\Phi}]$ of the amputated connected Green functions introduced in Sect. 6.1.3 satisfies a more useful boundary condition in the limit $\Lambda \to \infty$ where the Gaussian propagator is switched off. To see this, recall that according to Eq. (6.57) this generating functional can be written as

$$e^{\mathcal{G}_{\mathrm{ac},\Lambda}[\bar{\Phi}]} \equiv \frac{1}{\mathcal{Z}_{0,\Lambda}} \int \mathcal{D}[\Phi] e^{\frac{1}{2}(\Phi, \mathbf{G}_{0,\Lambda}^{-1}\Phi) - S_1[\Phi+\bar{\Phi}]}$$

$$= e^{-\frac{1}{2}\left(\frac{\delta}{\delta\bar{\Phi}}, \mathbf{G}_{0,\Lambda}^{T}\frac{\delta}{\delta\bar{\Phi}}\right)} e^{-S_1[\bar{\Phi}]} \,. \tag{7.36}$$

From the second line it is clear that for $\Lambda \to \infty$ where $\mathbf{G}_{0,\Lambda} \to 0$ the generating functional of the amputated connected Green functions is simply given by the negative of the interaction part of our initial action,

$$\mathcal{G}_{\mathrm{ac},\Lambda}[\bar{\Phi}] \to -S_1[\bar{\Phi}] \,, \quad \text{for } \Lambda \to \infty \,. \tag{7.37}$$

This is a more convenient starting point for approximations than the initial condition (7.35) for $\mathcal{G}_{c,\Lambda}[J]$. To obtain an exact FRG flow equation for $\mathcal{G}_{\mathrm{ac},\Lambda}[\bar{\Phi}]$, we simply differentiate the representation of $e^{\mathcal{G}_{\mathrm{ac},\Lambda}[\bar{\Phi}]}$ in the second line of Eq. (7.36) with respect to the cutoff Λ,

$$e^{\mathcal{G}_{\mathrm{ac},\Lambda}} \partial_\Lambda \mathcal{G}_{\mathrm{ac},\Lambda} = \partial_\Lambda e^{\mathcal{G}_{\mathrm{ac},\Lambda}} = -\frac{1}{2}\left(\frac{\delta}{\delta\bar{\Phi}}, [\partial_\Lambda \mathbf{G}_{0\Lambda}^{T}]\frac{\delta}{\delta\bar{\Phi}}\right) \underbrace{e^{-\frac{1}{2}\left(\frac{\delta}{\delta\bar{\Phi}}, \mathbf{G}_{0,\Lambda}^{T}\frac{\delta}{\delta\bar{\Phi}}\right)} e^{-S_1[\bar{\Phi}]}}_{e^{\mathcal{G}_{\mathrm{ac},\Lambda}}}$$

$$= e^{\mathcal{G}_{\mathrm{ac},\Lambda}}\left\{ -\frac{1}{2}\left(\frac{\delta}{\delta\bar{\Phi}}, [\partial_\Lambda \mathbf{G}_{0,\Lambda}^{T}]\frac{\delta}{\delta\bar{\Phi}}\right)\mathcal{G}_{\mathrm{ac},\Lambda} - \frac{1}{2}\left(\frac{\delta\mathcal{G}_{\mathrm{ac},\Lambda}}{\delta\bar{\Phi}}, [\partial_\Lambda \mathbf{G}_{0,\Lambda}^{T}]\frac{\delta\mathcal{G}_{\mathrm{ac},\Lambda}}{\delta\bar{\Phi}}\right)\right\} \,, \tag{7.38}$$

where in the second line we have carried out the functional differentiation as in Eq. (7.33). We thus conclude that the generating functional of the amputated connected Green functions satisfies the exact FRG flow equation

$$
\partial_\Lambda \mathcal{G}_{\mathrm{ac},\Lambda}[\bar{\Phi}] = -\frac{1}{2}\left(\frac{\delta \mathcal{G}_{\mathrm{ac},\Lambda}}{\delta \bar{\Phi}}, [\partial_\Lambda \mathbf{G}_{0,\Lambda}^T]\frac{\delta \mathcal{G}_{\mathrm{ac},\Lambda}}{\delta \bar{\Phi}}\right)
$$
$$
- \frac{1}{2}\mathrm{Tr}\left[[\partial_\Lambda \mathbf{G}_{0,\Lambda}^T]\left(\frac{\delta}{\delta \bar{\Phi}} \otimes \frac{\delta}{\delta \bar{\Phi}}\mathcal{G}_{\mathrm{ac},\Lambda}\right)^T\right].
\tag{7.39}
$$

As first pointed out by Keller and Kopper (1991) and further discussed by Morris (1994), the flow equation (7.39) for the generating functional of the amputated connected Green functions is identical with the flow equation for the cutoff-dependent Wilsonian effective action derived by Polchinski (1984). For this reason, the exact FRG flow equation (7.39) is sometimes called *Polchinski equation*. It turns out, however, that in approximate calculations based on Eq. (7.39) with a sharp momentum space cutoff, one encounters technical difficulties, because the first term on the right-hand side involving the combination $\left(\frac{\delta \mathcal{G}_{\mathrm{ac},\Lambda}}{\delta \bar{\Phi}}, [\partial_\Lambda \mathbf{G}_{0,\Lambda}^T]\frac{\delta \mathcal{G}_{\mathrm{ac},\Lambda}}{\delta \bar{\Phi}}\right)$ then generates a singular term involving a Dirac δ-function which is not integrated over. This is the reason why nowadays the formulation of the FRG in terms of the irreducible vertices is preferred for explicit calculations.[4] Nevertheless, the Polchinski equation (7.39) seems to be advantageous for gaining nonperturbative insights into the structure of the theory, such as renormalizability proofs (Polchinski 1984, Keller and Kopper 1991, Keller et al. 1992, Keller and Kopper 1996) or general properties of fixed points (Rosten 2009).

7.2.4 One-Line Irreducible Vertices

For our general class of models with cutoff-dependent Gaussian propagators $\mathbf{G}_{0,\Lambda}$ the generating functional of the one-line irreducible vertices is defined as in Eq. (6.61),

$$
\Gamma_\Lambda[\bar{\Phi}] = \mathcal{L}_\Lambda[\bar{\Phi}] + \frac{1}{2}\left(\bar{\Phi}, \mathbf{G}_{0,\Lambda}^{-1}\bar{\Phi}\right)
$$
$$
= (J_\Lambda[\bar{\Phi}], \bar{\Phi}) - \mathcal{G}_{c,\Lambda}[J_\Lambda[\bar{\Phi}]] + \frac{1}{2}\left(\bar{\Phi}, \mathbf{G}_{0,\Lambda}^{-1}\bar{\Phi}\right),
\tag{7.40}
$$

where the cutoff-dependent source field $J_\Lambda[\bar{\Phi}]$ is defined as a functional of the field $\bar{\Phi}$ via the usual relation[5]

---

[4] Note that an expansion of the generating functional $\mathcal{G}_{\mathrm{ac}}[\bar{\Phi}]$ of amputated connected Green functions in terms of normal-ordered monomials (defining so-called Wick-ordered vertex functions) also avoids terms without loop integrations (Salmhofer 1998, 1999); this version of the FRG hierarchy has been used by Halboth and Metzner (2000) to study the two-dimensional Hubbard model.

[5] In the case of spontaneous symmetry breaking the expectation value $\bar{\Phi}_\alpha^0 = \lim_{J\to 0} \langle \Phi_\alpha \rangle$ of at least one of the field components remains finite in the limit of vanishing sources. In this case it is

$$\bar{\Phi}_\alpha = \frac{\delta \mathcal{G}_{c,\Lambda}[J]}{\delta J_\alpha} \ . \tag{7.41}$$

It turns out that for $\Lambda \to \infty$, where $\mathbf{G}_{0,\Lambda}$ vanishes, the functional $\Gamma_\Lambda[\bar{\Phi}]$ as defined in (7.40) simply reduces to the interaction part of the bare action,

$$\Gamma_\Lambda[\bar{\Phi}] \to S_1[\bar{\Phi}] \ , \ \text{for} \ \Lambda \to \infty \ . \tag{7.42}$$

To prove Eq. (7.42), we follow Morris (1994) and first derive a relation between $\Gamma_\Lambda[\bar{\Phi}]$ and $\mathcal{G}_{ac,\Lambda}[\bar{\Phi}]$; the initial condition (7.42) for $\Gamma_\Lambda[\bar{\Phi}]$ follows then from the initial condition (7.37) for $\mathcal{G}_{ac,\Lambda}[\bar{\Phi}]$. Substituting Eq. (7.40) into the relation (6.49) between the generating functionals of the amputated connected and the connected Green functions, we may write

$$\mathcal{G}_{ac,\Lambda}[\bar{\Phi}] = \mathcal{G}_{c,\Lambda}[-\left(\mathbf{G}_{0,\Lambda}^T\right)^{-1}\bar{\Phi}] + \frac{1}{2}\left(\bar{\Phi}, \mathbf{G}_{0,\Lambda}^{-1}\bar{\Phi}\right)$$

$$= -\Gamma_\Lambda[\bar{\Phi}'] + (J[\bar{\Phi}], \bar{\Phi}') + \frac{1}{2}\left(\bar{\Phi}', \mathbf{G}_{0,\Lambda}^{-1}\bar{\Phi}'\right) + \frac{1}{2}\left(\bar{\Phi}, \mathbf{G}_{0,\Lambda}^{-1}\bar{\Phi}\right), \tag{7.43}$$

where $J[\bar{\Phi}] = -\left(\mathbf{G}_{0,\Lambda}^T\right)^{-1}\bar{\Phi}$, and the field $\bar{\Phi}'$ is defined as a functional of $\bar{\Phi}$ via Eq. (7.41),

$$\bar{\Phi}'_\alpha = \frac{\delta \mathcal{G}_{c,\Lambda}[J]}{\delta J_\alpha} = \int_\beta \frac{\delta \bar{\Phi}_\beta}{\delta J_\alpha} \frac{\delta \mathcal{G}_{c,\Lambda}[J]}{\delta \bar{\Phi}_\beta}$$

$$= -\int_\beta [\mathbf{G}_{0,\Lambda}]_{\alpha\beta} \frac{\delta}{\delta \bar{\Phi}_\beta}\left[\mathcal{G}_{ac,\Lambda}[\bar{\Phi}] - \frac{1}{2}\left(\bar{\Phi}, \mathbf{G}_{0,\Lambda}^{-1}\bar{\Phi}\right)\right] \ , \tag{7.44}$$

which in compact supervector notation can be written as

$$\bar{\Phi}' = \bar{\Phi} - \mathbf{G}_{0,\Lambda}\frac{\delta \mathcal{G}_{ac,\Lambda}[\bar{\Phi}]}{\delta \bar{\Phi}} \ . \tag{7.45}$$

Using this identity, the sum of the last three terms in the second line of Eq. (7.43) reduces to

$$(J[\bar{\Phi}], \bar{\Phi}') + \frac{1}{2}\left(\bar{\Phi}', \mathbf{G}_{0,\Lambda}^{-1}\bar{\Phi}'\right) + \frac{1}{2}\left(\bar{\Phi}, \mathbf{G}_{0,\Lambda}^{-1}\bar{\Phi}\right) = \frac{1}{2}\left(\frac{\delta \mathcal{G}_{ac,\Lambda}}{\delta \bar{\Phi}}, \mathbf{G}_{0,\Lambda}^T \frac{\delta \mathcal{G}_{ac,\Lambda}}{\delta \bar{\Phi}}\right) \ . \tag{7.46}$$

We thus obtain the following relation between the generating functionals of the amputated connected Green functions and of the irreducible vertices,

$$\boxed{\mathcal{G}_{ac,\Lambda}[\bar{\Phi}] = -\Gamma_\Lambda\left[\bar{\Phi} - \mathbf{G}_{0,\Lambda}\frac{\delta \mathcal{G}_{ac,\Lambda}}{\delta \bar{\Phi}}\right] + \frac{1}{2}\left(\frac{\delta \mathcal{G}_{ac,\Lambda}}{\delta \bar{\Phi}}, \mathbf{G}_{0,\Lambda}^T \frac{\delta \mathcal{G}_{ac,\Lambda}}{\delta \bar{\Phi}}\right) \ .} \tag{7.47}$$

---

more convenient to replace the last term on the right-hand side of Eq. (7.40) by $\frac{1}{2}\left(\delta\bar{\Phi}, \mathbf{G}_{0,\Lambda}^{-1}\delta\bar{\Phi}\right)$ where $\delta\bar{\Phi} = \bar{\Phi} - \bar{\Phi}^0$. We shall discuss this case separately in Sect. 7.4.

Finally, using Eq. (7.37) and the fact that by construction $\mathbf{G}_{0,\Lambda} \to 0$ for $\Lambda \to \infty$, we see that in this limit

$$\Gamma_\Lambda[\bar{\Phi}] \to -\mathcal{G}_{\mathrm{ac},\Lambda}[\bar{\Phi}] \to S_1[\bar{\Phi}] \ , \quad \text{for } \Lambda \to \infty \ , \tag{7.48}$$

which completes the proof of Eq. (7.42).

To obtain an exact FRG flow equation for the functional $\Gamma_\Lambda[\bar{\Phi}]$, we simply take the derivative of both sides of the definition (7.40) with respect to the cutoff Λ. Keeping in mind that now the fields $\bar{\Phi}$ are kept constant rather than the sources J, we have

$$\partial_\Lambda \Gamma_\Lambda[\bar{\Phi}] = \partial_\Lambda \mathcal{L}_\Lambda[\bar{\Phi}] + \frac{1}{2}\left(\bar{\Phi}, \partial_\Lambda \mathbf{G}_{0,\Lambda}^{-1} \bar{\Phi}\right) \ , \tag{7.49}$$

where

$$
\begin{aligned}
\partial_\Lambda \mathcal{L}_\Lambda[\bar{\Phi}] &= (\partial_\Lambda J_\Lambda[\bar{\Phi}], \bar{\Phi}) - \partial_\Lambda \mathcal{G}_{c,\Lambda}[J_\Lambda[\bar{\Phi}]] \\
&= \underbrace{(\partial_\Lambda J_\Lambda[\bar{\Phi}], \bar{\Phi}) - \left(\partial_\Lambda J_\Lambda[\bar{\Phi}], \frac{\delta \mathcal{G}_{c,\Lambda}[J]}{\delta J}\right)}_{\text{cancel due to } \bar{\Phi} = \delta \mathcal{G}_{c,\Lambda}/\delta J} - \partial_\Lambda \mathcal{G}_{c,\Lambda}[J]\Big|_{J=J_\Lambda[\bar{\Phi}]} \\
&= -\frac{1}{2}\mathrm{Tr}\left[[\partial_\Lambda \mathbf{G}_{0,\Lambda}^{-1}]\left(\frac{\delta}{\delta J} \otimes \frac{\delta}{\delta J}\mathcal{G}_{c,\Lambda}\right)^T\right]_{J=J_\Lambda[\bar{\Phi}]} + \partial_\Lambda \ln \mathcal{Z}_{0,\Lambda} \\
&\quad -\frac{1}{2}\left(\bar{\Phi}, [\partial_\Lambda \mathbf{G}_{0,\Lambda}^{-1}]\,\bar{\Phi}\right) \ .
\end{aligned}
\tag{7.50}
$$

In the last line we have substituted the exact FRG flow equation (7.34) for the generating functional $\mathcal{G}_{c,\Lambda}[J]$ of the connected Green function for constant sources and used the fact that $\bar{\Phi} = \delta \mathcal{G}_{c,\Lambda}/\delta J$. Substituting Eq. (7.50) into Eq. (7.49), we see that the last terms on the right-hand sides cancel. Finally, using the fact that according to Eq. (6.71) the second functional derivatives of $\mathcal{G}_{c,\Lambda}[J]$ and $\mathcal{L}_\Lambda[\bar{\Phi}]$ are inverse to each other, we obtain the following exact flow equation for the generating functional of the irreducible vertices,

$$\boxed{\partial_\Lambda \Gamma_\Lambda[\bar{\Phi}] = -\frac{1}{2}\mathrm{Tr}\left[[\partial_\Lambda \mathbf{G}_{0,\Lambda}^{-1}]\left(\frac{\delta}{\delta\bar{\Phi}} \otimes \frac{\delta}{\delta\bar{\Phi}}\mathcal{L}_\Lambda[\bar{\Phi}]\right)^{-1}\right] + \partial_\Lambda \ln \mathcal{Z}_{0,\Lambda} \ .} \tag{7.51}$$

Recall that the definition (7.40) implies

$$\frac{\delta}{\delta\bar{\Phi}} \otimes \frac{\delta}{\delta\bar{\Phi}}\mathcal{L}_\Lambda[\bar{\Phi}] = \frac{\delta}{\delta\bar{\Phi}} \otimes \frac{\delta}{\delta\bar{\Phi}}\Gamma_\Lambda[\bar{\Phi}] - \left[\mathbf{G}_{0,\Lambda}^T\right]^{-1} \ , \tag{7.52}$$

so that the right-hand side of Eq. (7.51) depends on the second functional derivative of $\Gamma_\Lambda[\bar{\Phi}]$. If we express the cutoff dependence of the Gaussian propagator in the form $\mathbf{G}_{0,\Lambda}^{-1} = \mathbf{G}_0^{-1} - \mathbf{R}_\Lambda$ (see Eq. (7.5)), then we may set $\partial_\Lambda \mathbf{G}_{0,\Lambda}^{-1} = -\partial_\Lambda \mathbf{R}_\Lambda$ in Eq. (7.51).

For classical field theories containing only bosonic fields, FRG equations equivalent to Eq. (7.51) have first been derived by Wetterich (1993) and slightly later by Bonini et al. (1993). Morris (1994) re-derived this equation and pointed out its advantages for vertex expansions. However, exact FRG flow equations for the irreducible vertices which are essentially equivalent to Eq. (7.51) have already been written down earlier by Nicoll et al. (1974), Weinberg (1976), and Nicoll and Chang (1977). The average effective action introduced by Wetterich (1993) corresponds in our notation to

$$
\begin{aligned}
\Gamma_\Lambda^{\mathrm{We}}[\bar{\Phi}] &= \mathcal{L}_\Lambda[\bar{\Phi}] - \frac{1}{2}(\bar{\Phi}, \mathbf{R}_\Lambda \bar{\Phi}) - \ln \mathcal{Z}_{0,\Lambda} \\
&= \Gamma_\Lambda[\bar{\Phi}] - \frac{1}{2}\left(\bar{\Phi}, \mathbf{G}_0^{-1}\bar{\Phi}\right) - \ln \mathcal{Z}_{0,\Lambda} .
\end{aligned}
\tag{7.53}
$$

The corresponding FRG flow equation can be written as

$$
\partial_\Lambda \Gamma_\Lambda^{\mathrm{We}}[\bar{\Phi}] = \frac{1}{2}\mathrm{Tr}\left[[\partial_\Lambda \mathbf{R}_\Lambda]\left(\frac{\delta}{\delta\bar{\Phi}} \otimes \frac{\delta}{\delta\bar{\Phi}} \Gamma_\Lambda^{\mathrm{We}}[\bar{\Phi}] + \mathbf{Z}\mathbf{R}_\Lambda \right)^{-1} \right] .
\tag{7.54}
$$

This equation is sometimes called the *Wetterich equation*. To completely conform with the notation used by Wetterich and coauthors for mixed Bose–Fermi theories (2004) let us define, consistent with Eq. (6.59) for $n = 2$,

$$
\boldsymbol{\Gamma}_\Lambda^{\mathrm{We}(2)}[\bar{\Phi}] = \frac{\delta}{\delta\bar{\Phi}} \otimes \frac{\delta}{\delta\bar{\Phi}} \Gamma_\Lambda^{\mathrm{We}}[\bar{\Phi}]\mathbf{Z} = \frac{\delta}{\delta\bar{\Phi}} \Gamma_\Lambda^{\mathrm{We}}[\bar{\Phi}]\frac{\overleftarrow{\delta}}{\delta\bar{\Phi}} ,
\tag{7.55}
$$

where $\frac{\overleftarrow{\delta}}{\delta\bar{\Phi}}$ is the row vector of right-handed derivatives. Equation (7.54) can then be written in the compact form

$$
\boxed{\;\partial_\Lambda \Gamma_\Lambda^{\mathrm{We}}[\bar{\Phi}] = \frac{1}{2}\mathrm{STr}\left[[\partial_\Lambda \mathbf{R}_\Lambda] \left(\boldsymbol{\Gamma}_\Lambda^{\mathrm{We}(2)}[\bar{\Phi}] + \mathbf{R}_\Lambda \right)^{-1} \right] .\;}
\tag{7.56}
$$

Here, $\mathrm{STr}[\ldots] = \mathrm{Tr}[\mathbf{Z}\ldots]$ denotes the so-called supertrace (Efetov 1983, 1997). Note that strictly speaking $\Gamma_\Lambda^{\mathrm{We}}[\bar{\Phi}]$ does not generate all irreducible vertices, because its second functional derivative is the inverse propagator and not the irreducible self-energy. Our functional $\Gamma_\Lambda[\bar{\Phi}]$ appearing in the exact FRG flow equation (7.51) (and also the corresponding functional $\Gamma_\Lambda^{\mathrm{Mo}}[\bar{\Phi}]$ introduced by Morris (1994)) generates the irreducible self-energy. The relation between our $\Gamma_\Lambda[\bar{\Phi}]$ and the corresponding functional used by Morris is

$$
\Gamma_\Lambda^{\mathrm{Mo}}[\bar{\Phi}] = \Gamma_\Lambda[\bar{\Phi}] - \ln \mathcal{Z}_{0,\Lambda} ,
\tag{7.57}
$$

so that in the FRG equation for $\Gamma_\Lambda^{\mathrm{Mo}}[\bar\Phi]$ the term $\partial_\Lambda \ln \mathcal{Z}_{0,\Lambda}$ in Eq. (7.51) should be omitted. The advantage of our normalization is that our functional $\Gamma_\Lambda[\bar\Phi]$ vanishes identically in the absence of interactions so that the field-independent part $\Gamma_\Lambda[0]$ can be identified with the interaction correction to the free energy.

7.3 Exact FRG Equations for the Irreducible Vertices

Mathematically, Eq. (7.51) is a very complicated functional integro-differential equation which in almost all physically interesting cases cannot be solved exactly. In order to make progress, we therefore have to rely on approximations. Roughly, the approximation strategies proposed so far can be divided into two classes. The first is based on the derivative expansion and generalizations thereof (Berges et al. 2002), which will be reviewed in Chap. 9. An alternative strategy, which has been advanced by Morris (1994), is to expand both sides of Eq. (7.51) in properly symmetrized powers of the fields and compare the coefficients of a given order in the field expansion. In this way Eq. (7.51) can be reduced to an infinite hierarchy of coupled integro-differential equations for the irreducible vertices. We shall refer to this hierarchy as the *vertex expansion*. To facilitate the derivation of this hierarchy, let us cast Eq. (7.51) into a more convenient form. As in Eq. (6.77), let us introduce the field-dependent supermatrix

$$\mathbf{U}_\Lambda[\bar\Phi] = \left(\frac{\delta}{\delta\bar\Phi} \otimes \frac{\delta}{\delta\bar\Phi}\Gamma_\Lambda[\bar\Phi]\right)^T - \left(\frac{\delta}{\delta\bar\Phi} \otimes \frac{\delta}{\delta\bar\Phi}\Gamma_\Lambda[\bar\Phi]\right)^T\Bigg|_{\bar\Phi=0}$$

$$= \left(\frac{\delta}{\delta\bar\Phi} \otimes \frac{\delta}{\delta\bar\Phi}\Gamma_\Lambda[\bar\Phi]\right)^T - \boldsymbol{\Sigma}_\Lambda\,, \tag{7.58}$$

so that the second functional derivative of $\mathcal{L}_\Lambda[\bar\Phi]$ can be written as (see Eq. (6.80)),

$$\frac{\delta}{\delta\bar\Phi} \otimes \frac{\delta}{\delta\bar\Phi}\mathcal{L}_\Lambda[\bar\Phi] = \mathbf{U}_\Lambda^T[\bar\Phi] - \left[\mathbf{G}_\Lambda^T\right]^{-1}\,, \tag{7.59}$$

where $\mathbf{G}_\Lambda^{-1} = \mathbf{G}_{0,\Lambda}^{-1} - \boldsymbol{\Sigma}_\Lambda$ is the cutoff-dependent exact inverse superfield propagator, and $\boldsymbol{\Sigma}_\Lambda$ is the corresponding irreducible self-energy. Moreover, using Eq. (7.31) we obtain for the last term on the right-hand side of Eq. (7.51),

$$\partial_\Lambda \ln \mathcal{Z}_{0,\Lambda} = -\frac{1}{2}\mathrm{Tr}\left[\mathbf{Z}\partial_\Lambda \ln\left(-\mathbf{G}_{0,\Lambda}^{-1}\right)\right] = -\frac{1}{2}\mathrm{Tr}\left[\left[\partial_\Lambda\mathbf{G}_{0,\Lambda}^{-1}\right]\mathbf{G}_{0,\Lambda}^T\right]\,. \tag{7.60}$$

Substituting Eqs. (7.59) and (7.60) into Eq. (7.51), we obtain

$$\partial_\Lambda\Gamma_\Lambda[\bar\Phi] = -\frac{1}{2}\mathrm{Tr}\left\{\left[\partial_\Lambda\mathbf{G}_{0,\Lambda}^{-1}\right]\left[\left(\mathbf{U}_\Lambda^T[\bar\Phi] - \left[\mathbf{G}_\Lambda^T\right]^{-1}\right)^{-1} + \mathbf{G}_{0,\Lambda}^T\right]\right\}\,. \tag{7.61}$$

Using the following chain of identities,

$$
\begin{aligned}
\left(\mathbf{U}^T - [\mathbf{G}^T]^{-1}\right)^{-1} + \mathbf{G}_0^{\cdot T} &= -[\mathbf{1} - \mathbf{G}^{\prime}\mathbf{U}^T]^{-1}\mathbf{G}^T + \mathbf{G}_0^T \\
&= -\left(\mathbf{G}^T + \mathbf{G}^T\mathbf{U}^T\mathbf{G}^T + \mathbf{G}^T\mathbf{U}^T\mathbf{G}^T\mathbf{U}^T\mathbf{G}^T + \ldots\right) + \mathbf{G}_0^T \\
&= -\mathbf{G}^T\mathbf{U}^T\left(\mathbf{1} - \mathbf{G}^T\mathbf{U}^T\right)^{-1}\mathbf{G}^T - \mathbf{G}^T + \mathbf{G}_0^T \\
&= -\mathbf{G}^T\mathbf{U}^T\left(\mathbf{1} - \mathbf{G}^T\mathbf{U}^T\right)^{-1}\mathbf{G}^T - \mathbf{G}_0^T\mathbf{\Sigma}^T\left(\mathbf{1} - \mathbf{G}_0^T\mathbf{\Sigma}^T\right)^{-1}\mathbf{G}_0^T \,,
\end{aligned}
\tag{7.62}
$$

and rearranging terms in the trace using its cyclic invariance, we finally obtain

$$
\boxed{
\begin{aligned}
\partial_\Lambda \Gamma_\Lambda[\bar{\Phi}] = -\frac{1}{2}\mathrm{Tr}\Big[& \dot{\mathbf{G}}_\Lambda \mathbf{U}_\Lambda^T[\bar{\Phi}]\left(\mathbf{1} - \mathbf{G}_\Lambda^T\mathbf{U}_\Lambda^T[\bar{\Phi}]\right)^{-1} \\
& + \dot{\mathbf{G}}_{0,\Lambda}\mathbf{\Sigma}_\Lambda^T\left(\mathbf{1} - \mathbf{G}_{0,\Lambda}^T\mathbf{\Sigma}_\Lambda^T\right)^{-1}\Big] \,,
\end{aligned}
}
\tag{7.63}
$$

where we have introduced the so-called *single-scale propagator*,

$$
\dot{\mathbf{G}}_\Lambda = -\mathbf{G}_\Lambda\left[\partial_\Lambda \mathbf{G}_{0,\Lambda}^{-1}\right]\mathbf{G}_\Lambda = \left(\mathbf{1} - \mathbf{G}_{0,\Lambda}\mathbf{\Sigma}_\Lambda\right)^{-1}\dot{\mathbf{G}}_{0,\Lambda}\left(\mathbf{1} - \mathbf{\Sigma}_\Lambda\mathbf{G}_{0,\Lambda}\right)^{-1} \,,
\tag{7.64}
$$

and its noninteracting limit

$$
\dot{\mathbf{G}}_{0,\Lambda} = \partial_\Lambda \mathbf{G}_{0,\Lambda} \,.
\tag{7.65}
$$

From Eq. (7.63) it is obvious that for a free field theory where $\mathbf{U}_\Lambda[\bar{\Phi}] = \mathbf{\Sigma}_\Lambda = 0$ the RG flow of our functional $\Gamma_\Lambda[\Phi]$ vanishes. If we are not interested in the interaction correction to the free energy the field-independent term in the second line of Eq. (7.63) can be dropped.

In order to reduce the FRG equation (7.63) for the generating functional $\Gamma_\Lambda[\bar{\Phi}]$ to an equivalent system of coupled integro-differential equations for the corresponding irreducible vertices $\Gamma^{(n)}_{\Lambda,\alpha_1\ldots\alpha_n}$ generated by $\Gamma_\Lambda[\bar{\Phi}]$ (see Eqs. (6.59) and (6.60)), we now expand both sides of Eq. (7.63) in powers of the fields and compare the coefficients of (properly symmetrized) powers of the fields on both sides. Expanding the right-hand side in powers of the functional $\mathbf{U}_\Lambda[\bar{\Phi}]$ we obtain

$$
\begin{aligned}
\partial_\Lambda \Gamma_\Lambda[\bar{\Phi}] &= \sum_{n=0}^{\infty} \frac{1}{n!} \int_{\alpha_1} \cdots \int_{\alpha_n} \left(\partial_\Lambda \Gamma^{(n)}_{\Lambda,\alpha_1\ldots\alpha_n}\right)\bar{\Phi}_{\alpha_1}\cdots\bar{\Phi}_{\alpha_n} \\
&= -\frac{1}{2}\mathrm{Tr}\left[\dot{\mathbf{G}}_\Lambda\mathbf{U}_\Lambda^T[\bar{\Phi}]\sum_{\nu=0}^{\infty}\left(\mathbf{G}_\Lambda^T\mathbf{U}_\Lambda^T[\bar{\Phi}]\right)^{\nu} + \dot{\mathbf{G}}_{0,\Lambda}\mathbf{\Sigma}_\Lambda^T\left(\mathbf{1} - \mathbf{G}_{0,\Lambda}^T\mathbf{\Sigma}_\Lambda^T\right)^{-1}\right] .
\end{aligned}
\tag{7.66}
$$

We now expand also the functional $\mathbf{U}_\Lambda[\bar{\Phi}]$ in powers of the fields, using the notation introduced in Eqs. (6.78) and (6.79),

$$\mathbf{U}_\Lambda[\bar\Phi] = \sum_{n=1}^\infty \frac{1}{n!} \int_{\alpha_1} \cdots \int_{\alpha_n} \boldsymbol{\Gamma}^{(n+2)}_{\Lambda,\alpha_1\ldots\alpha_n} \, \bar\Phi_{\alpha_1} \ldots \bar\Phi_{\alpha_n} \, , \tag{7.67}$$

where

$$\left[\boldsymbol{\Gamma}^{(n+2)}_{\Lambda,\alpha_1\ldots\alpha_n} \right]_{\alpha\alpha'} = \Gamma^{(n+2)}_{\Lambda,\alpha\alpha'\alpha_1\ldots\alpha_n} \, . \tag{7.68}$$

Comparing the field-independent parts on both sides of Eq. (7.66), we obtain the exact flow equation for the vertex $\Gamma^{(0)}_\Lambda$, which with our normalization can be identified with the interaction correction to the free energy,

$$\partial_\Lambda \Gamma^{(0)}_\Lambda = -\frac{1}{2}\mathrm{Tr}\left[\dot{\mathbf{G}}_{0,\Lambda} \boldsymbol{\Sigma}^T_\Lambda \left(1 - \mathbf{G}^T_{0,\Lambda} \boldsymbol{\Sigma}^T_\Lambda\right)^{-1} \right]. \tag{7.69}$$

After some rearrangements under the trace this can also be written as

$$\boxed{ \partial_\Lambda \Gamma^{(0)}_\Lambda = -\frac{1}{2}\mathrm{Tr}\left[\mathbf{Z}\dot{\mathbf{G}}_{0,\Lambda} \boldsymbol{\Sigma}_\Lambda \left(1 - \mathbf{G}_{0,\Lambda}\boldsymbol{\Sigma}_\Lambda\right)^{-1} \right]. } \tag{7.70}$$

The flow equations for the higher-order vertices $\Gamma^{(n)}_{\Lambda,\alpha_1\ldots\alpha_n}$ with $n \geq 1$ external legs can be obtained by comparing the coefficients of the monomials $\bar\Phi_{\alpha_1} \ldots \bar\Phi_{\alpha_n}$ on both sides of Eq. (7.66), taking into account that the vertices should be symmetrized for bosonic components and antisymmetrized for fermionic components. In fact, with the help of the notation introduced in Sect. 6.2.2 in the context of deriving an explicit algebraic formula for the tree expansion (see Eq. (6.90)), the resulting FRG flow equations for the vertices $\Gamma^{(n)}_{\Lambda,\alpha_1\ldots\alpha_n}$ with $n \geq 1$ can be written down in closed form,

$$\boxed{ \begin{aligned} \partial_\Lambda \Gamma^{(n)}_{\Lambda,\alpha_1\ldots\alpha_n} &= -\frac{1}{2}\sum_{\nu=1}^\infty \sum_{n_1=1}^\infty \cdots \sum_{n_\nu=1}^\infty \delta_{n,n_1+\ldots+n_\nu} \\ &\times \mathcal{S}_{\alpha_1\ldots\alpha_{n_1};\alpha_{n_1+1}\ldots\alpha_{n_1+n_2};\ldots;\alpha_{n-n_\nu+1}\ldots\alpha_n} \mathrm{Tr}\left[\mathbf{Z}\dot{\mathbf{G}}_\Lambda \boldsymbol{\Gamma}^{(n_\nu+2)}_{\Lambda,\alpha_{n-n_\nu+1}\ldots\alpha_n} \right. \\ &\left. \times \mathbf{G}_\Lambda \boldsymbol{\Gamma}^{(n_{\nu-1}+2)}_{\Lambda,\alpha_{n-n_\nu-n_{\nu-1}+1}\ldots\alpha_{n-n_\nu}} \ldots \mathbf{G}_\Lambda \boldsymbol{\Gamma}^{(n_1+2)}_{\Lambda,\alpha_1\ldots\alpha_{n_1}} \right], \end{aligned} } \tag{7.71}$$

where the symmetrization operator $\mathcal{S}_{\alpha_1\ldots\alpha_{n_1};\alpha_{n_1+1}\ldots\alpha_{n_2};\ldots;\alpha_{n-n_\nu+1}\ldots\alpha_n}$ has already been introduced in Eq. (6.88) in the context of the tree expansion.[6] As explained in Sect. 6.2.2 (see the text before Eq. (6.88)), we have divided the n labels $\alpha_1, \ldots, \alpha_n$

---

[6] Using the invariance of the trace under cyclic permutations and the fact that the transposition of a matrix does not change its trace, we have eliminated all transposition operators in Eqs. (7.70) and (7.71). Inverting the chain of manipulations leading to Eqs. (7.70) and (7.71), we conclude that Eqs. (7.61) and (7.63) can be alternatively written by omitting all transposition operators and

into $\nu \geq 1$ disjunct subsets s_i containing n_i elements as follows (where $i = 1, \ldots, \nu$ labels the subsets and $n = \sum_{i=1}^{\nu} n_i$),

$$\underbrace{\alpha_1 \ldots \alpha_{n_1}}_{s_1} \; ; \; \underbrace{\alpha_{n_1+1} \ldots \alpha_{n_1+n_2}}_{s_2} \; ; \; \ldots \; ; \; \underbrace{\alpha_{n-n_\nu+1}, \ldots, \alpha_n}_{s_\nu} \; . \tag{7.72}$$

The reason for the appearance of the symmetrization operator is that the right-hand side of Eq. (7.71) is already symmetric (for bosons) or antisymmetric (for fermions) with respect to the permutations which do not change the index set s_i, because all indices in a given index set s_i belong to the same symmetrized vertex. The operator $\mathcal{S}_{\alpha_1 \ldots \alpha_{n_1} ; \alpha_{n_1+1} \ldots \alpha_{n_2} ; \ldots ; \alpha_{n-n_\nu+1} \ldots \alpha_n}$ symmetrizes the right-hand side of Eq. (7.71) also with respect to arbitrary permutations which exchange indices belonging to different index groups.

The general flow equation (7.71) looks quite complicated, so let us explicitly write down the flow equations for the first few vertices. Note that for fixed n only a finite number of terms contribute on the right-hand side of Eq. (7.71). Recall that in the derivation of Eq. (7.71) we have implicitly assumed the absence of vacuum expectation values of all field components; the vertex $\Gamma_\alpha^{(1)}$ with a single external leg therefore vanishes identically.[7] Setting $n = 2$ in Eq. (7.71), we obtain the exact FRG flow equation for the vertex $\Gamma^{(2)}_{\Lambda,\alpha_1\alpha_2} = [\mathbf{\Sigma}_\Lambda]_{\alpha_1\alpha_2}$ with two external legs,

$$\partial_\Lambda \Gamma^{(2)}_{\Lambda,\alpha_1\alpha_2} = -\frac{1}{2}\mathrm{Tr}\left[\mathbf{Z}\dot{\mathbf{G}}_\Lambda \Gamma^{(4)}_{\Lambda,\alpha_1\alpha_2} + \mathcal{S}_{\alpha_1;\alpha_2}\left\{\mathbf{Z}\dot{\mathbf{G}}_\Lambda \Gamma^{(3)}_{\Lambda,\alpha_2} \mathbf{G}_\Lambda \Gamma^{(3)}_{\Lambda,\alpha_1}\right\}\right] . \tag{7.73}$$

A graphical representation of this equation is shown in Fig. 7.2. For $n = 3$ the right-hand side of our general flow equation (7.71) has already four terms, so that we obtain the following exact FRG flow equations for the irreducible vertex $\Gamma^{(3)}_{\alpha_1\alpha_2\alpha_3}$ with three external legs,

$$\partial_\Lambda \Gamma^{(3)}_{\Lambda,\alpha_1\alpha_2\alpha_3} = -\frac{1}{2}\mathrm{Tr}\left[\mathbf{Z}\dot{\mathbf{G}}_\Lambda \Gamma^{(5)}_{\Lambda,\alpha_1\alpha_2\alpha_3}\right.$$

$$\left. + \mathcal{S}_{\alpha_1\alpha_2;\alpha_3}\left\{\mathbf{Z}\dot{\mathbf{G}}_\Lambda \Gamma^{(4)}_{\Lambda,\alpha_2\alpha_3} \mathbf{G}_\Lambda \Gamma^{(3)}_{\Lambda,\alpha_1}\right\}\right.$$

---

simultaneously replacing $\mathrm{Tr}[\ldots] \to \mathrm{Tr}[\mathbf{Z}\ldots]$. For example, Eq. (7.61) can be written as

$$\partial_\Lambda \Gamma_\Lambda[\Phi] = -\frac{1}{2}\mathrm{Tr}\left\{\mathbf{Z}\left[\partial_\Lambda \mathbf{G}^{-1}_{0,\Lambda}\right]\left[\left(\mathbf{U}_\Lambda[\bar{\Phi}] - [\mathbf{G}_\Lambda]^{-1}\right)^{-1} + \mathbf{G}_{0,\Lambda}\right]\right\} .$$

Note that for fermions some of the elements of $\mathbf{U}_\Lambda[\bar{\Phi}]$ involve Grassmann fields, so that the operations of matrix inversion and transposition do in general not commute. Fortunately, in the manipulations leading from Eq. (7.61) to the equivalent equation given above this subtlety can be ignored because the matrix $\mathbf{U}_\Lambda[\bar{\Phi}]$ appears always under the trace.

[7] We shall discuss the modification of Eq. (7.71) in the presence of vacuum expectation values in Sect. 7.4.

$$+ \mathcal{S}_{\alpha_1;\alpha_2\alpha_3} \left\{ \mathbf{Z}\dot{\mathbf{G}}_\Lambda \mathbf{\Gamma}^{(3)}_{\Lambda,\alpha_3} \mathbf{G}_\Lambda \mathbf{\Gamma}^{(4)}_{\Lambda,\alpha_1\alpha_2} \right\}$$

$$+ \mathcal{S}_{\alpha_1;\alpha_2;\alpha_3} \left\{ \mathbf{Z}\dot{\mathbf{G}}_\Lambda \mathbf{\Gamma}^{(3)}_{\Lambda,\alpha_3} \mathbf{G}_\Lambda \mathbf{\Gamma}^{(3)}_{\Lambda,\alpha_2} \mathbf{G}_\Lambda \mathbf{\Gamma}^{(3)}_{\Lambda,\alpha_1} \right\} \Bigg] . \tag{7.74}$$

This equation is shown graphically in Fig. 7.3. Finally, for $n = 4$ the right-hand side of our general flow equation (7.71) contains in total 8 terms, involving the vertices $\Gamma^{(n)}$ with $n = 3, 4, 5, 6$. A graphical representation of this rather lengthy expression is shown in Fig. 7.4. For simplicity, let us here explicitly write down the FRG flow equation for $\Gamma^{(4)}_\Lambda$ only for the special case where all vertices with an odd number of external legs vanish (e.g., this is the case for theories involving only fermions or bosons in the absence of external fields and symmetry breaking). Then Eq. (7.71) with $n = 4$ reduces to

$$\partial_\Lambda \Gamma^{(4)}_{\Lambda,\alpha_1\alpha_2\alpha_3\alpha_4}$$

$$= -\frac{1}{2}\mathrm{Tr}\left[\mathbf{Z}\dot{\mathbf{G}}_\Lambda \mathbf{\Gamma}^{(6)}_{\Lambda,\alpha_1\alpha_2\alpha_3\alpha_4} + \mathcal{S}_{\alpha_1\alpha_2;\alpha_3\alpha_4} \left\{ \mathbf{Z}\dot{\mathbf{G}}_\Lambda \mathbf{\Gamma}^{(4)}_{\Lambda,\alpha_3\alpha_4} \mathbf{G}_\Lambda \mathbf{\Gamma}^{(4)}_{\Lambda,\alpha_1\alpha_2} \right\} \right]$$

$$= -\frac{\zeta}{2} \int_{\beta_1} \int_{\beta_2} [\dot{\mathbf{G}}_\Lambda]_{\beta_1\beta_2} \Gamma^{(6)}_{\Lambda,\beta_2\beta_1\alpha_1\alpha_2\alpha_3\alpha_4} - \frac{\zeta}{2} \int_{\beta_1} \int_{\beta_2} \int_{\beta_3} \int_{\beta_4} [\dot{\mathbf{G}}_\Lambda]_{\beta_1\beta_2} [\mathbf{G}_\Lambda]_{\beta_3\beta_4}$$

$$\times \bigg[\Gamma^{(4)}_{\Lambda,\beta_2\beta_3\alpha_3\alpha_4} \Gamma^{(4)}_{\Lambda,\beta_4\beta_1\alpha_1\alpha_2} + \Gamma^{(4)}_{\Lambda,\beta_2\beta_3\alpha_1\alpha_2} \Gamma^{(4)}_{\Lambda,\beta_4\beta_1\alpha_3\alpha_4}$$

$$+ \zeta\, \Gamma^{(4)}_{\Lambda,\beta_2\beta_3\alpha_3\alpha_1} \Gamma^{(4)}_{\Lambda,\beta_4\beta_1\alpha_4\alpha_2} + \zeta\, \Gamma^{(4)}_{\Lambda,\beta_2\beta_3\alpha_4\alpha_2} \Gamma^{(4)}_{\Lambda,\beta_4\beta_1\alpha_3\alpha_1}$$

$$+ \Gamma^{(4)}_{\Lambda,\beta_2\beta_3\alpha_3\alpha_2} \Gamma^{(4)}_{\Lambda,\beta_4\beta_1\alpha_4\alpha_1} + \Gamma^{(4)}_{\Lambda,\beta_2\beta_3\alpha_4\alpha_1} \Gamma^{(4)}_{\Lambda,\beta_4\beta_1\alpha_3\alpha_2} \bigg] . \tag{7.75}$$

This equation is shown graphically in Fig. 7.5.

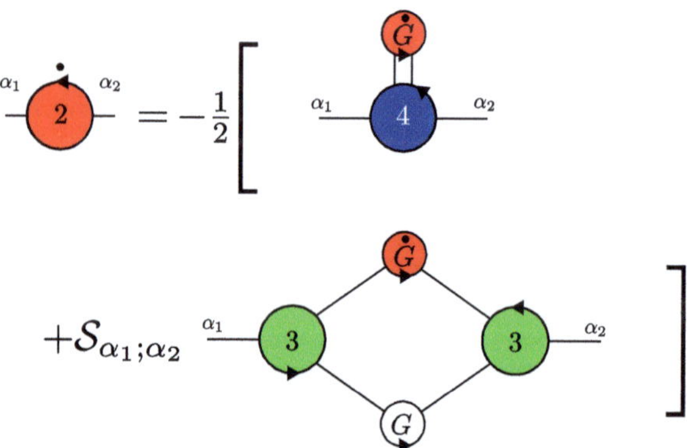

Fig. 7.2 Graphical representation of the exact FRG flow equation (7.73) for the irreducible vertex with two external legs (two-point vertex) in the absence of vacuum expectation values. The *dot* above the vertex on the *left-hand side* denotes the derivative with respect to the cutoff Λ. The single-scale propagator is represented by an *oriented circle* enclosing the symbol $\dot{G}$

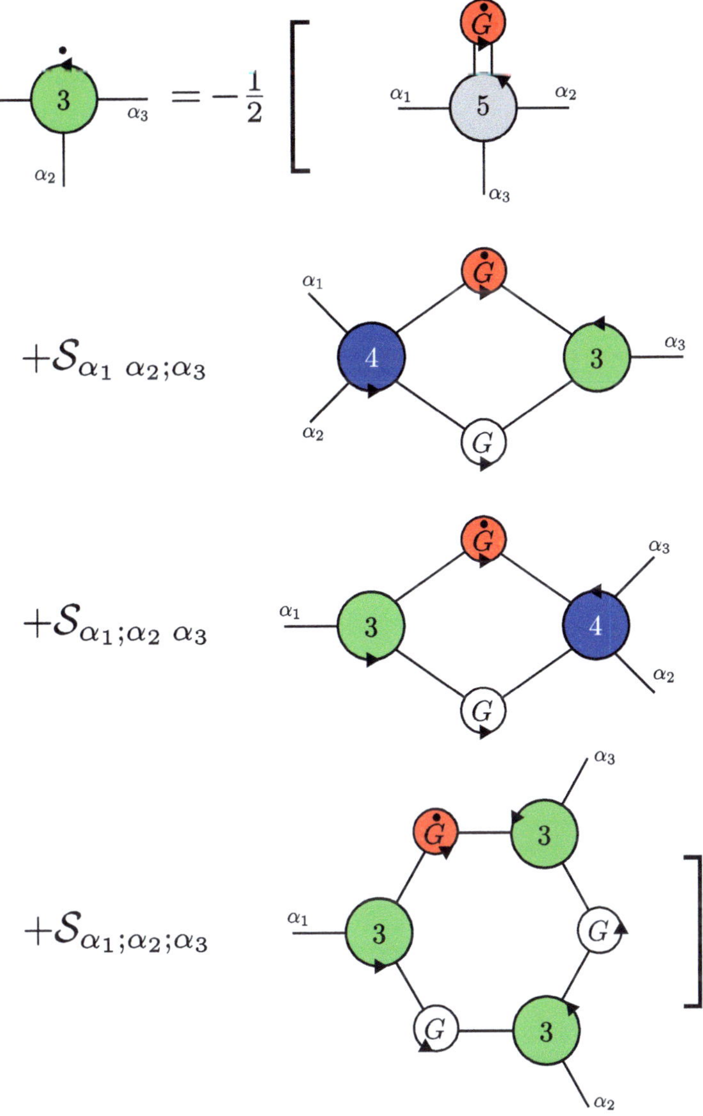

Fig. 7.3 Graphical representation of the exact FRG flow equation (7.74) for the irreducible vertex with three external legs in the absence of vacuum expectation values

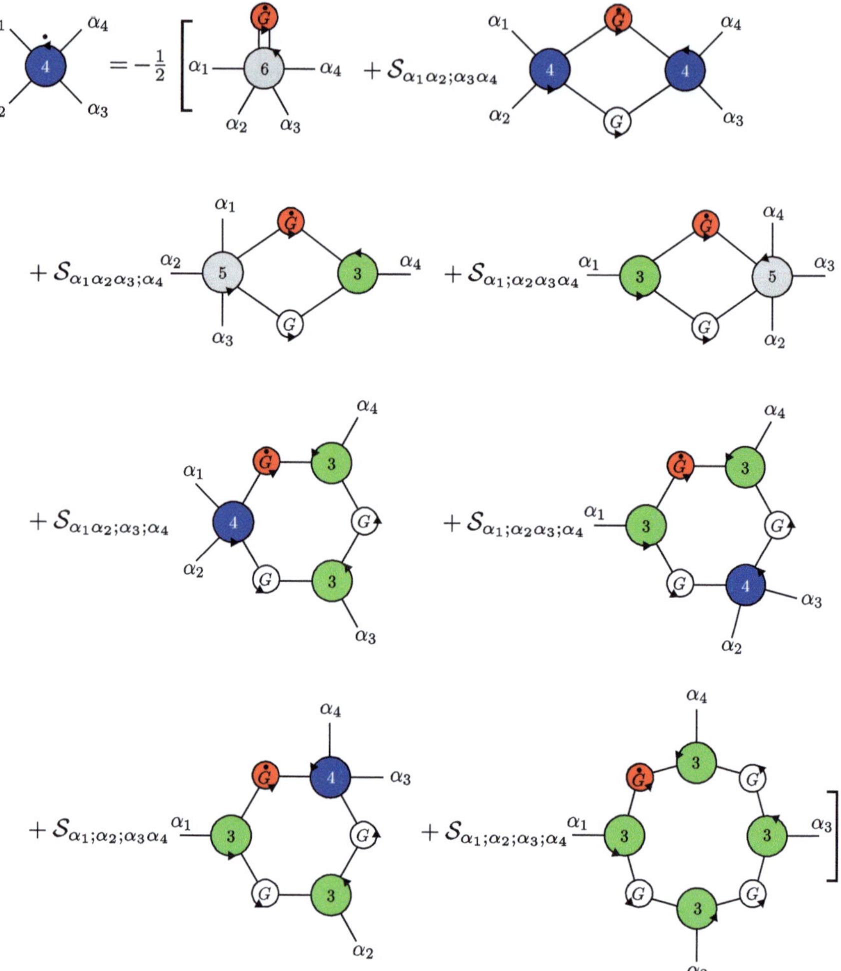

Fig. 7.4 Graphical representation of the exact FRG flow equation (7.71) for $n = 4$ in the absence of vacuum expectation values

7.4 Spontaneous Symmetry Breaking: The Vertex Expansion with Vacuum Expectation Values

As pointed out in the footnote at the beginning of Sect. 7.2.4, in the derivation of the general FRG flow equation (7.51) for the one-line irreducible vertices in Sect. 7.3 we have assumed that the expectation values $\bar{\Phi}_\alpha = \langle \Phi_\alpha \rangle$ of all field components vanish in the absence of external sources $J \to 0$. The hierarchy of FRG flow equations derived in Sect. 7.3 therefore does not describe the symmetry-broken phase of a given model system, where at least one field component has a finite vacuum expectation value in the absence of sources,

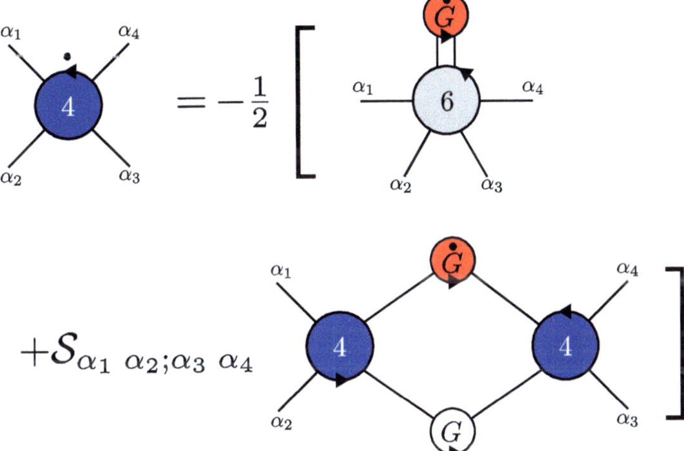

Fig. 7.5 Graphical representation of the exact FRG flow equation (7.75) for the irreducible vertex with four external legs for the special case where all vertices with an odd number of external legs vanish

$$\lim_{J \to 0} \frac{\delta \mathcal{G}_{c,\Lambda}[J]}{\delta J_\alpha} = \lim_{J \to 0} \langle \Phi_\alpha \rangle \equiv \bar{\Phi}_\alpha^0 \neq 0 \,. \tag{7.76}$$

The fact that $\bar{\Phi}_\alpha^0$ is finite does not necessarily mean that some symmetry is spontaneously broken; for example, if Φ_α represents the collective field which is conjugate to the bilinear fermion field representing the density of an electron system, then at any finite density the expectation value $\bar{\Phi}_\alpha^0$ is finite, but does not imply a spontaneously broken symmetry (Schütz and Kopietz 2006). On the other hand, as will be discussed in more detail in Chaps. 11 and 12, spontaneous symmetry breaking in Fermi systems can be conveniently described by introducing a suitable collective bosonic field which acquires a finite vacuum expectation value in the symmetry-broken phase.

It turns out that the generalization of the hierarchy of FRG flow equations for the irreducible vertices derived in Sect. 7.3 to include the possibility that one or several field components have finite vacuum expectation values is not entirely trivial due to several technical subtleties. Let us now carefully develop this generalization using the strategy developed by Schütz and Kopietz (2006). We start with the observation that in the presence of vacuum expectation values the Legendre transform

$$\mathcal{L}_\Lambda[\bar{\Phi}] = (J_\Lambda[\bar{\Phi}], \bar{\Phi}) - \mathcal{G}_{c,\Lambda}[J_\Lambda[\bar{\Phi}]] \tag{7.77}$$

has now an extremum at $\bar{\Phi} = \bar{\Phi}^0$, because according to Eq. (7.76) the limit $J \to 0$ corresponds to $\bar{\Phi} \to \bar{\Phi}^0$,

$$\frac{\delta \mathcal{L}_\Lambda[\bar{\Phi}]}{\delta \bar{\Phi}_\alpha}\bigg|_{\bar{\Phi}=\bar{\Phi}^0} = \zeta_\alpha J_{\Lambda,\alpha}[\bar{\Phi}^0] \to 0 \ . \tag{7.78}$$

Introducing the deviation of the field expectation value from its vacuum expectation value for vanishing sources,

$$\delta \bar{\Phi} = \bar{\Phi} - \bar{\Phi}^0 \ , \tag{7.79}$$

it is convenient to define the generating functional of the one-line irreducible vertices as follows,

$$\Gamma_\Lambda[\bar{\Phi}^0; \delta \bar{\Phi}] = \mathcal{L}_\Lambda[\bar{\Phi}^0 + \delta \bar{\Phi}] + \frac{1}{2} \left(\delta \bar{\Phi}, \mathbf{G}_{0,\Lambda}^{-1} \delta \bar{\Phi} \right)$$

$$= (J_\Lambda[\bar{\Phi}], \bar{\Phi}) - \mathcal{G}_{c,\Lambda}[J_\Lambda[\bar{\Phi}]] + \frac{1}{2} \left(\delta \bar{\Phi}, \mathbf{G}_{0,\Lambda}^{-1} \delta \bar{\Phi} \right) \ . \tag{7.80}$$

Note that on the right-hand side we have subtracted the Gaussian action with the field deviation $\delta \bar{\Phi}$, so that for $\bar{\Phi}^0 \neq 0$ the functional $\Gamma_\Lambda[\bar{\Phi}^0; \delta \bar{\Phi}]$ is not exactly the same as the functional $\Gamma_\Lambda[\bar{\Phi} \to \bar{\Phi}^0 + \delta \bar{\Phi}]$ defined in Eq. (7.40) with shifted argument. As before, we define the one-line irreducible vertices $\Gamma_{\Lambda,\alpha_1\ldots\alpha_n}^{(n)}[\bar{\Phi}^0]$ via the Taylor expansion of the functional (7.80),

$$\Gamma_\Lambda[\bar{\Phi}^0; \delta \bar{\Phi}] = \sum_{n=0}^\infty \frac{1}{n!} \int_{\alpha_1} \cdots \int_{\alpha_n} \Gamma_{\Lambda,\alpha_1\ldots\alpha_n}^{(n)}[\bar{\Phi}^0] \delta \bar{\Phi}_{\alpha_1} \ldots \delta \bar{\Phi}_{\alpha_n} \ . \tag{7.81}$$

The vertices now depend on the vacuum expectation value $\bar{\Phi}^0$ which we assume to be bosonic such that the $\Gamma_{\Lambda,\alpha_1\ldots\alpha_n}^{(n)}[\bar{\Phi}^0]$ commute with all fields.

In the derivation of the FRG flow equation for the functional $\Gamma_\Lambda[\bar{\Phi}^0; \delta \bar{\Phi}]$ defined in Eq. (7.80), one should take into account that the value of the vacuum expectation value $\bar{\Phi}^0$ depends on the cutoff Λ, so that we now obtain instead of Eqs. (7.49) and (7.50)

$$\partial_\Lambda \Gamma_\Lambda[\bar{\Phi}^0; \delta \bar{\Phi}] = \partial_\Lambda \mathcal{L}_\Lambda[\bar{\Phi}^0 + \delta \bar{\Phi}] + \frac{1}{2} \partial_\Lambda \left(\delta \bar{\Phi}, \mathbf{G}_{0,\Lambda}^{-1} \delta \bar{\Phi} \right)$$

$$= -\partial_\Lambda \mathcal{G}_{c,\Lambda}[J]\bigg|_{J=J_\Lambda[\bar{\Phi}]} + \frac{1}{2}(\delta \bar{\Phi}, \left[\partial_\Lambda \mathbf{G}_{0,\Lambda}^{-1} \right] \delta \bar{\Phi}) - (\delta \bar{\Phi}, \mathbf{G}_{0,\Lambda}^{-1} \partial_\Lambda \bar{\Phi}^0) \ . \tag{7.82}$$

As in Eq. (7.50), we now substitute on the right-hand side of Eq. (7.82) the exact FRG flow equation (7.34) for $\mathcal{G}_{c,\Lambda}[J]$, which is also valid in the presence of vacuum expectation values. We obtain

$$\partial_\Lambda \Gamma_\Lambda[\bar{\Phi}^0; \delta\bar{\Phi}] = -\frac{1}{2}\mathrm{Tr}\left[\left[\partial_\Lambda \mathbf{G}_{0,\Lambda}^{-1}\right]\left(\frac{\delta}{\delta\bar{\Phi}} \otimes \frac{\delta}{\delta\bar{\Phi}}\mathcal{L}_\Lambda[\bar{\Phi}]\right)^{-1}\right] + \partial_\Lambda \ln \mathcal{Z}_{0,\Lambda}$$

$$-\frac{1}{2}\left(\frac{\delta\mathcal{G}_{c,\Lambda}}{\delta J}, \left[\partial_\Lambda \mathbf{G}_{0,\Lambda}^{-1}\right]\frac{\delta\mathcal{G}_{c,\Lambda}}{\delta J}\right) + \frac{1}{2}(\delta\bar{\Phi}, \left[\partial_\Lambda \mathbf{G}_{0,\Lambda}^{-1}\right]\delta\bar{\Phi}) - \left(\delta\bar{\Phi}, \mathbf{G}_{0,\Lambda}^{-1}\partial_\Lambda \bar{\Phi}^0\right) \ . \quad (7.83)$$

Keeping in mind that by construction $\frac{\delta\mathcal{G}_{c,\Lambda}}{\delta J} = \bar{\Phi} = \bar{\Phi}^0 + \delta\bar{\Phi}$, we finally obtain

$$\partial_\Lambda \Gamma_\Lambda[\bar{\Phi}^0; \delta\bar{\Phi}] = -\frac{1}{2}\mathrm{Tr}\left[\left[\partial_\Lambda \mathbf{G}_{0,\Lambda}^{-1}\right]\left(\frac{\delta}{\delta\bar{\Phi}} \otimes \frac{\delta}{\delta\bar{\Phi}}\mathcal{L}_\Lambda[\bar{\Phi}]\right)^{-1}\right] + \partial_\Lambda \ln \mathcal{Z}_{0,\Lambda}$$

$$-\left(\delta\bar{\Phi}, \partial_\Lambda\left[\mathbf{G}_{0,\Lambda}^{-1}\bar{\Phi}^0\right]\right) - \frac{1}{2}(\bar{\Phi}^0, \left[\partial_\Lambda \mathbf{G}_{0,\Lambda}^{-1}\right]\bar{\Phi}^0) \ . \quad (7.84)$$

The two terms in the second line are not present in the corresponding FRG equation (7.51) without vacuum expectation values. To derive the vertex expansion with vacuum expectation values, we proceed as in Sect. 7.3. Similar to Eq. (7.58), we introduce the supermatrix

$$\mathbf{U}_\Lambda[\bar{\Phi}^0; \delta\bar{\Phi}] = \left(\frac{\delta}{\delta\bar{\Phi}} \otimes \frac{\delta}{\delta\bar{\Phi}}\Gamma_\Lambda[\bar{\Phi}^0; \delta\bar{\Phi}]\right)^T - \boldsymbol{\Sigma}_\Lambda[\bar{\Phi}^0] \ , \quad (7.85)$$

where the superfield self-energy

$$\boldsymbol{\Sigma}_\Lambda[\bar{\Phi}^0] = \left(\frac{\delta}{\delta\bar{\Phi}} \otimes \frac{\delta}{\delta\bar{\Phi}}\Gamma_\Lambda[\bar{\Phi}^0; \delta\bar{\Phi}]\right)^T\Bigg|_{\delta\bar{\Phi}=0} \quad (7.86)$$

now depends on the vacuum expectation value $\bar{\Phi}^0$. After the same manipulations as in Sect. 7.3, we can cast Eq. (7.84) into a similar form as Eq. (7.63),

$$\boxed{\begin{aligned}\partial_\Lambda \Gamma_\Lambda[\bar{\Phi}^0; \delta\bar{\Phi}] &= -\frac{1}{2}\mathrm{Tr}\left[\dot{\mathbf{G}}_\Lambda \mathbf{U}_\Lambda^T\left(1 - \mathbf{G}_\Lambda^T \mathbf{U}_\Lambda^T\right)^{-1}\right] - \left(\delta\bar{\Phi}, \partial_\Lambda[\mathbf{G}_{0,\Lambda}^{-1}\bar{\Phi}^0]\right) \\ &\quad -\frac{1}{2}\mathrm{Tr}\left[\dot{\mathbf{G}}_{0,\Lambda}\boldsymbol{\Sigma}_\Lambda^T\left(1 - \mathbf{G}_{0,\Lambda}^T \boldsymbol{\Sigma}_\Lambda^T\right)^{-1}\right] - \frac{1}{2}\left(\bar{\Phi}^0, [\partial_\Lambda \mathbf{G}_{0,\Lambda}^{-1}]\bar{\Phi}^0\right) \ .\end{aligned}}$$

$$(7.87)$$

To derive the hierarchy of FRG flow equations for the irreducible vertices associated with Eq. (7.87), one should take into account that the vacuum expectation values $\bar{\Phi}^0$ depend on the cutoff Λ, so that the differentiation of the functional Taylor expansion of $\Gamma[\bar{\Phi}^0; \delta\bar{\Phi}]$ in Eq. (7.81) generates an additional term arising from the cutoff dependence of $\bar{\Phi}^0$,

$$\partial_\Lambda \Gamma_\Lambda[\bar{\Phi}^0; \delta\bar{\Phi}] = \sum_{n=0}^{\infty} \frac{1}{n!} \int_{\alpha_1} \cdots \int_{\alpha_n} \left(\partial_\Lambda \Gamma^{(n)}_{\Lambda,\alpha_1\ldots\alpha_n}[\bar{\Phi}^0] \right) \delta\bar{\Phi}_{\alpha_1} \ldots \delta\bar{\Phi}_{\alpha_n}$$

$$- \sum_{n=0}^{\infty} \frac{1}{n!} \int_\alpha \int_{\alpha_1} \cdots \int_{\alpha_n} \Gamma^{(n+1)}_{\Lambda,\alpha\alpha_1\ldots\alpha_n}[\bar{\Phi}^0] \left(\partial_\Lambda \bar{\Phi}^0_\alpha \right) \delta\bar{\Phi}_{\alpha_1} \ldots \delta\bar{\Phi}_{\alpha_n} . \qquad (7.88)$$

We now substitute Eq. (7.88) into the left-hand side of our exact FRG equation (7.87), expand the right-hand side in powers of the fields $\delta\bar{\Phi}$, and compare the coefficients of properly symmetrized monomials, as discussed in Sect. 7.3. For the vertex without external legs (i.e., the interaction correction to the free energy) we obtain

$$\partial_\Lambda \Gamma^{(0)}_\Lambda[\bar{\Phi}^0] = -\frac{1}{2}\mathrm{Tr}\left[\mathbf{Z}\dot{\mathbf{G}}_{0,\Lambda}\mathbf{\Sigma}_\Lambda \left(1 - \mathbf{G}_{0,\Lambda}\mathbf{\Sigma}_\Lambda \right)^{-1} \right]$$

$$+ \int_\alpha \Gamma^{(1)}_{\Lambda,\alpha}[\bar{\Phi}^0]\left(\partial_\Lambda \bar{\Phi}^0_\alpha \right) - \frac{1}{2}\left(\bar{\Phi}^0, \left[\partial_\Lambda \mathbf{G}^{-1}_{0,\Lambda} \right] \bar{\Phi}^0 \right) , \qquad (7.89)$$

while the flow equation of the vertex $\Gamma^{(1)}_{\Lambda,\alpha}[\bar{\Phi}^0]$ with a single external leg can be written as

$$\partial_\Lambda \Gamma^{(1)}_{\Lambda,\alpha_1}[\bar{\Phi}^0] = -\frac{1}{2}\mathrm{Tr}\left[\mathbf{Z}\dot{\mathbf{G}}_\Lambda \mathbf{\Gamma}^{(3)}_{\Lambda,\alpha_1}[\bar{\Phi}^0] \right] + \int_\alpha \Gamma^{(2)}_{\Lambda,\alpha\alpha_1}[\bar{\Phi}^0]\left(\partial_\Lambda \bar{\Phi}^0_\alpha \right)$$

$$- \int_\alpha \left[\mathbf{G}^{-1}_{0,\Lambda}\right]_{\alpha_1\alpha}\left(\partial_\Lambda \bar{\Phi}^0_\alpha \right) - \int_\alpha \left[\partial_\Lambda \mathbf{G}^{-1}_{0,\Lambda}\right]_{\alpha_1\alpha} \bar{\Phi}^0_\alpha . \qquad (7.90)$$

Using the same notation as in Eq. (7.71), the flow equations for the irreducible vertices $\Gamma^{(n)}_{\Lambda,\alpha_1\ldots\alpha_n}[\bar{\Phi}^0]$ with $n \geq 2$ external legs can be written as

$$\partial_\Lambda \Gamma^{(n)}_{\Lambda,\alpha_1\ldots\alpha_n}[\bar{\Phi}^0] = \int_\alpha \Gamma^{(n+1)}_{\Lambda,\alpha\alpha_1\ldots\alpha_n}[\bar{\Phi}^0]\left(\partial_\Lambda \bar{\Phi}^0_\alpha \right)$$

$$- \frac{1}{2}\sum_{\nu=1}^{\infty} \sum_{n_1=1}^{\infty} \cdots \sum_{n_\nu=1}^{\infty} \delta_{n,n_1+\ldots+n_\nu}$$

$$\times \mathcal{S}_{\alpha_1\ldots\alpha_{n_1};\alpha_{n_1+1}\ldots\alpha_{n_1+n_2};\ldots;\alpha_{n-n_\nu+1}\ldots\alpha_n} \mathrm{Tr}\left[\mathbf{Z}\dot{\mathbf{G}}_\Lambda \mathbf{\Gamma}^{(n_\nu+2)}_{\Lambda,\alpha_{n-n_\nu+1}\ldots\alpha_n} \right.$$

$$\left. \times \mathbf{G}_\Lambda\mathbf{\Gamma}^{(n_{\nu-1}+2)}_{\Lambda,\alpha_{n-n_\nu-n_{\nu-1}+1}\ldots\alpha_{n-n_\nu}} \cdots \mathbf{G}_\Lambda\mathbf{\Gamma}^{(n_1+2)}_{\Lambda,\alpha_1\ldots\alpha_{n_1}} \right] . \qquad (7.91)$$

The nested sum on the right-hand side of this expression is precisely the same as in the corresponding flow equation (7.71) in the absence of vacuum expectation values, so that graphically the flow equation (7.91) can be represented as shown in Fig. 7.6.

To further simplify Eqs. (7.89) and (7.90), it is convenient to choose the cutoff-dependent inverse Gaussian propagator $\mathbf{G}^{-1}_{0,\Lambda}$ such that

$$\mathbf{G}^{-1}_{0,\Lambda}\bar{\Phi}^0 = 0 \quad , \quad \left[\partial_\Lambda \mathbf{G}^{-1}_{0,\Lambda}\right]\bar{\Phi}^0 = 0 . \qquad (7.92)$$

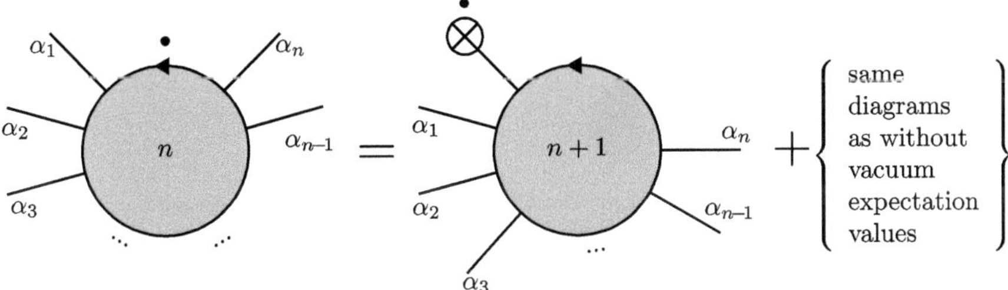

Fig. 7.6 Graphical representation of the exact FRG flow equation (7.91) for the irreducible vertices with $n \geq 2$ external legs. The diagrams in the curly bracket are the same as in the absence of vacuum expectation values. The *crossed circle* with a dot denotes the derivative of the vacuum expectation value with respect to the flow parameter, $\partial_\Lambda \bar{\Phi}^0$

If necessary one should add suitable counterterms to the inverse Gaussian propagator to satisfy these conditions.[8] With the conditions (7.92) the last term on the right-hand side of Eq. (7.89) and the two terms in the second line of Eq. (7.90) vanish. We now fix the flowing vacuum expectation value $\bar{\Phi}^0_\Lambda$ such that the vertex $\Gamma^{(1)}_\Lambda[\bar{\Phi}^0]$ with a single external leg vanishes identically,

$$\Gamma^{(1)}_\Lambda[\bar{\Phi}^0] = 0 \ . \tag{7.93}$$

Physically, this means that for any Λ the first functional derivative of the functional $\Gamma_\Lambda[\bar{\Phi}^0; \delta\bar{\Phi}]$ vanishes, so that we flow along an extremum of the effective potential.[9] Combining Eqs. (7.90) and (7.92) we see that Eq. (7.93) is equivalent with the following FRG flow equation for the vacuum expectation value,

$$\boxed{\int_\alpha \Gamma^{(2)}_{\Lambda,\alpha\alpha_1}[\bar{\Phi}^0](\partial_\Lambda \bar{\Phi}^0_\alpha) = \frac{1}{2}\mathrm{Tr}\Big[\mathbf{Z}\dot{\mathbf{G}}_\Lambda \Gamma^{(3)}_{\Lambda,\alpha_1}[\bar{\Phi}^0]\Big] \ ,} \tag{7.94}$$

which is shown graphically in Fig. 7.7. Recall that according to Eq. (6.75) $\Gamma^{(2)}_{\Lambda,\alpha_1\alpha_2} = [\mathbf{\Sigma}_\Lambda]_{\alpha_1\alpha_2}$. If the symmetry is spontaneously broken, then Eq. (7.94) determines the flowing order parameter as a function of the running cutoff Λ. Note that Eqs. (7.92) and (7.93) imply that both terms in the second line of our flow equation (7.89) for $\Gamma^{(0)}_\Lambda[\bar{\Phi}^0]$ vanish identically, so that the flow equation for the free energy in the presence of vacuum expectation values has precisely the same form as the corresponding flow equation (7.70) in the absence of vacuum expectation values.

---

[8] For example, in the ordered phase of classical φ^4-theory the $\mathbf{k} = 0$ Fourier component of the field $\varphi(\mathbf{k})$ has a finite expectation value, $\bar{\varphi}^0(\mathbf{k}) = (2\pi)^D \delta(\mathbf{k})\bar{\varphi}_0$, see Eq. (2.72). With the choice $G^{-1}_{0,\Lambda}(\mathbf{k}) = \Theta^{-1}_\Lambda(\mathbf{k})c_0 k^2$ we satisfy Eq. (7.92). The momentum-independent part $-2r_0$ appearing in the inverse Gaussian propagator in Eq. (2.78) should then be taken into account via the intial condition for the two-point vertex $\Gamma^{(2)}_{\Lambda_0}(\mathbf{k}) = -2r_0$.

[9] The effective potential is defined (up to an additive constant) by dividing the extensive quantity $\Gamma_\Lambda[\bar{\Phi}^0; 0]$ by the volume of the system, see Eq. (9.8) in Sect. 9.1.

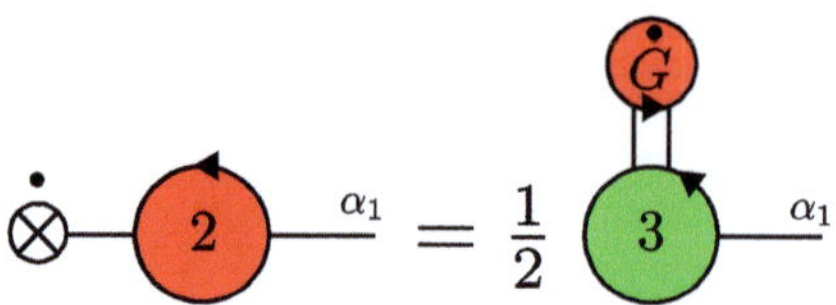

Fig. 7.7 Graphical representation of the exact FRG flow equation (7.94) relating the flow of the vacuum expectation value to the irreducible vertex with three external legs

Exercises

7.1 FRG Flow Equations for the Zero-Dimensional Field Theory Defined in Exercise 6.1

This exercise goes back to an idea by Schönhammer (2000) and is also discussed in unpublished lecture notes by Meden (2003). Let us assume that the "propagator" $G_0 = G_0^\Lambda$ depends on a parameter Λ, e.g., $G_0^\Lambda = -\Lambda$. Note that in contrast to the usual convention (where G_0^Λ approaches zero in the limit $\Lambda \to \infty$), the propagator $G_0^\Lambda = -\Lambda$ vanishes for $\Lambda = 0$. We will therefore start our flow at $\Lambda = 0$ and stop it at a finite Λ.

(a) Derive the flow equations

$$\partial_\Lambda g_c(j) = \frac{1}{2}\partial_\Lambda\left[G_0^{-1}\right]\left[\frac{\partial^2 g_c}{\partial j^2} + \left(\frac{\partial g_c}{\partial j}\right)^2\right] - \partial_\Lambda \ln Z_0 \,,$$

$$\partial_\Lambda g_{ac}(\bar\varphi) = -\frac{1}{2}\partial_\Lambda G_0\left[\frac{\partial^2 g_{ac}}{\partial\bar\varphi^2} + \left(\frac{\partial g_{ac}}{\partial\bar\varphi}\right)^2\right] \,,$$

$$\partial_\Lambda \gamma = -\frac{1}{2}\dot{G}U(1 - GU)^{-1} - \frac{1}{2}\dot{G}_0\Sigma(1 - G_0\Sigma)^{-1} \,,$$

where $\dot{G} = -G\partial_\Lambda\left[G_0^{-1}\right]G$, $G^{-1} = G_0^{-1} - \Sigma$, $\Sigma = \gamma^{(2)}$, and $U = \frac{\partial^2\gamma}{\partial\varphi^2} - \Sigma$.
Hint: It might turn out advantageous to derive the relation $e^{g_{ac}(\bar\varphi)} = e^{-\frac{1}{2}G_0\frac{\partial^2}{\partial\bar\varphi^2}}e^{-s_1(\bar\varphi)}$ to obtain the flow equation for $g_{ac}(\bar\varphi)$.

(b) Expand the flow equations for $g_{ac}(\bar\varphi)$ and $\gamma(\varphi)$ in powers of the sources to obtain flow equations for the vertices $\gamma^{(n)}$ and $g_{ac}^{(n)}$. Show for the specific choice $G_0^\Lambda = -\Lambda$,

$$\partial_\Lambda \gamma^{(0)} = \frac{1}{2}\frac{\gamma^{(2)}}{1+\Lambda\gamma^{(2)}} , \qquad\qquad \partial_\Lambda g_{ac}^{(0)} = \frac{1}{2}g_{ac}^{(2)} ,$$

$$\partial_\Lambda \gamma^{(2)} = \frac{1}{2}\frac{\gamma^{(4)}}{[1+\Lambda\gamma^{(2)}]^2} , \qquad\qquad \partial_\Lambda g_{ac}^{(2)} = \frac{1}{2}g_{ac}^{(4)} + \left[g_{ac}^{(2)}\right]^2 ,$$

$$\partial_\Lambda \gamma^{(4)} = \frac{1}{2}\frac{\gamma^{(6)}}{[1+\Lambda\gamma^{(2)}]^2} - 3\frac{\Lambda[\gamma^{(4)}]^2}{[1+\Lambda\gamma^{(2)}]^3} , \quad \partial_\Lambda g_{ac}^{(4)} = \frac{1}{2}g_{ac}^{(6)} + 4g_{ac}^{(2)}g_{ac}^{(4)} ,$$

$$\partial_\Lambda \gamma^{(6)} = \frac{1}{2}\frac{\gamma^{(8)}}{[1+\Lambda\gamma^{(2)}]^2} - 15\frac{\Lambda\gamma^{(4)}\gamma^{(6)}}{[1+\Lambda\gamma^{(2)}]^3} \quad \partial_\Lambda g_{ac}^{(6)} = \frac{1}{2}g_{ac}^{(8)} + 6g_{ac}^{(2)}g_{ac}^{(6)}$$
$$+45\frac{\Lambda^2[\gamma^{(4)}]^3}{[1+\Lambda\gamma^{(2)}]^4} , \qquad\qquad\qquad + 10\left[g_{ac}^{(4)}\right]^2 .$$

(c) If you have not already done so, derive the relation

$$e^{g_{ac}(\bar\varphi)} = e^{-\frac{1}{2}G_0\frac{\partial^2}{\partial\bar\varphi^2}} e^{-s_1(\bar\varphi)} ,$$

as well as

$$\gamma(\varphi) = \frac{(\bar\varphi(\varphi)-\varphi)^2}{2G_0} - g_{ac}(\bar\varphi(\varphi)) , \quad \text{where} \quad \bar\varphi - \varphi = G_0\frac{\partial g_{ac}}{\partial\bar\varphi} ,$$

and conclude the initial conditions $g_{ac}(\bar\varphi)|_{\Lambda=0} = -s_1(\bar\varphi)$ and $\gamma(\varphi)|_{\Lambda=0} = s_1(\varphi)$, when $G_0|_{\Lambda=0} = 0$. What are the initial conditions for the amputated connected and one-line irreducible vertices?

(d) Truncate the flow equations derived in b) by setting the vertices $\gamma^{(m)}$ and $g_{ac}^{(m)}$ with $m > m_c$ and $m_c = 2, 4, 6$ equal to their initial values. Integrate the remaining set of equations numerically (e.g., using *Mathematica*) from $\Lambda = 0$ to $\Lambda = 1$ and compare the change of the free energy $\gamma^{(0)} = -\ln[Z/Z_0]$, the self-energy Σ and the one-particle irreducible vertex $\gamma^{(4)}$ at the end of the flow in the different schemes [ac and irreducible, each with $m_c = 2, 4, 6$] with the perturbative result of Exercise 6.1 (for $G_0 = -1$) and the exact result obtained by numerically evaluating the integrals I_n defined in Exercise 6.1.

References

Baier, T., E. Bick, and C. Wetterich (2004), *Temperature dependence of antiferromagnetic order in the Hubbard model*, Phys. Rev. B **70**, 125111.

Berges, J., N. Tetradis, and C. Wetterich (2002), *Non-perturbative renormalization flow in quantum field theory and statistical physics*, Phys. Rep. **363**, 223.

Bonini, M., M. D'Attanasio, and G. Marchesini (1993), *Perturbative renormalization and infrared finiteness in the Wilson renormalization group: The massless scalar case*, Nucl. Phys. B **409**, 441.

Efetov, K. (1997), *Supersymmetry in Disorder and Chaos*, Cambridge University Press, Cambridge, UK.

Efetov, K. B. (1983), *Supersymmetry and theory of disordered metals*, Adv. Phys. **32**, 53.

Fisher, M. E. (1983), *Scaling, Universality and Renormalization Group Theory*, in F. J. W. Hahne, editor, *Lecture Notes in Physics*, volume 186, Springer, Berlin.

Halboth, C. J. and W. Metzner (2000), *Renormalization group analysis of the two-dimensional Hubbard model*, Phys. Rev. B **61**, 7364.

Hasselmann, N., S. Ledowski, and P. Kopietz (2004), *Critical behavior of weakly interacting bosons: A functional renormalization-group approach*, Phys. Rev. A **70**, 063621.

Honerkamp, C. and M. Salmhofer (2001), *Temperature-flow renormalization group and the competition between superconductivity and ferromagnetism*, Phys. Rev. B **64**, 184516.

Honerkamp, C., D. Rohe, S. Andergassen, and T. Enss (2004), *Interaction flow method for many-fermion systems*, Phys. Rev. B **70**, 235115.

Keller, G. and C. Kopper (1991), *Perturbative renormalization of QED via flow equations*, Phys. Lett. B **273**, 323.

Keller, G. and C. Kopper (1996), *Renormalizability proof for QED based on flow equations*, Commun. Math. Phys. **176**, 193.

Keller, G., C. Kopper, and M. Salmhofer (1992), *Perturbative renormalization and effective Lagrangians in* Φ_4^4, Helv. Phys. Acta **65**, 32.

Litim, D. F. (2001), *Optimized renormalization group flows*, Phys. Rev. D **64**, 105007.

Meden, V. (2003), *Lecture notes on the "Functional Renormalization Group"*, http://web.physik.rwth-aachen.de/~meden/funRG/.

Morris, T. R. (1994), *The Exact Renormalisation Group and Approximate Solutions*, Int. J. Mod. Phys. A **9**, 2411.

Nicoll, J. F. and T. S. Chang (1977), *An exact one-particle-irreducible renormalization-group generator for critical phenomena*, Phys. Lett. A **62**, 287.

Nicoll, J. F., T. S. Chang, and H. E. Stanley (1974), *Approximate Renormalization Group Based on the Wegner-Houghton Differential Generator*, Phys. Rev. Lett. **33**, 540.

Polchinski, J. (1984), *Renormalization and effective lagrangians*, Nucl. Phys. B **231**, 269.

Rosten, O. J. (2009), *Triviality from the exact renormalization group*, J. High Energy Phys. **07**, 019.

Salmhofer, M. (1998), *Continuous renormalization for fermions and Fermi liquid theory*, Comm. Math. Phys. **194**, 249.

Salmhofer, M. (1999), *Renormalization: An Introduction*, Springer, Berlin.

Sauli, F. and P. Kopietz (2006), *Low-density expansion for the twodimensional electron gas*, Phys. Rev. B **74**, 193106.

Schönhammer, K. (2000), *private communication*.

Schütz, F. and P. Kopietz (2006), *Functional renormalization group with vacuum expectation values and spontaneous symmetry breaking*, J. Phys. A: Math. Gen. **39**, 8205.

Schütz, F., L. Bartosch, and P. Kopietz (2005), *Collective fields in the functional renormalization group for fermions, Ward identities, and the exact solution of the Tomonaga-Luttinger model*, Phys. Rev. B **72**, 035107.

Sinner, A., N. Hasselmann, and P. Kopietz (2008), *Functional renormalization group in the broken symmetry phase: momentum dependence and twoparameter scaling of the self-energy*, J. Phys.: Condens. Matter **20**, 075208.

Weinberg, S. (1976), *Critical Phenomena for Field Theorists*, in A. Zichichi, editor, *Proc. Int. School of Subnuclear Physics 1*, Erice, Plenum, New York.

Wetterich, C. (1993), *Exact evolution equation for the effective potential*, Phys. Lett. B **301**, 90.

Chapter 8
Vertex Expansion

Our fundamental FRG flow equation (7.51) for the generating functional of the irreducible vertices and the equivalent hierarchy of integro-differential equations discussed in Sects. 7.3 and 7.4 are very complicated mathematical objects which usually cannot be solved exactly.[1] It is therefore important to develop reliable approximation strategies for solving these equations. Roughly, two different types of strategies have been developed. The first is based on a suitable truncation of the hierarchy of integro-differential equations for the vertices. This vertex expansion approach was pioneered by Morris (1994) and has been extensively used in the condensed matter community. The second approximation strategy, which has been preferentially used in field theory and statistical mechanics (see for example the review by Berges et al. (2002)), is based on the expansion of the functional $\Gamma_\Lambda[\bar{\Phi}]$ in powers of gradients of the field $\bar{\Phi}$. We first discuss in this chapter the vertex expansion, and then give in Chap. 9 a brief but self-contained introduction to the gradient expansion.

For simplicity, we shall explain these methods using the classical φ^4-theory with action $S_{\Lambda_0}[\varphi]$ defined in Eqs. (2.61) and (2.65) as an example. This is sufficient to understand the basic ideas underlying these two different approximation strategies.[2] Recall that in Sect. 2.2 we have derived the action $S_{\Lambda_0}[\varphi]$ microscopically from the Ising model. The FRG flow equations for the classical φ^4-theory can be obtained as a special case of the general flow equations derived in Chap. 7: our superfield Φ_α has then only a single component $\varphi(\mathbf{k})$, and the superlabel α corresponds to the wave vector $\mathbf{k}$, or alternatively to points $\mathbf{r}$ in D-dimensional space. As discussed in Sect. 7.4, in the ordered phase it is convenient to consider the momentum-independent part of the inverse Gaussian propagator as part of the interaction $S_1[\varphi]$. Introducing a cutoff Λ into the Gaussian propagator as discussed in Sect. 7.1, the action of our model can be written as

$$S[\varphi] = S_{0,\Lambda}[\varphi] + S_1[\varphi] , \tag{8.1}$$

---

[1] In Sect. 11.5 we shall give a nontrivial example where the exact hierarchy of FRG flow equations can be solved exactly using Ward identities.

[2] In Chap. 9 we shall also analyze the generalization of our scalar φ^4-theory to an N-component field theory.

Kopietz, P. et al.: *Vertex Expansion*. Lect. Notes Phys. **798**, 209–232 (2010)
DOI 10.1007/978-3-642-05094-7_8 © Springer-Verlag Berlin Heidelberg 2010

with the cutoff-dependent Gaussian part

$$S_{0,\Lambda}[\varphi] = \frac{1}{2} \int_k G_{0,\Lambda}^{-1}(k)\varphi(-k)\varphi(k) \,, \tag{8.2}$$

and the cutoff-independent interaction,

$$S_1[\varphi] = V f_0 + \frac{r_0}{2} \int_k \varphi(-k)\varphi(k)$$
$$+ \frac{u_0}{4!} \int_{k_1} \int_{k_2} \int_{k_3} \int_{k_4} (2\pi)^D \delta(k_1 + k_2 + k_3 + k_4)\varphi(k_1)\varphi(k_2)\varphi(k_3)\varphi(k_4) \,. \tag{8.3}$$

Here, we have used again the notation $\int_k = \int \frac{d^D k}{(2\pi)^D}$. For simplicity, we set the external magnetic field in Eq. (2.61) equal to zero. If we use a multiplicative cutoff with a smoothed step function $\Theta_\epsilon(|k| - \Lambda)$ as shown in Fig. 7.1, then

$$G_{0,\Lambda}(k) = \frac{\Theta_\epsilon(|k| - \Lambda)}{c_0 k^2} \,. \tag{8.4}$$

For $\epsilon \to 0$ we obtain a sharp momentum shell cutoff, $\lim_{\epsilon \to 0} \Theta_\epsilon(|k| - \Lambda) = \Theta(|k| - \Lambda)$. It turns out that in the symmetry-broken phase the nonanalyticity of the sharp cutoff can lead to technical complications (Berges et al. 2002, Sinner et al. 2008), so that in this case it is better to introduce the cutoff function additively in the inverse propagator as in Eq. (7.5),

$$G_{0,\Lambda}^{-1}(k) = c_0 k^2 + R_\Lambda(k) \,. \tag{8.5}$$

As already mentioned in the footnote after Eq. (7.20) in Sect. 7.1, a convenient choice for the cutoff function $R_\Lambda(k)$ is (Litim 2001)

$$R_\Lambda(k) = (1 - \delta_{k,0}) \frac{c_0 \Lambda^2}{Z_l} R(k^2/\Lambda^2) \,, \quad R(x) = (1 - x)\Theta(1 - x) \,, \tag{8.6}$$

where Z_l is the flowing wave function renormalization factor defined in Eqs. (4.57) and (4.62). The factor $1 - \delta_{k,0}$ in Eq. (8.6) is introduced to satisfy Eq. (7.92) in the ordered phase where $\varphi_{k=0}$ has a finite expectation value.

8.1 Vertex Expansion for Classical φ^4-Theory

To derive the vertex expansion for the classical φ^4-theory given above, it is convenient to work in momentum space. All superfield matrices of Chap. 7 should then be understood as infinite matrices in the momentum labels. The matrix elements of the free superfield Green function are given by

$$[G_{0,\Lambda}]_{kk'} = -(2\pi)^D \delta(k + k') G_{0,\Lambda}(k) \,, \tag{8.7}$$

where the minus sign on the right-hand side is due to the fact that the sign of the $G_{0,\Lambda}^{-1}(k)$ in the Gaussian action (8.1) is different from the sign convention in the definition of our general superfield propagator $\mathbf{G}_{0,\Lambda}^{1}$ in Eq. (7.23). By translational invariance, the Green function of the interacting system with running cutoff Λ has a similar structure,

$$[\mathbf{G}_\Lambda]_{kk'} = -(2\pi)^D \delta(k + k')G_\Lambda(k) , \tag{8.8}$$

and the self-energy matrix is of the form

$$[\mathbf{\Sigma}_\Lambda]_{kk'} = (2\pi)^D \delta(k + k')\Sigma_\Lambda(k') , \tag{8.9}$$

with

$$G_\Lambda^{-1}(k) = G_{0,\Lambda}^{-1}(k) + \Sigma_\Lambda(k) . \tag{8.10}$$

Our exact FRG flow equation (7.87) for the generating functional $\Gamma_\Lambda[\bar\varphi^0; \delta\bar\varphi]$ of the irreducible vertices (including the possibility of symmetry breaking) reduces then to

$$\begin{aligned}
\partial_\Lambda \Gamma_\Lambda[\bar\varphi^0; \delta\bar\varphi] &= \frac{1}{2} \int_k \left[\partial_\Lambda G_{0,\Lambda}^{-1}(k)\right] \left[\frac{\delta}{\delta\bar\varphi} \otimes \frac{\delta}{\delta\bar\varphi} \Gamma_\Lambda[\bar\varphi^0; \delta\bar\varphi] - [\mathbf{G}_{0,\Lambda}]^{-1}\right]^{-1}_{k,-k} \\
&\quad + \partial_\Lambda \ln \mathscr{Z}_{0,\Lambda} \\
&= \frac{1}{2} \int_k \dot{G}_\Lambda(k) \left[\mathbf{U}_\Lambda[\bar\varphi^0; \delta\bar\varphi]\left(1 - \mathbf{G}_\Lambda \mathbf{U}_\Lambda[\bar\varphi^0; \delta\bar\varphi]\right)^{-1}\right]_{k,-k} \\
&\quad + \frac{V}{2} \int_k \frac{\dot{G}_{0,\Lambda}(k)\Sigma_\Lambda(k)}{1 + G_{0,\Lambda}(k)\Sigma_\Lambda(k)} .
\end{aligned} \tag{8.11}$$

Here, the factor of volume V in last term arises from the regularized momentum space δ-function $(2\pi)^D \delta(k = 0) \to V$, the noninteracting single-scale propagator in momentum space is $\dot{G}_{0,\Lambda}(k) = \partial_\Lambda G_{0,\Lambda}(k)$, and the corresponding interacting single-scale propagator is given by (see Eq. (7.64)),

$$\dot{G}_\Lambda(k) = -G_\Lambda(k)\left[\partial_\Lambda G_{0,\Lambda}^{-1}(k)\right] G_\Lambda(k) = \frac{\dot{G}_{0,\Lambda}(k)}{[1 + G_{0,\Lambda}(k)\Sigma_\Lambda(k)]^2} . \tag{8.12}$$

The functional $\mathbf{U}_\Lambda$ is defined in general in Eq. (7.85); for our φ^4-theory the matrix elements of $\mathbf{U}_\Lambda[\bar\varphi^0; \delta\bar\varphi]$ in momentum space are

$$\left(\mathbf{U}_\Lambda[\bar\varphi^0; \delta\bar\varphi]\right)_{kk'} = \frac{\delta^2 \Gamma_\Lambda[\bar\varphi^0; \delta\bar\varphi]}{\delta\bar\varphi(k)\delta\bar\varphi(k')} - (2\pi)^D \delta(k + k')\Sigma_\Lambda(k) . \tag{8.13}$$

8.1.1 Exact FRG Flow Equations

Let us now explicitly write down the first few equations of the exact hierarchy of FRG flow equations for the irreducible vertices. Therefore, we simply specify the general flow equations derived in Sect. 7.4 to the case of the φ^4-theory considered in this chapter. The flow of the interaction correction to the free energy is given by the field-independent part of Eq. (8.11) (see also Eqs. (7.70) and (7.89)),

$$\partial_\Lambda \Gamma_\Lambda^{(0)}[\bar{\varphi}^0] = \frac{V}{2} \int_k \frac{\dot{G}_{0,\Lambda}(k)\Sigma_\Lambda(k)}{1 + G_{0,\Lambda}(k)\Sigma_\Lambda(k)} . \tag{8.14}$$

To write down the FRG flow equations for the higher-order vertices, it is convenient to explicitly factor out the momentum-conserving δ-function from the irreducible vertices with $n \geq 2$ external legs,

$$\Gamma_{\Lambda, k_1 \dots k_n}^{(n)}[\bar{\varphi}^0] = (2\pi)^D \delta(k_1 + \dots + k_n)\Gamma_\Lambda^{(n)}(k_1, \dots, k_n) , \tag{8.15}$$

where on the right-hand side we suppress the dependence on $\bar{\varphi}^0$. Similarly, we factor out the δ-function from the vacuum expectation value, defining

$$\bar{\varphi}_\Lambda^0(k) = (2\pi)^D \delta(k)\bar{\varphi}_\Lambda^0 . \tag{8.16}$$

Our general FRG equation (7.94) for the flowing vacuum expectation value then reduces to

$$\Sigma_\Lambda(0)\partial_\Lambda \bar{\varphi}_\Lambda^0 = -\frac{1}{2} \int_k \dot{G}_\Lambda(k)\Gamma_\Lambda^{(3)}(k, -k, 0) , \tag{8.17}$$

and the general flow equation (7.73) for the two-point vertex reduces to the following exact FRG equation for the self-energy $\Sigma_\Lambda(k) = \Gamma_\Lambda^{(2)}(k, -k)$,

$$\partial_\Lambda \Sigma_\Lambda(k) = \frac{1}{2} \int_{k'} \dot{G}_\Lambda(k')\Gamma_\Lambda^{(4)}(k', -k', k, -k) + (\partial_\Lambda \bar{\varphi}_\Lambda^0)\Gamma_\Lambda^{(3)}(k, -k, 0)$$
$$- \int_{k'} \dot{G}_\Lambda(k')G_\Lambda(k' + k)\Gamma_\Lambda^{(3)}(k, -k - k', k')\Gamma_\Lambda^{(3)}(-k', k + k', -k) . \tag{8.18}$$

A generalization of Eq. (8.18) to the $O(N)$-symmetric case will be given in Eq. (9.46). Recall that with our normalization of $G_{0,\Lambda}(k)$ in Eqs. (8.4) and (8.5), the self-energy at the initial scale $\Lambda = \Lambda_0$ is simply given by the momentum-independent part of the inverse Gaussian propagator (see Eqs. (2.75) and (2.78)),

$$\Sigma_{\Lambda_0}(k) = \begin{cases} r_0 & \text{for } T > T_c \\ -2r_0 & \text{for } T < T_c \end{cases} . \tag{8.19}$$

Obviously, the flow of the self-energy is driven by the three- and four-legged vertices, which satisfy even more complicated flow equations that can be obtained as special cases of our general FRG flow equation (7.91). For later reference, let us here explicitly write down the exact FRG flow equation for the four-legged vertex $\Gamma_\Lambda^{(4)}(k_1, k_2, k_3, k_4)$ in the symmetric phase, where all odd vertices and the vacuum expectation value $\bar{\varphi}_\Lambda^0$ vanish,

$$
\partial_\Lambda \Gamma_\Lambda^{(4)}(k_1, k_2, k_3, k_4) = \frac{1}{2} \int_k \dot{G}_\Lambda(k) \Gamma_\Lambda^{(6)}(k, -k, k_1, k_2, k_3, k_4)
$$
$$
- \int_k \left\{ \dot{G}_\Lambda(k) G_\Lambda(-k - k_1 - k_2) \Gamma_\Lambda^{(4)}(k_1, k_2, k, -k - k_1 - k_2) \right.
$$
$$
\left. \times \Gamma_\Lambda^{(4)}(-k - k_1 - k_2, k, k_3, k_4) + (k_2 \leftrightarrow k_3) + (k_2 \leftrightarrow k_4) \right\} . \tag{8.20}
$$

This equation is a special case of our general superfield FRG flow equation (7.75) for the effective interaction in the absence of vacuum expectation values shown graphically in Fig. 7.5.

8.1.2 Rescaled Flow Equations

In order to classify the vertices according to their relevance and to develop approximation strategies based on relevance, it is useful to properly rescale the above exact flow equations and rewrite them in dimensionless form. The rescaled form of the FRG flow equations is also most convenient to discuss fixed points of the RG. Let us therefore introduce the dimensionless vertices

$$
\tilde{\Gamma}_l^{(n)}(q_1, \ldots, q_n) = \Lambda^{D(\frac{n}{2}-1)-n} \left(\frac{Z_l}{c_0} \right)^{n/2} \Gamma_\Lambda^{(n)}(\Lambda q_1, \ldots, \Lambda q_n) , \tag{8.21}
$$

which are considered to be functions of the logarithmic flow parameter

$$
l = -\ln(\Lambda/\Lambda_0) , \tag{8.22}
$$

and the dimensionless rescaled momenta

$$
q_i = \frac{k_i}{\Lambda} . \tag{8.23}
$$

The flowing wave function renormalization factor Z_l is defined in analogy with Eq. (4.97) via

$$
Z_l = \frac{1}{1 + \left. \frac{\partial \Sigma_\Lambda(k)}{\partial(c_0 k^2)} \right|_{k=0}} . \tag{8.24}
$$

Note that the power of Λ on the right-hand side of Eq. (8.21) can be identified with the canonical dimension of the vertex which follows from simple dimensional analysis. In particular, the dimensionless two-point vertex is related to the flowing irreducible self-energy via the relation

$$\tilde{\Gamma}_l^{(2)}(\boldsymbol{q}, -\boldsymbol{q}) \equiv \tilde{\Gamma}_l^{(2)}(\boldsymbol{q}) = \frac{Z_l}{c_0 \Lambda^2} \Sigma_\Lambda(\Lambda \boldsymbol{q}) , \tag{8.25}$$

where we have used the fact that the vertex is multiplied by a momentum conserving δ-function so that we may set $\boldsymbol{q}_1 = -\boldsymbol{q}_2 = \boldsymbol{q}$. Combining Eqs. (8.24) and (8.25), the flowing wave function renormalization factor can also be written as

$$\boxed{Z_l = 1 - \left. \frac{\partial \tilde{\Gamma}_l^{(2)}(\boldsymbol{q})}{\partial q^2} \right|_{q=0}} . \tag{8.26}$$

In the symmetry-broken phase, we should complement Eq. (8.21) by the rescaled vacuum expectation value M_l, which we define via

$$\bar{\varphi}_\Lambda^0 = \Lambda^{\frac{D}{2}-1} \left(\frac{Z_l}{c_0} \right)^{1/2} M_l , \tag{8.27}$$

implying

$$\Lambda \partial_\Lambda \bar{\varphi}_\Lambda^0 = \Lambda^{\frac{D}{2}-1} \left(\frac{Z_l}{c_0} \right)^{1/2} \left[\frac{D-2+\eta_l}{2} - \partial_l \right] M_l , \tag{8.28}$$

where $\eta_l = -\partial_l \ln Z_l$ is the flowing anomalous dimension. Using Eq. (8.24) we can relate η_l to the FRG flow equation for the self-energy,

$$\boxed{\eta_l = -\frac{\partial_l Z_l}{Z_l} = -Z_l \Lambda \lim_{k \to 0} \frac{\partial [\partial_\Lambda \Sigma_\Lambda(k)]}{\partial(c_0 k^2)}} . \tag{8.29}$$

Introducing the rescaled propagator

$$\tilde{G}_l(\boldsymbol{q}) = \frac{c_0 \Lambda^2}{Z_l} G_\Lambda(\Lambda \boldsymbol{q}) , \tag{8.30}$$

and the corresponding single-scale propagator

$$\dot{\tilde{G}}_l(\boldsymbol{q}) = -\frac{c_0 \Lambda^3}{Z_l} \dot{G}_\Lambda(\Lambda \boldsymbol{q}) , \tag{8.31}$$

the rescaled version of the order-parameter flow equation (8.17) assumes the form

$$\partial_l M_l = \frac{D - 2 + \eta_l}{2} M_l - \frac{1}{2} \int_q \check{G}_l(q) \frac{\tilde{\Gamma}_l^{(3)}(q, -q, 0)}{\tilde{\Gamma}_l^{(2)}(0)} \; . \tag{8.32}$$

Similarly, we obtain for the rescaled version of the FRG flow equation (8.18) for the self-energy

$$\partial_l \tilde{\Gamma}_l^{(2)}(q) = (2 - \eta_l - q \partial_q) \tilde{\Gamma}_l^{(2)}(q) + \dot{\Gamma}_l^{(2)}(q) \, , \tag{8.33}$$

with

$$\dot{\Gamma}_l^{(2)}(q) = -\frac{Z_l}{c_0 \Lambda} \partial_\Lambda \Sigma_\Lambda(k)$$

$$= \frac{1}{2} \int_{q'} \dot{\check{G}}_l(q') \tilde{\Gamma}_l^{(4)}(q', -q', q, -q) - \tilde{\Gamma}_l^{(3)}(q, -q, 0) \left[\partial_l - \frac{D - 2 + \eta_l}{2} \right] M_l$$

$$- \int_{q'} \dot{\check{G}}_l(q') \tilde{G}_l(q' + q) \tilde{\Gamma}_\Lambda^{(3)}(q, -q - q', q') \tilde{\Gamma}_\Lambda^{(3)}(-q', q + q', -q) \, . \tag{8.34}$$

In the second term on the right-hand side of this expression, we may eliminate the flowing order parameter M_l using Eq. (8.32) and obtain

$$\dot{\Gamma}_l^{(2)}(q) = \frac{1}{2} \int_{q'} \dot{\check{G}}_l(q') \left[\tilde{\Gamma}_l^{(4)}(q', -q', q, -q) - \frac{\left[\tilde{\Gamma}_l^{(3)}(q, -q, 0) \right]^2}{\tilde{\Gamma}_l^{(2)}(0)} \right]$$

$$- \int_{q'} \dot{\check{G}}_l(q') \tilde{G}_l(q' + q) \tilde{\Gamma}_\Lambda^{(3)}(q, -q - q', q') \tilde{\Gamma}_\Lambda^{(3)}(-q', q + q', -q) \, . \tag{8.35}$$

In the symmetric phase the three-legged vertex vanishes so that only the first term on the right-hand side of Eq. (8.35) survives. Using the exact relation (8.29), we see that the coefficient of the quadratic term in the expansion of $\dot{\Gamma}_l^{(2)}(q)$ in powers of q can be identified with the flowing anomalous dimension,

$$\boxed{\eta_l = \lim_{q \to 0} \frac{\partial \dot{\Gamma}_l^{(2)}(q)}{\partial q^2} \, .} \tag{8.36}$$

From Eqs. (8.35) and (8.36) it is obvious that in the symmetric phase the anomalous dimension is determined by the momentum dependence of the four-point vertex, while in the symmetry-broken phase one obtains already a finite result for η_l if one neglects the momentum dependence of the three- and four-point vertices (Berges et al. 2002, Sinner et al. 2008).

For later reference, we also give the flow equation for the rescaled four-point vertex, which is defined by setting $n = 4$ in Eq. (8.21),

$$\tilde{\Gamma}_l^{(4)}(\boldsymbol{q}_1, \boldsymbol{q}_2, \boldsymbol{q}_3, \boldsymbol{q}_4) = \Lambda^{D-4} \left(\frac{Z_l}{c_0} \right)^2 \Gamma_\Lambda^{(4)}(\Lambda \boldsymbol{q}_1, \Lambda \boldsymbol{q}_2, \Lambda \boldsymbol{q}_3, \Lambda \boldsymbol{q}_4) . \tag{8.37}$$

In analogy with the rescaled flow equation (8.33) for the two-point vertex, we write the corresponding flow equation for the four-point vertex in the form

$$\partial_l \tilde{\Gamma}_l^{(4)}(\boldsymbol{q}_1, \boldsymbol{q}_2, \boldsymbol{q}_3, \boldsymbol{q}_4) = \left[4 - D - 2\eta_l - \sum_{i=1}^{4} \boldsymbol{q}_i \cdot \nabla_{\boldsymbol{q}_i} \right] \tilde{\Gamma}_l^{(4)}(\boldsymbol{q}_1, \boldsymbol{q}_2, \boldsymbol{q}_3, \boldsymbol{q}_4)$$
$$+ \dot{\tilde{\Gamma}}_l^{(4)}(\boldsymbol{q}_1, \boldsymbol{q}_2, \boldsymbol{q}_3, \boldsymbol{q}_4) . \tag{8.38}$$

In the symmetric phase, where all odd vertices and the order parameter vanish, the inhomogeneity in the second line of Eq. (8.38) is

$$\dot{\tilde{\Gamma}}_l^{(4)}(\boldsymbol{q}_1, \boldsymbol{q}_2, \boldsymbol{q}_3, \boldsymbol{q}_4) = -\Lambda^{D-4} \frac{Z_l^2}{c_0^2} \Lambda \partial_\Lambda \Gamma_\Lambda^{(4)}(\boldsymbol{k}_1, \boldsymbol{k}_2, \boldsymbol{k}_3, \boldsymbol{k}_4)$$
$$= \frac{1}{2} \int_q \dot{\tilde{G}}_l(\boldsymbol{q}) \tilde{\Gamma}_l^{(6)}(\boldsymbol{q}, -\boldsymbol{q}, \boldsymbol{q}_1, \boldsymbol{q}_2, \boldsymbol{q}_3, \boldsymbol{q}_4)$$
$$- \int_q \left\{ \dot{\tilde{G}}_l(\boldsymbol{q}) \tilde{G}_l(-\boldsymbol{q} - \boldsymbol{q}_1 - \boldsymbol{q}_2) \tilde{\Gamma}_l^{(4)}(\boldsymbol{q}_1, \boldsymbol{q}_2, \boldsymbol{q}, -\boldsymbol{q} - \boldsymbol{q}_1 - \boldsymbol{q}_2) \right.$$
$$\left. \times \tilde{\Gamma}_l^{(4)}(-\boldsymbol{q} - \boldsymbol{q}_1 - \boldsymbol{q}_2, \boldsymbol{q}, \boldsymbol{q}_3, \boldsymbol{q}_4) + (\boldsymbol{q}_2 \leftrightarrow \boldsymbol{q}_3) + (\boldsymbol{q}_2 \leftrightarrow \boldsymbol{q}_4) \right\} . \tag{8.39}$$

In general, the exact FRG flow equation for the rescaled n-point vertex $\tilde{\Gamma}_l^{(n)}(\boldsymbol{q}_1, \ldots, \boldsymbol{q}_n)$ defined in Eq. (8.21) is of the form

$$\partial_l \tilde{\Gamma}_l^{(n)}(\boldsymbol{q}_1, \ldots, \boldsymbol{q}_n) = \left[n - D \left(\frac{n}{2} - 1 \right) - \frac{n}{2} \eta_l - \sum_{i=1}^{n} \boldsymbol{q}_i \cdot \nabla_{\boldsymbol{q}_i} \right] \tilde{\Gamma}_l^{(n)}(\boldsymbol{q}_1, \ldots, \boldsymbol{q}_n)$$
$$+ \dot{\tilde{\Gamma}}_l^{(n)}(\boldsymbol{q}_1, \ldots, \boldsymbol{q}_n) , \tag{8.40}$$

where the function $\dot{\tilde{\Gamma}}_l^{(n)}(\boldsymbol{q}_1, \ldots, \boldsymbol{q}_n)$ depends on the vertices $\tilde{\Gamma}_l^{(m)}(\boldsymbol{q}_1, \ldots, \boldsymbol{q}_m)$ with $m \le n + 2$.

Finally, we note that first-order partial differential equations of the type given in Eqs. (8.33) and (8.38) can be converted into equivalent integral equations. By direct differentiation it is easy to see that the FRG flow equation (8.33) for the rescaled two-point vertex is equivalent to the following integral equation (Busche et al. 2002, Hasselmann et al. 2004),

$$\tilde{\Gamma}_l^{(2)}(\boldsymbol{q}) = e^{2l - \int_0^l d\tau \, \eta_\tau} \, \tilde{\Gamma}_{l=0}^{(2)}(e^{-l} \boldsymbol{q}) + \int_0^l dt \, e^{2t - \int_{l-t}^l d\tau \, \eta_\tau} \, \dot{\tilde{\Gamma}}_{l-t}^{(2)}(e^{-t} \boldsymbol{q})$$
$$= e^{2l - \int_0^l d\tau \, \eta_\tau} \left[\tilde{\Gamma}_{l=0}^{(2)}(e^{-l} \boldsymbol{q}) + \int_0^l dl' e^{-2l' + \int_0^{l'} d\tau \, \eta_\tau} \, \dot{\tilde{\Gamma}}_{l'}^{(2)}(e^{-(l-l')} \boldsymbol{q}) \right] ,$$
$$\tag{8.41}$$

where in the second line we have substituted $t = l - l'$. Similarly, the rescaled flow equation (8.38) for the four-point vertex is equivalent to

$$\tilde{\Gamma}_l^{(4)}(\boldsymbol{q}_1, \boldsymbol{q}_2, \boldsymbol{q}_3, \boldsymbol{q}_4) = e^{(4-D)l - 2\int_0^l d\tau \eta_\tau}\, \tilde{\Gamma}_{l=0}^{(4)}(e^{-l}\boldsymbol{q}_1, e^{-l}\boldsymbol{q}_2, e^{-l}\boldsymbol{q}_3, e^{-l}\boldsymbol{q}_4)$$

$$+ \int_0^l dt\, e^{(4-D)t - 2\int_{l-t}^l d\tau \eta_\tau}\, \dot{\Gamma}_{l-t}^{(4)}(e^{-t}\boldsymbol{q}_1, e^{-t}\boldsymbol{q}_2, e^{-t}\boldsymbol{q}_3, e^{-t}\boldsymbol{q}_4)\,. \tag{8.42}$$

In general, the partial differential equation (8.40) for the rescaled n-point vertex is equivalent to the integral equation (Hasselmann et al. 2004)

$$\tilde{\Gamma}_l^{(n)}(\boldsymbol{q}_1, \ldots, \boldsymbol{q}_n) = e^{[n - D(\frac{n}{2}-1)]l - \frac{n}{2}\int_0^l d\tau \eta_\tau}\, \tilde{\Gamma}_{l=0}^{(n)}(e^{-l}\boldsymbol{q}_1, \ldots, e^{-l}\boldsymbol{q}_n)$$

$$+ \int_0^l dt\, e^{[n - D(\frac{n}{2}-1)]t - \frac{n}{2}\int_{l-t}^l d\tau \eta_\tau}\, \dot{\Gamma}_{l-t}^{(n)}(e^{-t}\boldsymbol{q}_1, \ldots, e^{-t}\boldsymbol{q}_n)\,. \tag{8.43}$$

8.1.3 FRG Flow Equations for a Sharp Momentum Cutoff

So far we have not specified the cutoff procedure. In order to make contact with the simple one-loop momentum shell RG discussed in Sect. 4.2, we now choose a sharp multiplicative momentum shell cutoff, which amounts to setting

$$G_{0,\Lambda}(\boldsymbol{k}) = \frac{\Theta(|\boldsymbol{k}| - \Lambda)}{c_0 k^2}\,, \tag{8.44}$$

see Eqs. (7.3) and (7.12). For the corresponding rescaled propagator we obtain from Eq. (8.30)

$$\tilde{G}_{0,l}(\boldsymbol{q}) = \frac{\Theta(|\boldsymbol{q}| - 1)}{q^2}\,, \tag{8.45}$$

while the single-scale propagator associated with Eq. (8.44) is according to Eq. (7.64) given by the formal expression

$$\dot{G}_\Lambda(\boldsymbol{k}) = \frac{\partial_\Lambda G_{0,\Lambda}(\boldsymbol{k})}{[1 + G_{0,\Lambda}(\boldsymbol{k})\Sigma_\Lambda(\boldsymbol{k})]^2} = \frac{-\delta(|\boldsymbol{k}| - \Lambda)}{c_0 k^2 \left[1 + \frac{\Theta(|\boldsymbol{k}|-\Lambda)}{c_0 k^2}\Sigma_\Lambda(\boldsymbol{k})\right]^2}\,. \tag{8.46}$$

Obviously, this expression is ambiguous because the Dirac δ-function $\delta(|\boldsymbol{k}| - \Lambda)$ is multiplied by a factor containing the step function $\Theta(|\boldsymbol{k}| - \Lambda)$. To properly define Eq. (8.46), we should smooth out the step function $\Theta(|\boldsymbol{k}| - \Lambda) \to \Theta_\epsilon(|\boldsymbol{k}| - \Lambda)$ (as shown in Fig. 7.1), and then take the limit of vanishing step width $\epsilon \to 0$ only after we have solved the FRG equations for finite ϵ. It turns out, however, that in the symmetric phase the cutoff dependence of the self-energy $\Sigma_\Lambda(\boldsymbol{k})$ does not contain any additional step function, so that we explicitly know the dependence of the right-hand side of Eq. (8.46) on the step function. We can therefore use the Morris–Lemma (Morris 1994) given in Eq. (7.15) to unambiguously define the single-scale propagator for a sharp momentum cutoff,

$$
\dot{G}_\Lambda(\boldsymbol{k}) = -\frac{\delta(|\boldsymbol{k}| - \Lambda)}{c_0 k^2} \int_0^1 dt \, \frac{1}{\left[1 + t\frac{\Sigma_\Lambda(\boldsymbol{k})}{c_0 k^2}\right]^2}
$$
$$
= -\frac{\delta(|\boldsymbol{k}| - \Lambda)}{c_0 k^2} \frac{1}{1 + \frac{\Sigma_\Lambda(\boldsymbol{k})}{c_0 k^2}} = -\frac{\delta(|\boldsymbol{k}| - \Lambda)}{c_0 \Lambda^2 + \Sigma_\Lambda(\boldsymbol{k})} \, . \tag{8.47}
$$

For the corresponding rescaled single-scale propagator we then obtain from the definition (8.31),

$$
\dot{\tilde{G}}_l(\boldsymbol{q}) = \frac{\delta(|\boldsymbol{q}| - 1)}{Z_l + \tilde{\Gamma}_l^{(2)}(\boldsymbol{q})} \, . \tag{8.48}
$$

Note that the Dirac δ-function restricts the set of wave vectors where $\dot{G}_\Lambda(\boldsymbol{k})$ is nonzero to the cutoff scale Λ, which is the reason why $\dot{G}_\Lambda(\boldsymbol{k})$ deserves to be called the *single-scale propagator*.

The flow equation (8.14) for the free energy $\Gamma_\Lambda^{(0)}[\bar{\varphi}^0]$ involves a combination of δ- and Θ-functions which is different from the combination appearing in the single-scale propagator. Using

$$
\delta(x) A[1 + \Theta(x) A]^{-1} = \delta(x) \int_0^1 dt \, A[1 + tA]^{-1} = \delta(x) \ln(1 + A) \, , \tag{8.49}
$$

we obtain from Eq. (8.14) in the sharp cutoff limit

$$
\partial_\Lambda \Gamma_\Lambda^{(0)}[\bar{\varphi}^0] = -\frac{V}{2} \int_{\boldsymbol{k}} \delta(|\boldsymbol{k}| - \Lambda) \int_0^1 dt \, \frac{\Sigma_\Lambda(\boldsymbol{k})}{c_0 k^2} \frac{1}{\left[1 + t\frac{\Sigma_\Lambda(\boldsymbol{k})}{c_0 k^2}\right]}
$$
$$
= -\frac{V}{2} \int_{\boldsymbol{k}} \delta(|\boldsymbol{k}| - \Lambda) \ln\left[\frac{c_0 \Lambda^2 + \Sigma_\Lambda(\boldsymbol{k})}{c_0 \Lambda^2}\right] \, . \tag{8.50}
$$

An obvious advantage of working with a sharp momentum cutoff is that the δ-function can be used to get rid of the integration over the modulus of $\boldsymbol{k}$. On the other hand, the nonanalyticity of the sharp cutoff can give rise to unphysical singularities in the momentum dependence of the vertices. Although in principle one can take care of these terms by introducing appropriate coupling constants (Hasselmann et al. 2004), it is rather difficult to analyze FRG flow equations in the enlarged coupling space. Usually it is then better to work with a smooth cutoff, where the vertices are regular functions of the external momenta (Sinner et al. 2008).

8.2 Recovering the Momentum Shell Results from the FRG

Let us now show how to recover from the above exact FRG flow equations the one-loop flow equations (4.71) and (4.72) for the coupling constants $\bar{r}_l$ and $\bar{u}_l$, which we have derived in Chap. 4 within the framework of the simple Wilsonian momentum shell RG. Therefore, we identify the coupling constants $\bar{r}_l$ and $\bar{u}_l$ defined in Eq. (4.70) with the following zero-momentum limits of the flowing self-energy $\Sigma_\Lambda(\boldsymbol{k})$ and the flowing effective interaction $\Gamma_\Lambda^{(4)}(\boldsymbol{k}_1, \boldsymbol{k}_2, \boldsymbol{k}_3, \boldsymbol{k}_4)$,

$$\bar{r}_l = \frac{Z_l r_\Lambda}{c_0 \Lambda^2} \, , \tag{8.51}$$

$$\bar{u}_l = K_D \frac{Z_l^2 u_\Lambda}{c_0^2 \Lambda^{4-D}} \, , \tag{8.52}$$

where $\Lambda = \Lambda_0 e^{-l}$, and the dimensionful couplings r_Λ and u_Λ are

$$r_\Lambda = \Sigma_\Lambda(\boldsymbol{k} = 0) \, , \tag{8.53}$$

$$u_\Lambda = \Gamma_\Lambda^{(4)}(0, 0, 0, 0) \, . \tag{8.54}$$

The exact RG flow equations for r_Λ and u_Λ are easily obtained from Eqs. (8.18) and (8.20) by setting all external momenta equal to zero. In the symmetric phase where all odd vertices vanish these equations reduce to

$$\partial_\Lambda r_\Lambda = \frac{1}{2} \int_k \dot{G}_\Lambda(\boldsymbol{k}) \Gamma_\Lambda^{(4)}(\boldsymbol{k}, -\boldsymbol{k}, 0, 0) \, , \tag{8.55}$$

$$\partial_\Lambda u_\Lambda = \frac{1}{2} \int_k \dot{G}_\Lambda(\boldsymbol{k}) \Gamma_\Lambda^{(6)}(\boldsymbol{k}, -\boldsymbol{k}, 0, 0, 0, 0)$$
$$- 3 \int_k \dot{G}_\Lambda(\boldsymbol{k}) G_\Lambda(-\boldsymbol{k}) \Gamma_\Lambda^{(4)}(0, 0, \boldsymbol{k}, -\boldsymbol{k}) \Gamma_\Lambda^{(4)}(-\boldsymbol{k}, \boldsymbol{k}, 0, 0) \, . \tag{8.56}$$

The right-hand sides of these exact flow equations depend on the momentum-dependent vertices with four and six external legs. Moreover, the propagator $G_\Lambda(\boldsymbol{k})$ and the single-scale propagator $\dot{G}_\Lambda(\boldsymbol{k})$ depend on the flowing momentum-dependent self-energy $\Sigma_\Lambda(\boldsymbol{k})$, so that further approximations are necessary to obtain from Eqs. (8.55) and (8.56) a closed set of equations for r_Λ and u_Λ. To recover the one-loop momentum shell results derived in Sect. 4.2.1, it is sufficient to completely neglect the contribution from the six-point vertex $\Gamma_\Lambda^{(6)}$ in Eq. (8.56). Moreover, we may also neglect the momentum dependence of the self-energy and of the four-point vertices on the right-hand side of Eqs. (8.55) and (8.56), which amounts to approximating

$$\Sigma_\Lambda(\boldsymbol{k}) \approx \Sigma_\Lambda(0) = r_\Lambda \, , \tag{8.57}$$

$$\Gamma_\Lambda^{(4)}(\boldsymbol{k}, -\boldsymbol{k}, 0, 0) \approx \Gamma_\Lambda^{(4)}(0, 0, 0, 0) = u_\Lambda \, . \tag{8.58}$$

Using for simplicity a sharp momentum cutoff, we may approximate the single-scale propagator in Eq. (8.55) by

$$\dot{G}_\Lambda(k) \approx -\frac{\delta(|k| - \Lambda)}{c_0 k^2 + r_\Lambda} \, . \tag{8.59}$$

The momentum integration is then trivial and Eq. (8.55) reduces to

$$\partial_\Lambda r_\Lambda = -\frac{u_\Lambda}{2} \int \frac{d^D k}{(2\pi)^D} \frac{\delta(|k| - \Lambda)}{c_0 k^2 + r_\Lambda} = -\frac{K_D}{2} \frac{\Lambda^{D-1} u_\Lambda}{c_0 \Lambda^2 + r_\Lambda} \, . \tag{8.60}$$

To perform the momentum integration in Eq. (8.56) one should keep in mind that within our approximation the flowing propagator for sharp cutoff is of the form (see Eq. (8.44))

$$G_\Lambda(k) = \left[G_{0,\Lambda}^{-1}(k) + \Sigma_\Lambda(k) \right]^{-1} \approx \frac{\Theta(|k| - \Lambda)}{c_0 k^2 + \Theta(|k| - \Lambda) r_\Lambda} \, , \tag{8.61}$$

so that Eq. (8.56) reduces to

$$\partial_\Lambda u_\Lambda = -3 u_\Lambda^2 \int \frac{d^D k}{(2\pi)^D} \dot{G}_\Lambda(k) \frac{\Theta(|k| - \Lambda)}{c_0 k^2 + \Theta(|k| - \Lambda) r_\Lambda} \, . \tag{8.62}$$

To properly remove the ambiguities inherent in the product of the δ-function carried by $\dot{G}_\Lambda(k)$ and the Θ-function-dependent factor, we go back to the representation (8.46) of the single-scale propagator for a sharp cutoff and use the Morris–Lemma (7.15),

$$\partial_\Lambda u_\Lambda = 3 u_\Lambda^2 \int \frac{d^D k}{(2\pi)^D} \delta(|k| - \Lambda) \int_0^1 dt \, t \frac{c_0 \Lambda^2}{(c_0 \Lambda^2 + t \, r_\Lambda)^3}$$

$$= \frac{3 K_D}{2} \frac{\Lambda^{D-1} u_\Lambda^2}{(c_0 \Lambda^2 + r_\Lambda)^2} \, . \tag{8.63}$$

We now introduce the dimensionless couplings $\bar{r}_l$ and $\bar{u}_l$ defined in Eqs. (8.51) and (8.52). Keeping in mind that within our approximation $Z_l = 1$ and using the fact that $\Lambda = \Lambda_0 e^{-l}$ implies $\Lambda \partial_\Lambda = -\partial_l$, it is easy to see that Eq. (8.60) reduces to

$$\partial_l \bar{r}_l = 2 \bar{r}_l + \frac{1}{2} \frac{\bar{u}_l}{1 + \bar{r}_l} \, , \tag{8.64}$$

while Eq. (8.63) becomes

$$\partial_l \bar{u}_l = (4 - D) \bar{u}_l - \frac{3}{2} \frac{\bar{u}_l^2}{(1 + \bar{r}_l)^2} \, . \tag{8.65}$$

These approximate flow equations are identical with the flow equations (4.71) and (4.72) derived in Sect. 4.2 using the simple Wilsonian momentum shell method. Note that while we have worked here with the unrescaled flow equations and have introduced dimensionless rescaled variables only in the end, we could have obtained the same results directly from the rescaled flow equations given in Sect. 8.1.2. In fact, we will use the latter strategy in the next section to calculate the momentum dependence of the self-energy.

8.3 Momentum-Dependent Self-Energy in the Symmetric Phase

8.3.1 Scaling Functions

One advantage of the FRG is that the exact hierarchy of flow equations for the momentum-dependent vertices can be used to obtain all vertices at finite wave vectors (and for quantum systems also at finite frequencies). For example, at the critical point the self-energy $\Sigma(k)$ of φ^4-theory is expected to scale as $\Sigma(k) \propto |k|^{2-\eta}$ for $|k| \to 0$, where the exponent η is finite for $D < 4$. For sufficiently small $|k|$ the propagator then scales as $G(k) \sim 1/\Sigma(k) \propto |k|^{-2+\eta}$, see Eq. (1.14) and the discussion in Sect. 4.2.3. But how large is the regime in momentum space where the self-energy scales as $|k|^{2-\eta}$? Clearly, there must be a characteristic interaction-dependent crossover scale k_c where the k-dependence of $\Sigma(k)$ crosses over from the asymptotic long-wavelength regime to another short-wavelength regime characterized by some different momentum dependence which may not be perturbatively accessible.[3] In fact, if the crossover scale k_c is smaller than the microscopic scale $\Lambda_0 \approx 1/a$ set by the lattice spacing of the underlying lattice model, then it should be possible to describe the momentum dependence of the self-energy at the critical point in terms of a dimensionless universal scaling function $\sigma_*(y)$,

$$\Sigma(k) = c_0 k_c^2 \sigma_* \left(\frac{|k|}{k_c} \right) \quad , \quad T = T_c \; . \tag{8.66}$$

More generally, away from the critical point where the correlation length ξ is finite Eq. (8.66) should be replaced by a two-parameter scaling function (Hasselmann et al. 2007, Sinner et al. 2008),

$$\Sigma(k) = c_0 k_c^2 \sigma \left(|k|\xi, \frac{|k|}{k_c} \right) \; , \tag{8.67}$$

where $\sigma_*(y) = \sigma(\infty, y)$. Obviously, for $y \ll 1$ the critical scaling function $\sigma_*(y)$ must be proportional to $y^{2-\eta}$. In this asymptotic long-wavelength regime the

---

[3] We shall show shortly that the scale k_c can be identified with the Ginzburg scale discussed after Eq. (4.73) in Sect. 4.2.2.

momentum dependence of the self-energy is essentially determined by the scaling properties of the relevant couplings in the vicinity of the critical point. On the other hand, the behavior of the critical scaling function in the regime $y \gg 1$ is not determined by the scaling in the vicinity of the critical point, so that for a microscopic calculation of $\sigma_*(y)$ in this regime one has to take into account couplings which are irrelevant at the critical point. At the first sight it seems that for $|k| \gg k_c$ one can obtain $\Sigma(k)$ by simple perturbation theory. However, for $D < 4$ perturbation theory is still infrared divergent in this regime, so that one needs a nonperturbative approach (such as the FRG) to take into account also the irrelevant couplings which determine the behavior of the self-energy for $|k| \gtrsim k_c$.

It is worth pointing out that the problem of calculating the momentum dependence of $\Sigma(k)$ of momenta outside the asymptotic long-wavelength regime is not entirely academic: as shown by Baym et al. (1999, 2001), the interaction-induced shift of the critical temperature of the interacting Bose gas in three dimensions is essentially determined by the self-energy of the classical $O(2)$-model (i.e., a classical φ^4-theory with a two-component field) for momenta of the order of k_c. It is therefore important to have quantitatively accurate calculations of the self-energy of the classical $O(2)$-model in this regime. This problem has first been addressed within the framework of the FRG in Ledowski et al. (2004) and Hasselmann et al. (2004) using the truncation of the FRG vertex expansion described below. Later Blaizot et al. (2006a) proposed a more sophisticated truncation of the FRG flow equations which is based on the fact that the derivatives of the irreducible n-point vertices with respect to a uniform external field can be related to the irreducible vertices with one additional external leg carrying zero momentum. This truncation, which is sometimes called the BMW approximation, has been shown to give very accurate results for critical exponents and momentum-dependent correlation functions (Blaizot et al. 2006b,c, Benitez et al. 2009). On the other hand, the numerical value for the T_c-shift for the weakly interacting Bose gas obtained within the BMW approximation differs only by a few percent from the value obtained by means of the simpler truncation described in the following subsection.

8.3.2 Truncation Strategy Based on Relevance

We now develop a simple truncation of the FRG flow equations which yields the momentum dependence of the self-energy for all values of the external momenta. We focus in this section on the symmetric phase, where all odd vertices vanish. The symmetry-broken phase will be discussed in Sect. 8.4. To justify our truncation procedure, it is convenient to start from the rescaled version of the exact RG flow equations given in Sect. 8.1.2. Consider the exact RG flow equation for the rescaled four-point vertex $\tilde{\Gamma}_l^{(4)}(q_1, q_2, q_3, q_4)$ in the paramagnetic phase given in Eqs. (8.38) and (8.39). The advantage of working with the rescaled flow equations is that we can directly read off the canonical dimensions of the couplings and thus classify all couplings according to their relevance with respect to a given fixed point. For example, from the first term on the right-hand side of Eq. (8.38) it is obvious that the momentum-independent part of the rescaled four-point vertex,

$$\tilde{\Gamma}_l^{(4)}(0, 0, 0, 0) = \frac{Z_l^2 u_\Lambda}{c_0^2 \Lambda^{4-D}} \, , \tag{8.68}$$

has canonical dimension $4 - D$ and is therefore relevant at the Gaussian fixed point in $D < 4$. If we expand $\tilde{\Gamma}_l^{(4)}(\boldsymbol{q}_1, \boldsymbol{q}_2, \boldsymbol{q}_3, \boldsymbol{q}_4)$ in powers of the external momenta, then the operator $\sum_i \boldsymbol{q}_i \cdot \nabla_{\boldsymbol{q}_i}$ in Eq. (8.38) reduces the scaling dimension of couplings proportional to q^n by n, so that the coupling constants multiplying larger powers of the external momenta in the expansion of the vertices are increasingly irrelevant. Nevertheless, these irrelevant terms are important to describe the crossover from the short-wavelength regime $|\boldsymbol{k}| \gg k_c$ to the critical long-wavelength scaling regime $|\boldsymbol{k}| \ll k_c$. It is therefore important to retain infinite powers of momenta in the expansion of the vertices in order to correctly describe this crossover.

Our aim is to develop a truncation of the exact FRG flow equations (8.33) and (8.38) for the two-point vertex and the four-point vertex which gives a numerically accurate interpolation formula for the self-energy $\Sigma(\boldsymbol{k})$ for all $\boldsymbol{k}$. Consider first the inhomogeneity $\dot{\tilde{\Gamma}}_l^{(4)}(\boldsymbol{q}_1, \boldsymbol{q}_2, \boldsymbol{q}_3, \boldsymbol{q}_4)$ in the rescaled flow equation for the four-point vertex, see Eqs. (8.38) and (8.39). We truncate the exact expression for $\dot{\tilde{\Gamma}}_l^{(4)}(\boldsymbol{q}_1, \boldsymbol{q}_2, \boldsymbol{q}_3, \boldsymbol{q}_4)$ given in Eq. (8.39) by replacing the (a priori unknown) functions $\tilde{\Gamma}_l^{(2)}(\boldsymbol{q})$, $\tilde{\Gamma}_l^{(4)}(\boldsymbol{q}_1, \boldsymbol{q}_2, \boldsymbol{q}_3, \boldsymbol{q}_4)$ and $\tilde{\Gamma}_l^{(6)}(\boldsymbol{q}_1, \boldsymbol{q}_2, \boldsymbol{q}_3, \boldsymbol{q}_4, \boldsymbol{q}_5, \boldsymbol{q}_6)$ appearing on the right-hand side of Eq. (8.39) by their relevant and marginal parts (with respect to the Gaussian fixed point). We shall refer to this strategy as *truncation based on relevance*. In dimensions $3 < D < 4$ this amounts to the following approximations on the right-hand side of Eq. (8.39),

$$\tilde{\Gamma}_l^{(2)}(\boldsymbol{q}) \approx \bar{r}_l + (1 - Z_l)q^2 \, , \tag{8.69a}$$

$$\tilde{\Gamma}_l^{(4)}(\boldsymbol{q}_1, \boldsymbol{q}_2, \boldsymbol{q}_3, \boldsymbol{q}_4) \approx \tilde{\Gamma}_l^{(4)}(0, 0, 0, 0) \equiv \tilde{u}_l \, , \tag{8.69b}$$

$$\tilde{\Gamma}_l^{(6)}(\boldsymbol{q}_1, \boldsymbol{q}_2, \boldsymbol{q}_3, \boldsymbol{q}_4, \boldsymbol{q}_5, \boldsymbol{q}_6) \approx 0 \, . \tag{8.69c}$$

Here the flowing coupling $\bar{r}_l = \tilde{\Gamma}_l^{(2)}(0) = Z_l \Sigma_\Lambda(0)/(c_0 \Lambda^2)$ has already been introduced in Eq. (8.51), and the rescaled coupling constant $\tilde{u}_l$ is related to the coupling $\bar{u}_l$ defined in Eq. (8.52) via

$$\tilde{u}_l = \frac{\bar{u}_l}{K_D} = \frac{Z_l^2 u_\Lambda}{c_0^2 \Lambda^{4-D}} \, . \tag{8.70}$$

In $D = 3$ the momentum-independent part $\tilde{\Gamma}_l^{(6)}(0, 0, 0, 0, 0, 0) = \tilde{v}_l$ of the six-point vertex has a vanishing canonical dimension and is therefore marginal if we neglect the anomalous dimension. Moreover, if we work with a sharp momentum cutoff the expansion of $\tilde{\Gamma}_l^{(4)}(\boldsymbol{q}_1, \boldsymbol{q}_2, \boldsymbol{q}_3, \boldsymbol{q}_4)$ in powers of momenta contains linear terms whose canonical dimensions also vanish in $D = 3$. Fortunately, these additional couplings are marginally irrelevant in the sense that the finite flowing anomalous dimension forces these coupling to flow to finite limiting values which can be absorbed by re-defining the numerical values of the relevant couplings. It is therefore reasonable to use the truncation (8.69a)–(8.69c) even in three dimensions and neglect the above mentioned marginally irrelevant couplings.[4]

If we approximate $Z_l \approx 1$ and set all external momenta equal to zero we recover the flow equations (8.64) and (8.65) for the coupling constants $\bar{r}_l$ and $\bar{u}_l$. In our *truncation based on relevance* we retain in addition the flowing anomalous dimension $\eta_l = -\partial_l \ln Z_l$, so that the flow equation (8.64) for the momentum-independent part of the rescaled two-point vertex should be replaced by

$$\boxed{\partial_l \bar{r}_l = (2 - \eta_l)\bar{r}_l + \frac{1}{2}\frac{\bar{u}_l}{1 + \bar{r}_l}} \tag{8.71}$$

while the flow equation for the momentum-independent part of the rescaled interaction $\bar{u}_l = K_D \tilde{u}_l$ reads

$$\boxed{\partial_l \bar{u}_l = (4 - D - 2\eta_l)\bar{u}_l - \frac{3}{2}\frac{\bar{u}_l^2}{(1 + \bar{r}_l)^2}} \tag{8.72}$$

To obtain a closed system of flow equations, we need an additional equation for the flowing anomalous dimension η_l, which in the symmetric phase is determined by the momentum dependence of the four-point vertex $\Gamma_l^{(4)}(\boldsymbol{q}_1, \boldsymbol{q}_2, \boldsymbol{q}_3, \boldsymbol{q}_4)$, see Eqs. (8.35) and (8.36). With our truncation (8.69a)–(8.69c) the function $\dot{\Gamma}_l^{(4)}(\boldsymbol{q}_1, \boldsymbol{q}_2, \boldsymbol{q}_3, \boldsymbol{q}_4)$ appearing in the rescaled flow equation (8.39) for the four-point vertex is approximated by[5]

$$\dot{\Gamma}_l^{(4)}(\boldsymbol{q}_1, \boldsymbol{q}_2, \boldsymbol{q}_3, \boldsymbol{q}_4) \approx -\tilde{u}_l^2 \left[\dot{\chi}_l(\boldsymbol{q}_1 + \boldsymbol{q}_2) + \dot{\chi}_l(\boldsymbol{q}_1 + \boldsymbol{q}_3) + \dot{\chi}_l(\boldsymbol{q}_1 + \boldsymbol{q}_4)\right] , \tag{8.73}$$

---

[4] It turns out that this truncation strategy yields an accurate result for the critical temperature T_c of the weakly interacting Bose gas in three dimensions (Ledowski et al. 2004). On the other hand, the momentum-independent part of the six-point vertex $\tilde{v}_l$ and the linear terms in the expansion of the four-point vertex contribute substantially to the numerical value of the critical exponent η (Hasselmann et al. 2004). However, the interaction correction to T_c is mainly dominated by momenta of the order of the crossover scale k_c, so that it depends only weakly on the precise value of η. This seems to be the reason why the results for T_c based on the truncation (8.69a)–(8.69c) are numerically accurate in $D = 3$.

[5] In terms of unrescaled couplings the truncation (8.73) amounts to replacing the flow equation (8.20) for the four-point vertex by

$$\partial_\Lambda \Gamma_\Lambda^{(4)}(\boldsymbol{k}_1, \boldsymbol{k}_2, \boldsymbol{k}_3, \boldsymbol{k}_4) = -u_\Lambda^2 \left[I_\Lambda(\boldsymbol{k}_1 + \boldsymbol{k}_2) + I_\Lambda(\boldsymbol{k}_1 + \boldsymbol{k}_3) + I_\Lambda(\boldsymbol{k}_1 + \boldsymbol{k}_4)\right] ,$$

with

$$I_\Lambda(\boldsymbol{k}) = \int_{\boldsymbol{k}'} \dot{G}_\Lambda(\boldsymbol{k}')G_\Lambda(\boldsymbol{k}' + \boldsymbol{k}) ,$$

and

$$G_\Lambda(\boldsymbol{k}) = \frac{Z_l\Theta(|\boldsymbol{k}| - \Lambda)}{c_0 k^2 + r_\Lambda} , \quad \dot{G}_\Lambda(\boldsymbol{k}) = -\frac{Z_l\delta(|\boldsymbol{k}| - \Lambda)}{c_0\Lambda^2 + r_\Lambda} .$$

with

$$\dot{\chi}_l(q) = \int_{q'} \dot{\tilde{G}}_l(q')\tilde{G}_l(q'+q) \, , \tag{8.74}$$

and

$$\tilde{G}_l(q) = \frac{\Theta(|q|-1)}{q^2 + \bar{r}_l} \, , \quad \dot{\tilde{G}}_l(q) = \frac{\delta(|q|-1)}{1 + \bar{r}_l} \, . \tag{8.75}$$

A similar truncation has been used by Busche et al. (2002) to calculate the spectral function of the Tomonaga–Luttinger model using FRG methods. For $D = 3$ the integration in Eq. (8.74) can be carried out analytically (Hasselmann et al. 2004),

$$\dot{\chi}_l(q) = K_D \frac{\Theta(q_2 - q_1)}{4(1 + \bar{r}_l)|q|} \ln\left[\frac{1 + \bar{r}_l + q^2 + 2|q|q_2}{1 + \bar{r}_l + q^2 + 2|q|q_1}\right] \, , \tag{8.76}$$

where

$$q_1 = \begin{cases} -|q|/2 & \text{if } |q| < 2 \\ -1 & \text{if } |q| > 2 \end{cases} \, , \tag{8.77}$$

$$q_2 = \begin{cases} 1 & \text{if } |q| < e^l - 1 \\ \frac{e^{2l}-1}{2|q|} - \frac{|q|}{2} & \text{if } |q| > e^l - 1 \end{cases} \, . \tag{8.78}$$

In particular, for $|q| \le \min\{2, e^l - 1\}$ we have

$$\dot{\chi}_l(q) = \frac{K_D}{4(1 + \bar{r}_l)|q|} \ln\left[\frac{(1 + |q|)^2 + \bar{r}_l}{1 + \bar{r}_l}\right] \, . \tag{8.79}$$

We may now substitute our approximate Eq. (8.73) into the exact integral equation (8.42) for the rescaled four-point vertex to obtain an explicit formula relating the flowing four-point vertex $\tilde{\Gamma}_l^{(4)}(q_1, q_2, q_3, q_4)$ to the three running couplings $\bar{r}_l, \bar{u}_l,$ and Z_l,

$$\tilde{\Gamma}_l^{(4)}(q_1, q_2, q_3, q_4) \approx e^{(4-D)l - 2\int_0^l d\tau\, \eta_\tau}\tilde{u}_0 - \int_0^l dt\, e^{(4-D)t - 2\int_{l-t}^l d\tau\, \eta_\tau}\tilde{u}_{l-t}^2$$

$$\times \left[\dot{\chi}_{l-t}(e^{-t}(q_1 + q_2)) + \dot{\chi}_{l-t}(e^{-t}(q_1 + q_3)) + \dot{\chi}_{l-t}(e^{-t}(q_1 + q_4))\right] \, . \tag{8.80}$$

For the calculation of the self-energy it is useful to write Eq. (8.80) in a different way. Using the fact that by definition $\tilde{\Gamma}_l^{(4)}(0, 0, 0, 0) = \tilde{u}_l$, we may explicitly separate from the right-hand side of Eq. (8.80) the momentum-dependent part of the effective interaction,

$$\tilde{\Gamma}_l^{(4)}(q_1, q_2, q_3, q_4) = \tilde{u}_l + \gamma_l^{(4)}(q_1, q_2, q_3, q_4) \, , \tag{8.81}$$

where the flow of $\tilde{u}_l = \bar{u}_l/K_D$ is determined by Eq. (8.72), and the momentum-dependent part of the effective interaction is given by

$$
\gamma_l^{(4)}(\boldsymbol{q}_1, \boldsymbol{q}_2, \boldsymbol{q}_3, \boldsymbol{q}_4) = -\int_0^l dt \, e^{(4-D)t - 2\int_{l-t}^l d\tau \eta_\tau} \tilde{u}_{l-t}^2 \Big[\delta\dot{\chi}_{l-t}(e^{-t}(\boldsymbol{q}_1 + \boldsymbol{q}_2))
$$

$$
+ \delta\dot{\chi}_{l-t}(e^{-t}(\boldsymbol{q}_1 + \boldsymbol{q}_3)) + \delta\dot{\chi}_{l-t}(e^{-t}(\boldsymbol{q}_1 + \boldsymbol{q}_4)) \Big] , \quad (8.82)
$$

with

$$
\delta\dot{\chi}_l(\boldsymbol{q}) = \dot{\chi}_l(\boldsymbol{q}) - \dot{\chi}_l(0) . \tag{8.83}
$$

To calculate the rescaled two-point vertex, we need the flow function $\dot{\Gamma}_l^{(2)}(\boldsymbol{q})$ defined in Eq. (8.35). Using the fact that in the symmetric phase all odd vertices vanish and approximating the flowing effective interaction by Eqs. (8.81) and (8.82), we obtain from Eq. (8.35) in the symmetric phase

$$
\dot{\Gamma}_l^{(2)}(\boldsymbol{q}) = \frac{\tilde{u}_l}{2} \int_{q'} \check{\dot{G}}_l(\boldsymbol{q}') + \frac{1}{2} \int_{q'} \check{\dot{G}}_l(\boldsymbol{q}') \gamma_l^{(4)}(\boldsymbol{q}', -\boldsymbol{q}', \boldsymbol{q}, -\boldsymbol{q})
$$

$$
= \frac{\bar{u}_l}{2(1+\bar{r}_l)} - \int_{q'} \frac{\delta(|\boldsymbol{q}'| - 1)}{1 + \bar{r}_l} \int_0^l dt \, e^{(4-D)t - 2\int_{l-t}^l d\tau \eta_\tau} \tilde{u}_{l-t}^2 \delta\dot{\chi}_{l-t}(e^{-t}(\boldsymbol{q}' + \boldsymbol{q})) .
$$

$$
\tag{8.84}
$$

Next, we substitute our approximate result (8.84) into the exact relation (8.36) between the flowing anomalous dimension η_l and the function $\dot{\Gamma}_l^{(2)}(\boldsymbol{q})$ to obtain the following integral equation for η_l (Ledowski et al. 2004, Hasselmann et al. 2004),

$$
\boxed{\eta_l = \int_0^l dt \, K(l, t) \bar{u}_{l-t}^2 e^{-2\int_{l-t}^l d\tau \eta_\tau} , } \tag{8.85}
$$

where the kernel $K(l, t)$ is given by

$$
K(l, t) = -\frac{1}{2D(1+\bar{r}_l)} \Big[(D - 1)e^{-(D-3)t} \dot{\chi}_{l-t}'(e^{-t}) + e^{-(D-2)t} \dot{\chi}_{l-t}''(e^{-t}) \Big] . \tag{8.86}
$$

Here $\dot{\chi}_l'(\boldsymbol{q}) = \partial\dot{\chi}_l(\boldsymbol{q})/\partial q$ and $\dot{\chi}_l''(\boldsymbol{q}) = \partial^2\dot{\chi}_l(\boldsymbol{q})/\partial q^2$ are the derivatives of the function $\dot{\chi}_l(\boldsymbol{q})$ defined in Eq. (8.74) with respect to $q = |\boldsymbol{q}|$. Together with Eqs. (8.71) and (8.72), the integral equation (8.85) forms a closed system of equations for the three unknown functions $\bar{u}_l$, $\bar{r}_l$, and η_l, which can easily be solved numerically. If the initial value of $\bar{r}_0$ is fine-tuned such that the RG trajectory flows into the Wilson–Fisher fixed point (for $D < 4$), then the limits $\eta = \lim_{l\to\infty} \eta_l$ and $\bar{u}_* = \lim_{l\to\infty} \bar{u}_l$ are finite. In this case, Eq. (8.85) reduces to a self-consistency equation for the critical exponent η,

$$\eta = \bar{u}_*^2 \int_0^\infty dt\, K(\infty, t) e^{-2\eta t} \,. \qquad (8.87)$$

For small $\epsilon = 4 - D$ this yields the leading order result of the ϵ-expansion, $\eta \approx \frac{\epsilon^2}{54}$ (Zinn-Justin 2002), while a numerical solution in three dimensions gives $\eta \approx 0.101$ (Hasselmann et al. 2007). This is more than twice as large as the established value $\eta \approx 0.036$ for the three-dimensional Ising universality class (Pelissetto and Vicari 2002), indicating that our truncation is not sufficient to obtain quantitatively accurate results for the critical exponents.

On the other hand, our truncation yields a reasonable interpolation formula for the momentum-dependent self-energy $\Sigma(k)$ for all momenta which are small compared with the ultraviolet cutoff Λ_0. Let us therefore go back to the integrated form (8.41) of the exact FRG flow equation (8.33) for the rescaled two-point vertex $\tilde{\Gamma}_l^{(2)}(q) = Z_l \Sigma_\Lambda(\Lambda q)/(c_0 \Lambda^2)$. Introducing the momentum-dependent part of the function $\dot{\Gamma}_l^{(2)}(q)$ appearing on the right-hand side of Eq. (8.33),

$$\dot{\gamma}_l^{(2)}(q) = \dot{\Gamma}_l^{(2)}(q) - \dot{\Gamma}_l^{(2)}(0) \,, \qquad (8.88)$$

and keeping in mind that by definition $\tilde{\Gamma}_l^{(2)}(0) = \bar{r}_l$, it is easy to show from Eq. (8.41) that the exact self-energy can be expressed as the following integral over the entire RG trajectory (Hasselmann et al. 2007),

$$\boxed{\frac{\Sigma(k)}{\Lambda_0^2} = \lim_{l \to \infty} \frac{e^{-2l} \bar{r}_l}{Z_l} + \int_0^\infty dl\, e^{-2l + \int_0^l d\tau\, \eta_\tau}\, \dot{\gamma}_l^{(2)}(e^l k/\Lambda_0) \,.} \qquad (8.89)$$

Given our approximate result (8.84) for the function $\dot{\Gamma}^{(2)}(q)$, we have

$$\dot{\gamma}_l^{(2)}(q) = -\int_0^l dt\, e^{(4-D)t - 2\int_{l-t}^l d\tau\, \eta_\tau}\, \tilde{u}_{l-t}^2$$
$$\times \int_{q'} \delta(|q'| - 1) \left[\delta\dot{\chi}_{l-t}(e^{-t}(q' + q)) - \delta\dot{\chi}_{l-t}(e^{-t}q') \right] \,. \qquad (8.90)$$

Explicitly, we then obtain from Eq. (8.89),

$$\frac{\Sigma(k)}{\Lambda_0^2} = \lim_{l \to \infty} \frac{e^{-2l} \bar{r}_l}{Z_l} - \int_0^\infty dl\, \frac{e^{-2l + \int_0^l d\tau\, \eta_\tau}}{1 + \bar{r}_l} \int_0^l dt\, e^{(4-D)t - 2\int_{l-t}^l d\tau\, \eta_\tau}\, \tilde{u}_{l-t}^2$$
$$\times \int_{q'} \delta(|q'| - 1) \left[\delta\dot{\chi}_{l-t}(e^{-t}q' + e^{l-t}k/\Lambda_0)) - \delta\dot{\chi}_{l-t}(e^{-t}q') \right] \,. \qquad (8.91)$$

The calculation of the momentum-dependent self-energy is now reduced to a three-dimensional integration (two one-dimensional integrations over the RG flow parameters l and t, and one angular integration over the angle between q' and the external wave vector q), which can easily be performed numerically (Ledowski et al. 2004, Hasselmann et al. 2004, 2007).

8.3.3 FRG Results for the Self-Energy Scaling Function

In order to extract from Eq. (8.91) the scaling functions $\sigma_*(y)$ and $\sigma(x, y)$ defined in Eqs. (8.66) and (8.67), recall that slightly above the critical temperature the RG flow as a function of the logarithmic flow parameter l exhibits three characteristic regimes, as shown in Fig. 4.6 of Chap. 4. In the initial regime $0 < l \lesssim l_c$ the running couplings $\bar{r}_l$ and $\bar{u}_l$ rapidly flow toward the fixed point $(\bar{r}_*, \bar{u}_*)$; then there is an intermediate interval $l_c \lesssim l \lesssim l_*$ where the RG trajectory remains almost stationary in the vicinity of the fixed point; finally, for $l \gtrsim l_*$ the RG trajectory rapidly flows away from the fixed point so that the leading l-dependence of $\bar{r}_l$ in this regime is proportional to e^{2l}, as given by the canonical dimension of $\bar{r}_l$. Note that only in the vicinity of the critical point the asymptotic behavior is $\delta\bar{r} \propto e^{l/\nu}$. Keeping in mind that according to Eq. (4.98) the correlation length ξ of the system is related to the self-energy for vanishing wave vector via (see also the footnote in Sect. 3.3.1)

$$\frac{c_0}{\xi^2} = Z\Sigma(0) = Z \lim_{\Lambda \to 0} r_\Lambda = c_0 \Lambda_0^2 \lim_{l \to \infty} e^{-2l}\bar{r}_l \, , \tag{8.92}$$

we see that by construction the logarithmic scale l_* is directly related to the correlation length ξ of the system via (Hasselmann et al. 2007, Sinner et al. 2008)

$$e^{-2l_*} = \lim_{l \to \infty} e^{-2l}\bar{r}_l = \frac{1}{(\Lambda_0 \xi)^2} \, . \tag{8.93}$$

At the critical point where $\xi \to \infty$ the scale l_* moves to infinity. On the other hand, the scale l_c, remains finite at the critical point. Recall that l_c is defined in terms of the RG time needed for the initial point in coupling space to reach the vicinity of the fixed point. As already mentioned in Sect. 4.2.2, the corresponding scale $k_c = \Lambda_0 e^{-l_c}$ can be identified with the Ginzburg scale, which sets an upper limit to the regime in momentum space where the self-energy at the critical point exhibits the asymptotic scaling $\Sigma(k) \propto |k|^{2-\eta}$. We are now in a position to justify this identification by an explicit calculation of the momentum-dependent self-energy. Introducing the dimensionless scaling variables

$$x = |k|\xi \, , \quad y = |k|/k_c \, , \tag{8.94}$$

so that $x/y = k_c\xi = e^{l_*-l_c}$, we may rewrite our formally exact integral representation (8.89) for the self-energy $\Sigma(k)$ in the scaling form (8.67), with the scaling function $\sigma(x, y)$ explicitly given by

$$\sigma(x, y) = \frac{y^2}{Zx^2} + \int_0^\infty dl \, e^{-2l+\int_0^l d\tau\, \eta_\tau} \dot{\gamma}_l^{(2)}(e^{l-l_c}y) \, , \tag{8.95}$$

where we have written $\dot{\gamma}_l^{(2)}(|\mathbf{q}|) = \dot{\gamma}_l^{(2)}(\mathbf{q})$, using the fact that this function actually depends only on $|\mathbf{q}|$. The exact expression (8.95) is useful to discuss the scaling of the self-energy at or close to the critical point. For simplicity, let us focus here on the self-energy at the critical point $T = T_c$, where $x \to \infty$ and $\sigma(\infty, y) \equiv \sigma_*(y)$ can be identified with the critical scaling function defined in Eq. (8.66),

$$\sigma_*(y) = \int_0^\infty dl\, e^{-2l + \int_0^l d\tau\, \eta_\tau}\, \dot{\gamma}_l^{(2)}(e^{l-l_c} y) \,. \tag{8.96}$$

Substituting into this exact expression our approximate result (8.90) for the function $\dot{\gamma}_l^{(2)}(|\mathbf{q}|)$ we obtain an explicit expression for the critical scaling function $\sigma_*(y)$ which is easily evaluated numerically. The result for $D = 3$ is shown in Fig. 8.1. One clearly sees the crossover at $y \approx 1$ from the critical $y^{2-\eta}$-scaling to another short-wavelength regime where $\sigma_*(y) \propto \ln y$ in $D = 3$. Keeping in mind that $y \approx 1$ corresponds to $|\mathbf{k}| \approx k_c$ we have thus proven that the scale k_c can indeed be identified with the Ginzburg scale defining the upper boundary of the critical regime in momentum space. Recall that in Sect. 4.2.2 we have introduced k_c in terms of the scale $l_c = \ln(\Lambda_0/k_c)$ where the RG flow of the relevant couplings $\bar{u}_l$ and $\bar{r}_l$ approaches the vicinity of the critical fixed point, as shown in Fig. 4.6. That the scale $k_c = \Lambda_0 e^{-l_c}$ defined in this way in terms of the RG flow of $\bar{u}_l$ and $\bar{r}_l$ can also be identified with the upper boundary of the momentum regime where the self-energy exhibits critical scaling is a nontrivial result of our FRG calculation of the critical scaling function.

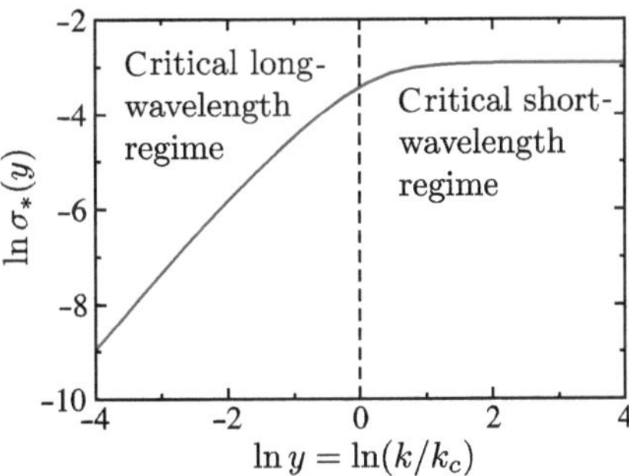

Fig. 8.1 Numerical evaluation of the critical scaling function $\sigma_*(y)$ defined in Eq. (8.96) in $D = 3$, using the approximation (8.90) for the function $\dot{\gamma}_l^{(2)}(q)$

8.4 Momentum-Dependent Self-Energy in the Symmetry-Broken Phase

Finally, let us derive a closed system of FRG flow equations for the momentum-dependent self-energy in the symmetry-broken phase $T < T_c$. Starting point is again the system of exact FRG flow equations (8.17), (8.18), and (8.20) given in Sect. 8.1.1. To motivate our truncation, it is useful to consider first the proper initial values of the vertices of our theory in the symmetry-broken phase at scale $\Lambda = \Lambda_0$. Recall that for $T < T_c$ we should shift the field according to $\varphi(\mathbf{k}) = \bar{\varphi}_\Lambda^0(\mathbf{k}) + \delta\varphi(\mathbf{k})$, where the average of $\delta\varphi(\boldsymbol{k})$ vanishes and $\bar{\varphi}_\Lambda^0(\boldsymbol{k}) = (2\pi)^D \delta(\boldsymbol{k})\bar{\varphi}_\Lambda^0$, see Eqs. (2.72) and (8.16). For simplicity we write $\bar{\varphi}_{\Lambda_0}^0 = \bar{\varphi}_0$ anticipating that the initial value of the flowing order parameter $\bar{\varphi}_\Lambda^0$ can be identified with the mean-field value of the order parameter which we have already introduced in the Gaussian approximation in Chapter 2.3, see Eq. (2.74). Substituting the above decomposition of the field into our bare action $S[\varphi] = S_{0,\Lambda}[\varphi] + S_1[\varphi]$ defined in Eqs. (8.1)–(8.3), we obtain for the interaction part,

$$
\begin{aligned}
S_1[\bar{\varphi}^0; \delta\varphi] &\equiv S_1[\varphi \to \bar{\varphi}^0 + \delta\varphi] \\
&= V\left[f_0 + \frac{r_0}{2}\bar{\varphi}_0^2 + \frac{u_0}{4!}\bar{\varphi}_0^4 \right] + \Gamma_{\Lambda_0}^{(1)}\delta\varphi_0 + \frac{1}{2}\int_k \Sigma_{\Lambda_0}(\boldsymbol{k})\delta\varphi_{-k}\delta\varphi_k \\
&\quad + \frac{1}{3!}\int_{k_1}\int_{k_2}\int_{k_3}(2\pi)^D\delta\left(\sum_{i=1}^3 \boldsymbol{k}_i\right)\Gamma_{\Lambda_0}^{(3)}(\boldsymbol{k}_1, \boldsymbol{k}_2, \boldsymbol{k}_3)\delta\varphi_{k_1}\delta\varphi_{k_2}\delta\varphi_{k_3} \\
&\quad + \frac{1}{4!}\int_{k_1}\cdots\int_{k_4}(2\pi)^D\delta\left(\sum_{i=1}^4 \boldsymbol{k}_i\right)\Gamma_{\Lambda_0}^{(4)}(\boldsymbol{k}_1, \boldsymbol{k}_2, \boldsymbol{k}_3, \boldsymbol{k}_4)\delta\varphi_{k_1}\delta\varphi_{k_2}\delta\varphi_{k_3}\delta\varphi_{k_4} \,,
\end{aligned}
\tag{8.97}
$$

where the bare vertices now depend on the vacuum expectation value $\bar{\varphi}_0$,

$$
\Gamma_{\Lambda_0}^{(1)} = r_0\bar{\varphi}_0 + \frac{u_0}{6}\bar{\varphi}_0^3 \,,
\tag{8.98a}
$$

$$
\Sigma_{\Lambda_0}(\boldsymbol{k}) = r_0 + \frac{u_0}{2}\bar{\varphi}_0^2 \,,
\tag{8.98b}
$$

$$
\Gamma_{\Lambda_0}^{(3)}(\boldsymbol{k}_1, \boldsymbol{k}_2, \boldsymbol{k}_3) = u_0\bar{\varphi}_0 \,,
\tag{8.98c}
$$

$$
\Gamma_{\Lambda_0}^{(4)}(\boldsymbol{k}_1, \boldsymbol{k}_2, \boldsymbol{k}_3, \boldsymbol{k}_4) = u_0 \,.
\tag{8.98d}
$$

We now require that $\Gamma_{\Lambda_0}^{(1)} = 0$, so that the initial value $\bar{\varphi}_0$ of the order parameter in our functional RG is the minimum of the effective potential in Landau approximation given in Eq. (2.74). Note that for $r_0 < 0$ we have $\bar{\varphi}_0^2 = -6r_0/u_0$, so that

$$
\Sigma_{\Lambda_0}(\boldsymbol{k}) = -2r_0 = \frac{u_0}{3}\bar{\varphi}_0^2 \,.
\tag{8.99}
$$

The interaction part of our initial action can then be written in real space as an integral of the form

$$
S_1[\varphi] = \int d^D r\, U_{\Lambda_0}(\rho(\boldsymbol{r})) \,,
\tag{8.100}
$$

where the so-called effective potential $U_{\Lambda_0}(\rho(\boldsymbol{r}))$ is the following local function of the density $\rho(\boldsymbol{r}) = \varphi^2(\boldsymbol{r})/2$,

$$
U_{\Lambda_0}(\rho(\boldsymbol{r})) = f_0 - \frac{3}{2}\frac{r_0^2}{u_0} + \frac{u_0}{4!}\left(\varphi^2(\boldsymbol{r}) - \bar{\varphi}_0^2\right)^2 = f_0 - \frac{3}{2}\frac{r_0^2}{u_0} + \frac{u_0}{6}(\rho - \rho_0)^2 \,.
\tag{8.101}
$$

Note that the first two terms on the right-hand side of Eq. (8.101) correspond to the free energy density in mean-field approximation, see Eq. (2.22). In the local potential approximation with a truncated quartic potential (which will be discussed in the following Chap. 9) one would now require that the renormalized effective potential is again of the form (8.101), but with the bare parameters u_0 and $\bar{\varphi}_0$ replaced by flowing couplings u_Λ and $\bar{\varphi}_\Lambda^0$. This approximation amounts to approximating the flowing vertices by

$$\Sigma_\Lambda(k) \approx \frac{u_\Lambda}{3}\left(\bar{\varphi}_\Lambda^0\right)^2 , \tag{8.102a}$$

$$\Gamma_\Lambda^{(3)}(k_1, k_2, k_3) \approx u_\Lambda \bar{\varphi}_\Lambda^0 , \tag{8.102b}$$

$$\Gamma_\Lambda^{(4)}(k_1, k_2, k_3, k_4) \approx u_\Lambda . \tag{8.102c}$$

The crucial observation is now that we may use these relations to obtain a non-trivial truncation of the exact flow equation (8.18) for the momentum-dependent self-energy $\Sigma_\Lambda(k)$: therefore we simply substitute Eqs. (8.102a)–(8.102c) on the right-hand sides of Eqs. (8.17) and (8.18), thus expressing the unknown vertices $\Gamma_\Lambda^{(3)}(k_1, k_2, k_3)$ and $\Gamma_\Lambda^{(4)}(k_1, k_2, k_3, k_4)$ in terms of known running couplings (Schütz and Kopietz 2006, Sinner et al. 2008). The order-parameter flow equation (8.17) then reduces to

$$\partial_\Lambda \left(\bar{\varphi}_\Lambda^0\right)^2 = -3 \int_k \dot{G}_\Lambda(k) , \tag{8.103}$$

while the flow equation (8.18) for the self-energy can be written as

$$\partial_\Lambda \Sigma_\Lambda(k) = \frac{u_\Lambda}{2} \int_{k'} \dot{G}_\Lambda(k') + \frac{u_\Lambda}{2} \partial_\Lambda \left(\bar{\varphi}_\Lambda^0\right)^2$$
$$- u_\Lambda^2 \left(\bar{\varphi}_\Lambda^0\right)^2 \int_{k'} \dot{G}_\Lambda(k')G_\Lambda(k' + k)$$
$$= -u_\Lambda \int_{k'} \dot{G}_\Lambda(k') - u_\Lambda^2 \left(\bar{\varphi}_\Lambda^0\right)^2 \int_{k'} \dot{G}_\Lambda(k')G_\Lambda(k' + k) . \tag{8.104}$$

By demanding that the flow of $\Sigma_\Lambda(0)$ is consistent with our truncation (8.102a), we obtain the flow equation for the effective interaction,

$$\partial_\Lambda u_\Lambda = -3u_\Lambda^2 \int_k \dot{G}_\Lambda(k)G_\Lambda(k) . \tag{8.105}$$

The above equations (8.103)–(8.105) form a closed system of coupled integro-differential equations for the order parameter $\bar{\varphi}_\Lambda^0$, the effective interaction u_Λ, and the momentum-dependent self-energy $\Sigma_\Lambda(k)$. Note that the right-hand side of Eq. (8.104) is momentum-dependent, implying that in the symmetry-broken phase a simple one-loop calculation already gives a finite estimate for the flowing anomalous dimension. For a detailed analysis of the above flow equations and an explicit calculation of the momentum-dependent self-energy in the symmetry broken phase see (Sinner et al. 2008). A similar truncation strategy has recently been used to calculate the momentum- and frequency-dependent single-particle spectral function in the condensed phase of the weakly interacting Bose gas in two dimensions (Sinner et al. 2009).

References

Baym, G., J. P. Blaizot, M. Holzmann, F. Laloë, and D. Vautherin (1999), *The transition temperature of the dilute interacting Bose gas*, Phys. Rev. Lett. **83**, 1703.

Baym, G., J. P. Blaizot, M. Holzmann, F. Laloë, and D. Vautherin (2001), *Bose-Einstein transition in a dilute interacting gas*, Eur. Phys. J. B **24**, 107.

Benitez, F., J.-P. Blaizot, H. Chaté, B. Delamotte, R. Méndez-Galain, and N. Wschebor (2009), *Solutions of renormalization-group flow equations with full momentum dependence*, Phys. Rev. E **80**, 030103(R).

Berges, J., N. Tetradis, and C. Wetterich (2002), *Non-perturbative renormalization flow in quantum field theory and statistical physics*, Phys. Rep. **363**, 223.

Blaizot, J. P., R. Méndez-Galain, and N. Wschebor (2006a), *A new method to solve the non-perturbative renormalization group equations*, Phys. Lett. B **632**, 571.

Blaizot, J. P., R. Méndez-Galain, and N. Wschebor (2006b), *Nonperturbative renormalization group and momentum dependence of n-point functions. I*, Phys. Rev. E **74**, 51116.

Blaizot, J. P., R. Méndez-Galain, and N. Wschebor (2006c), *Nonperturbative renormalization group and momentum dependence of n-point functions. II*, Phys. Rev. E **74**, 51117.

Busche, T., L. Bartosch, and P. Kopietz (2002), *Dynamic scaling in the vicinity of the Luttinger liquid fixed point*, J. Phys.: Condens. Matter **14**, 8513.

Hasselmann, N., S. Ledowski, and P. Kopietz (2004), *Critical behavior of weakly interacting bosons: A functional renormalization-group approach*, Phys. Rev. A **70**, 063621.

Hasselmann, N., A. Sinner, and P. Kopietz (2007), *Two-parameter scaling of correlation functions near continuous phase transitions*, Phys. Rev. E **76**, 040101.

Ledowski, S., N. Hasselmann, and P. Kopietz (2004), *Self-energy and critical temperature of weakly interacting bosons*, Phys. Rev. A **69**, 061601.

Litim, D. F. (2001), *Optimized renormalization group flows*, Phys. Rev. D **64**, 105007.

Morris, T. R. (1994), *The exact renormalisation group and approximate solutions*, Int. J. Mod. Phys. A **9**, 2411.

Pelissetto, A. and E. Vicari (2002), *Critical phenomena and renormalization group theory*, Phys. Rep. **368**, 549.

Schütz, F. and P. Kopietz (2006), *Functional renormalization group with vacuum expectation values and spontaneous symmetry breaking*, J. Phys. A: Math. Gen. **39**, 8205.

Sinner, A., N. Hasselmann, and P. Kopietz (2008), *Functional renormalization group in the broken symmetry phase: Momentum dependence and two-parameter scaling of the self-energy*, J. Phys.: Condens. Matter **20**, 075208.

Sinner, A., N. Hasselmann, and P. Kopietz (2009), *Spectral function and quasiparticle damping of interacting bosons in two dimensions*, Phys. Rev. Lett. **102**, 120601.

Zinn-Justin, J. (2002), *Quantum Field Theory and Critical Phenomena*, 4th ed., Clarendon Press, Oxford.

Chapter 9
Derivative Expansion

Apart from the vertex expansion discussed in Chap. 8 the other main strategy to obtain approximate solutions of the FRG flow equations is the derivative expansion. This method has been successfully applied to problems in statistical physics and field theory (see Bagnuls and Bervillier 2001, Berges et al. 2002, and Pawlowski 2007 for comprehensive reviews) where one is usually only interested in long-wavelength phenomena and an expansion of $\Gamma_\Lambda[\bar{\Phi}]$ in gradients of the fields $\bar{\Phi}$ seems reasonable. Such an expansion is justified by noting that, although the true generating functional of the irreducible vertices $\Gamma[\bar{\Phi}] = \Gamma_{\Lambda=0}[\bar{\Phi}]$ can contain non-analyticities, the flowing $\Gamma_\Lambda[\bar{\Phi}]$ is analytic for any finite value of the cutoff Λ.

To obtain the derivative expansion it is convenient to work with Wetterich's Legendre effective action $\Gamma_\Lambda^{\mathrm{We}}[\bar{\Phi}]$ defined in Eq. (7.53), whose second derivatives differ from the second derivatives of our generating functional of irreducible vertices $\Gamma_\Lambda[\bar{\Phi}]$ defined in Eq. (7.40) by the cutoff-independent inverse free propagator and a field-independent constant. For $\Lambda \to 0$, the functional $\Gamma_\Lambda^{\mathrm{We}}[\bar{\Phi}]$ is the true Legendre transform of the generating functional of connected Green functions (this is the reason why this functional deserves to be called *Legendre effective action*) and is therefore convex. As noted by Wetterich (1993), $\Gamma_\Lambda^{\mathrm{We}}[\bar{\Phi}]$ satisfies the initial condition

$$\lim_{\Lambda \to \Lambda_0} \Gamma_\Lambda^{\mathrm{We}}[\bar{\Phi}] = S_{\Lambda_0}[\bar{\Phi}] \,, \tag{9.1}$$

and during the flow of Λ nicely interpolates between the bare action $S_{\Lambda_0}[\bar{\Phi}] = S[\bar{\Phi}]$ and the Legendre effective action,

$$\lim_{\Lambda \to 0} \Gamma_\Lambda^{\mathrm{We}}[\bar{\Phi}] = \Gamma^{\mathrm{We}}[\bar{\Phi}] \,. \tag{9.2}$$

This follows directly from the relations established in Sect. 7.2.4.

9.1 Derivative Expansion for the $O(N)$-Symmetric Classical φ^4-Theory

For concreteness, let us study here the $O(N)$-symmetric classical φ^4-theory with bare action

Kopietz, P. et al.: *Derivative Expansion*. Lect. Notes Phys. **798**, 233–247 (2010)
DOI 10.1007/978-3-642-05094-7_9   © Springer-Verlag Berlin Heidelberg 2010

$$S_{\Lambda_0}[\boldsymbol{\varphi}] = \int d^D r \left[\frac{r_0}{2} \boldsymbol{\varphi}^2(\boldsymbol{r}) + \frac{c_0}{2} \left(\nabla \boldsymbol{\varphi}(\boldsymbol{r}) \right)^2 + \frac{u_0}{4!} \left(\boldsymbol{\varphi}^2(\boldsymbol{r}) \right)^2 \right] , \qquad (9.3)$$

where $\boldsymbol{\varphi} = (\varphi_1, \ldots, \varphi_N)$ is an N-component classical field. This model is a straightforward generalization of our scalar φ^4-theory given in Eq. (2.65) to the case of N real fields; the case $N = 2$ describes the XY and the case $N = 3$ describes the Heisenberg universality class. The generating functional $\Gamma_\Lambda^{\mathrm{We}}[\bar{\boldsymbol{\varphi}}]$ satisfies the Wetterich equation (7.56) which for our model can be written as

$$\partial_\Lambda \Gamma_\Lambda^{\mathrm{We}}[\bar{\boldsymbol{\varphi}}] = \frac{1}{2} \mathrm{Tr} \left[\frac{\partial_\Lambda \mathbf{R}_\Lambda}{\Gamma_\Lambda^{\mathrm{We}\,(2)}[\bar{\boldsymbol{\varphi}}] + \mathbf{R}_\Lambda} \right] , \qquad (9.4)$$

where the trace is over momentum space and the flavor space associated with the N field components. Here, we have introduced the matrix of second-order derivatives,

$$\boldsymbol{\Gamma}_\Lambda^{\mathrm{We}\,(2)}[\bar{\boldsymbol{\varphi}}] \equiv \left(\frac{\delta}{\delta \bar{\boldsymbol{\varphi}}} \otimes \frac{\delta}{\delta \bar{\boldsymbol{\varphi}}} \right) \Gamma_\Lambda^{\mathrm{We}}[\bar{\boldsymbol{\varphi}}] . \qquad (9.5)$$

Expanding the generating functional $\Gamma_\Lambda^{\mathrm{We}}[\bar{\boldsymbol{\varphi}}]$ in terms of gradients of the field $\bar{\boldsymbol{\varphi}}$ and introducing the density

$$\rho(\boldsymbol{r}) \equiv \frac{1}{2} \bar{\boldsymbol{\varphi}}^2(\boldsymbol{r}) , \qquad (9.6)$$

we have[1]

$$\Gamma_\Lambda^{\mathrm{We}}[\bar{\boldsymbol{\varphi}}] = \int d^D r \left[U_\Lambda(\rho(\boldsymbol{r})) + \frac{c_0}{2} Z_\Lambda^{-1}(\rho(\boldsymbol{r})) \sum_{i=1}^{N} (\nabla \bar{\varphi}_i(\boldsymbol{r}))^2 \right.$$
$$\left. + \frac{c_0}{4} Y_\Lambda(\rho(\boldsymbol{r})) (\nabla \rho(\boldsymbol{r}))^2 + \ldots \right] . \qquad (9.7)$$

It should be noted that by choosing the same Z-factor for all field components and by assuming that Z_Λ^{-1} is a function of $\rho(\boldsymbol{r})$ instead of $\bar{\boldsymbol{\varphi}}(\boldsymbol{r})$, we have already made two approximations. The function $U_\Lambda(\bar{\boldsymbol{\varphi}}) = U_\Lambda(\rho)$ in the first term of the above expansion of the Legendre effective action is known as the *effective potential*. It can be obtained from $\Gamma_\Lambda^{\mathrm{We}}[\bar{\boldsymbol{\varphi}}]$ by choosing a space-independent configuration of the field $\bar{\boldsymbol{\varphi}}$, implying

---

[1] In the literature on the derivative expansion the factor Z^{-1} introduced above is often called just Z. However, to be consistent with the rest of this book and to have the interpretation of the Z-factor as the quasiparticle weight in the context of quantum many-body physics we will stick to the above definition.

$$U_\Lambda(\bar{\varphi}) \equiv \frac{1}{V} \Gamma_\Lambda^{\mathrm{We}}[\bar{\varphi}]\Big|_{\bar{\varphi}(r)=\bar{\varphi}} \ , \tag{9.8}$$

where V is the volume of the system. As we have discussed in Chap. 7, in the limit of vanishing external sources, the vacuum expectation value $\bar{\varphi}_0 \equiv \lim_{J\to 0}\langle\varphi\rangle$ (describing the spontaneous magnetization in applications of φ^4-theory to magnetic systems) is determined by

$$\frac{\delta \mathcal{L}_\Lambda[\bar{\varphi}]}{\delta\bar{\varphi}} = \frac{\delta \Gamma_\Lambda^{\mathrm{We}}[\bar{\varphi}]}{\delta\bar{\varphi}} + R_\Lambda(0)\bar{\varphi} = 0 \ , \tag{9.9}$$

which follows from the general relation (7.53) between $\mathcal{L}_\Lambda$ and $\Gamma_\Lambda^{\mathrm{We}}$. Using the fact that[2] $\lim_{\Lambda\to 0} R_\Lambda(k^2) = 0$, we see that the vacuum expectation value $\bar{\varphi}_0$ can be obtained by minimizing $\lim_{\Lambda\to 0} \Gamma_\Lambda^{\mathrm{We}}[\bar{\varphi}]$. Similarly, within mean-field theory the vacuum expectation value $\bar{\varphi}_0^{\mathrm{MF}}$ is obtained from the minimum of the classical action $S_{\Lambda_0}[\bar{\varphi}] = \lim_{\Lambda\to\Lambda_0} \Gamma_\Lambda^{\mathrm{We}}[\bar{\varphi}]$. The crossover of the vacuum expectation value from its mean-field value $\bar{\varphi}_0^{\mathrm{MF}}$ to the vacuum expectation value $\bar{\varphi}_0$ including all fluctuations can therefore be described by the location of the running minimum of $\Gamma_\Lambda^{\mathrm{We}}[\bar{\varphi}]$.[3] Assuming a translationally invariant vacuum expectation value $\bar{\varphi}_{0,\Lambda}(r) = \bar{\varphi}_{0,\Lambda}$, we obtain the flowing vacuum expectation value from the minimum of the effective potential,

$$U_\Lambda'(\rho) \equiv \frac{dU_\Lambda(\rho)}{d\rho} = 0 \ . \tag{9.10}$$

To derive a flow equation for $U_\Lambda(\rho)$, we insert Eq. (9.8) into the exact FRG flow equation (9.4), resulting in

$$\partial_\Lambda U(\bar{\varphi}) = \frac{1}{2} \sum_{i=1}^{N} \int_k \left[\partial_\Lambda R_\Lambda(k^2)\right] G_{i,\Lambda}(\bar{\varphi}; k) \ , \tag{9.11}$$

where we have used again the notation $\int_k = \int \frac{d^D k}{(2\pi)^D}$, and the inverse of the field-dependent propagator for a translationally invariant field configuration $G_{i,\Lambda}^{-1}(\bar{\varphi}; k)$ is

---

[2] The relation $\lim_{\Lambda\to 0} R_\Lambda(k^2) = 0$ is not satisfied for the $k = 0$-component of the sharp cutoff function given in Eq. (7.12); although with Eq. (7.9) one obtains with a sharp momentum cutoff $\lim_{k\to 0}\lim_{\Lambda\to 0} R_\Lambda(k^2) = 0$, the opposite order of limits gives an infinite result.

[3] The reader might wonder why we are taking the minimum of $\Gamma_\Lambda^{\mathrm{We}}[\bar{\varphi}]$ instead of $\mathcal{L}_\Lambda[\bar{\varphi}]$. Formally both expressions only give rise to the same minimum for $\Lambda \to 0$. However, as the regulator R_Λ introduces an artificial gap into the low-energy modes for finite Λ, it is important to remove this gap for the $k = 0$-mode before discussing symmetry breaking. In other words, to obtain a finite vacuum expectation value which smoothly interpolates between the mean-field result for $\Lambda \to \Lambda_0$ and the exact result for $\Lambda \to 0$ we ignore the fluctuations in the $k = 0$-mode which are not integrated out yet. For $\Lambda \to \Lambda_0$ we thereby recover mean-field theory.

given by

$$G_{i,\Lambda}^{-1}(\bar{\varphi}; k) = \frac{1}{V} \frac{\delta}{\delta\bar{\varphi}_i(k)} \frac{\delta}{\delta\bar{\varphi}_i(-k)} \Gamma_\Lambda^{\mathrm{We}}[\bar{\varphi}]\Big|_{\bar{\varphi}(r)=\bar{\phi}} + R_\Lambda(k^2) \,. \qquad (9.12)$$

Using the elementary functional derivatives

$$\frac{\delta\rho(r)}{\delta\bar{\varphi}_j(k)} = \bar{\varphi}_j(r)e^{ik\cdot r} \quad , \quad \frac{\delta(\nabla\bar{\varphi}_i(r))}{\delta\bar{\varphi}_j(k)} = i\delta_{ij}ke^{ik\cdot r} \,, \qquad (9.13)$$

we can perform the functional derivatives in Eq. (9.12) and obtain

$$G_{i,\Lambda}^{-1}(\bar{\varphi}; k) = U_\Lambda'(\rho) + \varphi_i^2 U_\Lambda''(\rho) + c_0 Z_\Lambda^{-1}(\rho)k^2 + \frac{c_0}{2}Y_\Lambda(\rho)k^2\varphi_i^2 + R_\Lambda(k^2) \,. \quad (9.14)$$

Let us now evaluate the trace over the field components, i.e., perform the sum over the index i. To do so, it is convenient to choose a coordinate system in which $\bar{\phi}_0$ is orientated along one of the eigendirections. Identifying the inverse longitudinal and transverse propagators,

$$G_{\ell,\Lambda}^{-1}(\rho, k^2) = c_0\left[Z_\Lambda^{-1}(\rho) + \rho Y_\Lambda(\rho)\right]k^2 + U_\Lambda'(\rho) + 2\rho U_\Lambda''(\rho) + R_\Lambda(k^2) \,,$$
$$\qquad (9.15a)$$
$$G_{t,\Lambda}^{-1}(\rho, k^2) = c_0 Z_\Lambda^{-1}(\rho)k^2 + U_\Lambda'(\rho) + R_\Lambda(k^2) \,, \qquad (9.15b)$$

the flow equation for the effective potential can then be written as

$$\partial_\Lambda U_\Lambda(\rho) = \frac{1}{2}\int_k \left[\partial_\Lambda R_\Lambda(k^2)\right]\left[G_{\ell,\Lambda}(\rho, k^2) + (N-1)G_{t,\Lambda}(\rho, k^2)\right] \,. \qquad (9.16)$$

The angular part of the integration can now be easily carried out, resulting in

$$\boxed{\begin{aligned} \partial_\Lambda U_\Lambda(\rho) = \frac{K_D}{2}\int_0^\infty dk\, k^{D-1}\left[\partial_\Lambda R_\Lambda(k^2)\right] \\ \times\left[G_{\ell,\Lambda}(\rho, k^2) + (N-1)G_{t,\Lambda}(\rho, k^2)\right] \,. \end{aligned}} \qquad (9.17)$$

It should be noted that while in the symmetry-broken phase the longitudinal mode is gapped with $\rho_\Lambda \neq 0$ for $\Lambda \to 0$, the $N-1$ transverse modes become gapless. These gapless modes correspond to massless particles which are commonly known as Goldstone bosons.

Although the above flow equation for the effective potential is exact and implies an exact proof for the existence of Goldstone bosons in the $O(N)$-model (with $N \geq 2$), the functions $Z_\Lambda^{-1}(\rho)$ and $Y_\Lambda(\rho)$ cannot be derived from the effective potential alone and some approximations are needed to close the hierarchy of flow equations.

9.2 Local Potential Approximation

A very convenient approximation which closes the hierarchy of flow equations for the effective potential is commonly known as the local potential approximation (LPA). Within this approximation, one simply sets $Z_\Lambda^{-1}(\rho)$ and $Y_\Lambda(\rho)$ equal to their initial values $Z_{\Lambda_0}^{-1}(\rho) = 1$ and $Y_{\Lambda_0}(\rho) = 0$ and neglects the flow of these couplings. The LPA amounts to setting the anomalous dimension η equal to zero and can only be expected to give accurate results as long as this is a good approximation.

9.2.1 RG Equation for the Effective Potential

To calculate the effective potential $U(\rho) = \lim_{\Lambda \to 0} U_\Lambda(\rho)$ for the $O(N)$-symmetric field theory considered here we have to solve the partial differential equation (9.17) for $U_\Lambda(\rho)$ as a function of Λ and ρ, starting from the initial condition (see Eqs. (2.67) and (8.101))

$$U_{\Lambda_0}(\rho) = f_0 + \frac{r_0}{2}\bar{\varphi}^2 + \frac{u_0}{4!}\left(\bar{\varphi}^2\right)^2 . \tag{9.18}$$

It is convenient to choose the zero of energy such that $f_0 = u_0\rho_0^2/6 = \frac{3}{2}r_0^2/u_0$ where $\rho_0 \equiv -3r_0/u_0$. The initial condition of the effective potential can then be written as

$$U_{\Lambda_0}(\rho) = \frac{u_0}{6}(\rho - \rho_0)^2 . \tag{9.19}$$

An especially convenient cutoff function which works fine within the LPA is the Litim cutoff (Litim 2001) given in Eq. (7.21),

$$R_\Lambda(k^2) = c_0(\Lambda^2 - k^2)\Theta(\Lambda^2 - k^2) , \tag{9.20}$$

which for $k \leq \Lambda$ implies the k-independent inverse propagators,

$$G_{\ell,\Lambda}^{-1}(\rho, k^2) = c_0\Lambda^2 + U_\Lambda'(\rho) + 2\rho U_\Lambda''(\rho) , \tag{9.21a}$$

$$G_{t,\Lambda}^{-1}(\rho, k^2) = c_0\Lambda^2 + U_\Lambda'(\rho) . \tag{9.21b}$$

Using

$$\partial_\Lambda R_\Lambda(k^2) = 2c_0\Lambda\Theta(\Lambda^2 - k^2) , \tag{9.22}$$

the integral over k in Eq. (9.17) reduces to $\int_0^\Lambda dk\, k^{D-1} = \Lambda^D/D$, such that within the LPA and with the Litim cutoff the flow of the effective potential is determined by

$$\partial_\Lambda U_\Lambda(\rho) = \frac{c_0 K_D \Lambda^{D+1}}{D} \left[\frac{1}{c_0 \Lambda^2 + U'_\Lambda(\rho) + 2\rho U''_\Lambda(\rho)} + \frac{N-1}{c_0 \Lambda^2 + U'_\Lambda(\rho)} \right].$$

$$(9.23)$$

This is a simple partial differential equation which can easily be solved numerically, using for instance *Mathematica* or standard numerical routines. For simplicity, let us study once again the case $N = 1$ in $D = 3$, describing the Ising universality class in three dimensions. Choosing $\rho_0 = 0.07\,\Lambda_0/c_0$ and $u_0 = 0.01\,c_0^2\Lambda_0$ one obtains the flow of the effective potential $U_\Lambda(\rho)$ shown in Fig. 9.1 for representative values of Λ. For clarity we have shifted each curve by the running minimum of $U_\Lambda(\rho)$. As expected, fluctuations reduce the vacuum expectation value $\bar{\varphi}_0 = \lim_{\Lambda\to 0} \sqrt{2\rho_0(\Lambda)}$ from its mean-field value $\bar{\varphi}_0^{\mathrm{MF}}$. The fact that we do not get a truly flat plateau for $\bar{\varphi}$ between $\pm\lim_{\Lambda\to 0}\bar{\varphi}_0(\Lambda)$ turning the effective potential convex is due to the approximations involved in the LPA. Of course, having a finite vacuum expectation value indicates that the parameters chosen correspond to the low-temperature broken symmetry phase. Changing the parameter ρ_0, we can now tune our system across the expected phase transition. In Fig. 9.2 we show the effective potential $U_\Lambda(\rho)$ for $\rho_0 = 0.03\,\Lambda_0/c_0$ and $u_0 = 0.01\,c_0^2\Lambda_0$. Clearly, the location of the running minimum of $U_\Lambda(\rho)$ becomes equal to zero for a finite value of Λ, leading to a vanishing vacuum expectation value and the restoration of the spontaneously broken $O(N)$-symmetry. It is reasonable to expect that the value Λ_* where this happens is of the order of the inverse correlation length ξ^{-1}: as long as $\Lambda \gtrsim \xi^{-1}$ fluctuations on

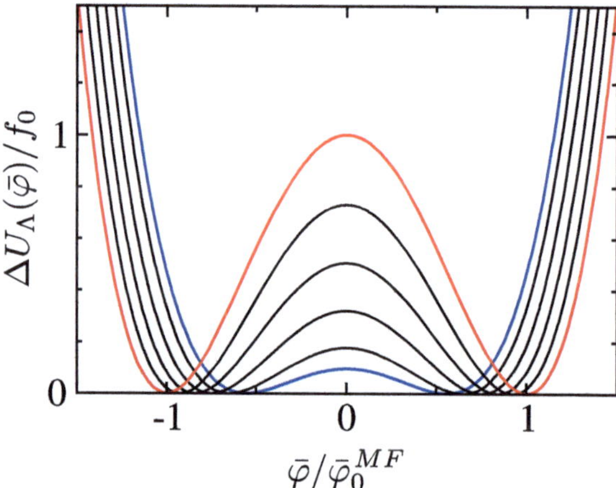

Fig. 9.1 Shifted effective potential $\Delta U_\Lambda(\rho) \equiv U_\Lambda(\rho) - U_\Lambda(\rho_0(\Lambda))$ in units of $f_0 = u_0\rho_0^2/6 = \frac{3}{2}r_0^2/u_0$ for a one-component φ^4-theory in $D = 3$ with Λ ranging from its initial value Λ_0 (*red line*) to $\Lambda = 0$ (*blue line*) in steps of $0.2\,\Lambda_0$. In terms of Λ_0, the coupling constants are given by $u_0 = 0.01\,c_0^2\Lambda_0$ and $\rho_0 = 0.07\,\Lambda_0/c_0$. Decreasing Λ, the effective potential becomes almost convex, but has a minimum at $\lim_{\Lambda\to 0}\bar{\varphi}_0(\Lambda) \approx 0.56$, indicating that fluctuations have reduced the vacuum expectation value (spontaneous magnetization) $\bar{\varphi}_0(\Lambda)$ to almost half the mean-field value $\bar{\varphi}_0^{\mathrm{MF}}$

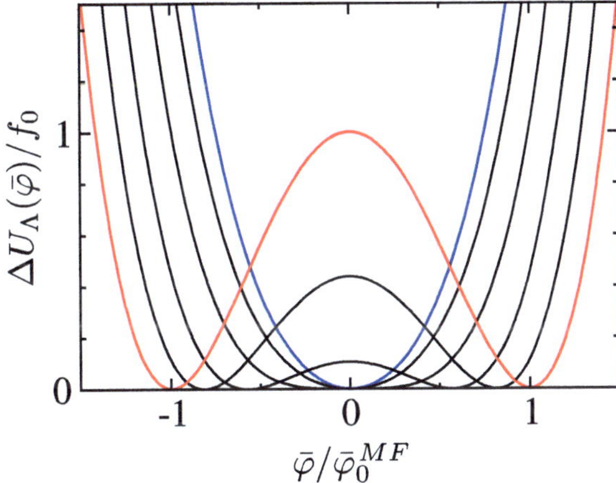

Fig. 9.2 Shifted effective potential $\Delta U_\Lambda(\rho) \equiv U_\Lambda(\rho) - U_\Lambda(\rho_0(\Lambda))$ for the one-component φ^4-theory in $D = 3$ as in Fig. 9.1 but with $\rho_0 = 0.03\,\Lambda_0/c_0$. Fluctuations now drive the vacuum expectation value $\bar{\varphi}_0(\Lambda)$ to zero at a finite value of Λ, restoring the spontaneously broken symmetry

length scales larger than the correlation length have not been included yet, leaving all fields in a coarse-grained patch correlated and thereby leading to a finite magnetization (which in applications of φ^4-theory to magnetic systems is the physical meaning of the vacuum expectation value $\bar{\varphi}_0$). However, as the length scale Λ_*^{-1} of the coarse-grained patch is increased beyond ξ, fluctuations succeed in destroying the spontaneous magnetization, pushing $\bar{\varphi}_0(\Lambda)$ to zero for $\Lambda < \Lambda_*$.

9.2.2 Fixed Points and Critical Exponents

To discuss the properties of the RG fixed point associated with the critical point lying between the high and low temperature phases, it is advantageous to introduce again dimensionless rescaled variables. As before, we express energies and inverse length scales in terms of the running cutoff $\Lambda = \Lambda_0 e^{-l}$. Introducing the dimensionless variables and functions

$$q = k/\Lambda \,, \tag{9.24a}$$

$$x = \Lambda r \,, \tag{9.24b}$$

$$\tilde{\varphi}(x) = \sqrt{c_0}\,\Lambda^{(2-D)/2}\bar{\varphi}(r) \,, \tag{9.24c}$$

$$\tilde{\rho}(x) = c_0 \Lambda^{2-D}\rho(r) \,, \tag{9.24d}$$

$$\tilde{U}_l(\tilde{\varphi}) = \tilde{f}_0 + \frac{\tilde{r}_l}{2}\tilde{\varphi}^2 + \frac{\tilde{u}_l}{4!}(\tilde{\varphi}^2)^2 + \cdots = U_\Lambda(\bar{\varphi})/\Lambda^D \,, \tag{9.24e}$$

the FRG flow equation (9.23) for the effective potential turns into the scale-invariant form

$$\partial_l \tilde{U}_l(\tilde{\rho}) = D\tilde{U}_l(\tilde{\rho}) - (D-2)\tilde{\rho}\tilde{U}_l'(\tilde{\rho})$$
$$- \frac{K_D}{D}\left[\frac{1}{1 + \tilde{U}_l'(\tilde{\rho}) + 2\tilde{\rho}\tilde{U}_l''(\tilde{\rho})} + \frac{N-1}{1 + \tilde{U}_l'(\tilde{\rho})}\right] \,. \tag{9.25}$$

While the first two terms on the right-hand side are due to the rescaling of U_Λ and ρ, the term in the second line is the nontrivial interaction contribution. In contrast to Eq. (9.23), this rescaled flow equation allows for a fixed point $\tilde{U}^*(\tilde{\rho})$ which satisfies $\partial_l \tilde{U}_l(\tilde{\rho}) = 0$. To obtain this fixed point, it is convenient to introduce

$$\tilde{W}_l(\tilde{\rho}) = \tilde{U}'_l(\tilde{\rho}) \,, \tag{9.26}$$

which obeys

$$\boxed{\begin{aligned} \partial_l \tilde{W}_l(\tilde{\rho}) = 2\tilde{W}_l(\tilde{\rho}) - (D-2)\tilde{\rho}\,\tilde{W}'_l(\tilde{\rho}) \\ + \frac{K_D}{D}\left[\frac{3\tilde{W}'_l(\tilde{\rho}) + 2\tilde{\rho}\,\tilde{W}''_l(\tilde{\rho})}{[1 + \tilde{W}_l(\tilde{\rho}) + 2\tilde{\rho}\,\tilde{W}'_l(\tilde{\rho})]^2} + \frac{(N-1)\tilde{W}'_l(\tilde{\rho})}{[1 + \tilde{W}_l(\tilde{\rho})]^2} \right] \,. \end{aligned}} \tag{9.27}$$

The location of the running minimum $\tilde{\rho}_l$ of $\tilde{U}_\Lambda(\tilde{\rho})$ is simply given by

$$W_l(\tilde{\rho}) = 0 \,. \tag{9.28}$$

Before we actually solve the above flow equation for $W_l(\tilde{\rho})$, let us briefly summarize what we should expect. As we have discussed in Sect. 3.3.3, in the zero-field Ising model, or – more generally – in the $O(N)$-symmetric φ^4-theory considered here, apart from the Gaussian fixed point $\tilde{U}_l(\tilde{\rho}) = const$ implying $\tilde{W}_l(\tilde{\rho}) = 0$, we can distinguish three fixed points of the RG flow (see Delamotte (2007) for a similar discussion):

(a) *Ferromagnetic (low temperature) fixed point $\mathcal{F}$:* In this case the $O(N)$-symmetry is spontaneously broken and we have a finite vacuum expectation value $\bar{\varphi}_0$ (which in a magnetic system represents the spontaneous magnetization). In terms of dimensionless variables we therefore expect the rescaled field $\tilde{\varphi}_0(l)$ to diverge within the LPA as

$$\tilde{\varphi}_0(l) = \sqrt{c_0}\bar{\varphi}_0 e^{(D-2)l/2} \propto e^{(D-2)l/2} \,. \tag{9.29}$$

(b) *Paramagnetic (high temperature) fixed point $\mathcal{P}$:* In the paramagnetic phase the $O(N)$-symmetry is not broken and both the unrescaled and the rescaled vacuum expectation value are equal to zero. For $\rho_0^{\mathrm{MF}} > 0$ this is in contrast to mean-field theory which predicts a symmetry-broken phase. As discussed above, the vanishing of the order parameter is fluctuation-driven and occurs at a cutoff scale Λ_* which is expected to be of the order of the inverse correlation length ξ^{-1}.

(c) *Critical fixed point $\mathcal{C}$:* Finally, at criticality $\xi \to \infty$ and the system becomes self-similar such that $\partial_l \tilde{W}_l(\tilde{\rho}) = 0$. This is the Wilson–Fisher fixed point. As Λ is sent to zero, $\tilde{\varphi}_0$ approaches a finite value, implying that $\bar{\varphi}_0(\Lambda)$ vanishes as

$$\bar{\varphi}_0(\Lambda) = (c_0)^{-1/2}\tilde{\varphi}_0 \,\Lambda^{(D-2)/2} \propto \Lambda^{(D-2)/2} \,. \tag{9.30}$$

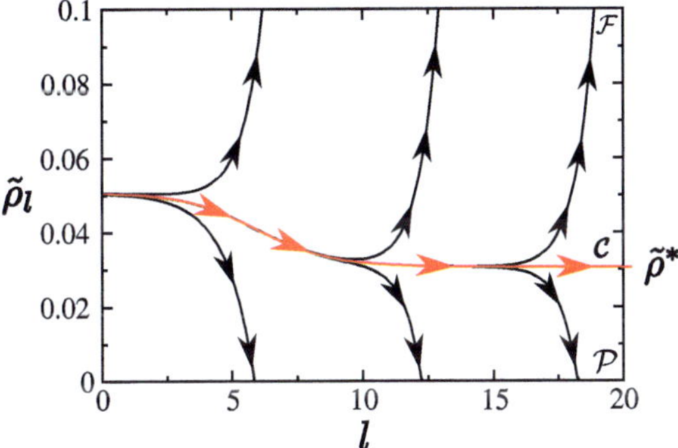

Fig. 9.3 Flow of $\tilde{\rho}_l$ as a function of l for $\tilde{u}_0 = 0.01$ and $\tilde{\rho}_0 = \tilde{\rho}_0^* = 0.050459\ldots$ (*red line*) as well as $\tilde{\rho}_0 = \tilde{\rho}_0^* \pm 10^{-4}$, $\tilde{\rho}_0^* \pm 10^{-8}$, $\tilde{\rho}_0^* \pm 10^{-12}$. For $\tilde{\rho}_0$ very close to $\tilde{\rho}_0^*$ the flow of $\tilde{\rho}_l$ is almost critical and, as we have seen in Chap. 4, stays close to the critical value $\tilde{\rho}^* \neq \tilde{\rho}_0^*$ for some RG time before being attracted by either the high-temperature (paramagnetic, $\mathcal{P}$) fixed point for $\tilde{\rho}_0 < \tilde{\rho}_0^*$, or low-temperature (ferromagnetic, $\mathcal{F}$) fixed point for $\tilde{\rho}_0 > \tilde{\rho}_0^*$

Let us now explicitly solve the partial differential equation (9.27). Fine-tuning the initial value for $\tilde{\rho}_0$ with $\tilde{u}_0$ fixed, our FRG analysis does indeed give rise to flows which are attracted by one of the three fixed points discussed above. This is shown in Fig. 9.3, where we have plotted $\tilde{\rho}_l$ as a function of l for $\tilde{u}_0 = 0.01$ and different values of $\tilde{\rho}_0$. The flow is close to criticality for $\tilde{\rho}_0 \approx \tilde{\rho}_0^* = 0.050459\ldots$. While for $\tilde{\rho}_0 < \tilde{\rho}_0^*$ the flow reaches the paramagnetic fixed point $\mathcal{P}$, for $\tilde{\rho}_0 > \tilde{\rho}_0^*$ it is attracted by the ferromagnetic fixed point $\mathcal{F}$.

Within the LPA, the anomalous dimension η is set equal to zero such that by computing one additional scaling exponent all other scaling exponents can be obtained by using the scaling relations discussed in Sect. 1.3. As in our analysis of the Wilson–Fisher fixed point in Sect. 4.2.2, it seems to be most convenient to calculate ν. Having $\eta = 0$ and an estimate for ν, the thermodynamic exponents α, β, γ, and δ can then be obtained from the scaling relations (1.33a), (1.33b), (1.33c), and (1.33d). To actually calculate ν it is possible to proceed as in Sect. 4.2.2 and calculate ν as the inverse of the eigenvalue y_t of the relevant scaling variable t_l defined in Eq. (4.87). As t_l is a linear combination of $\delta\tilde{r}_l$ and $\delta\tilde{u}_l$, we can write the flow of the effective potential in the vicinity of the critical point as

$$\tilde{U}_l(\tilde{\rho}) = \tilde{U}^*(\tilde{\rho}) + e^{l/\nu}\delta\tilde{U}(\tilde{\rho}) \, , \tag{9.31}$$

where $\delta\tilde{U}(\tilde{\rho})$ is a quadratic function in $\tilde{\rho}$. Working with $\tilde{W}_l$, we can write this as

$$\tilde{W}_l(\tilde{\rho}) = \tilde{W}^*(\tilde{\rho}) + e^{l/\nu}\delta\tilde{W}(\tilde{\rho}) \, , \tag{9.32}$$

where $\delta\tilde{W}(\tilde{\rho})$ is a linear function in $\tilde{\rho}$.

Alternatively, one can calculate the correlation length ξ for different initial values of $\tilde{\rho}_0$ and extract ν by using the scaling relation

$$\xi \propto |\tilde{r}_0 - \tilde{r}_0^*|^{-\nu} \, , \tag{9.33}$$

where $\tilde{r}_0^*$ is the fine-tuned value of $\tilde{r}_0$ (with $\tilde{u}_0$ fixed) flowing into the critical point with $\tilde{r} = \tilde{r}^*$. In fact, using this approach we can even get the nonuniversal prefactor. To obtain ξ, we note that in an expansion of the effective potential $\tilde{U}_l(\tilde{\rho})$ the coefficient of the term linear in $\tilde{\rho}$ equals the square of the inverse rescaled correlation length,

$$\tilde{W}_l(0) = \tilde{r}_l = \bar{r}_l = \frac{1}{(\Lambda_l \xi)^2} \ . \tag{9.34}$$

Eq. (9.34) directly follows from Eq. (8.92) and implies

$$\xi^{-2} = \Lambda_0^2 \lim_{l \to \infty} e^{-2l} \tilde{W}_l(0) \ . \tag{9.35}$$

Since close to the critical point we expect ξ to obey Eq. (9.33), we can obtain the critical exponent ν by plotting $\ln\left[(\Lambda_0 \xi)^{-1}\right]$ as a function of $\ln\left|\tilde{r}_0 - \tilde{r}_0^*\right|$. For small $\tilde{r}_0 - \tilde{r}_0^*$, we expect

$$\ln\left[(\Lambda_0 \xi)^{-1}\right] = \nu \ln\left|\tilde{r}_0 - \tilde{r}_0^*\right| + const \ , \tag{9.36}$$

so that ν can be obtained from the slope of the expected straight line, while the constant contains information about the nonuniversal prefactor. Finally, by noticing over which region the linearity holds we obtain information about the range of the scaling regime. The approach outlined above has been used by Berges et al. (2002) to calculate the critical exponent ν. Within the LPA and with the Litim regulator implemented as above, one finds for the critical exponent ν of the three-dimensional Ising universality class the value $\nu \approx 0.65$, which should be compared with the best numerical result, $\nu \approx 0.63$, as compiled in Table 1.1 in Sect. 1.2.

9.3 Beyond the Local Potential Approximation

To go beyond the simple LPA we need to include additional flow equations for $Z_\Lambda^{-1}(\rho)$ and $Y_\Lambda(\rho)$. In the simplest extension to the LPA, called the LPA′, one includes a field-independent wave function renormalization factor. Setting $Z_\Lambda^{-1}(\rho) = Z_\Lambda^{-1}$ and also $Y_\Lambda(\rho) = 0$ the effective action reads

$$\Gamma_\Lambda^{\mathrm{LPA}'}[\bar{\varphi}] = \frac{1}{2} \int_k Z_\Lambda^{-1} c_0 k^2 \sum_i \bar{\varphi}_i(-k)\bar{\varphi}_i(k) + \int d^D r\, U_\Lambda\left(\rho(r)\right) \ . \tag{9.37}$$

Since in comparison with the LPA there is now an additional factor of Z_Λ^{-1} in front of the term proportional to $c_0 k^2$, it is advantageous to include this prefactor in the definition of the cutoff function as well. Let us therefore replace the regulator (9.20) by

$$R_\Lambda(k^2) = c_0 Z_\Lambda^{-1}(\Lambda^2 - k^2)\Theta(\Lambda^2 - k^2) \ . \tag{9.38}$$

Anticipating that the flowing anomalous dimension will be given by (see also Eq. (8.29)),

$$\eta_\Lambda = \Lambda \frac{\partial_\Lambda Z_\Lambda}{Z_\Lambda} = -\Lambda \frac{\partial_\Lambda Z_\Lambda^{-1}}{Z_\Lambda^{-1}} ,$$ (9.39)

we can write the derivative of $R_\Lambda(k^2)$ with respect to Λ as

$$\partial_\Lambda R_\Lambda(k^2) = c_0 Z_\Lambda^{-1} 2\Lambda \left(1 - \eta_\Lambda \frac{\Lambda^2 - k^2}{2\Lambda^2}\right) \Theta(\Lambda^2 - k^2) .$$ (9.40)

Evaluating the general flow equation (9.17) for the effective potential with the modified Litim regulator, we obtain

$$\partial_\Lambda U_\Lambda(\rho) = \frac{c_0 Z_\Lambda^{-1} K_D \Lambda^{D+1}}{D} \left(1 - \frac{\eta_\Lambda}{D+2}\right)$$
$$\times \left[\frac{1}{c_0 Z_\Lambda^{-1} \Lambda^2 + U_\Lambda'(\rho) + 2\rho U_\Lambda''(\rho)} + \frac{N-1}{c_0 Z_\Lambda^{-1} \Lambda^2 + U_\Lambda'(\rho)}\right] .$$ (9.41)

To close our set of flow equations we need an additional flow equation for Z_Λ^{-1}. This can be obtained by noting that Z_Λ^{-1} is the coefficient in front of the term proportional to $c_0 k^2$ in our truncated effective action (9.37). Let us define the irreducible two-point vertex

$$\Gamma_{i,\Lambda}^{(2)}(\bar{\boldsymbol{\varphi}}; \boldsymbol{k}) = \frac{1}{V} \frac{\delta}{\delta\varphi_i(-\boldsymbol{k})} \frac{\delta}{\delta\varphi_i(\boldsymbol{k})} \Gamma_\Lambda^{\text{LPA}'}[\bar{\boldsymbol{\varphi}}(\boldsymbol{r})]\bigg|_{\bar{\varphi}(\boldsymbol{r})=\bar{\varphi}}$$ (9.42)

for a homogeneous field configuration $\bar{\boldsymbol{\varphi}}(\boldsymbol{r}) = \bar{\boldsymbol{\varphi}}$. Within the LPA′, Eq. (9.42) is then given by

$$\Gamma_{i,\Lambda}^{(2)}(\bar{\boldsymbol{\varphi}}; \boldsymbol{k}) = Z_\Lambda^{-1} c_0 k^2 + U_\Lambda'(\rho) + \varphi_i^2 U_\Lambda''(\rho) .$$ (9.43)

Z_Λ^{-1} can now be defined by

$$Z_\Lambda^{-1} = \frac{\partial \Gamma_{t,\Lambda}^{(2)}(\bar{\boldsymbol{\varphi}}_0; \boldsymbol{k})}{\partial(c_0 k^2)}\bigg|_{k=0} ,$$ (9.44)

so that the flow of Z_Λ^{-1} is determined by the flow of $\Gamma_{i,\Lambda}^{(2)}(\bar{\boldsymbol{\varphi}}; \boldsymbol{k})$, which is usually evaluated for a transverse mode (e.g., $i = 2$) at the minimum of the potential. Alternative definitions of Z_Λ^{-1} which involve finite values of k are discussed in Berges et al. (2002). Comparing Eqs. (9.43) and (9.44) with Eq. (8.24) we see that our factor Z_Λ defined above is indeed the wave function renormalization factor, which in quantum many-body systems would correspond to the quasiparticle weight in the single-particle Green function. This confirms our definition of the flowing anomalous dimension η_Λ in Eq. (9.39).

To derive a flow equation for $\Gamma^{(2)}_{t,\Lambda}(\bar{\varphi}; k)$ let us use here an alternative method to the ones used before which is particularly suitable for the calculation of the field-dependent vertices occurring within the derivative expansion. Following (Berges et al. 2002) we differentiate the exact FRG flow equation with respect to $\bar{\varphi}_i(-k)$ and $\bar{\varphi}_i(k)$ and subsequently set the field configuration $\bar{\varphi}(r)$ equal to $\bar{\varphi}$. Using the fact that the derivative of the inverse of a matrix $\mathbf{A}(x)$ is given by

$$\frac{d}{dx}\mathbf{A}^{-1}(x) = -\mathbf{A}^{-1}(x)\frac{d\mathbf{A}(x)}{dx}\mathbf{A}^{-1}(x) , \tag{9.45}$$

we find

$$\partial_\Lambda \Gamma^{(2)}_{i,\Lambda}(\bar{\varphi}; k) = -\frac{1}{2}\int_{k'}\sum_{j=1}^{N}\left[\partial_\Lambda R_\Lambda(k'^2)\right]G^2_{j,\Lambda}(\varphi; k')\Gamma^{(4)}_{jjii,\Lambda}(\bar{\varphi}; k', -k', k, -k)$$

$$+\int_{k'}\sum_{i_1,i_2=1}^{N}\left[\partial_\Lambda R_\Lambda(k'^2)\right]\left[G_{i_1,\Lambda}(\bar{\varphi}; k')\right]^2 G_{i_2,\Lambda}(\bar{\varphi}; k'+k)$$

$$\times\; \Gamma^{(3)}_{ii_2i_1,\Lambda}(\bar{\varphi}; k, -k-k', k')\Gamma^{(3)}_{i_1i_2i,\Lambda}(\bar{\varphi}; -k', k+k', -k) , \tag{9.46}$$

which generalizes Eq. (8.18) to the $O(N)$-symmetric case. Within the LPA$'$ the vertices $\Gamma^{(3)}_{i_1i_2i_3,\Lambda}(\bar{\varphi}; k_1, k_2, k_3)$ and $\Gamma^{(4)}_{iijj,\Lambda}(\bar{\varphi}; k, -k, k', -k')$ do not have any momentum dependence and are given by

$$\Gamma^{(3)}_{i_1i_2i_3,\Lambda}(\bar{\varphi}) = U''_\Lambda(\rho)\left[\delta_{i_1,i_2}\bar{\varphi}_{i_3} + \delta_{i_2,i_3}\bar{\varphi}_{i_1} + \delta_{i_3,i_1}\bar{\varphi}_{i_2}\right] + U'''_\Lambda(\rho)\bar{\varphi}_{i_1}\bar{\varphi}_{i_2}\bar{\varphi}_{i_3} , \tag{9.47}$$

$$\Gamma^{(4)}_{iijj,\Lambda}(\bar{\varphi}) = \left[1 + 2\delta_{i,j}\right]\left[U''_\Lambda(\rho) + U'''_\Lambda(\rho)(\varphi_i^2 + \varphi_j^2)\right] + U''''_\Lambda(\rho)\bar{\varphi}_i^2\bar{\varphi}_j^2 . \tag{9.48}$$

First of all let us note that since we are only interested in the k-dependence of $\Gamma^{(2)}_{t,\Lambda}(\bar{\varphi}; k)$ and since the four-point vertex $\Gamma^{(4)}_{iijj,\Lambda}(\bar{\varphi})$ is entirely k-independent we can safely drop the first term on the right-hand side of Eq. (9.46). To isolate the k-dependent part of $\Gamma^{(2)}_{t,\Lambda}(\bar{\varphi}; k)$, it is convenient to introduce

$$\delta\Gamma^{(2)}_{t,\Lambda}(\bar{\varphi}; k) \equiv \Gamma^{(2)}_{t,\Lambda}(\bar{\varphi}; k) - \Gamma^{(2)}_{t,\Lambda}(\bar{\varphi}; 0) . \tag{9.49}$$

Now, since by definition $\bar{\varphi}_i = 0$ for $i > 1$, the only nonvanishing $\Gamma^{(3)}$-vertex apart from $\Gamma^{(3)}_{\ell\ell\ell,\Lambda}$ is

$$\Gamma^{(3)}_{tt\ell,\Lambda}(\rho) = \sqrt{2\rho}\, U''_\Lambda(\rho) . \tag{9.50}$$

Our flow equation (9.46) can therefore be rewritten as

$$\partial_\Lambda \delta \Gamma_{t,\Lambda}^{(2)}(\rho;k) = 2\rho \left[U_\Lambda''(\rho)\right]^2 \int_{k'} \partial_\Lambda R_\Lambda(k'^2)[G_{t,\Lambda}^2(\rho;k')G_{\ell,\Lambda}(\rho;k'+k)$$

$$+ G_{\ell,\Lambda}^2(\rho;k')G_{t,\Lambda}(\rho;k'+k)] - \{k \to 0\}$$

$$= -2\rho \left[U_\Lambda''(\rho)\right]^2 \tilde{\partial}_\Lambda \int_{k'} G_{\ell,\Lambda}(\rho;k')\left[G_{t,\Lambda}(\rho;k'+k) - G_{t,\Lambda}(\rho;k')\right] , \quad (9.51)$$

where the partial derivative $\tilde{\partial}_\Lambda$ acts only on $R_\Lambda(k^2)$ and derivatives thereof. To obtain the anomalous dimension, we need to expand the last term in Eq. (9.51) up to order k^2. To this end, let us write

$$G_{t,\Lambda}(\rho;k'+k) - G_{t,\Lambda}(\rho;k')$$

$$= -G_{t,\Lambda}^2(\rho;k')\delta G_{t,\Lambda}^{-1}(\rho;k',k) + G_{t,\Lambda}^3(\rho;k')\delta G_{t,\Lambda}^{-2}(\rho;k',k) - \dots , \quad (9.52)$$

where

$$\delta G_{t,\Lambda}^{-1}(\rho;k',k) \equiv G_{t,\Lambda}^{-1}(\rho;k'+k) - G_{t,\Lambda}^{-1}(\rho;k')$$

$$= \left[c_0 Z_\Lambda^{-1} + R_\Lambda'(k'^2)\right](k^2 + 2k \cdot k') + \frac{1}{2} R_\Lambda''(k'^2)(2k \cdot k')^2 + \mathcal{O}(k^3) . \quad (9.53)$$

Inserting this into Eq. (9.51) and using

$$G_{t,\Lambda}^3(\rho;k')\left[c_0 Z_\Lambda^{-1} + R_\Lambda'(k'^2)\right] = -\frac{1}{2}\frac{\partial}{\partial k'^2}G_{t,\Lambda}^2(\rho;k') , \quad (9.54)$$

where the derivative with respect to k'^2 can also be replaced by the derivative with respect to $k_t'^2$ (with the coordinate system chosen such that $k = ke_i$), we obtain after integrating by parts and finally replacing $(k \cdot k')^2$ by $k^2 k'^2/D$,

$$\partial_\Lambda \delta \Gamma_{t,\Lambda}^{(2)}(\rho;k) = c_0 k^2 \frac{4\rho \left[U_\Lambda''(\rho)\right]^2}{D} \tilde{\partial}_\Lambda \int_{k'} c_0 k'^2 G_{\ell,\Lambda}^2(\rho;k')G_{t,\Lambda}^2(\rho;k')$$

$$\times \left[Z_\Lambda^{-1} + c_0^{-1} R_\Lambda'(k'^2)\right]^2 + \mathcal{O}(k^4) . \quad (9.55)$$

Taking the derivative with respect to $c_0 k^2$ and setting $\rho = \rho_0$, we finally obtain the flow of Z_Λ^{-1}. For the Litim regulator (9.38), we have

$$R_\Lambda'(k^2) = \frac{d R_\Lambda(k^2)}{d(k^2)} = -c_0 Z_\Lambda^{-1} \Theta(\Lambda^2 - k^2) , \quad (9.56)$$

such that

$$Z_\Lambda^{-1} + c_0^{-1} R_\Lambda'(k'^2) = Z_\Lambda^{-1} - Z_\Lambda^{-1}\Theta(\Lambda^2 - k'^2) = Z_\Lambda^{-1}\Theta(k'^2 - \Lambda^2) . \quad (9.57)$$

Combining the two terms Z_Λ^{-1} and $Z_\Lambda^{-1}\Theta(\Lambda^2-k^2)$ in the last step of Eq. (9.57) is not as innocuous as it looks because when taking the derivative $\tilde\partial_\Lambda$ with respect to Λ this derivative should only act on the regulator or its derivatives, i.e., $Z_\Lambda^{-1}\Theta(\Lambda^2-k'^2)$. We therefore have

$$\tilde\partial_\Lambda\left[Z_\Lambda^{-1}+c_0^{-1}R_\Lambda'(k'^2)\right]^2 = -2Z_\Lambda^{-1}\Theta(k'^2-\Lambda^2)\partial_\Lambda\left[Z_\Lambda^{-1}\Theta(\Lambda^2-k'^2)\right]$$
$$= -2Z_\Lambda^{-1}\Theta(k'^2-\Lambda^2)\left[\partial_\Lambda Z_\Lambda^{-1}\Theta(\Lambda^2-k'^2)+Z_\Lambda^{-1}\delta(k'-\Lambda)\right]$$
$$= -\left(Z_\Lambda^{-1}\right)^2\delta(k'-\Lambda)\,. \tag{9.58}$$

It should be noted that ignoring derivatives of Z_Λ^{-1} on the right-hand side of this flow equation and using $\left[Z_\Lambda^{-1}+c_0^{-1}R_\Lambda'(k'^2)\right]^2 = \left(Z_\Lambda^{-1}\right)^2\Theta(k'-\Lambda)$ would have given us exactly the same result. The k'-integral in Eq. (9.55) is now easily done analytically. First of all, as we have seen before, the angular integration gives us a factor K_D defined in Eq. (2.86). Next we note that since the Θ-function in the numerator of Eq. (9.57) restricts the integral in Eq. (9.55) to values $k' \geq \Lambda$ and since $R_\Lambda(k'^2)$ vanishes for $k' \geq \Lambda$, the only contribution to the k'-integral comes from the δ-function in Eq. (9.58). Recalling that $U_\Lambda'(\rho_0) = 0$, we obtain for the flowing anomalous dimension,

$$\eta_\Lambda = -\Lambda\frac{\partial_\Lambda Z_\Lambda^{-1}}{Z_\Lambda^{-1}} = \frac{4K_D\Lambda^{D-2}\rho_0\left[U_\Lambda''(\rho_0)\right]^2}{Dc_0Z_\Lambda^{-1}\left[c_0Z_\Lambda^{-1}\Lambda^2+2\rho_0U_\Lambda''(\rho_0)\right]^2}\,. \tag{9.59}$$

In combination with the flow equation (9.41) for the effective potential $U_\Lambda(\rho)$, we now have a complete set of flow equations.

To discuss fixed-point properties and to calculate the critical exponents it is again advantageous to introduce dimensionless variables and $\tilde W_l = \tilde U_l'$. This can be accomplished as in the discussion of the LPA in the previous section with the only difference being that in Eqs. (9.24a), (9.24b), (9.24c), (9.24d), and (9.24e) we should now replace c_0 by $c_0Z_\Lambda^{-1}$. In rescaled form our complete set of flow equations reads (see also (Blaizot et al. 2006a)) as follows,

$$\boxed{\begin{aligned}\partial_l\tilde W_l(\tilde\rho) &= (2-\eta_l)\tilde W_l(\tilde\rho)-(D+\eta_l-2)\tilde\rho\,\tilde W_l'(\tilde\rho)\\ &+\frac{K_D}{D}\left(1-\frac{\eta_l}{D+2}\right)\left[\frac{3\tilde W_l'(\tilde\rho)+2\tilde\rho\,\tilde W_l''(\tilde\rho)}{\left[1+\tilde W_l(\tilde\rho)+2\tilde\rho\,\tilde W_l'(\tilde\rho)\right]^2}+\frac{(N-1)\tilde W_l'(\tilde\rho)}{\left[1+\tilde W_l(\tilde\rho)\right]^2}\right]\,,\end{aligned}}$$
$$\tag{9.60}$$

where

$$\boxed{\eta_l = \frac{K_D}{D}\frac{4\tilde\rho_0\left[\tilde W_l'(\tilde\rho_0)\right]^2}{\left[1+2\tilde\rho_0\tilde W_l'(\tilde\rho_0)\right]^2}\,.}$$
$$\tag{9.61}$$

Since the anomalous dimension can be simply expressed analytically, these LPA$'$ flow equations are almost as easy to solve as the corresponding LPA equations. While the flow of the effective potential and the flow of $\tilde{\rho}_l$ look qualitatively similar to the flow within the LPA, the flowing anomalous dimension now picks up a finite value and stays finite if we fine-tune our system towards criticality. For more details we refer the reader to Berges et al. (2002) or Blaizot et al. (2006a) who find an anomalous dimension $\eta = 0.044$ for $N = 2$ and $D = 3$ which should be compared with the more accurate value $\eta = 0.038$ compiled in Table 1.1 in Sect. 1.2.

References

Bagnuls, C. and C. Bervillier (2001), *Exact renormalization group equations: An introductory review*, Phys. Rep. **348**, 91.

Berges, J., N. Tetradis, and C. Wetterich (2002), *Non-perturbative renormalization flow in quantum field theory and statistical physics*, Phys. Rep. **363**, 223.

Blaizot, J. P., R. Méndez-Galain, and N. Wschebor (2006a), *A new method to solve the non-perturbative renormalization group equations*, Phys. Lett. B **632**, 571.

Delamotte, B. (2007), *An Introduction to the Nonperturbative Renormalization Group*, arXiv:cond-mat/0702365.

Litim, D. F. (2001), *Optimized renormalization group flows*, Phys. Rev. D **64**, 105007.

Pawlowski, J. M. (2007), *Aspects of the functional renormalisation group*, Ann. Phys. **322**, 2831.

Wetterich, C. (1993), *Exact evolution equation for the effective potential*, Phys. Lett. B **301**, 90.

Part III
Functional Renormalization Group Approach to Fermions

Lattice models for strongly correlated electrons in solids are of central interest in condensed matter physics, because it is believed that the physical mechanism underlying collective phenomena such as superconductivity or magnetism can be qualitatively understood using simplified fermionic lattice models such as the Hubbard model. In the condensed matter community, the renormalization group has therefore been used extensively to analyze the low-energy phase diagram of interacting many-electron systems, especially in reduced spatial dimensionality where correlation effects play an important role. A perturbative treatment of these systems can lead to infrared divergencies due to low-energy excitations such as particle–hole excitations or various types of collective modes. Often these divergencies signal a tendency toward the onset of long-range order. In cases where these instabilities of the normal state are dominated by scattering processes in a single channel, the physical properties of the system can be obtained within simple mean-field theory or the Gaussian approximation, which in this context is called *random phase approximation* or *ladder approximation*, as will be discussed in Sect. 10.3. However, in one and two spatial dimensions, different instabilities often compete and need to be treated on equal footing. The renormalization group then provides an unbiased means of resumming the perturbation theory. The leading order renormalization group equations for one-dimensional electrons have been written down a long time ago (Sólyom 1979). After the discovery of the cuprate high-temperature superconductors, the RG approach to interacting fermions has been extended to higher dimensions using the modern language of functional integration over anticommuting fields (see, e.g., Shankar 1994). In this context, different approximation schemes have been developed and applied to various versions of the Hubbard model during the last decade.[1] Because the single-particle propagator of normal fermions is singular on the entire Fermi surface which forms a continuum in $D > 1$, the low-energy

---

[1] It should be mentioned that a few years before Shankar's didactic review made the formulation of the Wilsonian RG via fermionic functional integrals popular in the condensed matter community, most of the ideas and techniques described in Shankar (1994) had already been published in the mathematical physics literature (Benfatto and Gallavotti 1990a,b, Feldman and Trubowitz 1990, 1991, Feldman et al. 1992, 1993a,b). We shall not discuss these works further in this book and

classification of couplings according to relevance (to be discussed in Sect. 10.4) leads to infinitely many marginal couplings associated with all possible two-body scattering processes where the momenta of all particles lie on the Fermi surface. To formulate a consistent RG theory for interacting fermions, one should therefore keep track of entire coupling functions, which is most conveniently done using the functional renormalization group (FRG) methods developed in this book.

Early FRG studies of the Hubbard model with a pure nearest-neighbor hopping were based on a fermionic version of the Polchinski equation (Zanchi and Schulz 1998, 2000, Zanchi 2001). These authors found a dominating antiferromagnetic instability in the vicinity of half filling that is replaced by d-wave superconductivity at sufficient hole doping. Halboth and Metzner (2000a,b) implemented an analogous scheme using an expansion in normal-ordered monomials (which defines so-called Wick-ordered vertex functions) for a model including also next-nearest-neighbor hopping. Apart from d-wave superconductivity and commensurate antiferromagnetism, they also found regions with incommensurate spin-density waves. Most authors now use the one-particle irreducible version of the fermionic FRG (Kopietz and Busche 2001, Salmhofer and Honerkamp 2001, Honerkamp and Salmhofer 2001b, Honerkamp 2003, Honerkamp and Salmhofer 2005, Honerkamp 2001, Honerkamp et al. 2004, Kampf and Katanin 2003, Katanin and Kampf 2003, 2004, Katanin 2004, 2009, Ossadnik et al. 2008). Honerkamp et al. (2001) first applied this one-particle irreducible scheme to the two-dimensional Hubbard model and also discussed the possibility of a spin-liquid phase stabilized by Umklapp scattering.

At low energies and independent of the type of vertices used, the one-loop FRG flow exhibits a divergence of some particular coupling constants at a finite value of the infrared cutoff Λ. The type and symmetry of the couplings which diverge first then yield useful information about the dominating instability of the Fermi liquid. Yet, with the strong increase of the coupling constants, the justification of the one-loop truncation breaks down. For the case of only one dominating instability, the flow to strong coupling is then interpreted as an indication of a broken symmetry. Yet, especially for parameter regions with several divergent couplings of similar strength, states without long-range order are also conceivable. Strictly speaking, a continuous symmetry in two dimensions can only be broken in the ground state due to the Mermin–Wagner theorem. At finite temperatures a quasi long-range order with algebraic decay of correlation functions can occur. Furthermore, a weak coupling in the third spatial direction can suffice to stabilize long-range three-dimensional order in the channel of the dominating two-dimensional instability.

To be able to continue the FRG flow down to $\Lambda \rightarrow 0$, some workers currently explore possibilities to incorporate symmetry breaking into the formalism. In the purely fermionic language, this can be achieved by introducing a small symmetry-breaking initial source field. During the flow, this generates anomalous, gap-creating

---

refer the reader to the book by Salmhofer (1999) for a more mathematical introduction to the fermionic functional renormalization group.

self-energy terms and also regularizes the flow of the two-particle vertex (Salmhofer et al. 2004, Gersch et al. 2005, 2006, 2008). In an alternative approach which will be discussed in Chaps. 11 and 12, one directly introduces bosonic order parameter fields via a Hubbard–Stratonovich transformation and derives FRG flow equations for the coupled field theory (Baier et al. 2004, 2005, Schütz et al. 2005, Schütz 2005). Symmetry breaking then becomes manifest in finite vacuum expectation values of some bosonic field components (Schütz and Kopietz 2006). This partially bosonized FRG flow has the advantage that a bosonic four-point vertex is still tractable (Baier et al. 2004, 2005) and phenomena that are controlled by a non-Gaussian fixed point in the bosonic sector are accessible. In the purely fermionic language, the corresponding eight-point vertex is extremely difficult to treat. A two-stage approach is also possible which first uses an FRG calculation in the fermionic language and analyzes the resulting low-energy action with the help of bosonic techniques. A mean-field analysis has recently been carried out for regions with equally strong superconducting as well as antiferromagnetic instabilities (Reiss et al. 2007).

Further interesting developments in the fermionic FRG include the possibility of alternative flow parameters, such as the strength of the interaction (Honerkamp et al. 2004), a cutoff in the momentum transfer of the interaction (Schütz et al. 2005), or the temperature. The latter approach allows for the treatment of ferromagnetic instabilities (Honerkamp and Salmhofer 2001a,b, Honerkamp 2001, Katanin and Kampf 2003). Furthermore, an explicit calculation of the self-energy is of great interest. Zanchi first calculated the quasiparticle residue (Zanchi 2001) and found a strong suppression in the vicinity of the Van Hove points. Recently, a calculation of the frequency dependence of the spectral function has been achieved (Katanin and Kampf 2004, Rohe and Metzner 2005) leading to results that suggest a pseudo-gap behavior at certain points on the Fermi surface. Using the field-theoretical renormalization group in a two-loop truncation, Ferraz and coworkers found a spin liquid with vanishing quasiparticle weight for a Fermi surface with parallel sections (Ferraz 2003, Freire et al. 2005) and later on also for the two-dimensional Hubbard model (Freire et al. 2008). In connection with self-energy effects, one should also mention the problem of self-consistently determining the interacting Fermi surface (Ledowski and Kopietz 2003, Ledowski et al. 2005, Ledowski and Kopietz 2007). It is an open question to what extend a change of the form of the Fermi surface has an influence on the interplay of different instabilities. Up to now all calculations in two dimensions have neglected the renormalization of the dispersion, which is consistent with the one-loop truncation. The one-loop patching approximation has also been used to study phase diagrams of two-dimensional models with more complicated geometries such as triangular lattices (Tsai and Marston 2001, Honerkamp and Salmhofer 2003, Mathey et al. 2006) as well as the Honeycomb lattice (Raghu et al. 2008). The existence of more exotic phases, for example, staggered flux phases, has been addressed for the extended Hubbard model on the square lattice (Kampf and Katanin 2003). Finally, let us mention here very recent applications of the FRG to the study of the pairing symmetry and the pairing mechanism of the iron arsenide-based superconductors (Wang et al. 2009, Platt et al. 2009, Zhai et al. 2009).

The third part of this book is divided into three chapters. We begin in Chap. 10 with a detailed discussion of the fermionic FRG. After a careful derivation of the exact FRG flow equations for the irreducible vertices, we discuss various truncations of the fermionic vertex expansion. We also present a detailed discussion of the rescaling problem in the presence of a Fermi surface, and show how standard single-channel approximations such as the random phase approximation and the ladder approximation can be recovered within the FRG. Moreover, in Sect. 10.5 we give a self-contained introduction to the one-loop patching approximation where all scattering channels are treated on equal footing.

In the last two chapters, Chaps. 11 and 12, we then introduce the reader to the subtleties of partially bosonized FRG flows for interacting fermions. The basic idea is to represent certain types of interaction processes as the exchange of suitable bosonic fields, which are formally introduced via Hubbard–Stratonovich transformations. Unfortunately, this can be done in many ways, so that in practice one needs some a priori knowledge about the dominant scattering processes. In Chap. 11 we shall discuss normal fermions with dominant forward scattering, where it is natural to decouple the interaction in the zero-sound channel describing particle–hole scattering with small momentum transfers. Finally, in Chap. 12 we shall discuss superfluid fermions with an attractive short-range interaction—in this case it is natural to use a Hubbard–Stratonovich decoupling in the particle–particle channel. We present a new truncation strategy of the exact hierarchy of FRG flow equations using skeleton equations and Ward identities (Bartosch et al. 2009).

References

Baier, T., E. Bick, and C. Wetterich (2004), *Temperature dependence of antiferromagnetic order in the Hubbard model*, Phys. Rev. B **70**, 125111.

Baier, T., E. Bick, and C. Wetterich (2005), *Antiferromagnetic gap in the Hubbard model*, Phys. Lett. B **605**, 144.

Bartosch, L., P. Kopietz, and A. Ferraz (2009), *Renormalization of the BCS-BEC crossover by order-parameter fluctuations*, Phys. Rev. B **80**, 104504.

Benfatto, G. and G. Gallavotti (1990a), *Perturbation theory of the Fermi surface in a quantum liquid. A general quasiparticle formalism and one-dimensional systems*, J. Stat. Phys. **59**, 541.

Benfatto, G. and G. Gallavotti (1990b), *Renormalization group approach to the theory of the Fermi surface*, Phys. Rev. B **42**, 9967.

Feldman, J. and E. Trubowitz (1990), *Perturbation theory for many fermion systems*, Helv. Phys. Acta **63**, 156.

Feldman, J. and E. Trubowitz (1991), *The flow of an electron-phonon system to the superconducting state*, Helv. Phys. Acta **64**, 213.

Feldman, J., J. Magnen, V. Rivasseau, and E. Trubowitz (1992), *An infinite volume expansion for many fermion Green's functions*, Helv. Phys. Acta **65**, 679.

Feldman, J., J. Magnen, V. Rivasseau, and E. Trubowitz (1993a), *An intrinsic 1/N expansion for many-fermion systems*, Europhys. Lett. **24**, 437.

Feldman, J., J. Magnen, V. Rivasseau, and E. Trubowitz (1993b), *Two-dimensional many-fermion systems as vector models*, Europhys. Lett. **24**, 521.

Ferraz, A. (2003), *Non-Fermi liquid in a truncated two-dimensional Fermi surface*, Phys. Rev. B **68**, 075115.

Freire, H., E. Corrêa, and A. Ferraz (2005), *Field theoretical renormalization group for a flat two-dimensional Fermi surface*, Phys. Rev. B **71**, 165113.

Freire, H., E. Corrêa, and A. Ferraz (2008), *Breakdown of the Fermi-liquid regime in the two-dimensional Hubbard model from a two-loop field-theoretical renormalization group approach*, Phys. Rev. B **78**, 125114.

Gersch, R., C. Honerkamp, D. Rohe, and W. Metzner (2005), *Fermionic renormalization group flow into phases with broken discrete symmetry: charge-density wave mean-field model*, Eur. Phys. J. B **48**, 349.

Gersch, R., J. Reiss, and C. Honerkamp (2006), *Fermionic functional renormalization-group for first-order phase transitions: a mean-field model*, New J. Phys. **8**, 320.

Gersch, R., C. Honerkamp, and W. Metzner (2008), *Superconductivity in the attractive Hubbard model: functional renormalization group analysis*, New J. Phys. **10**, 045003.

Halboth, C. J. and W. Metzner (2000a), *d-Wave superconductivity and Pomeranchuk instability in the two-dimensional Hubbard model*, Phys. Rev. Lett. **85**, 5162.

Halboth, C. J. and W. Metzner (2000b), *Renormalization group analysis of the two-dimensional Hubbard model*, Phys. Rev. B **61**, 7364.

Honerkamp, C. (2001), *Electron-doping versus hole-doping in the 2D t-t' Hubbard model*, Eur. Phys. J. B **21**, 81.

Honerkamp, C. (2003), *Instabilities of interacting electrons on the triangular lattice*, Phys. Rev. B **68**, 104510.

Honerkamp, C. and M. Salmhofer (2001a), *Magnetic and superconducting instabilities of the Hubbard model at the Van Hove filling*, Phys. Rev. Lett. **87**, 187004.

Honerkamp, C. and M. Salmhofer (2001b), *Temperature-flow renormalization group and the competition between superconductivity and ferromagnetism*, Phys. Rev. B **64**, 184516.

Honerkamp, C. and M. Salmhofer (2003), *Flow of the quasiparticle weight in the N-patch renormalization group scheme*, Phys. Rev. B **67**, 174504.

Honerkamp, C. and M. Salmhofer (2005), *Eliashberg equations derived from the functional renormalization group*, Progr. Theoret. Phys. **113**, 1145.

Honerkamp, C., M. Salmhofer, N. Furukawa, and T. M. Rice (2001), *Breakdown of the Landau-Fermi liquid in two dimensions due to Umklapp scattering*, Phys. Rev. B **63**, 035109.

Honerkamp, C., D. Rohe, S. Andergassen, and T. Enss (2004), *Interaction flow method for many-fermion systems*, Phys. Rev. B **70**, 235115.

Kampf, A. P. and A. A. Katanin (2003), *Competing phases in the extended U-V-J Hubbard model near the Van Hove fillings*, Phys. Rev. B **67**, 125104.

Katanin, A. A. (2004), *Fulfillment of Ward identities in the functional renormalization group approach*, Phys. Rev. B **70**, 115109.

Katanin, A. A. (2009), *The two-loop functional renormalization group approach to the one- and two-dimensional Hubbard model*, Phys. Rev. B **79**, 235119.

Katanin, A. A. and A. P. Kampf (2003), *Renormalization group analysis of magnetic and superconducting instabilities near Van Hove band fillings*, Phys. Rev. B **68**, 195101.

Katanin, A. A. and A. P. Kampf (2004), *Quasiparticle anisotropy and pseudogap formation from the weak-coupling renormalization group point of view*, Phys. Rev. Lett. **93**, 106406.

Kopietz, P. and T. Busche (2001), *Exact renormalization group flow equations for nonrelativistic fermions: Scaling toward the Fermi surface*, Phys. Rev. B **64**, 155101.

Ledowski, S. and P. Kopietz (2003), *An exact integral equation for the renormalized Fermi surface*, J. Phys.: Condens. Matter **15**, 4779.

Ledowski, S. and P. Kopietz (2007), *Fermi surface renormalization and confinement in two coupled metallic chains*, Phys. Rev. B **75**, 045134.

Ledowski, S., P. Kopietz, and A. Ferraz (2005), *Self-consistent Fermi surface renormalization of two coupled Luttinger liquids*, Phys. Rev. B **71**, 235106.

Mathey, L., S.-W. Tsai, and A. H. Castro Neto (2006), *Competing types of order in two-dimensional bose-fermi mixtures*, Phys. Rev. Lett. **97**, 030601.

Ossadnik, M., C. Honerkamp, T. M. Rice, and M. Sigrist (2008), *Breakdown of Landau theory in overdoped cuprates near the onset of superconductivity*, Phys. Rev. Lett. **101**, 256405.

Platt, C., C. Honerkamp, and W. Hanke (2009), *Pairing in the iron arsenides: a functional RG treatment*, New J. Phys. **11**, 055058.

Raghu, S., X.-L. Qi, C. Honerkamp, and S.-C. Zhang (2008), *Topological Mott insulators*, Phys. Rev. Lett. **100**, 156401.

Reiss, J., D. Rohe, and W. Metzner (2007), *Renormalized mean-field analysis of antiferromagnetism and d-wave superconductivity in the two-dimensional Hubbard model*, Phys. Rev. B **75**, 075110.

Rohe, D. and W. Metzner (2005), *Pseudogap at hot spots in the two-dimensional Hubbard model at weak coupling*, Phys. Rev. B **71**, 115116.

Salmhofer, M. (1999), *Renormalization: An Introduction*, Springer, Berlin.

Salmhofer, M. and C. Honerkamp (2001), *Fermionic renormalization group flows*, Progr. Theoret. Phys. 105, 1.

Salmhofer, M., C. Honerkamp, W. Metzner, and O. Lauscher (2004), *Renormalization group flows into phases with broken symmetry*, Progr. Theoret. Phys. **112**, 943.

Schütz, F. (2005), *Aspects of Strong Correlations in Low Dimensions*, Doktorarbeit, Goethe Universität Frankfurt.

Schütz, F. and P. Kopietz (2006), *Functional renormalization group with vacuum expectation values and spontaneous symmetry breaking*, J. Phys. A: Math. Gen. **39**, 8205.

Schütz, F., L. Bartosch, and P. Kopietz (2005), *Collective fields in the functional renormalization group for fermions, Ward identities, and the exact solution of the Tomonaga-Luttinger model*, Phys. Rev. B **72**, 035107.

Shankar, R. (1994), *Renormalization group approach to interacting fermions*, Rev. Mod. Phys. **66**, 129.

Sólyom, J. (1979), *The Fermi gas model of one-dimensional conductors*, Adv. Phys. **28**, 201.

Tsai, S.-W. and J. B. Marston (2001), *κ–(BEDT–TTF)₂X organic crystals: Superconducting versus anti-ferromagnetic instabilities in the Hubbard model on an anisotropic triangular lattice*, Can. J. Phys. **79**, 1463.

Wang, F., H. Zhai, Y. Ran, A. Vishwanath, and D.-H. Lee (2009), *Functional renormalization group study of the pairing mechanism of the FeAs-based high-temperature superconductors*, Phys. Rev. Lett. **102**, 047005.

Zanchi, D. (2001), *Angle-resolved loss of Landau quasiparticles in 2D Hubbard model*, Europhys. Lett. **55**, 376.

Zanchi, D. and H. J. Schulz (1998), *Weakly correlated electrons on a square lattice: A renormalization group theory*, Europhys. Lett. **44**, 235.

Zanchi, D. and H. J. Schulz (2000), *Weakly correlated electrons on a square lattice: Renormalization group theory*, Phys. Rev. B **61**, 13609.

Zhai, H., F. Wang, and D.-H. Lee (2009), *Antiferromagnetically driven electronic correlation in iron pnictides and cuprates*, Phys. Rev. B **80**, 064519.

Chapter 10
Fermionic Functional Renormalization Group

In this chapter we shall derive the fundamental FRG flow equations for one-band models of nonrelativistic fermions. Although these flow equations follow as a special case of the general flow equations derived in Chap. 7, it is still useful to write down these equations explicitly in order to identify the various terms in the vertex expansion with familiar types of scattering processes (Kopietz and Busche 2001, Salmhofer and Honerkamp 2001). Moreover, for fermions with $SU(2)$ spin rotational symmetry it is useful to take the constraints imposed by this symmetry via a proper parameterization of the vertices into account. We shall derive the corresponding exact FRG flow equations in Sect. 10.2. We then show in Sect. 10.3 how to recover standard single-channel approximations such as the random phase approximation or the ladder approximation from our FRG equations. We proceed in Sect. 10.4 with a discussion of the rescaling problem for normal fermions, which leads to a nonperturbative definition of the Fermi surface. Finally, we discuss in Sect. 10.5 the one-loop patching approximation and present numerical results for the square-lattice Hubbard model.

10.1 Symmetries of the Two-Fermion Interaction

The correlation functions of interacting fermions can be represented as functional integrals over pairs of anticommuting Grassmann fields ψ and $\bar{\psi}$ (see, e.g., Negele and Orland 1988). In terms of these fields, we can write the Euclidean action of an interacting many-fermion system with general two-body interaction in the form

$$S[\bar{\psi}, \psi] = S_0[\bar{\psi}, \psi] + S_1[\bar{\psi}, \psi] \,, \tag{10.1}$$

with the Gaussian part

$$S_0[\bar{\psi}, \psi] = -\sum_{\sigma} \int_K G_0^{-1}(K\sigma)\bar{\psi}_{K\sigma}\psi_{K\sigma} = -\sum_{\sigma} \int_K (i\omega - \xi_{k\sigma})\bar{\psi}_{K\sigma}\psi_{K\sigma} \,. \tag{10.2}$$

Kopietz, P. et al.: *Fermionic Functional Renormalization Group*. Lect. Notes Phys. **798**, 255–303 (2010)
DOI 10.1007/978-3-642-05094-7_10

Here, $\xi_{k\sigma} = \epsilon_{k\sigma} - \mu$, where μ is the chemical potential and $\epsilon_{k\sigma}$ is some energy dispersion which may also depend on the spin projection σ. The collective label $K = (i\omega, k)$ consists of fermionic Matsubara frequencies $i\omega$ and momenta k, and the symbol $\int_K$ denotes the corresponding integrations or summations. This notation has already been introduced in the introductory paragraph of Chap. 6, see Eqs. (6.6)–(6.8). Although in this chapter we are mainly interested in Fermi systems where $\zeta = -1$, we shall retain the statistics factor ζ so that by setting $\zeta = +1$ we obtain the corresponding FRG flow equations for interacting bosons.

For fermions with spin, the interaction part of the bare action will mostly be assumed to possess an $SU(2)$ spin rotational symmetry, or at least to be invariant under spin rotations around the quantization axis. Thus, according to Eq. (6.119) the two-body interaction must be of the form

$$S_1[\bar{\psi}, \psi] = \frac{1}{2} \sum_{\sigma_1\sigma_2} \int_{K_1'} \int_{K_2'} \int_{K_2} \int_{K_1} \delta_{K_1'+K_2', K_2+K_1}$$

$$\times U^{(4)}_{\sigma_1\sigma_2} \left(K_1', K_2'; K_2, K_1 \right) \bar{\psi}_{K_1'\sigma_1} \bar{\psi}_{K_2'\sigma_2} \psi_{K_2\sigma_2} \psi_{K_1\sigma_1} . \tag{10.3}$$

The vertex $U^{(4)}_{\sigma_1\sigma_2} \left(K_1', K_2'; K_2, K_1 \right)$ is symmetric under a simultaneous permutation of its ingoing and outgoing labels,

$$U^{(4)}_{\sigma_1\sigma_2} \left(K_1', K_2'; K_2, K_1 \right) = U^{(4)}_{\sigma_2\sigma_1} \left(K_2', K_1'; K_1, K_2 \right) , \tag{10.4}$$

which is a special case of the general relation (6.120). Diagrammatically, we represent the vertex $U^{(4)}_{\sigma_1\sigma_2}$ by an elongated rectangle, signaling that the spin is conserved along continuous particle lines at each end,

$$U^{(4)}_{\sigma_1\sigma_2} \left(K_1', K_2'; K_2, K_1 \right) = \tag{10.5}$$

The permutation symmetry in Eq. (10.4) then corresponds to flipping the diagram such that the upper and lower ends are exchanged. If furthermore the interaction depends only on the momentum transfer $q = k_1' - k_1$, i.e., $U^{(4)}_{\sigma_1\sigma_2} \left(K_1', K_2'; K_2, K_1 \right) = U_{\sigma_1\sigma_2} \left(k_1' - k_1 \right)$ (such as the bare Coulomb interaction), it is conveniently represented as follows:

$$U_{\sigma_1\sigma_2}(q) = \tag{10.6}$$

The wavy line can be thought of as an effective particle mediating the interaction; we shall come back to this interpretation in Chap. 11.

In order to derive exact FRG flow equations for the irreducible vertices, we will also need properly antisymmetrized vertices. Therefore, we write Eq. (10.3) in the form

$$S_1[\bar{\psi}, \psi] = \frac{1}{(2!)^2} \int_{K_1'\sigma_1'} \int_{K_2'\sigma_2'} \int_{K_2\sigma_2} \int_{K_1\sigma_1} \delta_{K_1'+K_2', K_2+K_1}$$

$$\times \Gamma_0^{(4)} \left(K_1'\sigma_1', K_2'\sigma_2'; K_2\sigma_2, K_1\sigma_1 \right) \bar{\psi}_{K_1'\sigma_1'} \bar{\psi}_{K_2'\sigma_2'} \psi_{K_2\sigma_2} \psi_{K_1\sigma_1} , \quad (10.7)$$

where $\int_{K\sigma} \equiv \int_K \sum_\sigma$ and the vertex $\Gamma_0^{(4)} \left(K_1'\sigma_1', K_2'\sigma_2'; K_2\sigma_2, K_1\sigma_1 \right)$ is antisymmetric with respect to the exchange of its first two and its second two labels. Diagrammatically, we represent $\Gamma^{(4)}$ by a square,

$$\Gamma_0^{(4)} \left(K_1'\sigma_1', K_2'\sigma_2'; K_2\sigma_2, K_1\sigma_1 \right) = \quad \vcenter{\hbox{(diagram)}} \quad . \qquad (10.8)$$

The relation between this antisymmetrized vertex and the symmetric vertex $U_{\sigma_1\sigma_2}^{(2)} \left(K_1', K_2'; K_2, K_1 \right)$ in Eq. (10.3) follows as a special case of the general expression (6.121),

$$\Gamma_0^{(4)} \left(K_1'\sigma_1', K_2'\sigma_2'; K_2\sigma_2, K_1\sigma_1 \right) = \delta_{\sigma_1'\sigma_1} \delta_{\sigma_2'\sigma_2} U_{\sigma_1\sigma_2}^{(4)} \left(K_1', K_2'; K_2, K_1 \right)$$

$$+ \zeta \delta_{\sigma_1'\sigma_2} \delta_{\sigma_2'\sigma_1} U_{\sigma_2\sigma_1}^{(4)} \left(K_1', K_2'; K_1, K_2 \right) . \quad (10.9)$$

Note that for bosons ($\zeta = 1$) this vertex is symmetric under the exchange of the first two or the second two labels. Sometimes it is useful to decompose the interaction into a spin-singlet part and a spin-triplet part,

$$\Gamma_0^{(4)} \left(K_1'\sigma_1', K_2'\sigma_2'; K_2\sigma_2, K_1\sigma_1 \right) =$$

$$\left[\delta_{\sigma_1'\sigma_1} \delta_{\sigma_2'\sigma_2} - \delta_{\sigma_1'\sigma_2} \delta_{\sigma_2'\sigma_1} \right] U_{\sigma_1\sigma_2}^{\perp} \left(K_1', K_2'; K_2, K_1 \right)$$

$$+ \left[\delta_{\sigma_1'\sigma_1} \delta_{\sigma_2'\sigma_2} + \delta_{\sigma_1'\sigma_2} \delta_{\sigma_2'\sigma_1} \right] U_{\sigma_1\sigma_2}^{\parallel} \left(K_1', K_2'; K_2, K_1 \right) , \qquad (10.10)$$

where

$$U_{\sigma_1\sigma_2}^{\perp} \left(K_1', K_2'; K_2, K_1 \right) =$$

$$\frac{1}{2} \left[U_{\sigma_1\sigma_2}^{(4)} \left(K_1', K_2'; K_2, K_1 \right) - \zeta U_{\sigma_2\sigma_1}^{(4)} \left(K_1', K_2'; K_1, K_2 \right) \right] , \quad (10.11a)$$

and

$$U_{\sigma_1\sigma_2}^{\parallel} \left(K_1', K_2'; K_2, K_1 \right) =$$

$$\frac{1}{2} \left[U_{\sigma_1\sigma_2}^{(4)} \left(K_1', K_2'; K_2, K_1 \right) + \zeta U_{\sigma_2\sigma_1}^{(4)} \left(K_1', K_2'; K_1, K_2 \right) \right] . \quad (10.11b)$$

If the system exhibits full $SU(2)$ spin rotation invariance, the function $U_{\sigma_1\sigma_2}^{(4)}$ is independent of the spin labels, see Eq. (6.124). Consequently, the spin-singlet and spin-triplet functions $U^{\perp}$ and $U^{\parallel}$ are also independent of the spin labels. Note that for fermions (bosons) the functions $U^{\perp}$ and $U^{\parallel}$ are then symmetric (antisymmetric) and antisymmetric (symmetric) under independent exchange of the ingoing or outgoing labels, respectively. In contrast, $U^{(4)}$ is only symmetric under simultaneous permutation of both incoming and outgoing momentum labels. The pair $(U^\perp, U^\parallel)$ thus contains the same information as the single function $U^{(4)}$. In the following, we will mostly work directly with $U^{(4)}$, but Eqs. (10.11a) and (10.11b) can always be used to obtain the corresponding singlet and triplet components of the interaction.

As an example, consider the Hubbard model with general spin-dependent hopping $t_{ij,\sigma}$ connecting lattice sites $\boldsymbol{r}_i$ and $\boldsymbol{r}_j$. The Euclidean action in imaginary time is then given by

$$S[\bar{\psi}, \psi] = \int_0^\beta d\tau \left[\sum_{ij\sigma} \bar{\psi}_{i\sigma}(\tau)[\delta_{ij}(\partial_\tau - \mu) + t_{ij,\sigma}]\psi_{j\sigma}(\tau) \right.$$
$$\left. +U \sum_i n_{i\uparrow}(\tau)n_{i\downarrow}(\tau) \right], \tag{10.12}$$

where $n_{i\sigma}(\tau) = \bar{\psi}_{i\sigma}(\tau)\psi_{i\sigma}(\tau)$. The on-site interaction U is only effective if two electrons with opposite spin occupy the same lattice site. For an infinite lattice or for periodic boundary conditions, the system is invariant under discrete lattice translations. The hoppings $t_{ij,\sigma}$ can then only depend on the distance $\boldsymbol{\delta} = \boldsymbol{r}_j - \boldsymbol{r}_i$ between the two lattice sites. For a single-band model, we can thus write $t_{ij,\sigma} = t_\sigma(\boldsymbol{r}_j - \boldsymbol{r}_i)$ and the energy dispersion in Eq. (10.2) is

$$\epsilon_{k\sigma} = \sum_\delta e^{-ik\cdot\delta} t_\sigma(\boldsymbol{\delta}) , \tag{10.13}$$

where the sum is over all translational vectors of the Bravais lattice. For short-range hoppings, only a small number of these terms contributes. Note that if the unit cell contains more than one site, multiple bands occur and a band index has to be included in addition to the spin index. We will not consider the multiband situation further here. For the Hubbard model on a D-dimensional hypercubic lattice with lattice spacing a, the interaction $U^{(4)}_{\sigma_1\sigma_2}$ in Eq. (10.3) is given by[1]

$$U^{(4)}_{\sigma_1\sigma_2}\left(K_1', K_2'; K_2, K_1\right) \equiv U^{(4)}\left(K_1', K_2'; K_2, K_1\right) = a^D U , \tag{10.14}$$

which is a special case of Eq. (6.124). Recall that for lattice models momentum is only conserved up to vectors of the reciprocal lattice, as discussed in Sect. 6.3. All momentum summations are thus over the first Brillouin zone and $\delta_{K,K'} = \delta^{(G)}_{K,K'}$ is the periodic delta-function as defined in Eq. (6.115). The momentum conservation modulo a vector of the reciprocal lattice in the presence of discrete translational invariance has been formally derived in Sect. 6.3.2 (see Eq. (6.112)).

---

[1] The factor of a^D in Eq. (10.14) is due to the fact that in Eq. (10.3) we use the continuum normalization of the Fourier-transformed fields $\psi_{K\sigma}$, which is related to the lattice fields appearing in Eq. (10.12) via $\psi_{i\sigma}(\tau) = \frac{a^{D/2}}{\beta V} \sum_K e^{i(k\cdot r_i - \omega\tau)} \psi_{K\sigma}$.

10.2 Exact FRG Flow Equations for the Irreducible Vertices

10.2.1 From Superfield to Partially Symmetrized Notation

Nonrelativistic Fermi or Bose systems with Euclidean action $S[\bar{\psi}, \psi]$ given by
Eqs. (10.1)–(10.3) belong to the class of models introduced in Chap. 6, so that
the exact FRG flow equations for these systems can be obtained as a special case
of the general flow equations derived in Chap. 7. The superfield Φ_α has in this
case two field components ψ and $\bar{\psi}$, whose configuration is specified by the labels
$(K\sigma) = (i\omega, \boldsymbol{k}, \sigma)$. Our superfield label α is therefore of the form $\alpha = (i, K, \sigma)$,
where the field-type label i assumes the values $i = \psi, \bar{\psi}$. For fixed values of K and
σ, we have therefore a two-component superfield,

$$\Phi_{K\sigma} \equiv \begin{pmatrix} \Phi_{\psi K\sigma} \\ \Phi_{\bar{\psi} K\sigma} \end{pmatrix} = \begin{pmatrix} \psi_{K\sigma} \\ \bar{\psi}_{K\sigma} \end{pmatrix} . \tag{10.15}$$

The symbol $\int_\alpha$ introduced in Eq. (6.3) should then be understood as

$$\int_\alpha f(\Phi_\alpha) = \sum_{i=\psi,\bar{\psi}} \int_{K\sigma} f(\Phi_{iK\sigma}) , \tag{10.16}$$

where $f(\Phi_\alpha)$ is any function of the field component Φ_α. The Gaussian part (10.2)
of our bare action can then be written as

$$S_0[\Phi] = -\frac{1}{2} \int_{K\sigma} \Phi_{K\sigma}^T \begin{pmatrix} 0 & \zeta G_0^{-1}(K\sigma) \\ G_0^{-1}(K\sigma) & 0 \end{pmatrix} \Phi_{K\sigma} , \tag{10.17}$$

which for fermions ($\zeta = -1$) reduces to Eq. (6.6).

To relate the two-body interaction to the completely symmetrized interaction
vertices introduced in superfield notation, we note that according to Eq. (6.60) the
interaction in superfield notation should be written as

$$S_1[\Phi] = \frac{1}{4!} \int_{\alpha_1} \cdots \int_{\alpha_4} \Gamma^{(4)}_{\alpha_1\alpha_2\alpha_3\alpha_4} \Phi_{\alpha_1} \Phi_{\alpha_2} \Phi_{\alpha_3} \Phi_{\alpha_4} . \tag{10.18}$$

Comparing this expression with Eq. (10.7), we conclude that

$$\Gamma^{(4)}_{\alpha_1=(\bar{\psi},K_1',\sigma_1'),\alpha_2=(\bar{\psi},K_2',\sigma_2'),\alpha_3=(\psi,K_2,\sigma_2),\alpha_4=(\psi,K_1,\sigma_1)}$$
$$= \delta_{K_1'+K_2',K_2+K_1} \Gamma_0^{(4)} \left(K_1'\sigma_1', K_2'\sigma_2'; K_2\sigma_2, K_1\sigma_1 \right) . \tag{10.19}$$

which is a special case of the general relation (6.114).

For simplicity, we shall assume in this chapter that our system is not superfluid. The only nonvanishing elements of the flowing two-point vertex $\Gamma^{(2)}_{\Lambda,\alpha_1\alpha_2}$ are then[2]

$$\Gamma^{(2)}_{\Lambda,\alpha_1=(\bar\psi K_1\sigma_1),\alpha_2=(\psi K_2\sigma_2)} = \zeta\,\Gamma^{(2)}_{\Lambda,\alpha_1=(\psi K_1\sigma_1),\alpha_2=(\bar\psi K_2\sigma_2)}$$
$$= \delta_{K_1 K_2}\delta_{\sigma_1\sigma_2}\Sigma_\Lambda(K_1\sigma_1)\,, \tag{10.20}$$

where $\Sigma_\Lambda(K\sigma)$ is the flowing irreducible self-energy. Note that here and in the following subsection we assume conservation of the spin projection onto the quantization axis. The flowing inverse matrix Green function $\mathbf{G}^{-1}_\Lambda$ and the self-energy matrix $\mathbf{\Sigma}_\Lambda$ have the same block structure as the inverse free propagator in Eq. (10.17),

$$\mathbf{G}^{-1}_\Lambda = \begin{pmatrix} 0 & (\mathbf{G}^{-1}_\Lambda)_{\psi\bar\psi} \\ (\mathbf{G}^{-1}_\Lambda)_{\bar\psi\psi} & 0 \end{pmatrix} = \begin{pmatrix} 0 & \zeta\hat{G}^{-1}_\Lambda \\ \hat{G}^{-1}_\Lambda & 0 \end{pmatrix} = \mathbf{G}^{-1}_0 - \mathbf{\Sigma}_\Lambda\,, \tag{10.21}$$

$$\mathbf{\Sigma}_\Lambda = \begin{pmatrix} 0 & (\mathbf{\Sigma}_\Lambda)_{\psi\bar\psi} \\ (\mathbf{\Sigma}_\Lambda)_{\bar\psi\psi} & 0 \end{pmatrix} = \begin{pmatrix} 0 & \zeta\hat{\Sigma}_\Lambda \\ \hat{\Sigma}_\Lambda & 0 \end{pmatrix}\,, \tag{10.22}$$

where $\hat{G}^{-1}_\Lambda$ and $\hat{\Sigma}_\Lambda$ are infinite diagonal matrices in momentum-frequency and spin space, with matrix elements given by

$$\left[\hat{G}^{-1}_\Lambda\right]_{K\sigma,K'\sigma'} = \delta_{KK'}\delta_{\sigma\sigma'}G^{-1}_\Lambda(K\sigma)\,, \tag{10.23}$$

$$\left[\hat{\Sigma}_\Lambda\right]_{K\sigma,K'\sigma'} = \delta_{KK'}\delta_{\sigma\sigma'}\Sigma_\Lambda(K\sigma)\,. \tag{10.24}$$

Here, the flowing single-particle Green function $G^{-1}_\Lambda(K\sigma)$ is related to the flowing self-energy via

$$G^{-1}_\Lambda(K\sigma) = G^{-1}_0(K\sigma) - \Sigma_\Lambda(K\sigma)\,. \tag{10.25}$$

The matrix Green function $\mathbf{G}_\Lambda$ and the corresponding single-scale propagator $\dot{\mathbf{G}}_\Lambda$ which appear in the exact FRG flow equations have then the block structure

$$\mathbf{G}_\Lambda = \begin{pmatrix} 0 & (\mathbf{G}_\Lambda)_{\psi\bar\psi} \\ (\mathbf{G}_\Lambda)_{\bar\psi\psi} & 0 \end{pmatrix} = \begin{pmatrix} 0 & \hat{G}_\Lambda \\ \zeta\hat{G}_\Lambda & 0 \end{pmatrix}\,, \tag{10.26a}$$

$$\dot{\mathbf{G}}_\Lambda = \begin{pmatrix} 0 & (\dot{\mathbf{G}}_\Lambda)_{\psi\bar\psi} \\ (\dot{\mathbf{G}}_\Lambda)_{\bar\psi\psi} & 0 \end{pmatrix} = \begin{pmatrix} 0 & \dot{\hat{G}}_\Lambda \\ \zeta\dot{\hat{G}}_\Lambda & 0 \end{pmatrix}\,, \tag{10.26b}$$

where $\dot{\hat{G}}_\Lambda$ is again a diagonal matrix whose matrix elements are given by the single-scale propagator in momentum–frequency space,

$$[\dot{\hat{G}}_\Lambda]_{K\sigma,K'\sigma'} = \delta_{KK'}\delta_{\sigma\sigma'}\dot{G}_\Lambda(K\sigma)\,. \tag{10.27}$$

---

[2] In Chap. 12 we shall generalize the FRG approach to the superfluid state of neutral fermions. In this case the matrix elements $\Gamma^{(2)}_{\psi\psi}$ and $\Gamma^{(2)}_{\bar\psi\bar\psi}$ are also nonzero.

Note that according to Eqs. (6.12), (6.22c), and (6.31) the cutoff-dependent propagator can be represented as the following functional average,

$$G_\Lambda(K\sigma) = -(\beta V)^{-1} \langle \psi_{K\sigma} \bar\psi_{K\sigma} \rangle = -\zeta (\beta V)^{-1} \langle \bar\psi_{K\sigma} \psi_{K\sigma} \rangle \,, \tag{10.28}$$

where the factor of $(\beta V)^{-1}$ is due to the fact that for $K = K'$ the Kronecker delta-symbol $\delta_{K,K'}$ should be understood as βV, (see Eq. (6.11)).

In the normal state all vertices $\Gamma^{(n)}_{\alpha_1\dots\alpha_n}$ where the number of external legs corresponding to $\bar\psi$ is not equal to the number of legs corresponding to ψ vanish, so that our theory has only even vertices with an equal number of $\bar\psi$- and ψ-legs. Similar to Eq. (10.19), it is useful to factor out the overall energy–momentum conserving δ-function and define the flowing, partially symmetrized[3] vertices

$$\Gamma^{(2n)}_{\Lambda,(\bar\psi,K'_1,\sigma'_1)\dots(\bar\psi,K'_n,\sigma'_n)(\psi,K_n,\sigma_n)\dots(\psi,K_1,\sigma_1)}$$
$$= \delta_{K'_1+\dots+K'_n,K_n+\dots+K_1} \Gamma^{(2n)}_\Lambda \left(K'_1\sigma'_1, \dots, K'_n\sigma'_n; K_n\sigma_n, \dots, K_1\sigma_1 \right) \,. \tag{10.29}$$

The corresponding functional Taylor expansion of the generating functional $\Gamma_\Lambda[\bar\psi, \psi]$ is given in Eq. (6.113). To visualize the various terms in the flow equations, it is then useful to switch from the graphical representation of the completely symmetrized vertices used in Chaps. 6 and 7 to a new graphical notation where the two different field types of our theory are represented by incoming or outgoing arrows, which enter or leave vertices which are now represented by squares, as shown in Fig. 10.1. A leg associated with a $\bar\psi$ field is represented by an arrow pointing outward, while a leg associated with ψ is represented by an arrow pointing inward. Note that on each side of the square only fields of the same type are attached. As usual, we represent the flowing propagator $G_\Lambda(K\sigma) = -(\beta V)^{-1} \langle \psi_{K\sigma} \bar\psi_{K\sigma} \rangle$ by a thick arrow pointing from $\bar\psi$ to ψ, as shown on the right-hand side of Fig. 10.1. The corresponding single-scale propagators are marked by an additional slash in front of the arrow. Because according to Eq. (10.29) the totally symmetric superfield vertices $\Gamma^{(2n)}_{\alpha_1,\dots,\alpha_n}$ agree, up to the energy and momentum conserving δ-function, with the partially symmetrized vertices $\Gamma^{(2n)}_\Lambda \left(K'_1\sigma'_1, \dots, K'_n\sigma'_n; K_n\sigma_n, \dots, K_1\sigma_1 \right)$ if we choose the same ordering of the indices, we can obtain the flow equations for the partially symmetrized vertices by choosing a definite realization of the external legs and by carrying out the intermediate sums over the different field species in our general FRG flow equations for the totally symmetrized vertices derived in Sect. 7.2. Graphically, we simply draw lines with all possible orientations of arrows in the intermediate loop and keep only those diagrams that due to particle conservation

---

[3] For simplicity we call vertices for both bosons and fermions *symmetrized* if they have the proper permutation symmetry of the labels. For fermions the partially symmetrized vertices $\Gamma^{(2n)}_\Lambda \left(K'_1\sigma'_1, \dots, K'_n\sigma'_n; K_n\sigma_n, \dots, K_1\sigma_1 \right)$ are therefore antisymmetric with respect to the exchange of any pair of labels within the first n and the last n group of indices.

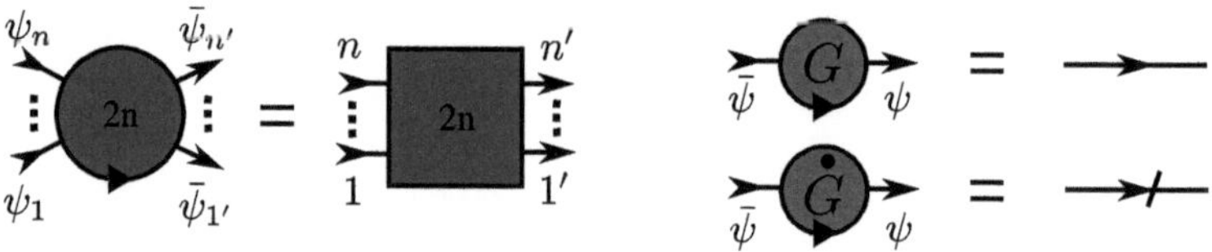

Fig. 10.1 Graphical elements of the diagrams representing the FRG flow equations for a theory containing only Fermi or Bose fields. A *square-shaded box* with $2n$ external legs represents the flowing irreducible vertex $\Gamma_\Lambda^{(2n)}\left(K_1'\sigma_1', \ldots, K_n'\sigma_n'; K_n\sigma_n, \ldots, K_1\sigma_1\right)$ which is related to the fully symmetrized vertex by Eq. (10.29). *Arrows* pointing into the vertex represent ψ, while *arrows* pointing out of the vertex represent $\bar{\psi}$. The exact propagator is represented by an *arrow* pointing from the point associated with $\bar{\psi}$ to the point associated with ψ, while the single-scale propagator is marked by an additional slash

have an equal number of incoming and outgoing arrows at each vertex and propagator. On the right-hand side one then has to appropriately order all legs attached to a given vertex, keeping in mind that an exchange of neighboring legs leads to a statistical factor ζ. Having done so, we can use the pictorial dictionary in Fig. 10.1 to replace the fully symmetrized vertices and propagators by their partially symmetrized energy–momentum conserving counterparts. The resulting diagrams can then be labeled with momentum and frequency indices just as ordinary Feynman diagrams.

10.2.2 Exact FRG Flow Equations

Let us now explicitly write down the exact FRG flow equations for the first three vertices: the interaction correction to the grand canonical potential $\Gamma_\Lambda^{(0)}$, the irreducible self-energy $\Gamma_\Lambda^{(2)}(K\sigma, K\sigma) \equiv \Sigma_\Lambda(K\sigma)$, and the effective interaction $\Gamma_\Lambda^{(4)}\left(K_1'\sigma_1', K_2'\sigma_2'; K_2\sigma_2, K_1\sigma_1\right)$. In Sect. 10.2.3 we shall further specify our general FRG flow equations for $SU(2)$-symmetric models. The general flow equations considered here are directly applicable to spinless fermions if all spin indices and spin summations are simply dropped.

For our field theory containing only nonrelativistic fermions or bosons, the FRG flow equation (7.70) for the interaction correction to the grand canonical potential reduces to

$$\partial_\Lambda \Gamma_\Lambda^{(0)} = -\zeta\beta V \int_{K\sigma} \frac{\dot{G}_{0,\Lambda}(K\sigma)\Sigma_\Lambda(K\sigma)}{1 - G_{0,\Lambda}(K\sigma)\Sigma_\Lambda(K\sigma)} . \tag{10.30}$$

For bosons ($\zeta = 1$) in the classical limit, corresponding to high temperatures where it is allowed to retain only the zeroth Matsubara frequency in the sum $\beta V \int_{K\sigma} = \sum_{k\omega\sigma}$, this expression reduces to the classical result (8.14) (up to a factor of 2, because here we consider now a two-component theory). Note, however the different convention for the sign of the Green function for classical systems, i.e., we should replace $G \to -G$ in Eq. (10.30) to recover Eq. (8.14).

Next, consider the exact FRG flow equation for the two-point vertex $\Gamma_\Lambda^{(2)}(K\sigma, K\sigma) \equiv \Sigma_\Lambda(K\sigma)$. Keeping in mind that in our case the vertices with three external legs vanish, the general FRG flow equation (7.73) reduces to

$$\partial_\Lambda \Gamma^{(2)}_{\Lambda,\alpha_1\alpha_2} = -\frac{1}{2}\int_\alpha \int_{\alpha'} (\mathbf{Z}\dot{\mathbf{G}}_\Lambda)_{\alpha\alpha'}\,\Gamma^{(4)}_{\Lambda,\alpha'\alpha\alpha_1\alpha_2}\,. \tag{10.31}$$

Setting the external labels in this expression equal to $\alpha_1 = (\bar{\psi}K_1\sigma_1)$ and $\alpha_2 = (\psi K_2\sigma_2)$ and using Eq. (10.20), this reduces in superfield notation to

$$\begin{aligned}
\delta_{K_1 K_2}&\delta_{\sigma_1\sigma_2}\partial_\Lambda\Sigma_\Lambda(K_1\sigma_1)\\
&= -\frac{\zeta}{2}\int_{K\sigma}\int_{K'\sigma'}\Big[(\dot{\mathbf{G}}_\Lambda)_{\psi K\sigma,\bar{\psi}K'\sigma'}\,\Gamma^{(4)}_{\Lambda,\bar{\psi}K'\sigma',\psi K\sigma,\bar{\psi}K_1\sigma_1,\psi K_2\sigma_2}\\
&\qquad +(\dot{\mathbf{G}}_\Lambda)_{\bar{\psi}K\sigma,\psi K'\sigma'}\,\Gamma^{(4)}_{\Lambda,\psi K'\sigma',\bar{\psi}K\sigma,\bar{\psi}K_1\sigma_1,\psi K_2\sigma_2}\Big]\,. \tag{10.32}
\end{aligned}$$

Substituting for the matrix elements of the single-scale propagator the expressions in Eqs. (10.26b) and (10.27), and using the relation (10.29) with $n = 2$ to express the totally symmetrized four-point vertex in terms of the partially symmetrized energy–momentum conserving four-point vertex, we finally obtain the exact FRG flow equation for the irreducible self-energy,

$$\partial_\Lambda \Sigma_\Lambda(K\sigma) = -\zeta \int_{K'\sigma'} \dot{G}_\Lambda(K'\sigma')\Gamma^{(4)}_\Lambda(K\sigma, K'\sigma'; K'\sigma', K\sigma)\,. \tag{10.33}$$

A diagrammatic representation of Eq. (10.33) and its derivation using the graphical elements defined in Fig. 10.1 is shown in Fig. 10.2.

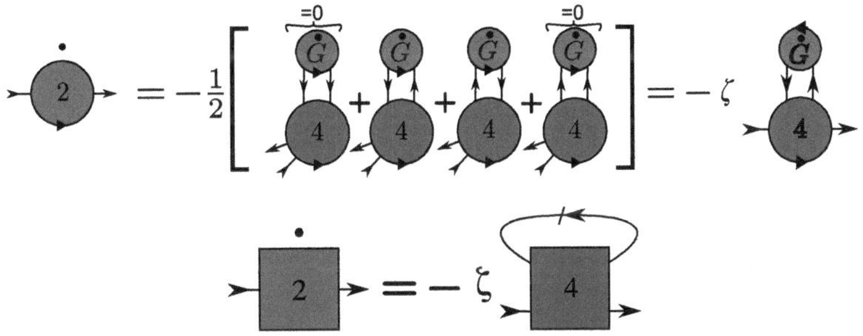

Fig. 10.2 *Upper panel*: Diagrammatic derivation of the flow equation for the self-energy. The first and last diagrams in the rectangular brackets vanish, as they do not conserve particle number. Note that the diagrams in Fig. 7.2 containing vertices with three external legs have been dropped directly, because in a theory with only particle fields three-legged vertices cannot conserve particle number. *Lower panel*: Resulting graphical representation of the exact flow equation (10.33). The *dot* over the vertex on the left-hand side denotes the derivative with respect to the cutoff Λ. The other symbols are defined in Fig. 10.1

Finally, consider the exact FRG flow equation for the effective interaction $\Gamma_\Lambda^{(4)}\left(K_1'\sigma_1', K_2'\sigma_2'; K_2\sigma_2, K_1\sigma_1\right)$ which appears on the right-hand side of the FRG flow equation (10.33) for the self-energy. The general flow equation for the effective interaction in superfield notation is given in Eq. (7.75) and shown graphically in Fig. 7.5. To obtain the flow equation for the corresponding partially symmetrized vertex $\Gamma_\Lambda^{(4)}\left(K_1'\sigma_1', K_2'\sigma_2'; K_2\sigma_2, K_1\sigma_1\right)$, let us explicitly write down all terms which are generated by the symmetrization operator $\mathcal{S}_{\alpha_1\alpha_2;\alpha_3\alpha_4}$ on the right-hand side of Eq. (7.75). From the general definition of the operator $\mathcal{S}$ given in Eq. (6.88), it is easy to see that the application of $\mathcal{S}_{\alpha_1\alpha_2;\alpha_3\alpha_4}$ to any function $f(\alpha_1, \alpha_2; \alpha_3, \alpha_4)$ which is already (anti)-symmetric with respect to the first and second pair of labels generates the following $6 = \frac{4!}{(2!)^2}$ terms,

$$\mathcal{S}_{1\,2;3\,4}\Big[f(1, 2; 3, 4)\Big] = f(1, 2; 3, 4) + f(3, 4; 1, 2) + \zeta f(1, 3; 2, 4)$$
$$+ \zeta f(2, 4; 1, 3) + f(1, 4; 2, 3) + f(2, 3; 1, 4) , \quad (10.34)$$

where for simplicity we have written 1 for α_1 and similarly for the other labels. Fixing the external labels in Eq. (7.75) to $\alpha_1 = \left(\bar\psi K_1'\sigma_1'\right)$, $\alpha_2 = \left(\bar\psi K_2'\sigma_2'\right)$, $\alpha_3 = (\psi K_2\sigma_2)$, and $\alpha_4 = (\psi K_1\sigma_1)$, and using the relation (10.29) between totally symmetrized and partially symmetrized vertices, our general FRG flow equation (7.75) for the four-point vertex reduces to

$$\delta_{K_1'+K_2',K_2+K_1}\partial_\Lambda\Gamma_\Lambda^{(4)}\left(K_1'\sigma_1', K_2'\sigma_2'; K_2\sigma_2, K_1\sigma_1\right)$$
$$= -\frac{1}{2}\int_{K\sigma}\int_{K'\sigma'}\Big[(\mathbf{Z}\dot{\mathbf{G}}_\Lambda)_{\psi K\sigma,\bar\psi K'\sigma'}\Gamma_{\Lambda,\bar\psi K'\sigma',\psi K\sigma,\bar\psi_{1'}\bar\psi_{2'}\psi_2\psi_1}^{(6)}$$
$$+(\mathbf{Z}\dot{\mathbf{G}}_\Lambda)_{\bar\psi K\sigma,\psi K'\sigma'}\Gamma_{\Lambda,\psi K'\sigma',\bar\psi K\sigma,\bar\psi_{1'}\bar\psi_{2'}\psi_2\psi_1}^{(6)}\Big]$$
$$-\frac{\zeta}{2}\mathrm{Tr}\Big\{\dot{\mathbf{G}}_\Lambda\boldsymbol{\Gamma}_{\Lambda,\bar\psi_{1'},\bar\psi_{2'}}^{(4)}\,\mathbf{G}_\Lambda\boldsymbol{\Gamma}_{\Lambda,\psi_2\psi_1}^{(4)} + \dot{\mathbf{G}}_\Lambda\boldsymbol{\Gamma}_{\Lambda,\psi_2\psi_1}^{(4)}\,\mathbf{G}_\Lambda\boldsymbol{\Gamma}_{\Lambda,\bar\psi_{1'}\bar\psi_{2'}}^{(4)}$$
$$+\zeta\dot{\mathbf{G}}_\Lambda\boldsymbol{\Gamma}_{\Lambda,\bar\psi_{1'}\psi_2}^{(4)}\,\mathbf{G}_\Lambda\boldsymbol{\Gamma}_{\Lambda,\bar\psi_{2'}\psi_1}^{(4)} + \zeta\dot{\mathbf{G}}_\Lambda\boldsymbol{\Gamma}_{\Lambda,\bar\psi_{2'}\psi_1}^{(4)}\,\mathbf{G}_\Lambda\boldsymbol{\Gamma}_{\Lambda,\bar\psi_{1'}\psi_2}^{(4)}$$
$$+\dot{\mathbf{G}}_\Lambda\boldsymbol{\Gamma}_{\Lambda,\bar\psi_{1'},\psi_1}^{(4)}\,\mathbf{G}_\Lambda\boldsymbol{\Gamma}_{\Lambda,\bar\psi_{2'}\psi_2}^{(4)} + \dot{\mathbf{G}}_\Lambda\boldsymbol{\Gamma}_{\Lambda,\bar\psi_{2'}\psi_2}^{(4)}\,\mathbf{G}_\Lambda\boldsymbol{\Gamma}_{\Lambda,\bar\psi_{1'}\psi_1}^{(4)}\Big\} , \quad (10.35)$$

where we have used the notation $\psi_1 \equiv (\psi K_1\sigma_1)$ and analogously for ψ_2, $\bar\psi_{1'}$, and $\bar\psi_{2'}$. Recall that according to Eq. (7.68) quantities like $\boldsymbol{\Gamma}_{\Lambda,\bar\psi_{1'},\bar\psi_{2'}}^{(4)}$ should be understood as matrices in the superlabels, with matrix elements given by

$$\left[\boldsymbol{\Gamma}_{\Lambda,\bar\psi_{1'}\bar\psi_{2'}}^{(4)}\right]_{\alpha\alpha'} = \Gamma_{\Lambda,\alpha\alpha'\bar\psi_{1'}\bar\psi_{2'}}^{(4)} , \quad (10.36)$$

and analogously for the other matrices in Eq. (10.35). Keeping in mind that in the normal state two of the external legs of the effective interaction $\Gamma_{\Lambda,\alpha_1\alpha_2\alpha_3\alpha_4}^{(4)}$ must correspond to $\bar\psi$-fields while the other two legs must correspond to ψ-fields, it is easy to see that each of the terms in the last three lines of Eq. (10.35) is nonzero only for one or two choices of the internal fields. For example, the vertex in Eq. (10.36) is only nonzero if the internal labels α and α' are both ψ-fields. The sum over the internal field configurations in Eq. (10.35) is therefore easily carried out. Using the explicit form of the matrix elements of $\dot{\mathbf{G}}_\Lambda$ and $\mathbf{G}_\Lambda$ given in Eqs. (10.26a) and (10.26b), we finally obtain the flow equation for the effective interaction in the following explicit form,

$$
\begin{aligned}
&\partial_\Lambda \Gamma_\Lambda^{(4)}\left(K_1'\sigma_1', K_2'\sigma_2'; K_2\sigma_2, K_1\sigma_1\right)\\
&= -\zeta \int_K \sum_\sigma \dot{G}_\Lambda(K\sigma)\Gamma_\Lambda^{(6)}\left(K_1'\sigma_1', K_2'\sigma_2', K\sigma; K\sigma, K_2\sigma_2, K_1\sigma_1\right)\\
&\quad - \int_K \sum_{\sigma\sigma'} \dot{G}_\Lambda(K\sigma)G_\Lambda(K_1+K_2-K\sigma')\\
&\qquad \times \Gamma_\Lambda^{(4)}\left(K_1'\sigma_1', K_2'\sigma_2'; K_1+K_2-K\sigma', K\sigma\right)\\
&\qquad \times \Gamma_\Lambda^{(4)}(K\sigma, K_1+K_2-K\sigma'; K_2\sigma_2, K_1\sigma_1)\\
&\quad - \zeta \int_K \sum_{\sigma\sigma'} \left[\dot{G}_\Lambda(K\sigma)G_\Lambda\left(K+K_1-K_1'\sigma'\right) + G_\Lambda(K\sigma)\dot{G}_\Lambda\left(K+K_1-K_1'\sigma'\right)\right]\\
&\qquad \times \Gamma_\Lambda^{(4)}\left(K_1'\sigma_1', K+K_1-K_1'\sigma'; K\sigma, K_1\sigma_1\right)\\
&\qquad \times \Gamma_\Lambda^{(4)}\left(K_2'\sigma_2', K\sigma; K+K_1-K_1'\sigma', K_2\sigma_2\right)\\
&\quad - \int_K \sum_{\sigma\sigma'} \left[\dot{G}_\Lambda(K\sigma)G_\Lambda\left(K+K_2-K_1'\sigma'\right) + G_\Lambda(K\sigma)\dot{G}_\Lambda\left(K+K_2-K_1'\sigma'\right)\right]\\
&\qquad \times \Gamma_\Lambda^{(4)}\left(K_1'\sigma_1', K+K_2-K_1'\sigma'; K\sigma, K_2\sigma_2\right)\\
&\qquad \times \Gamma_\Lambda^{(4)}\left(K_2'\sigma_2', K\sigma, K+K_2-K_1'\sigma', K_1\sigma_1\right).
\end{aligned}
$$

$$\tag{10.37}$$

This equation was derived independently by Kopietz and Busche (2001) and by Salmhofer and Honerkamp (2001) (for fermions where $\zeta = -1$); see also the appendix of Honerkamp et al. (2001). Note that the last term on the right-hand side of Eq. (10.37) can be obtained from the term above it by exchanging the labels $K_1\sigma_1 \leftrightarrow K_2\sigma_2$ of the two incoming fields. To exhibit the structure of the various terms appearing on the right-hand side of Eq. (10.37), the diagrammatic representation in Fig. 10.3 is very useful. Note that the right-hand side of Eq. (10.37) depends on the flowing irreducible six-point vertex $\Gamma_\Lambda^{(6)}$, so that we need an additional flow equation for $\Gamma_\Lambda^{(6)}$ to determine the effective interaction $\Gamma_\Lambda^{(4)}$. The exact FRG flow equation for the six-point vertex can be explicitly found in the appendix of Kopietz and Busche (2001) and involves the irreducible eight-point vertex $\Gamma_\Lambda^{(8)}$. In general, the exact FRG flow equation for $\Gamma_\Lambda^{(2n)}$ depends on all vertices up to $\Gamma_\Lambda^{(2n+2)}$. Salmhofer and Honerkamp (2001) have given a detailed mathematical analysis of the effect of $\Gamma_\Lambda^{(6)}$ on the FRG flow of $\Gamma_\Lambda^{(4)}$. Using rigorous bounds for overlapping fermion loops (Feldman et al. 1996, Salmhofer 1998) they showed that if the bare interaction $\Gamma_{\Lambda_0}^{(4)}$ is sufficiently small then the contributions from $\Gamma_\Lambda^{(6)}$ can indeed

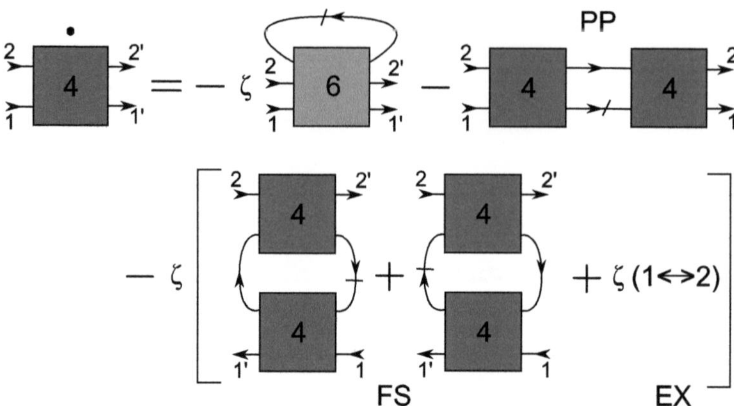

Fig. 10.3 Graphical representation of the exact FRG flow equation (10.37) for the effective interaction of nonrelativistic fermions or bosons. The symbols are defined in Fig. 10.1; see also Fig. 10.2. The diagram labeled PP involving parallel arrows in the intermediate loop describes particle–particle scattering. The diagrams in the second line have antiparallel arrows in the intermediate loop and are therefore called particle–hole diagrams. There are two distinct particle–hole scattering channels: the diagram labeled FS (forward scattering channel) involves particle–hole scattering with small momentum transfers, while the diagram labeled EX (exchange scattering) differs from FS by the exchange of the labels of the two incoming particles

be neglected in Eq. (10.37) as long as the running cutoff Λ remains larger than a certain scale Λ_6. On the other hand, for $\Lambda \lesssim \Lambda_6$ the six-point vertex can become large so that its effect on the FRG flow of $\Gamma_\Lambda^{(4)}$ can be important. Yet, even if we restrict ourselves to the regime $\Lambda > \Lambda_6$ and drop the contribution from $\Gamma_\Lambda^{(6)}$ on the right-hand side of Eq. (10.37), we still have to deal with a complicated nonlinear integro-differential equation for $\Gamma_\Lambda^{(4)}$ which cannot be solved exactly.

Each of the three terms on the right-hand side of Eq. (10.37) involving two powers of the flowing four-point vertex has a distinct physical meaning, as indicated by our labeling of the corresponding diagrams in Fig. 10.3. In the first term labeled PP (particle–particle scattering), the arrows in the intermediate loop point in the same direction, corresponding to the simultaneous excitation of two virtual particles (or two holes). If we truncate Eq. (10.37) by retaining only this term, the solution of our FRG flow equation amounts to approximating the effective interaction by the infinite series of particle–particle ladder diagrams, as will be shown in Sect. 10.3.1. In diagrammatic perturbation theory, these diagrams are usually summed by solving the corresponding Bethe–Salpeter equation (Fetter and Walecka 1971). For attractive interactions between fermions, the divergence of the series of particle–particle ladder diagrams indicates the BCS (Bardeen-Cooper-Schrieffer) instability.

The arrows attached to the intermediate loops in the other two contributions labeled FS (forward scattering) and EX (exchange scattering) in Fig. 10.3 point in opposite directions, corresponding to the simultaneous excitation of a virtual particle and a virtual hole. These processes are therefore called particle–hole scattering processes. The term labeled FS corresponds to the so-called forward scattering process, which describes the renormalization of the effective interaction by particle–hole scattering involving small momentum transfers $k_1' - k_1$. For fermions this scattering channel is also called zero-sound or Landau channel (Shankar 1994), because this component of the interaction determines the spectrum of the collective zero-sound mode in neutral Fermi liquids, describing collective long-wavelength density fluctuations consisting of an infinite coherent superposition of particle–hole excitations (Pines and Nozières 1966, Lifshitz and Pitaevskii 1980).

In the other particle–hole channel labeled EX in Fig. 10.3, the external labels K_1 and K_2 of the two incoming arrows are exchanged relative to the FS channel. Note that for fermions the contribution of the exchange channel to the flow of the effective interaction has the opposite sign as the contribution from the FS channel, so that both FS and EX channels are needed to preserve the antisymmetry of the effective interaction (Dupuis and Chitov 1996, Chitov and Sénéchal 1998).

In many-body theory one often argues that for the physical phenomenon of interest only the interaction in one particular channel is important, so that it is sufficient to solve the corresponding Bethe–Salpeter equation in this channel to obtain a reasonable estimate for the effective interaction. Such a procedure requires some a priori knowledge about the physical behavior of the system and is therefore biased. On the other hand, in the exact FRG equation (10.37) for the effective interaction all scattering channels appear on equal footing, so that this equation can be used as an unbiased starting point to investigate the physical behavior of a system with a given bare interaction. Of course, as will be explained in more detail in Sect. 10.5, the analysis of the FRG equation (10.37) including all channels can only be performed numerically and with some drastic approximations.

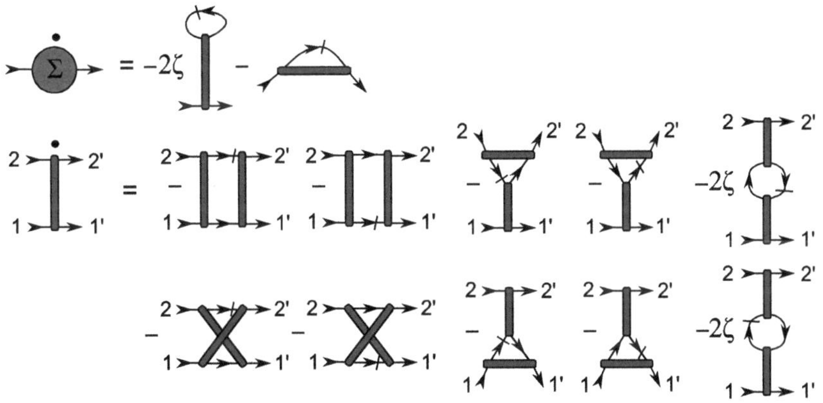

Fig. 10.4 Graphical representation of the flow equation for the self-energy and the four-point vertex for a system with $SU(2)$ spin rotational symmetry. The flow of the self-energy contains the usual Hartree and Fock contributions. In the flow equation of the four-point vertex, we have dropped the contribution from the six-point vertex, consistent with the truncations discussed in the rest of this chapter

10.2.3 *SU*(2)-Invariant Flow Equations

For a system with spin rotational symmetry, we now rewrite the flow equations in terms of the spin-conserving vertices defined in Eq. (6.119), using the relation (6.121). The explicit version of this relation for the four-point vertex given in Eq. (10.9) can be represented diagrammatically as

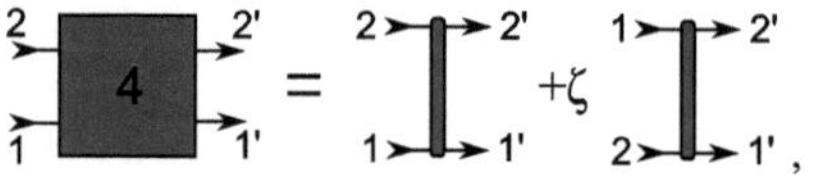

where the spin is now conserved along continuous particle lines. Using this replacement in Figs. 10.2 and 10.3, and then equating contributions on both sides that have continuous spin-conserving lines between the same external legs, we obtain the flow equations depicted in Fig. 10.4. For the flow of the self-energy, we obtain from Eq. (10.33),

$$\partial_\Lambda \Sigma_\Lambda(K\sigma) = -\zeta \int_{K'} \sum_{\sigma'} \dot{G}_\Lambda(K'\sigma') U^{(4)}_{\Lambda,\sigma\sigma'}(K, K'; K', K) - \int_{K'} \dot{G}_\Lambda(K'\sigma) U^{(4)}_{\Lambda,\sigma\sigma}(K, K'; K, K') \,. \tag{10.38}$$

If we approximate the flowing interaction in this expression by the bare interaction and drop the self-energy in the propagators, the first term on the right-hand side can be identified with the Hartree correction to the self-energy, while the second term is the Fock (or exchange) correction to the self-energy. Note that for the bare Hubbard interaction (10.14) only the Hartree term (with $\sigma' = -\sigma$) contributes.

In the presence of full $SU(2)$ spin rotational invariance, all vertices in Eq. (10.38) are independent of spin indices (see Eq. (6.124)), so that we obtain

$$\partial_\Lambda \Sigma_\Lambda(K) = -\zeta \int_{K'} \dot{G}_\Lambda(K') \left[2U^{(4)}_\Lambda(K, K'; K', K) + \zeta\, U^{(4)}_\Lambda(K, K'; K, K') \right] . \tag{10.39}$$

Neglecting in the exact FRG flow equation (10.37) for the effective interaction the contribution from the six-point vertex (as this will not contribute to the truncation discussed below), the flow of the four-point spin-conserving vertex for full $SU(2)$-invariance is given by

$$\partial_\Lambda U^{(4)}\left(K_1', K_2'; K_2, K_1\right) =$$

$$-\int_K [\dot{G}(K)G(-K + Q_{\mathrm{pp}}) + G(K)\dot{G}(-K + Q_{\mathrm{pp}})]$$

$$\times U^{(4)}\left(K_1', K_2'; -K + Q_{\mathrm{pp}}, K\right) U^{(4)}(K, -K + Q_{\mathrm{pp}}; K_2, K_1)$$

$$-\zeta \int_K [\dot{G}(K)G(K - Q_{\mathrm{fs}}) + G(K)\dot{G}(K - Q_{\mathrm{fs}})]$$

$$\times [2U^{(4)}\left(K_1', K - Q_{\mathrm{fs}}; K, K_1\right) U^{(4)}\left(K_2', K; K - Q_{\mathrm{fs}}, K_2\right)$$

$$+ \zeta\, U^{(4)}\left(K_1', K - Q_{\mathrm{fs}}; K_1, K\right) U^{(4)}\left(K_2', K; K - Q_{\mathrm{fs}}, K_2\right)$$

$$+ \zeta\, U^{(4)}\left(K_1', K - Q_{\mathrm{fs}}; K, K_1\right) U^{(4)}\left(K_2', K; K_2, K - Q_{\mathrm{fs}}\right)]$$

$$-\int_K [\dot{G}(K)G(K - Q_{\mathrm{ex}}) + G(K)\dot{G}(K - Q_{\mathrm{ex}})]$$

$$\times U^{(4)}\left(K_1', K - Q_{\mathrm{ex}}; K_2, K\right) U^{(4)}\left(K, K_2'; K - Q_{\mathrm{ex}}, K_1\right) , \qquad (10.40)$$

where we have omitted for simplicity the cutoff labels Λ and have defined the characteristic energy–momentum combinations associated with the three channels PP, FS, and EX,

$$Q_{\mathrm{pp}} = K_1 + K_2 = K_1' + K_2' , \qquad (10.41\mathrm{a})$$

$$Q_{\mathrm{fs}} = K_1' - K_1 = K_2 - K_2' , \qquad (10.41\mathrm{b})$$

$$Q_{\mathrm{ex}} = K_1' - K_2 = K_1 - K_2' . \qquad (10.41\mathrm{c})$$

We have used energy–momentum conservation to write these expressions in two different ways.

10.3 Single-Channel Truncations

We now show how the usual particle–particle ladder approximation and the so-called random phase approximation can be recovered from Eq. (10.40) if one retains only one particular interaction channel and neglects self-energy corrections to single-particle Green functions on the right-hand side.

10.3.1 Ladder Approximation in the Particle–Particle Channel

It is well known (Lifshitz and Pitaevskii 1980) that the effective interaction between two particles in vacuum is given by the infinite series of particle–particle ladder diagrams shown in Fig. 10.5. For two particles this series represents the Lippmann–Schwinger equation for the T-matrix of elementary scattering theory (Sakurai 1994). Assuming that the zero-density limit is continuously connected to the low-density regime, we expect that the same series of ladder diagrams is also a good

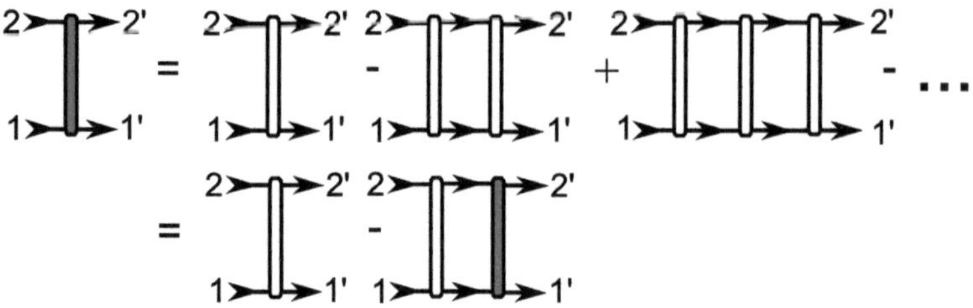

Fig. 10.5 Graphical representation of the effective interaction in the particle–particle ladder approximation. The *empty bars* represent the bare interaction $U_0(Q', Q; P)$ in Eq. (10.43)

approximation for the effective interaction at finite but sufficiently low densities. Carrying out the infinite ladder summation amounts mathematically to solving a linear integral equation, the so-called Bethe–Salpeter equation in the particle–particle channel (Galitskii 1958, Fetter and Walecka 1971). Therefore it is convenient to introduce relative and total energy momenta as independent variables, $Q = (K_1 - K_2)/2$, $Q' = \left(K'_1 - K'_2\right)/2$ and $P \equiv Q_{\mathrm{pp}} = K_1 + K_2 = K'_1 + K'_2$. The labels of the outgoing fields are then $K'_1 = \frac{P}{2} + Q'$, $K'_2 = \frac{P}{2} - Q'$, while the incoming fields are labeled by $K_1 = \frac{P}{2} + Q$, $K_2 = \frac{P}{2} - Q$. In terms of these variables the effective interaction can be written as

$$T(Q', Q; P) = U^{(4)} \left(\frac{P}{2} + Q', \frac{P}{2} - Q'; \frac{P}{2} - Q, \frac{P}{2} + Q \right) . \tag{10.42}$$

Denoting the bare interaction by $U_0(Q', Q; P)$ and the noninteracting Green function by $G_0(K)$, the Bethe–Salpeter equation in the particle–particle channel shown diagrammatically in Fig. 10.5 can be written as

$$T(Q', Q; P) = U_0(Q', Q; P)$$
$$- \int_K U_0(Q', K; P) G_0 \left(\frac{P}{2} - K \right) G_0 \left(\frac{P}{2} + K \right) T(K, Q; P) . \tag{10.43}$$

The notation indicates that in this approximation the effective interaction is sometimes called the many-body T-matrix. In the special case where the bare interaction $U_0(P)$ depends only on the total energy momentum $P = (i\bar{\omega}, \mathbf{p})$ this linear integral equation reduces to an algebraic one,

$$T(P) = U_0(P) - U_0(P)\Pi_0^{\mathrm{pp}}(P)T(P) , \tag{10.44}$$

which is easily solved for the effective interaction,

$$T(P) = \frac{U_0(P)}{1 + \Pi_0^{\mathrm{pp}}(P)U_0(P)} . \tag{10.45}$$

Here, the particle–particle susceptibility $\Pi_0^{\mathrm{pp}}(P)$ can be written as

$$\Pi_0^{\mathrm{pp}}(P) = \int_K G_0\left(\frac{P}{2} + K\right) G_0\left(\frac{P}{2} - K\right)$$

$$= \int_k \int \frac{d\omega}{2\pi} \frac{1}{\left[i\left(\frac{\bar{\omega}}{2} + \omega\right) - \xi_{\frac{P}{2}+k}\right]\left[i\left(\frac{\bar{\omega}}{2} - \omega\right) - \xi_{\frac{P}{2}-k}\right]}$$

$$= -\int_k \frac{1 - f\left(\xi_{\frac{P}{2}+k}\right) - f\left(\xi_{\frac{P}{2}-k}\right)}{i\bar{\omega} - \xi_{\frac{P}{2}+k} - \xi_{\frac{P}{2}-k}} , \tag{10.46}$$

where $f(\xi) = \Theta(-\xi)$ is the Fermi function at zero temperature and we use again the notation $\int_k = \int \frac{d^D k}{(2\pi)^D}$.

We now show how to recover the above results from our FRG flow equation (10.40). Therefore we simply retain only the first term involving the particle–particle channel on the right-hand side of Eq. (10.40). Using the same variables as in Eq. (10.42), we obtain the truncated FRG flow equation

$$\partial_\Lambda T_\Lambda(Q', Q; P) = -\int_K \left[\dot{G}_\Lambda(\frac{P}{2} + K) G_\Lambda(\frac{P}{2} - K) + G_\Lambda(\frac{P}{2} + K)\dot{G}_\Lambda(\frac{P}{2} - K)\right]$$

$$\times T_\Lambda(Q', K; P) T_\Lambda(K, Q; P) . \tag{10.47}$$

Note that this integro-differential equation is nonlinear, whereas the Bethe–Salpeter equation (10.43) is a linear integral equation for the effective interaction. The fact that the summation of diagrams described by linear Bethe–Salpeter equations can alternatively be described by quadratic differential equations has first been noticed by Sudakov (1956). To exhibit the precise connection between Eqs. (10.43) and (10.47), we replace the flowing Green functions in Eq. (10.47) by noninteracting ones. Using a multiplicative sharp momentum–space cutoff $\Theta_\Lambda(k) = \Theta(|k| - \Lambda)$, we have

$$G_\Lambda(K) \approx \frac{\Theta_\Lambda(k)}{i\omega - \xi_k} , \quad \dot{G}_\Lambda(K) \approx \frac{\partial_\Lambda \Theta_\Lambda(k)}{i\omega - \xi_k} . \tag{10.48}$$

Assuming that the initial interaction $T_{\Lambda_0}(Q', Q; P) = U_0(q', q; P)$ is independent of the frequency part of Q and Q', the flowing $T_\Lambda(Q', Q; P)$ has also this property so that we may write $T_\Lambda(Q', Q; P) = T_\Lambda(q', q; P)$. The frequency sum in Eq. (10.47) is then easily carried out,

$$\partial_\Lambda T_\Lambda(q', q; P) = \int_k \partial_\Lambda \left[\Theta_\Lambda\left(\frac{P}{2} + k\right) \Theta_\Lambda\left(\frac{P}{2} - k\right)\right]$$

$$\times \frac{1 - f\left(\xi_{\frac{P}{2}+k}\right) - f\left(\xi_{\frac{P}{2}-k}\right)}{i\bar{\omega} - \xi_{\frac{P}{2}+k} - \xi_{\frac{P}{2}-k}} T_\Lambda(q', k; P) T_\Lambda(k, q; P) . \tag{10.49}$$

For simplicity, let us further assume that $T_\Lambda(q', q; P) = T_\Lambda(P)$ is independent of q and q', in which case Eq. (10.49) reduces to a simple first-order differential equation,

$$\partial_\Lambda T_\Lambda(P) = -\dot{\Pi}_\Lambda^{\mathrm{pp}}(P)T_\Lambda^2(P)\,, \tag{10.50}$$

where

$$\dot{\Pi}_\Lambda^{\mathrm{pp}}(P) = \int_K \left[\dot{G}_\Lambda\left(\frac{P}{2}+K\right)G_\Lambda\left(\frac{P}{2}-K\right) + G_\Lambda\left(\frac{P}{2}+K\right)\dot{G}_\Lambda\left(\frac{P}{2}-K\right) \right]$$

$$= -\int_k \partial_\Lambda\left[\Theta_\Lambda\left(\frac{p}{2}+k\right)\Theta_\Lambda\left(\frac{p}{2}-k\right)\right]\frac{1-f\left(\xi_{\frac{p}{2}+k}\right)-f\left(\xi_{\frac{p}{2}-k}\right)}{i\bar{\omega}-\xi_{\frac{p}{2}+k}-\xi_{\frac{p}{2}-k}}\,. \tag{10.51}$$

Writing Eq. (10.50) as

$$\partial_\Lambda T_\Lambda^{-1}(P) = \dot{\Pi}_\Lambda^{\mathrm{pp}}(P)\,, \tag{10.52}$$

and integrating both sides of this equation over the flow parameter Λ with boundary condition $T_{\Lambda=\Lambda_0}(P) = U_0(P)$, we obtain for the effective interaction

$$T(P) \equiv \lim_{\Lambda\to 0} T_\Lambda(P) = \frac{U_0(P)}{1 + \Pi_{0,\Lambda_0}^{\mathrm{pp}}(P)U_0(P)}\,, \tag{10.53}$$

where

$$\Pi_{0,\Lambda_0}^{\mathrm{pp}}(P) = -\int_0^{\Lambda_0} d\Lambda\,\dot{\Pi}_\Lambda^{\mathrm{pp}}(P)$$

$$= -\int_k \left[1-\Theta_{\Lambda_0}\left(\frac{p}{2}+k\right)\Theta_{\Lambda_0}\left(\frac{p}{2}-k\right)\right]\frac{1-f\left(\xi_{\frac{p}{2}+k}\right)-f\left(\xi_{\frac{p}{2}-k}\right)}{i\bar{\omega}-\xi_{\frac{p}{2}+k}-\xi_{\frac{p}{2}-k}}\,. \tag{10.54}$$

For $\Lambda_0 \to \infty$ the Θ-functions in the second line vanish, so that $\Pi_{0,\Lambda_0}^{pp}(P)$ reduces in this limit to the noninteracting particle–particle pair susceptibility given in Eq. (10.46). We conclude that with the approximation (10.48) the solution (10.53) of our FRG flow equation (10.47) agrees precisely with the solution (10.45) of the Bethe–Salpeter equation (10.43).

In the limit of vanishing density, the Bethe–Salpeter equation (10.43) reduces to the Lippmann–Schwinger equation for the T-matrix in vacuum; the corresponding FRG flow equation (10.47) for vanishing total momentum $\boldsymbol{p} = 0$ assumes this limit in the form

$$\partial_\Lambda T_\Lambda(\boldsymbol{q}',\boldsymbol{q},i\bar{\omega}) = -\int_k \frac{\delta(|\boldsymbol{k}|-\Lambda)}{i\bar{\omega}-\Lambda^2/m} T_\Lambda(\boldsymbol{q}',\boldsymbol{k},i\bar{\omega})T_\Lambda(\boldsymbol{k},\boldsymbol{q},i\bar{\omega})\,, \tag{10.55}$$

which is the RG version of the off-shell Lippmann–Schwinger equation of elementary scattering theory (Sauli and Kopietz 2006).

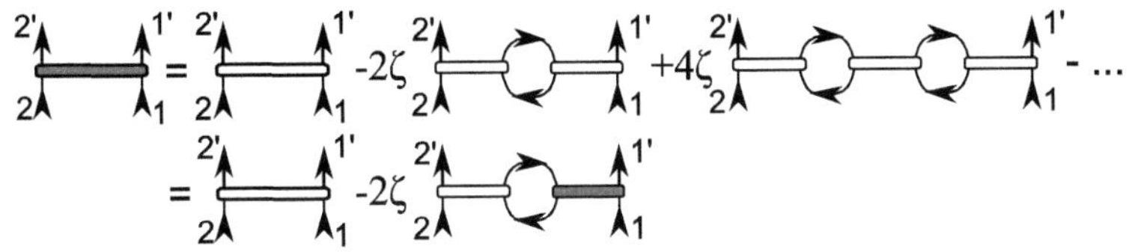

Fig. 10.6 Graphical representation of the effective interaction in random phase approximation

10.3.2 Random Phase Approximation in the Forward Scattering Channel

At high densities and for interactions whose Fourier transform has a strong maximum for small momentum transfers $q = k_1' - k_1$ (giving rise to enhanced forward scattering), the effective interaction is strongly screened at long distances. A simple diagrammatic method to take the physics of screening into account is to sum the infinite series of bubble diagrams shown in Fig. 10.6. This approximation is usually called the random phase approximation, which is abbreviated by RPA (Fetter and Walecka 1971, Negele and Orland 1988). To derive the RPA from our FRG flow equation (10.40), let us try an ansatz where the flowing interaction depends only on the energy–momentum transfer $Q_{\mathrm{fs}} = K_1' - K_1$,

$$U_\Lambda^{(4)}(K_1', K_2'; K_2, K_1) = f_\Lambda(K_1' - K_1) = f_\Lambda(Q_{\mathrm{fs}}) \, . \tag{10.56}$$

Substituting this into Eq. (10.40) and neglecting the contribution from the particle–particle channel, we obtain

$$
\begin{aligned}
\partial_\Lambda f_\Lambda(Q_{\mathrm{fs}}) \approx -\zeta \int_K &[\dot{G}_\Lambda(K)G_\Lambda(K - Q_{\mathrm{fs}}) + G_\Lambda(K)\dot{G}_\Lambda(K - Q_{\mathrm{fs}})] \\
&\times \big[2 f_\Lambda(Q_{\mathrm{fs}}) f_\Lambda(-Q_{\mathrm{fs}}) \\
&\quad + \zeta f_\Lambda\left(K_1' - K\right) f_\Lambda(-Q_{\mathrm{fs}}) + \zeta f_\Lambda(Q_{\mathrm{fs}}) f_\Lambda(K_2 - K)\big] \\
&- \int_K [\dot{G}_\Lambda(K)G_\Lambda(K - Q_{\mathrm{ex}}) + G_\Lambda(K)\dot{G}_\Lambda(K - Q_{\mathrm{ex}})] \\
&\quad \times f_\Lambda\left(K_1' - K\right) f_\Lambda(K - K_1) \, .
\end{aligned}
\tag{10.57}
$$

Note that the external labels on the right-hand side of this expression appear not only in the assumed combination $Q_{\mathrm{fs}} = K_1' - K_1$, so that our ansatz (10.56) seems to be inconsistent. However, for fermions the typical loop momentum k contributing to the K-integral is of the order of the Fermi momentum k_F. If the range of the interaction in momentum space is small compared with k_F and its strength is strongly enhanced for small energy–momentum transfers $Q_{\mathrm{fs}} = K_1' - K_1$, then the term in the second line of Eq. (10.57) involving the combination $f_\Lambda(Q_{\mathrm{fs}}) f_\Lambda(-Q_{\mathrm{fs}})$ dominates the integral, so that we may neglect the other terms. In this approximation Eq. (10.57) simplifies to

$$\partial_\Lambda f_\Lambda(Q) = -\dot{\Pi}^{\mathrm{ph}}_\Lambda(Q) f_\Lambda^2(Q) \,, \tag{10.58}$$

where we have set $Q = Q_{\mathrm{fs}} = K_1' - K_1$ and assumed that $f_\Lambda(-Q) = f_\Lambda(Q)$. The function $\dot{\Pi}^{\mathrm{ph}}_\Lambda(Q)$ is given by the derivative of the particle–hole bubble with respect to the RG cutoff,

$$\dot{\Pi}^{\mathrm{ph}}_\Lambda(Q) = 2\zeta \int_K \left[\dot{G}_\Lambda(K) G_\Lambda(K-Q) + G_\Lambda(K) \dot{G}_\Lambda(K-Q) \right] \,, \tag{10.59}$$

where the factor of 2 is due to two spin species. Equation (10.58) has the same form as the corresponding flow equation (10.50) for the T-matrix, so that the solution is similar to Eq. (10.53),

$$f(Q) \equiv \lim_{\Lambda \to 0} f_\Lambda(Q) = \frac{f_{\Lambda_0}(Q)}{1 + \Pi^{\mathrm{ph}}_{0,\Lambda_0}(Q) f_{\Lambda_0}(Q)} \,, \tag{10.60}$$

where

$$\Pi^{\mathrm{ph}}_{0,\Lambda_0}(Q) = -\int_0^{\Lambda_0} d\Lambda \, \dot{\Pi}^{\mathrm{ph}}_\Lambda(Q) \,. \tag{10.61}$$

If we replace again the flowing Green functions in Eq. (10.59) by the noninteracting ones and take the limit $\Lambda_0 \to \infty$, Eq. (10.61) reduces to the bare particle–hole susceptibility (or polarization), which for fermions ($\zeta = -1$) can be written as

$$\begin{aligned}
\Pi^{\mathrm{ph}}_0(Q) &= -2 \int_k \int \frac{d\omega}{2\pi} \frac{1}{[i\omega - \xi_k][i(\omega - \bar{\omega}) - \xi_{k-q}]} \\
&= 2 \int_k \frac{f(\xi_k) - f(\xi_{k-q})}{i\bar{\omega} - \xi_k + \xi_{k-q}} \,.
\end{aligned} \tag{10.62}$$

The interaction (10.60) then reduces to the RPA for the effective interaction (Fetter and Walecka 1971, Negele and Orland 1988).

10.4 Rescaled Flow Equations and Definition of the Fermi Surface

In the FRG flow equations given so far in this chapter, the rescaling step, which according to the general discussion in Chap. 3 is an essential part of the Wilsonian RG procedure, is still missing. Recall that in general a rescaling of the fields and the coupling constants is necessary to obtain critical fixed points of the RG. Of course, the proper rescaling depends on the nature of the fixed point under consideration. We now assume that our system remains in the normal state at zero temperature and has a sharp Fermi surface. In this case the exact propagator $G(i\omega, \boldsymbol{k}) = [i\omega - \epsilon_k + \mu - \Sigma(i\omega, \boldsymbol{k})]^{-1}$ exhibits singularities for $\omega = 0$ and for all wave vectors $\boldsymbol{k}_F$ satisfying

$$\epsilon_{k_F} + \Sigma(i0, \boldsymbol{k}_F) = \mu \ . \tag{10.63}$$

For a D-dimensional system this equation defines a $(D-1)$-dimensional manifold in momentum space, which is called the Fermi surface. Note that the definition (10.63) involves the exact self-energy $\Sigma(i0, \boldsymbol{k}_F)$ of the many-body system, which is a priori unknown. Equation (10.63) should therefore be considered as a complicated self-consistency equation for the true Fermi surface of the many-body system. If the self-energy is calculated perturbatively to some finite order in the interaction, one has to be careful to include the corrections to the noninteracting Fermi surface from the beginning in the propagators in order to avoid unphysical divergencies (Luttinger 1960, Nozières 1964). In this section we shall derive a rescaled version of the exact FRG flow equations given in Sect. 10.2, which allows us to give a nonperturbative definition of the Fermi surface as a fixed point of the RG (Kopietz and Busche 2001, Ledowski and Kopietz 2003).

10.4.1 Scaling Toward the Fermi Surface

The Wilsonian RG is based on the iterative elimination of the high-energy degrees of freedom, resulting in an effective action for the low-energy degrees of freedom. Keeping in mind that the zero-frequency propagator $G(i0, \boldsymbol{k})$ of normal fermions is singular on the entire Fermi surface, in fermionic many-body systems the normal state has a continuum of low-energy degrees of freedom described by fermion fields with momenta in a thin shell around the Fermi surface. To carry out the Wilsonian RG procedure for normal fermions,[4] one should therefore successively integrate over Fermi fields with momenta outside an increasingly thin shell around the Fermi surface. After the mode elimination, one should then rescale the distances $\delta k_\parallel$ of the momenta from the Fermi surface (see Fig. 10.7) in order to bring the thickness of the momentum shell back to the value before mode elimination (Shankar 1994).

Since only the true Fermi surface of an interacting Fermi system has a physical meaning, it is useful to formulate the FRG in such a way that all momenta are measured relative to the interacting Fermi surface. To implement this technically, we add the following interaction-dependent quadratic counterterm to the noninteracting part $S_0[\bar{\psi}, \psi]$ of the bare action (10.2),

---

[4] Actually, there exist also classical systems whose fluctuation spectrum has a minimum on a higher-dimensional manifold (Brazovskii universality class), for example, cholesteric liquid crystals (Brazovskii 1975, Hohenberg and Swift 1995). It is therefore not surprising that the rescaled version of the FRG equations for interacting fermions discussed in this section (Kopietz and Busche 2001) has also been used to study the Brazovskii universality class (Shiwa 2006).

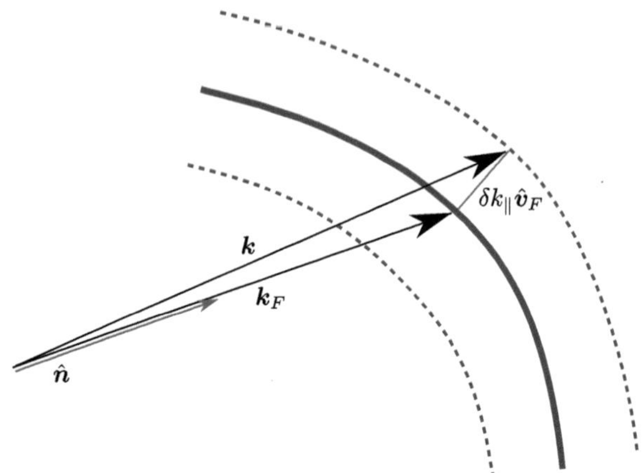

Fig. 10.7 Decomposition of a momentum $\boldsymbol{k} = \hat{\boldsymbol{n}} k_F(\hat{\boldsymbol{n}}) + \delta k_\| \hat{\boldsymbol{v}}_F$ into a component $\boldsymbol{k}_F = \hat{\boldsymbol{n}} k_F(\hat{\boldsymbol{n}})$ on the Fermi surface and a component $\delta k_\| \hat{\boldsymbol{v}}_F$ in the direction $\hat{\boldsymbol{v}}_F$ of the local Fermi velocity $\boldsymbol{v}_F$. The *thick solid line* is a sector of the Fermi surface. This construction defines $\boldsymbol{k}_F$ and $\hat{\boldsymbol{n}}$ as a function of $\boldsymbol{k}$. Here, $\boldsymbol{v}_F = \nabla_{\boldsymbol{k}} \epsilon_{\boldsymbol{k}}|_{\boldsymbol{k}=\boldsymbol{k}_F}$ is defined in terms of the gradient of the free dispersion at the true Fermi surface, so that $\boldsymbol{v}_F$ is not necessarily perpendicular to the Fermi surface

$$S_c[\bar{\psi}, \psi] = \sum_\sigma \int_K \Sigma(i0, \boldsymbol{k}_F) \bar{\psi}_{K\sigma} \psi_{K\sigma} \,, \tag{10.64}$$

where $\boldsymbol{k}_F$ is a vector on the true Fermi surface which is defined as a function of $\boldsymbol{k}$ via the geometric construction shown in Fig. 10.7. We then write the fermionic action (10.1) as $S[\bar{\psi}, \psi] = \tilde{S}_0[\bar{\psi}, \psi] + \tilde{S}_1[\bar{\psi}, \psi]$, where $\tilde{S}_0 = S_0 + S_c$ and the counterterm is again subtracted from the two-body part, $\tilde{S}_1 = S_1 - S_c$. Explicitly, the new Gaussian part of the action can be written as

$$\begin{aligned}
\tilde{S}_0[\bar{\psi}, \psi] &= -\sum_\sigma \int_K [i\omega - \epsilon_{\boldsymbol{k}} + \mu - \Sigma(i0, \boldsymbol{k}_F)] \bar{\psi}_{K\sigma} \psi_{K\sigma} \\
&= -\sum_\sigma \int_K [i\omega - \epsilon_{\boldsymbol{k}} + \epsilon_{\boldsymbol{k}_F}] \bar{\psi}_{K\sigma} \psi_{K\sigma} \,,
\end{aligned} \tag{10.65}$$

where in the last line we have used the definition (10.63) of the interacting Fermi surface. Suppose now that we have eliminated all fields with momenta outside a thin shell of thickness Λ_0 around the Fermi surface, so that the momentum integration in Eq. (10.65) is restricted to the regime

$$\frac{|\epsilon_{\boldsymbol{k}} - \epsilon_{\boldsymbol{k}_F}|}{|\boldsymbol{v}_F|} < \Lambda_0 \,, \tag{10.66}$$

as shown in Fig. 10.7. Here, $\boldsymbol{v}_F = \nabla_{\boldsymbol{k}} \epsilon_{\boldsymbol{k}}|_{\boldsymbol{k}_F}$ is the Fermi velocity of the noninteracting system at the true $\boldsymbol{k}_F$. Assuming now that Λ_0 is sufficiently small, we may expand the energy dispersion $\epsilon_{\boldsymbol{k}}$ to linear order around the reference points $\boldsymbol{k}_F$

on the Fermi surface. Therefore we perform in Eq. (10.65) a nonlinear coordinate transformation in momentum space by setting[5]

$$\boldsymbol{k} = \boldsymbol{k}_F + \delta\boldsymbol{k} = \hat{\boldsymbol{n}}k_F(\hat{\boldsymbol{n}}) + \delta k_\| \hat{\boldsymbol{v}}_F(\hat{\boldsymbol{n}}) \, , \tag{10.67}$$

where $k_F(\hat{\boldsymbol{n}})$ is the length of $\boldsymbol{k}_F$ parameterized by a unit vector $\hat{\boldsymbol{n}}$, and $\delta k_\| = \hat{\boldsymbol{v}}_F \cdot (\boldsymbol{k} - \boldsymbol{k}_F)$ is the component of $\boldsymbol{k} - \boldsymbol{k}_F$ parallel to the direction $\hat{\boldsymbol{v}}_F(\hat{\boldsymbol{n}}) = \boldsymbol{v}_F/|\boldsymbol{v}_F|$ of the local Fermi velocity $\boldsymbol{v}_F$, as shown in Fig. 10.7. We now eliminate $\boldsymbol{k}$ in favor of the component $\delta k_\|$ and the angular variables specifying the vector $\hat{\boldsymbol{n}}$ as new integration variables. This nonlinear coordinate transformation of the D-dimensional momentum integration gives rise to a nontrivial Jacobian, which is conveniently written down using D-dimensional spherical coordinates,

$$\int \frac{d^D k}{(2\pi)^D} = K_D \int \frac{d\Omega_{\hat{\boldsymbol{n}}}}{\Omega_D} \int d\delta k_\| \left| \hat{\boldsymbol{n}}k_F(\hat{\boldsymbol{n}}) + \delta k_\| \hat{\boldsymbol{v}}_F(\hat{\boldsymbol{n}}) \right|^{D-1}$$

$$= 2\pi \nu_0 v_0 \int \frac{d\Omega_{\hat{\boldsymbol{n}}}}{\Omega_D} \int \frac{d\delta k_\|}{2\pi} J_D(\hat{\boldsymbol{n}}, \delta k_\|) \, , \tag{10.68}$$

where $d\Omega_{\hat{\boldsymbol{n}}}$ is the surface element and $\Omega_D = (2\pi)^D K_D$ is the surface area of the D-dimensional unit sphere (see also Eq. (2.86)). To write the Jacobian in dimensionless form, we have pulled out in the last line of Eq. (10.68) a factor of $2\pi \nu_0 v_0$, where

$$\nu_0 = K_D \frac{k_0^{D-1}}{v_0} \tag{10.69}$$

is the density of states of free electrons with Fermi momentum k_0 and Fermi velocity v_0. Note that in one dimension $\nu_0 = 1/(\pi v_0)$ so that $2\pi \nu_0 v_0 = 2$. The dimensionless Jacobian is

$$J_D(\hat{\boldsymbol{n}}, \delta k_\|) = \left| \frac{\hat{\boldsymbol{n}}k_F(\hat{\boldsymbol{n}}) + \delta k_\| \hat{\boldsymbol{v}}_F(\hat{\boldsymbol{n}})}{k_0} \right|^{D-1} \, . \tag{10.70}$$

For a spherical Fermi surface, where $\hat{\boldsymbol{n}} = \hat{\boldsymbol{v}}_F$ and $k_F(\hat{\boldsymbol{n}})$ is independent of the direction $\hat{\boldsymbol{n}}$, the Jacobian $J_D(\delta k_\|) = |1 + \delta k_\|/k_0|^{D-1}$ depends only on $\delta k_\|$.

To derive the proper scaling of momenta, frequencies, and fields, we assume that the cutoff Λ_0 defining the width of the momentum shell in Eq. (10.66) is small compared with the minimal value of $k_F(\hat{\boldsymbol{n}})$. Then we may approximate

$$\epsilon_{\boldsymbol{k}} \approx \epsilon_{\boldsymbol{k}_F} + \boldsymbol{v}_F \cdot (\boldsymbol{k} - \boldsymbol{k}_F) = \epsilon_{\boldsymbol{k}_F} + v_F(\hat{\boldsymbol{n}})\delta k_\| \, , \tag{10.71}$$

---

[5] If we neglect interaction corrections to the direction $\hat{\boldsymbol{v}}_F$ of the local Fermi velocity, then $\delta k_\|$ measures the distance of a given momentum $\boldsymbol{k}$ from the true Fermi surface. Note that for a nonspherical Fermi surface $\hat{\boldsymbol{v}}_F(\hat{\boldsymbol{n}})$ is in general different from the direction $\hat{\boldsymbol{n}} = \boldsymbol{k}_F/|\boldsymbol{k}_F|$.

so that Eq. (10.66) reduces to the condition $|\delta k_\parallel| \lesssim \Lambda_0$. Moreover, for small $\delta k_\parallel$ we may approximate

$$J_D(\hat{\boldsymbol{n}}, \delta k_\parallel) \approx J_D(\hat{\boldsymbol{n}}, 0) = \left(\frac{k_F(\hat{\boldsymbol{n}})}{k_0}\right)^{D-1} . \tag{10.72}$$

Relabeling the fermion fields,

$$\psi_{\hat{\boldsymbol{n}}\sigma}(i\omega, \delta k_\parallel) \equiv \psi_\sigma(i\omega, \boldsymbol{k} \to \hat{\boldsymbol{n}} k_F(\hat{\boldsymbol{n}}) + \delta k_\parallel \hat{\boldsymbol{v}}_F(\hat{\boldsymbol{n}})) , \tag{10.73}$$

the low-energy version of the subtracted Gaussian action (10.65) becomes

$$\tilde{S}_0[\bar{\psi}, \psi] = -2\pi v_0 v_0 \sum_\sigma \int_{\hat{\boldsymbol{n}}} \int \frac{d\omega d\delta k_\parallel}{(2\pi)^2} \Theta(\Lambda_0 - |\delta k_\parallel|)[i\omega - v_F(\hat{\boldsymbol{n}})\delta k_\parallel]$$
$$\times \bar{\psi}_{\hat{\boldsymbol{n}}\sigma}(i\omega, \delta k_\parallel)\psi_{\hat{\boldsymbol{n}}\sigma}(i\omega, \delta k_\parallel) , \tag{10.74}$$

where

$$\int_{\hat{\boldsymbol{n}}} = \int \frac{d\Omega_{\hat{\boldsymbol{n}}}}{\Omega_D} J_D(\hat{\boldsymbol{n}}, 0) \tag{10.75}$$

denotes weighted angular averaging. Let us now rewrite the Gaussian action (10.74) in terms of dimensionless rescaled integration variables and rescaled fields which have the proper scaling dimensions. Therefore we introduce a momentum scale $\Lambda < \Lambda_0$ and define, for a given direction $\hat{\boldsymbol{n}}$, the dimensionless variables

$$q_\parallel = \frac{\delta k_\parallel}{\Lambda} \ , \quad \epsilon = \frac{\omega}{v_0 \Lambda} , \tag{10.76}$$

and the dimensionless fields

$$\psi'_{\hat{\boldsymbol{n}}\sigma}(i\epsilon, q_\parallel) = \sqrt{2\pi v_0 (v_0 \Lambda)^3} \ \psi_{\hat{\boldsymbol{n}}\sigma}(i\omega, \delta k_\parallel) . \tag{10.77}$$

Then Eq. (10.74) can be written as

$$\tilde{S}_0[\bar{\psi}', \psi'] = -\sum_\sigma \int_{\hat{\boldsymbol{n}}} \int \frac{d\epsilon dq_\parallel}{(2\pi)^2} \Theta\left(\frac{\Lambda_0}{\Lambda} - |q_\parallel|\right) [i\epsilon - c_0(\hat{\boldsymbol{n}})q_\parallel]$$
$$\times \bar{\psi}'_{\hat{\boldsymbol{n}}\sigma}(i\epsilon, q_\parallel)\psi'_{\hat{\boldsymbol{n}}\sigma}(i\epsilon, q_\parallel) , \tag{10.78}$$

with

$$c_0(\hat{\boldsymbol{n}}) = v_F(\hat{\boldsymbol{n}})/v_0 . \tag{10.79}$$

Note that the unit vector $\hat{\boldsymbol{n}}$ plays the role of a continuous flavor index which labels the different fields. Obviously, for $\Lambda/\Lambda_0 \to 0$ the action (10.78) is independent of the scale factor Λ and becomes scale invariant. The associated RG fixed point describes the normal metallic state and is called the Fermi liquid fixed point in dimensions $D > 1$.

To implement the above scaling within the framework of the FRG, let us now introduce properly rescaled vertices and write down the corresponding exact FRG flow equations (Kopietz and Busche 2001). Given the flowing self-energy $\Sigma_\Lambda(K\sigma)$ which satisfies the exact FRG flow equation (10.33), we introduce its rescaled and subtracted counterpart as a function of the logarithmic flow parameter $l = \ln(\Lambda_0/\Lambda)$,

$$\tilde{\Gamma}_l^{(2)}(i\epsilon, q_\parallel, \hat{n}, \sigma) = \frac{Z_l(\hat{n}, \sigma)}{v_0 \Lambda} \left[\Sigma_\Lambda(i\omega, k, \sigma) - \Sigma(i0, k_F, \sigma) \right] , \qquad (10.80)$$

where on the right-hand side it is understood that we should express $\omega = v_F(\hat{n})\Lambda\epsilon$ and $k = k_F(\hat{n}) + \hat{v}_F(\hat{n})\Lambda q_\parallel$ in terms of the dimensionless scaling variables $i\epsilon$ and $q_\parallel$, and the subtraction $\Sigma(i0, k_F, \sigma)$ is the (a priori unknown) exact self-energy at the true Fermi surface k_F. The flowing wave function renormalization factor is defined by

$$Z_l(\hat{n}, \sigma) = \frac{1}{1 - \left.\frac{\partial \Sigma_\Lambda(i\omega, k_F, \sigma)}{\partial(i\omega)}\right|_{\omega=0}} = 1 + \left.\frac{\partial \tilde{\Gamma}_l^{(2)}(i\epsilon, 0, \hat{n}, \sigma)}{\partial(i\epsilon)}\right|_{\epsilon=0} , \qquad (10.81)$$

where the second identity follows from the definition (10.80). We also introduce the rescaled propagator,

$$\tilde{G}_l(i\epsilon, q_\parallel, \hat{n}, \sigma) = \frac{v_0 \Lambda}{Z_l(\hat{n}, \sigma)} G_\Lambda(i\omega, k, \sigma) , \qquad (10.82)$$

and the corresponding rescaled single-scale propagator,

$$\dot{\tilde{G}}_l(i\epsilon, q_\parallel, \hat{n}, \sigma) = -\frac{v_0 \Lambda^2}{Z_l(\hat{n}, \sigma)} \dot{G}_\Lambda(i\omega, k, \sigma) . \qquad (10.83)$$

If we work with a multiplicative cutoff $\Theta_\Lambda(K)$ as in Eq. (7.10), then the flowing propagator can be written as

$$G_\Lambda(K\sigma) = \frac{\Theta_\Lambda(K)}{i\omega - \epsilon_k + \mu - \Sigma_\Lambda(K\sigma)}$$
$$= \frac{\Theta_\Lambda(K)}{i\omega - \epsilon_k + \epsilon_{k_F} - [\Sigma_\Lambda(K\sigma) - \Sigma(i0, k_F, \sigma)]} , \qquad (10.84)$$

and the corresponding rescaled propagator (10.82) is

$$\tilde{G}_l(i\epsilon, q_\parallel, \hat{n}, \sigma) = \frac{\Theta_\Lambda(K)}{Z_l(\hat{n}, \sigma)[i\epsilon - \xi_{\hat{n}}(q_\parallel)] - \tilde{\Gamma}_l^{(2)}(i\epsilon, q_\parallel, \hat{n}, \sigma)} , \qquad (10.85)$$

where

$$\xi_{\hat{n}}(q_{\parallel}) = \frac{\epsilon_{k_F+k} - \epsilon_{k_F}}{v_0 \Lambda} \approx c_0(\hat{n}) q_{\parallel} + \mathcal{O}\left(q_{\parallel}^2\right) \tag{10.86}$$

and $c_0(\hat{n}) = v_F(\hat{n})/v_0$ (see Eq. (10.79)). The function $\Theta_\Lambda(K)$ in Eq. (10.85) should be expressed in terms of rescaled variables; for example, for sharp momentum shell cutoff we have

$$\Theta_\Lambda(K) = \Theta\left(\Lambda - \frac{|\epsilon_k - \epsilon_{k_F}|}{|v_F|}\right) \approx \Theta(1 - |q_{\parallel}|) \ . \tag{10.87}$$

With these definitions the rescaled version of the exact FRG flow equation (10.33) for the self-energy assumes the form

$$\boxed{\begin{aligned}\partial_l \tilde{\Gamma}_l^{(2)}(i\epsilon, q_{\parallel}, \hat{n}, \sigma) &= [1 - \eta_l(\hat{n}, \sigma) - q_{\parallel}\partial_{q_{\parallel}} - \epsilon\partial_\epsilon]\tilde{\Gamma}_l^{(2)}(i\epsilon, q_{\parallel}, \hat{n}, \sigma) \\ &\quad + \dot{\Gamma}_l^{(2)}(i\epsilon, q_{\parallel}, \hat{n}, \sigma) \ ,\end{aligned}} \tag{10.88}$$

where the flowing anomalous dimension is defined by

$$\eta_l(\hat{n}, \sigma) = -\partial_l \ln Z_l(\hat{n}, \sigma) = -\frac{\partial_l Z_l(\hat{n}, \sigma)}{Z_l(\hat{n}, \sigma)} \ , \tag{10.89}$$

and the inhomogeneity $\dot{\Gamma}_l^{(2)}(\epsilon, q_{\parallel}, \hat{n}, \sigma)$ is determined by the right-hand side of the unrescaled FRG flow equation (10.33),

$$\begin{aligned}\dot{\Gamma}_l^{(2)}(i\epsilon, q_{\parallel}, \hat{n}, \sigma) &= \frac{Z_l(\hat{n}, \sigma)}{v_0 \Lambda}[-\Lambda\partial_\Lambda \Sigma_\Lambda(K\sigma)] \\ &= \sum_{\sigma'}\int_{\hat{n}'}\int \frac{d\epsilon' dq_{\parallel}'}{(2\pi)^2}\dot{\tilde{G}}_l\left(i\epsilon', q_{\parallel}', \hat{n}', \sigma'\right)\tilde{\Gamma}_l^{(4)}(Q\sigma, Q'\sigma'; Q'\sigma', Q\sigma) \ .\end{aligned} \tag{10.90}$$

We have now set $\zeta = -1$ in Eq. (10.33) (because the above rescaling makes only sense for fermions) and have introduced the collective labels

$$Q = (i\epsilon, q_{\parallel}, \hat{n}) \ . \tag{10.91}$$

The rescaled effective interaction is defined in terms of the dimensionful effective interaction $\Gamma_\Lambda^{(4)}$ appearing in Eq. (10.37) as

$$\begin{aligned}\tilde{\Gamma}_l^{(4)}\left(Q_1'\sigma_1', Q_2'\sigma_2'; Q_2\sigma_2, Q_1\sigma_1\right) &= \left[Z_l\left(\hat{n}_1', \sigma_1'\right)Z_l\left(\hat{n}_2', \sigma_2'\right)Z_l(\hat{n}_2, \sigma_2)Z_l(\hat{n}_1, \sigma_1)\right]^{1/2} \\ &\quad \times 2\pi v_0 \Gamma_\Lambda^{(4)}\left(K_1'\sigma_1', K_2'\sigma_2'; K_2\sigma_2, K_1\sigma_1\right) \ .\end{aligned} \tag{10.92}$$

Substituting Eqs. (10.81) and (10.88) into the definition (10.89), we see that the flowing anomalous dimension is directly related to the frequency dependence of the rescaled flowing effective interaction (Busche et al. 2002, Ledowski and Kopietz 2007),

$$
\eta_l(\hat{\boldsymbol{n}}, \sigma) = - \left. \frac{\dot{\Gamma}_l^{(2)}(i\epsilon, 0, \hat{\boldsymbol{n}}, \sigma)}{\partial(i\epsilon)} \right|_{\epsilon=0}
$$

$$
= - \sum_{\sigma'} \int_{\hat{\boldsymbol{n}}'} \int \frac{d\epsilon' dq'_\parallel}{(2\pi)^2} \dot{\tilde{G}}_l(Q'\sigma') \left. \frac{\partial \tilde{\Gamma}_l^{(4)}(i\epsilon, 0, \hat{\boldsymbol{n}}, \sigma, Q'\sigma'; Q'\sigma', i\epsilon, 0, \hat{\boldsymbol{n}}, \sigma)}{\partial(i\epsilon)} \right|_{\epsilon=0} .
$$

$$(10.93)$$

The FRG flow equation for the rescaled effective interaction $\tilde{\Gamma}_l^{(4)}$ is given by the rescaled version of the exact flow equation (10.37), which depends on the flowing irreducible six-point vertex $\Gamma_\Lambda^{(6)}$. For general n, we define rescaled dimensionless irreducible vertices with $2n$ external legs via

$$
\tilde{\Gamma}_l^{(2n)}\left(Q'_1\sigma'_1, \ldots, Q'_n\sigma'_n; Q_n\sigma_n, \ldots, Q_1\sigma_1\right) =
$$

$$
\left[Z_l\left(\hat{\boldsymbol{n}}'_1\sigma'_1\right) \cdots Z_l\left(\hat{\boldsymbol{n}}'_n\sigma'_n\right) Z_l(\hat{\boldsymbol{n}}_n\sigma_n) \cdots Z_l(\hat{\boldsymbol{n}}_1\sigma_1)\right]^{1/2}
$$

$$
\times (2\pi v_0)^{n-1}(v_0\Lambda)^{n-2} \Gamma_\Lambda^{(2n)}\left(K'_1\sigma'_1, \ldots, K'_n\sigma'_n; K_n\sigma_n, \ldots, K_1\sigma_1\right) . \quad (10.94)
$$

The rescaled version of Eq. (10.37) can then be written as (Kopietz and Busche 2001)

$$
\partial_l \tilde{\Gamma}_l^{(4)}\left(Q'_1\sigma'_1, Q'_2\sigma'_2; Q_2\sigma_2, Q_1\sigma_1\right) = -\sum_{i=1}^{2}\left[\frac{\eta_l\left(\hat{\boldsymbol{n}}'_i\sigma'_i\right) + \eta_l(\hat{\boldsymbol{n}}_i\sigma_i)}{2}\right.
$$

$$
\left. +q'_{i\parallel}\partial_{q'_{i\parallel}} + \epsilon'_i\partial_{\epsilon'_i} + q_{i\parallel}\partial_{q_{i\parallel}} + \epsilon_i\partial_{\epsilon_i}\right]\tilde{\Gamma}_l^{(4)}\left(Q'_1\sigma'_1, Q'_2\sigma'_2; Q_2\sigma_2, Q_1\sigma_1\right)
$$

$$
+\sum_{\sigma}\int_Q \dot{\tilde{G}}_l(Q\sigma)\tilde{\Gamma}_l^{(6)}\left(Q'_1\sigma'_1, Q'_2\sigma'_2, Q\sigma; Q\sigma, Q_2\sigma_2, Q_1\sigma_1\right)
$$

$$
-\sum_{\sigma\sigma'}\int_Q \left[\dot{\tilde{G}}_l(Q\sigma)\tilde{G}_l(Q'\sigma') + \tilde{G}_l(Q\sigma)\dot{\tilde{G}}_l(Q'\sigma')\right]
$$

$$
\times \left\{\frac{1}{2}\left[\tilde{\Gamma}_l^{(4)}\left(Q'_1\sigma'_1, Q'_2\sigma'_2; Q'\sigma', Q\sigma\right)\tilde{\Gamma}_l^{(4)}(Q\sigma, Q'\sigma'; Q_2\sigma_2, Q_1\sigma_1)\right]_{K'=K_1+K_2-K}\right.
$$

$$
-\left[\tilde{\Gamma}_l^{(4)}\left(Q'_1\sigma'_1, Q'\sigma'; Q\sigma, Q_1\sigma_1\right)\tilde{\Gamma}_l^{(4)}\left(Q'_2\sigma'_2, Q\sigma; Q'\sigma', Q_2\sigma_2\right)\right]_{K'=K+K_1-K'_1}
$$

$$
\left.+\left[\tilde{\Gamma}_l^{(4)}\left(Q'_1\sigma'_1, Q'\sigma'; Q\sigma, Q_2\sigma_2\right)\tilde{\Gamma}_l^{(4)}\left(Q'_2\sigma'_2, Q\sigma; Q'\sigma', Q_1\sigma_1\right)\right]_{K'=K+K_2-K'_1}\right\},
$$

$$(10.95)$$

where $\int_Q = \int_{\hat{\boldsymbol{n}}} \int \frac{d\epsilon dq_\parallel}{(2\pi)^2}$, and $K = (i\omega, \boldsymbol{k})$, K', K_i, and K'_i should be considered as functions of the dimensionless variables $Q = (i\epsilon, q_\parallel, \hat{\boldsymbol{n}})$, Q', Q_i, and Q'_i defined via Eqs. (10.67) and (10.76). Due to the highly nonlinear character of the transformation (10.67) from absolute momenta $\boldsymbol{k}$ to the local variables $(q_\parallel, \hat{\boldsymbol{n}})$, the momentum conservation enforced by identities like $K' = K_1 + K_2 - K$ implies a very complicated functional dependence of $q'_\parallel$ and $\hat{\boldsymbol{n}}'$ on $q_{1\parallel}, q_{2\parallel}, q_\parallel$ as well as on the three directions $\hat{\boldsymbol{n}}_1, \hat{\boldsymbol{n}}_2$, and $\hat{\boldsymbol{n}}$. In order to detect the leading instabilities of normal Fermi systems in dimensions $D > 1$, the rescaled flow equation (10.95) is therefore not very useful and it is better to analyze directly its unrescaled counterpart (10.37). On the other hand, rescaling is essential to obtain fixed points of the RG and to classify the couplings according to their relevance at a given fixed point, which we shall do in the following section. The rescaled version of the FRG flow equation for the irreducible six-point vertex can be found in the appendix of Kopietz and Busche (2001).

10.4.2 Classification of Couplings

From the rescaled FRG flow equations (10.88) and (10.95), we can now read off the scaling dimensions of all coupling constants. Consider first the flow equation (10.88) for the rescaled two-point vertex $\tilde{\Gamma}_l^{(2)}(i\epsilon, q_\parallel, \hat{n}, \sigma)$. For $\epsilon = q_\parallel = 0$ this equation reduces to

$$\partial_l r_l(\hat{n}, \sigma) = [1 - \eta_l(\hat{n}, \sigma)] r_l(\hat{n}, \sigma) + \dot{\Gamma}_l^{(2)}(\hat{n}, \sigma) , \tag{10.96}$$

where we have defined

$$r_l(\hat{n}, \sigma) = \tilde{\Gamma}_l^{(2)}(0, 0, \hat{n}, \sigma) , \tag{10.97a}$$

$$\dot{\Gamma}_l^{(2)}(\hat{n}, \sigma) = \dot{\Gamma}_l^{(2)}(0, 0, \hat{n}, \sigma) . \tag{10.97b}$$

Obviously, for a normal Fermi system with finite quasiparticle residue where $\lim_{l\to\infty} \eta_l(\hat{n}, \sigma) = 0$, the dimensionless couplings $r_l(\hat{n}, \sigma)$ are all relevant with scaling dimension $+1$. In dimensions $D > 1$, the Fermi surface is a $D - 1$-dimensional continuum whose points are labeled by the unit vectors $\hat{n}$. In $D > 1$, a normal Fermi system is therefore characterized by infinitely many relevant couplings $r_l(\hat{n}, \sigma)$. We shall show below that the true Fermi surface of the interacting system can be defined self-consistently by demanding that all relevant couplings $r_l(\hat{n}, \sigma)$ flow into a fixed point (Kopietz and Busche 2001, Ledowski and Kopietz 2003, Ledowski et al. 2005, Ledowski and Kopietz 2007). As already discussed in Sect. 3.3.3, the Fermi surface can thus be viewed as a multicritical RG fixed point of infinite order, which is characterized by the fixed-point values of infinitely many relevant couplings.

The first derivatives of the rescaled two-point vertex $\tilde{\Gamma}_l^{(2)}(i\epsilon, q_\parallel, \hat{n}, \sigma)$ with respect to ϵ and $q_\parallel$ define two marginal couplings whose scaling dimension vanishes. To define these couplings, we write the expansion of $\tilde{\Gamma}_l^{(2)}(i\epsilon, q_\parallel, \hat{n}, \sigma)$ in powers of ϵ and $q_\parallel$ as

$$\tilde{\Gamma}_l^{(2)}(i\epsilon, q_\parallel, \hat{n}, \sigma) = r_l(\hat{n}, \sigma) + [Z_l(\hat{n}, \sigma) - 1] i\epsilon + [Y_l(\hat{n}, \sigma) - Z_l(\hat{n}, \sigma)] q_\parallel$$
$$+ \mathcal{O}\left(\epsilon^2, q_\parallel^2, \epsilon q_\parallel\right) . \tag{10.98}$$

Note that to this order in the expansion the rescaled propagator (10.85) can be written as

$$\tilde{G}_l(i\epsilon, q_\parallel, \hat{n}, \sigma) \approx \frac{\Theta_\Lambda(K)}{i\epsilon - c_l(\hat{n}, \sigma) q_\parallel - r_l(\hat{n}, \sigma)} , \tag{10.99}$$

with

$$c_l(\hat{\boldsymbol{n}}, \sigma) = Z_l(\hat{\boldsymbol{n}}, \sigma)[c_0(\hat{\boldsymbol{n}}) - 1] + Y_l(\hat{\boldsymbol{n}}, \sigma) \ . \tag{10.100}$$

By definition, $Z_l(\hat{\boldsymbol{n}}, \sigma)$ is the flowing wave function renormalization factor, which satisfies the flow equation

$$\partial_l Z_l(\hat{\boldsymbol{n}}, \sigma) = -\eta_l(\hat{\boldsymbol{n}}, \sigma) Z_l(\hat{\boldsymbol{n}}, \sigma) \ , \tag{10.101}$$

see Eqs. (10.81) and (10.89). From the exact FRG flow equation (10.88) it is easy to show that the other marginal coupling related to the first derivatives of the two-point vertex,

$$Y_l(\hat{\boldsymbol{n}}, \sigma) = Z_l(\hat{\boldsymbol{n}}, \sigma) + \left. \frac{\partial \tilde{\Gamma}_l^{(2)}(0, q_{\|}, \hat{\boldsymbol{n}}, \sigma)}{\partial q_{\|}} \right|_{q_{\|}=0} , \tag{10.102}$$

satisfies the flow equation (Busche et al. 2002, Ledowski and Kopietz 2007)

$$\partial_l Y_l(\hat{\boldsymbol{n}}, \sigma) = -\eta_l(\hat{\boldsymbol{n}}, \sigma) Y_l(\hat{\boldsymbol{n}}, \sigma) + \left. \frac{\partial \dot{\Gamma}_l^{(2)}(0, q_{\|}, \hat{\boldsymbol{n}}, \sigma)}{\partial q_{\|}} \right|_{q_{\|}=0} . \tag{10.103}$$

Next, let us classify the couplings associated with the rescaled four-point vertex $\tilde{\Gamma}_l^{(4)}\left(Q_1'\sigma_1', Q_2'\sigma_2'; Q_2\sigma_2, Q_1\sigma_1\right)$. From our rescaled FRG flow equation (10.95), we see that the dependence of the four-point vertex on the frequencies and projected momenta parallel to the local Fermi velocity gives rise to negative scaling dimensions, so that the marginal part of $\tilde{\Gamma}_l^{(4)}\left(Q_1'\sigma_1', Q_2'\sigma_2'; Q_2\sigma_2, Q_1\sigma_1\right)$ is obtained by setting $\epsilon_1' = \epsilon_2' = \epsilon_2 = \epsilon_1 = 0$ and $q_{1\|}' = q_{2\|}' = q_{2\|} = q_{1\|} = 0$. Obviously, there are infinitely many marginal couplings, labeled by the unit vectors $\hat{\boldsymbol{n}}_1', \hat{\boldsymbol{n}}_2', \hat{\boldsymbol{n}}_2, \hat{\boldsymbol{n}}_1$. But if both incoming and outgoing momenta lie on the Fermi surface, then momentum conservation imposes the constraint

$$\boldsymbol{k}_{F1}' + \boldsymbol{k}_{F2}' = \boldsymbol{k}_{F2} + \boldsymbol{k}_{F1} \ . \tag{10.104}$$

In terms of the unit vectors pointing in the directions of the Fermi momenta, Eq. (10.104) reads as

$$k_F\left(\hat{\boldsymbol{n}}_1'\right)\hat{\boldsymbol{n}}_1' + k_F\left(\hat{\boldsymbol{n}}_2'\right)\hat{\boldsymbol{n}}_2' = k_F(\hat{\boldsymbol{n}}_2)\hat{\boldsymbol{n}}_2 + k_F(\hat{\boldsymbol{n}}_1)\hat{\boldsymbol{n}}_1 \ . \tag{10.105}$$

The crucial point is now that in dimensions $D > 1$ the geometric constraint imposed by these relations allows only to choose freely two of the Fermi momenta (or directions), say $\boldsymbol{k}_{F1}$ and $\boldsymbol{k}_{F2}$ (or $\hat{\boldsymbol{n}}_1$ and $\hat{\boldsymbol{n}}_2$). If we arbitrarily fix three momenta on the Fermi surface, for example, $\boldsymbol{k}_{F1}, \boldsymbol{k}_{F2}$, and $\boldsymbol{k}_{F1}'$, then in general the difference $\boldsymbol{k}_{F2} + \boldsymbol{k}_{F1} - \boldsymbol{k}_{F1}'$ does not lie on the Fermi surface so that we cannot satisfy Eq. (10.104). The set of allowed marginal couplings related to the four-point vertex can therefore be classified in terms of two independent unit vectors. For a more detailed discussion of the classification of the various scattering processes, we refer the reader to the reviews by Shankar (1994) and by Metzner et al. (1998).

Finally, the rescaled vertices $\tilde{\Gamma}_l^{(2n)}$ defined in Eq. (10.94) with six and more external legs ($n \geq 3$) have all negative scaling dimensions and are therefore irrelevant. For example, according to Eq. (10.94) the most important part of $\tilde{\Gamma}_l^{(6)}$ $\left(Q_1'\sigma_1', Q_2'\sigma_2', Q_3'\sigma_3'; Q_3\sigma_3, Q_2\sigma_2, Q_1\sigma_1 \right)$, which is obtained by setting all external frequencies and distances from the Fermi surface equal to zero, is proportional to Λ, so that this contribution has scaling dimension -1. It is therefore reasonable to neglect this vertex in the FRG flow equation (10.95) if one is only interested in the infrared behavior of the system at the Fermi liquid fixed point.

10.4.3 Exact Integral Equation for the Fermi Surface

In order to calculate the true Fermi surface of an interacting Fermi system, we need to know the exact self-energy $\Sigma(0, k)$ at vanishing frequency as a function of k and look for solutions of the defining equation (10.63). The rescaled version of the exact FRG flow equations discussed above allows us to define the Fermi surface nonperturbatively in terms of a fixed point condition for the relevant part $r_l(\hat{n})$ of the rescaled two-point vertex $\tilde{\Gamma}_l^{(2)}(Q)$ defined in Eq. (10.97a) (Ledowski and Kopietz 2003). To derive this equation, let us transform the differential RG equation (10.96) into an equivalent integral equation similar to Eq. (8.41),

$$r_l(\hat{n}) = e^{l - \int_0^l d\tau \, \eta_\tau(\hat{n})} \left[r_0(\hat{n}) + \int_0^l dl' \, e^{-l' + \int_0^{l'} d\tau \, \eta_\tau(\hat{n})} \dot{\Gamma}_{l'}^{(2)}(\hat{n}) \right] . \tag{10.106}$$

For simplicity, we omit in this section the spin labels. Suppose now that we have adjusted the initial couplings such that for $l \rightarrow \infty$ the flowing couplings $r_l(\hat{n})$ indeed approach finite fixed point values. Assuming that the associated anomalous dimensions $\eta_\infty(\hat{n})$ are smaller than unity, Eq. (10.106) implies that the limit $r_\infty(\hat{n}) = \lim_{l \rightarrow \infty} r_l(\hat{n})$ can only be finite if the initial values $r_0(\hat{n})$ are chosen such that

$$\begin{aligned} r_0(\hat{n}) &= -\int_0^\infty dl \, e^{-l + \int_0^l d\tau \, \eta_\tau(\hat{n})} \dot{\Gamma}_l^{(2)}(\hat{n}) \\ &= \int_0^\infty dl \, e^{-l + \int_0^l d\tau \, \eta_\tau(\hat{n})} \int_{Q'} \dot{G}_l(Q') \tilde{\Gamma}_l^{(4)}(0, 0, \hat{n}, Q'; Q', 0, 0, \hat{n}) , \end{aligned} \tag{10.107}$$

where in the second line we have substituted Eq. (10.90). This is an implicit condition for $r_0(\hat{n})$, relating it to the flowing two-point vertex and the four-point vertex on the entire RG trajectory. Keeping in mind that the right-hand side of Eq. (10.107) implicitly depends on $r_l(\hat{n})$ and that according to Eq. (10.80)

$$\Sigma(0, \boldsymbol{k}_F) - \Sigma_{\Lambda_0}(0, \boldsymbol{k}_F) = \frac{v_0 \Lambda_0}{Z_0(\hat{\boldsymbol{n}})} r_0(\hat{\boldsymbol{n}}) , \tag{10.108}$$

we see that Eq. (10.107) can be regarded as an integral equation for the counterterm $\Sigma(0, \boldsymbol{k}_F)$, which is needed in order to calculate the true shape of the Fermi surface.

Finally, let us transform Eq. (10.107) back to unrescaled variables, choosing for simplicity the initial conditions $\Sigma_{\Lambda_0}(0, \boldsymbol{k}_F) = 0$ and $Z_0(\hat{\boldsymbol{n}}) = 1$. Using a sharp momentum shell cutoff of the form (7.12) it is easy to show that Eq. (10.107) is equivalent with

$$\Sigma(0, \boldsymbol{k}_F) = \int \frac{d^D k'}{(2\pi)^D} \frac{d\omega'}{2\pi} \frac{\Theta(\Lambda_0 - \Lambda_{k'})}{i\omega' - \epsilon_{k'} + \mu - \Sigma_{\Lambda_{k'}}(K')}$$
$$\times \Gamma^{(4)}_{\Lambda_{k'}}(0, \boldsymbol{k}_F, i\omega', \boldsymbol{k}'; i\omega', \boldsymbol{k}', 0, \boldsymbol{k}_F) , \tag{10.109}$$

where $\Lambda_{k'} = |\epsilon_{k'} - \epsilon_{k'_F}| / |\boldsymbol{v}'_F|$. Note that the right-hand side of Eq. (10.109) involves the flowing self-energy and four-point vertex at the scales $\Lambda = \Lambda_{k'}$ which depend on the distance from the true Fermi surface. The exact integral equation (10.109) and the equivalent rescaled equation (10.107) determine the counterterm $\Sigma(0, \boldsymbol{k}_F)$ by the requirement that in the limit $l \to \infty$ all rescaled couplings approach finite fixed point values. For an explicit calculation of the Fermi surface of quasi one-dimensional metals using this nonperturbative method (see Ledowski et al. 2005, Ledowski and Kopietz 2007).

10.5 One-Loop Patching Approximations

During the last decade, many workers have applied renormalization group methods to contribute to an understanding of the electronic phase diagram of layered materials such as the cuprate high-temperature superconductors. The Fermi surface of these two-dimensional systems is a continuous line. Thus functional aspects of the RG are important, since in this case the marginal couplings depend on a continuous position on the Fermi surface, as discussed in Sect. 10.4.2. For practical calculations, the continuous dependence is discretized, and the marginal couplings related to the effective interaction now depend only on discrete patch labels. A numerical integration of the flow equations for the resulting finite number of couplings is then performed. Typically, one finds divergencies at a finite energy scale. Although this approach thus formally breaks down at this scale, the divergencies can be taken as an indication for either a broken symmetry or at least dominating correlations in a particular interaction channel. An analysis of the nature and the symmetry of the diverging coupling provides a numerical tool to investigate the phase diagram of two-dimensional models. A short review of this one-loop patching approximation is given in Metzner (2005).

As a minimal model for the copper-oxide planes of the cuprate high-temperature superconductors, Hubbard models on a square lattice (see Eq. (10.90)) and possible extensions have been intensively studied. Depending on the strength of microscopic parameters such as hopping amplitudes, Hubbard interactions, and the filling of the band, one finds regions in the phase diagram with spin- or charge-density waves, d-wave superconductivity, ferromagnetism, or even exotic phases, such as staggered-flux phases, for example. Starting from the limit of a vanishing interaction strength, the occurrence of these phases can be understood as instabilities of the Fermi liquid. In this context, the geometry of the Fermi surface is essential for determining the nature of the dominant instability. Of special interest are regions in parameter space which are close to a half-filled band in a model with pure nearest-neighbor hopping, having a square Fermi surface. In this case, the parallel sections of the Fermi surface lead to nesting, and the points $\left(0, \pm\frac{\pi}{a}\right)$, $\left(\pm\frac{\pi}{a}, 0\right)$ (where a is the lattice spacing) on the boundary of the Brillouin zone have a vanishing Fermi velocity resulting in Van Hove singularities in the density of states. Also, Umklapp scattering processes are important at half filling. For the strictly half-filled case, it is well known that an insulating antiferromagnetic phase emerges. Yet, slightly away from half filling, or in the presence of next-nearest-neighbor hopping, the nesting instability can be suppressed and other ordering tendencies dominate. This interplay of different instabilities is often discussed in the context of an experimentally found non-Fermi-liquid behavior of the normal phase of the high-temperature supercon-ductors below optimal doping.

Some aspects of the physics in these two-dimensional models can be captured by simple toy models for which analytical treatments of the renormalization group equations are still possible. The importance of Van Hove singularities can be analyzed using a model dispersion that includes only momenta close to the two saddle points at $\left(0, \frac{\pi}{a}\right)$ and $\left(\frac{\pi}{a}, 0\right)$ (Lederer et al. 1987, Furukawa et al. 1998). The renormalization group flow is particularly strong in this case as the perturbation expansion exhibits a $\log^2$-divergence due to the Van Hove singularity in the density of states leading to an enhancement of d-wave pairing tendencies. The role of nesting and Umklapp scattering can be investigated with a model containing only patches on the Fermi surface along or close to the diagonals of the Brillouin zone (Houghton and Marston 1993, Tam et al. 2006). However, for a treatment of the interplay between these two tendencies, the number of patches on the Fermi surface has to be increased and a numerical solution of the FRG flow equations is necessary.

The FRG flow equations derived so far are part of an infinite hierarchy which for practical calculations has to be truncated. For weak interactions a systematic truncation is possible: If the bare interaction contains only a four-point vertex proportional to a small coupling constant U, the perturbative corrections for the irreducible vertex $\Gamma^{(2n)}$ contains only terms that are at least of order U^n. Hence, to analyze the leading renormalization of the four-point vertex, we have to keep terms up to order U^2. We will thus neglect the contribution of the six-point vertex and all higher vertices to the RG flow. Equations (10.39) and (10.40) then constitute a closed set of integro-differential equations. However, as the effective interaction $U^{(4)}$ depends on three independent momentum and frequency variables, these equations are still too complicated to be amenable to a numerical solution. Rescaling arguments can then be used to identify the important contributions at low energies.

In this section we shall derive explicit flow equations for the marginal couplings in the one-loop patching approximation and apply them to the simplest case of a single one-dimensional chain as well as to the two-dimensional square lattice Hubbard model with nearest-neighbor hopping close to half filling.

10.5.1 Flow of Marginal Couplings

In Sect. 10.4.2, we have used rescaling arguments to show that the frequency dependencies of the four-point vertex as well as its momentum dependencies perpendicular to the Fermi surface have negative scaling dimensions at the Fermi-liquid fixed point and are thus irrelevant at low energies. To obtain the leading instabilities around the Fermi-liquid fixed point, it is thus sufficient to keep track of couplings with the momenta of all four legs on the Fermi surface.[6] Throughout Sect. 10.4 the position of the momenta on the Fermi surface has been parameterized by a unit vector $\hat{n}$; the dependence of the vertex functions on these unit vectors is expected to be sufficiently smooth such that for a numerical treatment it can be discretized. In the arguments of the four-point vertex, we thus effectively replace the energy–momentum indices K by integer patch indices n. Momentum space is divided exhaustively into these nonoverlapping patches, usually by subdividing the Fermi surface into discrete regions and defining an appropriate procedure to project an arbitrary momentum onto the Fermi surface. Each patch has a central momentum k_n on the Fermi surface. We only treat systems with full spin rotational symmetry in this section. We thus replace the interaction function in the relevant FRG flow equation (10.40) by[7]

$$U^{(4)}(K_1, K_2; K_3, K_4) \longrightarrow u(n_1, n_2; n_3) , \qquad (10.110)$$

where the discrete couplings $u(n_1, n_2; n_3)$ depend only on three integer patch indices n_1, n_2, and n_3, while the fourth momentum

$$k_4 = k_{n_1} + k_{n_2} - k_{n_3} \qquad (10.111)$$

is determined by momentum conservation. The integer n_4 is then the index of the patch that contains k_4. Note that for many of these couplings, the fourth momentum

---

[6] For a one-dimensional extended Hubbard model Tam et al. (2006) have proposed a truncation of the FRG flow equations which include scattering processes involving momenta far away from the Fermi points.

[7] Note that in the context of the discrete couplings, we simply enumerate momenta using integers n_1, n_2, n_3 instead of marking them as incoming and outgoing by using primed variable. Thus compared to the notation for continuous momenta, we replace $U^{(4)}(K_1', K_2'; K_2, K_1) \rightarrow U^{(4)}(K_1, K_2; K_3; K_4) \rightarrow u(n_1, n_2; n_3)$, where the new K_1 and K_2 denote the two outgoing energy-momenta. Hence, $u(n_1, n_2; n_3)$ becomes a tensor of rank 3 which has some advantages for the following numerical calculations.

does not actually lie on the Fermi surface as in general only two momenta can be chosen freely on a surface in order to satisfy momentum conservation. However, in the general formulation of the flow equations we shall keep these terms; the reason is that couplings where the fourth leg is in some sense close to the Fermi surface can still dominate the flow at intermediate energy scales and these couplings are thus important to describe the crossover from the bare microscopic model to the very low energy regime.

In this section, we shall work with unrescaled vertex functions, as is usually done in the literature in this context; within the one-loop approximation, rescaled and unrescaled flow equations are equivalent because anomalous dimensions only appear at two-loop order. The feedback of the flow of the self-energy and thus the shape of the interacting Fermi surface on the flow of the couplings $u(n_1, n_2; n_3)$ is neglected. This is inconsistent from a pure scaling point of view as discussed in Sect. 10.4. For a full two-dimensional problem the inclusion of self-energy effects is a very complicated numerical problem that has not been fully addressed yet, although an attempt in this direction has recently been made by Katanin (2009). The propagators in Eq. (10.40) are thus taken as cutoff-dependent bare propagators without self-energy corrections. After the replacement (10.110), the momentum and frequency integration in Eq. (10.40) can be carried out. We define the derivatives of the particle–particle and particle–hole bubble integrals[8] with respect to Λ for a particular patch centered at k_n,

$$\dot{\Pi}^{\pm}(n, q) = \pm \Lambda \int_{K \in K_n} \dot{G}(K) G(\pm(K - Q_0)) , \qquad (10.112)$$

where $K \in K_n$ means that the momentum part of the integration is over the area of the patch with the index n, and $Q_0 = (i0, q)$. As usual, the logarithmic RG flow parameter l (which we shall also call *RG time*) is related to the cutoff $\Lambda = \Lambda_0 e^{-l}$ and thus $\partial_l = -\Lambda \partial_\Lambda$. The integro-differential equation (10.40) then reduces to the following system of coupled ordinary differential equations,

$$\partial_l u(n_1, n_2; n_3) = -\sum_n \left\{ \dot{\Pi}^-(n, q_{\mathrm{pp}}) \left[u(n_2, n_1; n) u(n_3, n_4; n) \right. \right.$$
$$\left. + u(n_1, n_2; n) u(n_4, n_3; n) \right]$$
$$+ \dot{\Pi}^+(n, q_{\mathrm{fs}}) \left[2u(n, n_4; n_1) u(n, n_2; n_3) \right.$$
$$\left. - u(n_4, n; n_1) u(n, n_2; n_3) - u(n, n_4; n_1) u(n_2, n; n_3) \right]$$

---

[8] Up to a different normalization, the function $\dot{\Pi}^-(n, q)$ is a simplified version of the single-scale particle–particle bubble $\dot{\Pi}^{\mathrm{pp}}(P)$ defined in Eq. (10.51), while the function $\dot{\Pi}^+(n, q)$ corresponds to the single-scale particle–hole bubble $\dot{\Pi}^{\mathrm{ph}}(Q)$ given in Eq. (10.59).

$$
\begin{aligned}
&+ \dot{\Pi}^{+}(n, -\boldsymbol{q}_{\mathrm{fs}}) \Big[2u(n, n_1; n_4)u(n, n_3; n_2) \\
&\qquad - u(n_1, n; n_4)u(n, n_3; n_2) - u(n, n_1; n_4)u(n_3, n; n_2) \Big] \\
&- \dot{\Pi}^{+}(n, \boldsymbol{q}_{\mathrm{ex}})u(n_3, n; n_1)u(n_2, n; n_4) \\
&- \dot{\Pi}^{+}(n, -\boldsymbol{q}_{\mathrm{ex}})u(n_1, n; n_3)u(n_4, n; n_2) \Big\} \; .
\end{aligned}
\tag{10.113}
$$

Here, the fourth index n_4 is determined by (approximate) momentum conservation as described below Eq. (10.111). The momentum combinations appearing in the bubble integrals are

$$
\boldsymbol{q}_{\mathrm{pp}} = \boldsymbol{k}_{n_1} + \boldsymbol{k}_{n_2} \, , \qquad \boldsymbol{q}_{\mathrm{fs}} = \boldsymbol{k}_{n_3} - \boldsymbol{k}_{n_2} \, , \qquad \boldsymbol{q}_{\mathrm{ex}} = \boldsymbol{k}_{n_1} - \boldsymbol{k}_{n_3} \, .
\tag{10.114}
$$

This notation is equivalent to the notation in Eqs. (10.41a)–(10.41c) if one takes momentum conservation and the different labeling of the momenta in this section into account. In the derivation of Eq. (10.113), we have made use of the following symmetries of the coupling function (see Eqs. (6.120) and (6.128)),

$$
\begin{aligned}
U^{(4)}\left(K_1', K_2'; K_2, K_1\right) &= U^{(4)}\left(K_2', K_1'; K_1, K_2\right) \\
&= U^{(4)}\left(K_1, K_2; K_1', K_2'\right) \, ,
\end{aligned}
\tag{10.115}
$$

which allows us to order the indices such that the shifted integration momentum $\boldsymbol{k} - \boldsymbol{q}$ never appears as an index in the coupling functions. These symmetries follow from Eqs. (6.120) and (6.128). Note that for discrete couplings $u(n_1, n_2; n_3)$ that conserve momentum only approximately (so that $\boldsymbol{k}_{n_1} + \boldsymbol{k}_{n_2} - \boldsymbol{k}_{n_3} - \boldsymbol{k}_{n_4} \neq 0$), the symmetries (10.115) are in general not fulfilled as for example $u(n_1, n_2; n_3)$ describes a slightly different scattering process than $u(n_2, n_1; n_4)$. However, the pure initial Hubbard interaction $u(n_1, n_2; n_3) = U$ trivially fulfills the symmetries (10.115) for all combinations of patch indices. For strictly momentum–conserving couplings, e.g., for particle–particle processes on the Fermi surface, the symmetries hold and are conserved exactly by the flow equations (10.113).

Let us now explicitly evaluate the bubble integrals $\dot{\Pi}^{\pm}$ using a momentum cutoff. The propagators in Eq. (10.112) are then given by Eq. (10.48), which we write as

$$
G(K) = \frac{\Theta_\Lambda(\boldsymbol{k})}{i\omega - \xi_k} \quad , \quad \dot{G}(K) = -\frac{\delta_\Lambda(\boldsymbol{k})}{i\omega - \xi_k} \, ,
\tag{10.116}
$$

where $\Theta_\Lambda(\boldsymbol{k})$ is a function that suppresses modes in a shell of width Λ around the Fermi surface, and we have defined $\delta_\Lambda(\boldsymbol{k}) = -\partial_\Lambda \Theta(\boldsymbol{k})$. The frequency sum in Eq. (10.112) can be carried out via contour integration leading to

$$\frac{1}{\beta}\sum_{\omega}\left[\frac{1}{i\omega - \xi_k}\right]\left[\frac{1}{\pm i\omega - \xi_{k'}}\right] = \pm\frac{f(\xi_k) - f(\pm\xi_{k'})}{\xi_k \mp \xi_{k'}} = \mp\frac{\Theta(\mp\xi_k\xi_{k'})}{|\xi_k| + |\xi_{k'}|} \,.$$

$$(10.117)$$

Here, $f(x) = 1/[e^{\beta x} + 1]$ is the Fermi function and the last equality is valid for $T = 0$. We thus have

$$\dot{\Pi}^{\pm}(n, q) = -\Lambda \int_{k \in K_n} \frac{f(\xi_k) - f(\pm\xi_{k-q})}{\xi_k - \xi_{k-q}} \delta_\Lambda(k)\Theta_\Lambda(k - q)$$

$$= \Lambda \int_{k \in K_n} \frac{\Theta(\mp\xi_k\xi_{k-q})}{|\xi_k| + |\xi_{k-q}|} \delta_\Lambda(k)\Theta_\Lambda(k - q) \,, \qquad (10.118)$$

where the last line is again valid for $T = 0$. For the rest of this chapter, we will work with a sharp energy cutoff, i.e., we set

$$\Theta_\Lambda(k) = \Theta(\Lambda < |\xi_k| < \Lambda_0) \,, \qquad (10.119)$$

where $\Theta(X) = 1$ if the logical expression X is true and $\Theta(X) = 0$ if X is wrong. The derivative of this cutoff is $\delta_\Lambda(k) = \delta(|\xi_k| - \Lambda)$. For $T = 0$, Eq. (10.118) can then be simplified further leading to

$$\dot{\Pi}^s(n, q) = \sum_{s'=\pm} \int_{k \in K_n} \frac{\delta(\xi_k - s'\Lambda)\Theta(-ss'\xi_{k-q} - \Lambda)}{1 - ss'\xi_{k-q}/\Lambda} \,, \qquad (10.120)$$

where $s = \pm$ labels the two types of bubbles ($s = -$ for particle–particle and $s = +$ for particle–hole).

10.5.2 Spinless Fermions

To obtain FRG flow equations for spinless fermions, we may simply drop the spin indices in our general FRG flow equation (10.37) for the effective interaction. For a numerical treatment, this equation can be discretized as in the previous section. We thus replace[9]

$$\Gamma^{(4)}(K_1, K_2; K_3, K_4) \longrightarrow \gamma(n_1, n_2; n_3) \,, \qquad (10.121)$$

and obtain from Eq. (10.37) the following system of coupled ordinary differential equations,

---

[9] Note that for spinless fermions the discretized interactions $\gamma(n_1, n_2; n_3)$ have different symmetries than the corresponding interactions $u(n_1, n_2; n_3)$ introduced in Eq. (10.110) for spinful fermions with $SU(2)$-invariance.

$$\partial_l \gamma(n_1, n_2; n_3) = -\Lambda \partial_\Lambda \gamma(n_1, n_2; n_3)$$

$$= -\sum_n \Big\{ \dot{\Pi}^-(n, \boldsymbol{q}_{\mathrm{pp}}) \gamma(n_1, n_2; n) \gamma(n_4, n_3; n)$$

$$+ \dot{\Pi}^+(n, \boldsymbol{q}_{\mathrm{fs}}) \gamma(n, n_4; n_1) \gamma(n, n_2; n_3)$$
$$+ \dot{\Pi}^+(n, -\boldsymbol{q}_{\mathrm{fs}}) \gamma(n, n_1; n_4) \gamma(n, n_3; n_2)$$
$$- \dot{\Pi}^+(n, \boldsymbol{q}_{\mathrm{ex}}) \gamma(n, n_3; n_1) \gamma(n, n_2; n_4)$$
$$- \dot{\Pi}^+(n, -\boldsymbol{q}_{\mathrm{ex}}) \gamma(n, n_1; n_3) \gamma(n, n_4; n_2) \Big\} . \qquad (10.122)$$

Here, the definitions of the bubble integrals and the associated momenta $\boldsymbol{q}$ are identical to the spinful case in the previous section. To derive Eq. (10.122), we have again used the symmetries in Eq. (10.115) which also hold for $\Gamma^{(4)}$. Additionally, $\Gamma^{(4)}$ is also antisymmetric with respect to an exchange of the ingoing or outgoing momenta,

$$\Gamma^{(4)}\left(K_1', K_2'; K_2, K_1\right) = -\Gamma^{(4)}\left(K_2', K_1'; K_2, K_1\right) = -\Gamma^{(4)}\left(K_1', K_2'; K_1, K_2\right) .$$
$$(10.123)$$

Note that the antisymmetry with respect to an exchange of the first two indices also holds for all discrete couplings $\gamma(n_1, n_2; n_3)$ and is preserved during the flow, whereas the relation $\gamma(n_1, n_2; n_3) = -\gamma(n_1, n_2; n_4)$ is only valid for couplings that strictly conserve momentum.

The simplest application of Eq. (10.122) is to a one-dimensional chain of spinless fermions. In this case, the patch index n can take the two values $n = R = +$ and $n = L = -$ for patches around the right and left Fermi points. Due to the antisymmetry, only a single nonvanishing coupling g needs to be considered,

$$\pi v_F g = \gamma(R, L; L) = \gamma(L, R; R) = -\gamma(L, R; L) = -\gamma(R, L; R) . \qquad (10.124)$$

Carrying out the summation over the intermediate patch index n explicitly, Eq. (10.122) then reduces to

$$\pi v_F \dot{g} = -\Lambda \partial_\Lambda \gamma(R, L; L)$$
$$= -\dot{\Pi}^-(R, 0) \gamma(R, L; R) \gamma(R, L; R) - \dot{\Pi}^-(L, 0) \gamma(R, L; L) \gamma(R, L; L)$$
$$+ \dot{\Pi}^+(R, 2k_F) \gamma(R, L; R) \gamma(R, L; R) + \dot{\Pi}^+(L, -2k_F) \gamma(L, R; L) \gamma(L, R; L)$$
$$= [\dot{\Pi}^+(R, 2k_F) + \dot{\Pi}^+(R, -2k_F) - \dot{\Pi}^+(R, 0) - \dot{\Pi}^+(R, 0)](\pi v_F g)^2$$
$$= 0 , \qquad (10.125)$$

where we have dropped terms containing the vanishing couplings $\gamma(R, R; R)$, $\gamma(L, L; L)$, $\gamma(R, R; L)$, or $\gamma(L, L; R)$. For the last equality in Eq. (10.125), we have used the result

$$\dot{\Pi}^+(R, 2k_F) = \dot{\Pi}^+(L, -2k_F) = \dot{\Pi}^+(R, 0) = \dot{\Pi}^+(L, 0) = \frac{1}{4\pi v_F} , \qquad (10.126)$$

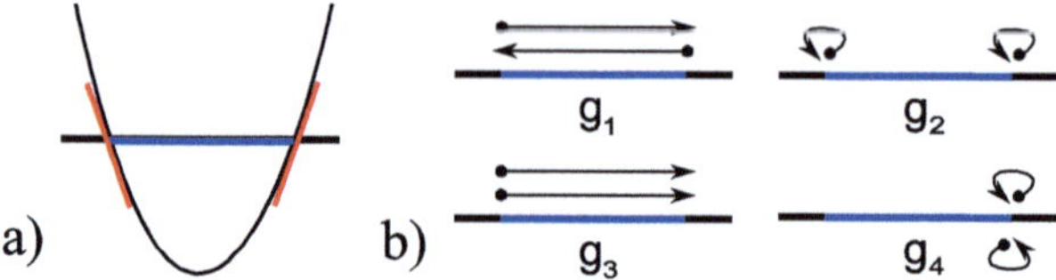

Fig. 10.8 (a) Linearization of the energy dispersion of a one-dimensional chain around the two Fermi points. The *parabola* represents the full energy dispersion while the *red lines* represents its linear approximation. The *thick horizontal line* is the Fermi energy. (b) Classification of (lattice) momentum-conserving two-particle interaction processes in g-ology notation. The *arrows* denote the transfered momentum of the two particles

which is valid for a linearized dispersion relation around the Fermi points as will be shown in detail in the next section. In one spatial dimension, the coupling g thus remains marginal in the one-loop approximation and we obtain a line of fixed points for arbitrary small g. Hence, the system remains a conductor, a so-called Luttinger liquid, at small finite interaction strength. Note that this behavior is due to cancellations of particle–particle and particle–hole contributions in Eq. (10.125) which, taken individually, would lead to a runaway flow to strong coupling corresponding to a charge-density-wave or pairing instability. In higher dimensions, this cancellation does not occur any longer, and charge-density waves or paired states can form at an arbitrary small interaction strength (Shankar 1994).

10.5.3 One-Dimensional g-ology for SU(2)-Invariant Models

An example of a system of interacting spinful fermions with a nontrivial RG flow that can be solved analytically is a single one-dimensional chain. The patch index n can again take the two values $n = R = +$ and $n = L = -$ for patches around the right and left Fermi points, respectively, as shown in Fig. 10.8a. The $2^3 = 8$ couplings $u(n_1, n_2; n_3)$ are related in pairs by the symmetries in Eq. (10.115). Following the standard notation in the literature (Sólyom 1979), we thus define the couplings g_1, g_2, g_3, and g_4 as follows,

$$\pi v_F g_1 = u(L, R; L) = u(R, L; R) \,, \tag{10.127a}$$

$$\pi v_F g_2 = u(L, R; R) = u(R, L; L) \,, \tag{10.127b}$$

$$\pi v_F g_3 = u(L, L; R) = u(R, R; L) \,, \tag{10.127c}$$

$$\pi v_F g_4 = u(L, L; L) = u(R, R; R) \,. \tag{10.127d}$$

Note, that the coupling g_3 represents Umklapp scattering, which does not conserve momentum in a strict sense, but can conserve lattice momentum for a lattice model at a commensurate band filling with $2k_F$ equal to a vector of the reciprocal lattice. Here, we will only consider the asymptotic low energy regime, where we may drop the coupling g_3 altogether if the filling of the lattice is incommensurate with the reciprocal lattice.

For sufficiently weak initial interactions, the couplings are only weakly renormalized during the initial stages of the flow and we can analyze the asymptotic behavior by starting with a small ultraviolet cutoff $\Lambda_0 \ll \epsilon_F$. We may then linearize the dispersion relation around the Fermi points,

$$\xi_k \approx n v_F (k - n k_F) \,, \tag{10.128}$$

where v_F is the Fermi velocity and we have assumed that k is in the vicinity of the Fermi point $n k_F$. Recall that in one dimension the patch index assumes only the two values $n = \pm$ which label the left and right Fermi points. The momentum $s(k - q)$ on the second leg of the particle–particle ($s = -$) or particle–hole ($s = +$) bubble also has to be close to a Fermi point which we will denote by $n' k_F$. The momentum q is then given by $q = (n - sn')k_F$ and we may also linearize ξ_{k-q},

$$\xi_{(k-q)} = \xi_{s(k-q)} \approx s n' v_F (k - n k_F) \,. \tag{10.129}$$

With these preparations, the integrations in Eq. (10.120) can be carried out analytically,

$$\dot{\Pi}^s(n, (n - sn')k_F) \approx \sum_{s'=\pm} \int_{-\Lambda_0}^{\Lambda_0} \frac{d\epsilon}{2\pi v_F} \frac{\delta(n\epsilon - s'\Lambda)\Theta(-s'n'\epsilon - \Lambda)}{1 - s'n'\epsilon/\Lambda} = \frac{\delta_{n',-n}}{4\pi v_F} \,, \tag{10.130}$$

where we have used the Morris–Lemma Eq. (7.15) in the last equality, since the occurring product of the δ- and Θ-function is only defined via the limit from a smooth to a progressively sharper cutoff. For the momentum combination occurring in Eq. (10.113), we thus have,

$$\dot{\Pi}^-(n, q_{\mathrm{pp}}) = \dot{\Pi}^-(n, (n_1 + n_2)k_F) = \frac{\delta_{n_2,-n_1}}{4\pi v_F} \,, \tag{10.131a}$$

$$\dot{\Pi}^+(n, q_{\mathrm{fs}}) = \dot{\Pi}^+(n, (n_3 - n_2)k_F) = \frac{\delta_{n_2,-n_3}\delta_{n,n_3}}{4\pi v_F} \,, \tag{10.131b}$$

$$\dot{\Pi}^+(n, -q_{\mathrm{fs}}) = \dot{\Pi}^+(n, (n_2 - n_3)k_F) = \frac{\delta_{n_2,-n_3}\delta_{n,-n_3}}{4\pi v_F} \,, \tag{10.131c}$$

$$\dot{\Pi}^+(n, q_{\mathrm{ex}}) = \dot{\Pi}^+(n, (n_1 - n_3)k_F) = \frac{\delta_{n_3,-n_1}\delta_{n,n_1}}{4\pi v_F} \,, \tag{10.131d}$$

$$\dot{\Pi}^+(n, -q_{\mathrm{ex}}) = \dot{\Pi}^+(n, (n_3 - n_1)k_F) = \frac{\delta_{n_3,-n_1}\delta_{n,-n_1}}{4\pi v_F} \,. \tag{10.131e}$$

For the particular choice $n_1 = R$, $n_2 = L$, $n_3 = R$, we thus obtain the flow equation

$$
\begin{aligned}
4\pi v_F \partial_l u(R, L; R) = {}& \\
-\Big[& u(L, R; R)u(R, L; R) + u(R, L; R)u(L, R; R) \\
& + u(L, R; L)u(R, L; L) + u(R, L; L)u(L, R; L) \Big] \\
-\Big[& 2u(R, L; R)u(R, L; R) - u(L, R; R)u(R, L; R) - u(R, L; R)u(L, R; R) \Big] \\
-\Big[& 2u(L, R; L)u(L, R; L) - u(R, L; L)u(L, R; L) - u(L, R; L)u(R, L; L) \Big] \\
= {}& -4[u(R, L; R)]^2 \; .
\end{aligned}
\tag{10.132}
$$

Here, each term in square brackets corresponds to one of the channels in Eq. (10.113); note that the exchange terms in the last two sums in Eq. (10.113) do not contribute to the flow of this coupling. Note also the large amount of cancellations between the particle–particle and particle–hole terms. Proceeding in a similar way for all other combinations of external labels n_i and using the shorthand notations in Eqs. (10.127a)–(10.127d), we obtain the well-known flow equations (Sólyom 1979)

$$
\begin{aligned}
\partial_l g_1 &= -g_1^2 \, , \\
\partial_l g_2 &= -(g_1^2 - g_3^2)/2 \, , \\
\partial_l g_3 &= -g_3(g_1 - 2g_2) \, , \\
\partial_l g_4 &= 0 \, .
\end{aligned}
\tag{10.133}
$$

To discuss these equations, let us first consider a continuum model or a lattice model with a band that is not exactly half filled. In the low energy and asymptotically small coupling regime, the coupling g_3 should be dropped from Eq. (10.133) altogether. Defining the spin coupling $g_s = g_1$ and the charge coupling $g_c = g_1 - 2g_2$, the flow equations (10.133) reduce to

$$
\partial_l g_s = -g_s^2 \, , \qquad \partial_l g_c = \partial_l g_4 = 0 \, .
\tag{10.134}
$$

Here, the decoupling of the equations for spin- and charge-couplings is a manifestation of spin-charge separation of interacting electrons in one spatial dimension. However, the consequences for correlation functions can be better understood using complementary bosonization techniques (see, e.g., Sachdev 1999, Schönhammer 2003). The solution of Eq. (10.134) is given by

$$
g_c(l) = g_c(0) \, , \qquad g_4(l) = g_4(0) \, , \qquad g_s(l) = \frac{1}{l + [g_s(0)]^{-1}} \, .
\tag{10.135}
$$

For a repulsive initial spin interaction $g_s(0) > 0$, the spin coupling g_s vanishes for $l \to \infty$ and the continuum of fixed points with $g_s = 0$ parameterized by g_c and g_4 is approached. The absence of any divergence signals that no ordering occurs and the resulting state is the one-dimensional analogon of a Fermi liquid, i.e., a Luttinger liquid (Haldane 1981). In terms of the original couplings, the model at the fixed point only contains the forward scattering terms g_2 and g_4, as the back scattering g_1 flows to zero. As a consequence, the asymptotic model can be solved exactly using bosonization techniques. Chap. 11 describes how the results from bosonization can also be obtained within the FRG framework.

On the contrary, for an attractive spin interaction $g_s(0) < 0$, the spin-coupling $g_s(l)$ diverges at a finite energy scale $l \to l_c = -[g_s(0)]^{-1}$; the weak-coupling RG approach breaks down well before this energy scale is reached. It turns out that for a particular coupling strength the model can be solved by refermionization (Luther and Emery 1974). The resulting state is a Luther–Emery liquid with a gap to spin excitations (Sólyom 1979). It has strong pairing and spin correlations although no true long range order can occur in one spatial dimension.

At half filling where the Umklapp coupling g_3 is finite, it is still useful to rewrite the flow equations in terms of spin and charge couplings,

$$\partial_l g_s = -g_s^2 , \qquad \partial_l g_c = -g_3^2 , \qquad \partial_l g_3 = -g_3 g_c . \qquad (10.136)$$

Note that the flow of g_s is still decoupled from the other couplings. For the flow of g_c and g_3, one finds the following conservation rule

$$\partial_l \left(g_3^2 - g_c^2 \right) = 0 . \qquad (10.137)$$

Thus, the flow is along the hyperbolas defined by

$$g_3^2 - g_c^2 = \text{const} , \qquad (10.138)$$

as sketched in the flow diagram in Fig. 10.9. The line $g_3 = 0$ is a line of fixed points which for $g_c > 0$ is stable against small perturbations $g_3 \neq 0$, whereas for $g_c < 0$ the fixed line is unstable and the flow diverges at a finite scale l_c along the diagonal rays with $g_c = \pm g_3 < 0$. For very small initial interaction strengths, the asymptotic diagonal rays are approached right before the weak coupling approach breaks down, suggesting a universal behavior of the system in the same basin of attraction. Thus, the asymptotics of the one-loop RG flow equations provide a classification of the occurring states or phases of the system, even though the approach itself breaks down when a runaway flow to strong coupling occurs. A description of the nature of the phases, for example, their gaps and correlation functions, is not directly accessible within the present approach.

For a one-dimensional Hubbard chain with one electron per site and with only an on-site repulsion U, the initial couplings are identical, i.e., $g_1 = g_2 = g_3 = g_4 \propto U$. The RG flow is thus toward $g_1(l) \to 0$ and along the ray $g_3(l) \sim -g_c(l) \to \infty$ shown as a red line in Fig. 10.9. The divergence of the charge coupling and the vanishing spin coupling signals a gap in the charge sector and no gap in the spin sector. In the strong coupling limit, the half-filled Hubbard model can be mapped onto an antiferromagnetic Heisenberg spin-chain; the exact solution of this model

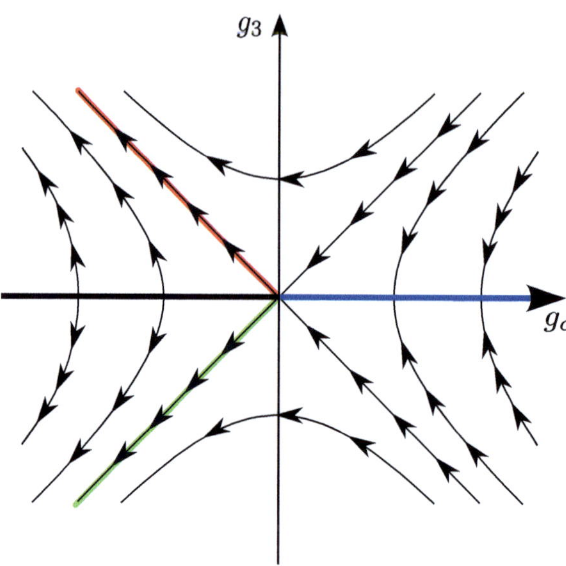

Fig. 10.9 Schematic RG flow of the charge coupling g_c and the Umklapp coupling g_3 for a one-dimensional chain with a half-filled single band. The flow of the spin coupling g_s is independent of the flow shown here. Note that the flow is formally identical with the RG flow of the two-dimensional classical XY-model shown in Fig. 3.9 if we replace $g_c \to x = \pi \beta J - 2$ and $g_3 \to y = 2\pi e^{-\beta E_c}$

via the Bethe Ansatz shows that there is indeed no gap to spin excitations. For other choices of microscopic couplings, the flow starts at a different position in the flow diagram in Fig. 10.9. As an example, consider a one-dimensional Hubbard chain with on-site as well as nearest neighbor repulsion. At half filling, the g-ology couplings are then given by

$$g_1 = u - v \,, \qquad g_2 = u + v \,, \qquad g_3 = u - v \,, \qquad g_4 = u + v \,, \qquad (10.139)$$

and thus

$$g_s = g_1 = u - v \,, \qquad g_c = g_1 - 2g_2 = -u - 3v \,. \qquad (10.140)$$

Here, u and v denote the strength of the on-site and nearest-neighbor repulsion in appropriate units, respectively. When only a repulsive nearest-neighbor coupling is present, i.e., $u = 0$ and $v > 0$, the asymptotic flow is then toward $g_1(l) \to -\infty$ and $g_3(l) \sim g_c(l) \to -\infty$, which is marked by a green line in Fig. 10.9. From a microscopic perspective and in the strong coupling limit, it is straightforward to see that the coupling v leads to the formation of a commensurate charge density wave with every other lattice site doubly occupied and gapped spin or charge excitations.

10.5.4 Many-Patch RG for the Square-Lattice Hubbard Model

Let us now apply the one loop patching approximation to the two-dimensional Hubbard model with nearest-neighbor hopping t and next-nearest-neighbor hopping t'. The single-particle dispersion relation is given by

$$\epsilon_k = -2t[\cos(k_x a) + \cos(k_y a)] - 4t'\cos(k_x a)\cos(k_y a) , \qquad (10.141)$$

and is shown graphically in Figs. 10.10a and 10.10b. At $t' = 0$ and $\mu = 0$, the Fermi surface is a perfect square touching the boundary of the first Brillouin zone at the points $ka = (\pm\pi, 0)$ and $ka = (0, \pm\pi)$ which are saddle points of ϵ_k and lead to logarithmic Van Hove singularities in the density of states. The perfect square is also called the Umklapp surface, since the parallel sides of the square are shifted by a vector Q with $Qa = (\pi, \pi)$. As Q is just half of a reciprocal lattice vector, this geometry allows for Umklapp scatterings with all four legs on the Fermi surface. An example of such a coupling is shown graphically in Fig. 10.10d. Apart from this Umklapp property, the square Fermi surface is also perfectly nested, since a translation by the vector Q maps the Fermi surface onto itself (up to shifts by a reciprocal lattice vector). The band filling n can be read off graphically as the ratio of the surface area enclosed by the Fermi surface to the area of the complete Brillouin zone. Thus the square Fermi surface for $\mu = 0$ corresponds to a half-filled band. Away from half filling or for $t' \neq 0$, the nesting is only approximate and Umklapp processes exactly on the Fermi surface occur only for isolated points at the intersection of the Fermi surface with the Umklapp surface as shown in Fig. 10.10b. However, for Fermi surface geometries slightly away from the perfect square, Umklapp and nesting processes are expected to still be important at higher energy scales.

Some aspects of this two-dimensional dispersion relation can be addressed by applying the RG to simplified toy models with only a very limited number of patches. A two-patch model containing the regions around the saddle points has been used by several authors (Lederer et al. 1987, Furukawa et al. 1998). Due to the Van Hove singularities in the density of states, the bare susceptibilities contain $\log^2$-singularities and, consequently, the bubble integrals $\Pi^\pm$ have a contribution proportional to the RG time l. The relative importance of nesting depends on the strength of the nearest-neighbor hopping t', since only for $t' = 0$ the Fermi surface is perfectly nested. A solution of the RG equations shows strong tendencies to pairing with a d-wave symmetry. However, due to nesting a gap can also appear in the charge sector, leading not to a superconducting state, but rather to a spin-liquid phase, similar to the d-Mott phase for the half-filled two-leg ladder (Balents and Fisher 1996). The role of nesting and Umklapp scatterings for regions away from the saddle points has been analyzed by placing patches around the diagonals of the Brillouin zone, i.e., near the points $ka = (\pm\pi, \pm\pi)$ (Houghton and Marston 1993, Furukawa and Rice 1998). It has been found that Umklapp scattering can lead to the opening of a charge gap which, however, does not necessarily lead to long-range magnetic ordering, but can also imply a spin-liquid behavior.

These toy models contribute important insights to a qualitative understanding of the low-energy behavior of two-dimensional fermionic lattice models. However, to properly analyze the interplay of the different instabilities, the continuous nature of the Fermi surface needs to be captured. This can be achieved by successively increasing the number of patches in a numerical treatment of Eq. (10.113). The spirit of such an approach is very similar to the one-dimensional model discussed in Sect. 10.5.3, yet details of the approximation scheme differ as discussed in the following. The focus is not so much on asymptotic low-energy scales and vanishing coupling strength, but rather on crossover phenomena at a finite coupling strength starting from the full bandwidth of a given lattice model.

To subdivide the Brillouin zone into the patches used in Eq. (10.113), one usually first defines the projection of an arbitary point k in the Brillouin zone onto the Fermi surface. This can be done by simply determining the intersection of the radial section between the origin and k itself with the Fermi surface (Tsai and Marston 2001). A little more elaborate projection that also respects the particle–hole symmetry for $t' = 0$ and $\mu = 0$ uses a line composed of two straight sections, from the origin to the Umklapp surface and from there to the nearest corner of the Brillouin zone (Zanchi and Schulz 2000, Honerkamp and Salmhofer 2001). Here, we use a more general scheme and project along a line following the path of steepest descent (ascent) whose tangent points in the direction of the gradient of the dispersion (Halboth and Metzner 2000). An example of such a projection is shown in Fig. 10.10c. The original point k now belongs to the patch whose central point is closest to the projection of k according to an appropriate measure. Here, we simply place the central points at a constant angular distance on the Fermi surface and use the Euclidean distance between the projection and the central point as a measure. In

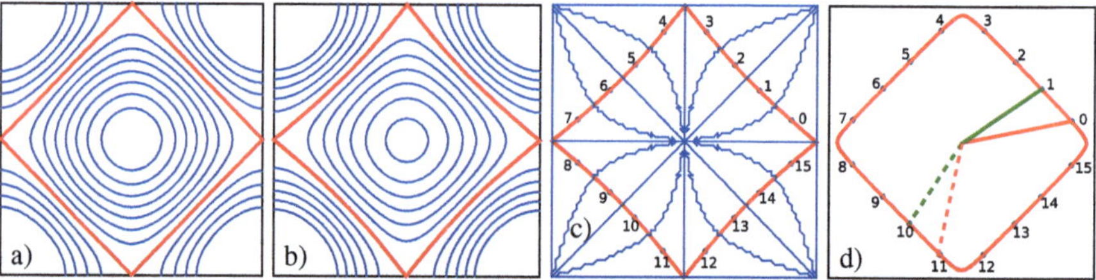

Fig. 10.10 Fermi surface and patching scheme for the square-lattice Hubbard model. (**a**) Dispersion relation for $t' = 0$. The *red square* is the Fermi surface at $\mu = 0$ (also called Umklapp surface); the *blue lines* are lines of constant energy; moving outward the energy is increasing in steps of $0.5t$. (**b**) Same as (a) for $t' = -0.1t$. The red line is for Van Hove filling $\mu = 4t'$. (**c**) Subdivision of the Brillouin zone into 16 patches for $t' = -0.1t$ and $\mu = 4t'$. The central points of the patches marked by *small circles* are placed at regular angular intervals, an arbitrary point is projected onto the Fermi surface along the direction of the gradient of the dispersion and belongs to the patch whose central point is closest to the projection. (**d**) Graphical representation of the Umklapp coupling $u(n_1 = 0, n_2 = 1; n_3 = 10)$ with all four legs on the Fermi surface so that $n_4 = 11$. Legs for outgoing particles are marked by *solid lines* and incoming particles by *dashed lines*. The color coding marks pairs of incoming and outgoing legs which belong together: $(n_2, n_3) = green, (n_1, n_4) = red$

Fig. 10.10c we show 16 patches for $t' = -0.1t$ at Van Hove filling. To determine the shape of the patches, we have divided the Brillouin zone into a triangular mesh. The center of each triangle is then projected onto the Fermi surface to determine to which patch it belongs to. The union of all such triangles leading to the same central point $\boldsymbol{k}_n$ then forms the patch n. The outer boundaries of these regions shown in Fig. 10.10c are a little raggedy due to the finite size of the underlying triangles. The flow equations (10.113) are then solved numerically with a standard Runge–Kutta algorithm with adaptive step size control (Press et al. 2007). At each step, the integrals $\dot{\Pi}^{\pm}(n, \boldsymbol{q})$ in Eq. (10.118) for particle–hole and particle–particle bubbles have to be determined numerically. We use a sharp energy cutoff

$$\Theta_\Lambda(\boldsymbol{k}) = \Theta(|\xi_{\boldsymbol{k}}| - \Lambda) , \qquad (10.142)$$

leading to $\delta_\Lambda(\boldsymbol{k}) = \delta(|\xi_{\boldsymbol{k}}| - \Lambda)$. This δ-function effectively reduces our expression (10.118) for the single-scale bubbles $\dot{\Pi}^{\pm}$ to integrals along lines of constant energy $\pm\Lambda$ in momentum space. We evaluate these numerically with an algorithm designed for contour plotting (Bourke 1987) using the triangular mesh mentioned above. Tabulating the energies at the corners of the triangles once at the beginning, we can easily determine the triangles with edges that intersect the lines at the energies $\pm\Lambda$. Line searches along these edges then yield the points at the given energy. The line of constant energy is then replaced by straight lines inside the triangles. For the Hubbard model, the initial condition for the couplings is given by $u(n_1, n_2; n_3) = U$ for all combinations of n_i. The integration is started at the full band with, i.e. $\Lambda = B = 8t$ for $t' = 0$. All following results are for $U = t$, i.e., $U/B = 1/8$.

Results of the numerical integration for the perfectly nested Fermi surface are shown in Fig. 10.11. At a finite renormalization group time l_c corresponding to a cutoff scale Λ_c, one encounters a runaway flow where some of the couplings diverge. As the truncation of the hierarchy of RG equations is based on a weak

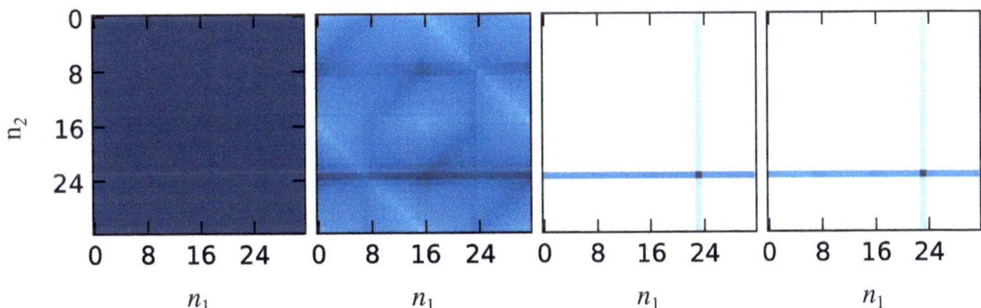

Fig. 10.11 Renormalization group flow for the exactly half-filled Hubbard model on a square lattice with nearest-neighbor hopping t and on-site repulsion U with $U/t = 1$. The coupling strength $u(n_1, n_2; n_3)$ for fixed $n_3 = 0$ is color coded as a function of n_1 and n_2. *Blue* denotes positive couplings; *red* would denotes negative couplings, but does not occur here. From left to right the renormalization group time is given by $l = 0.00, 2.42, 5.28, 5.33$. Results are shown for a Brillouin zone divided into 32 patches

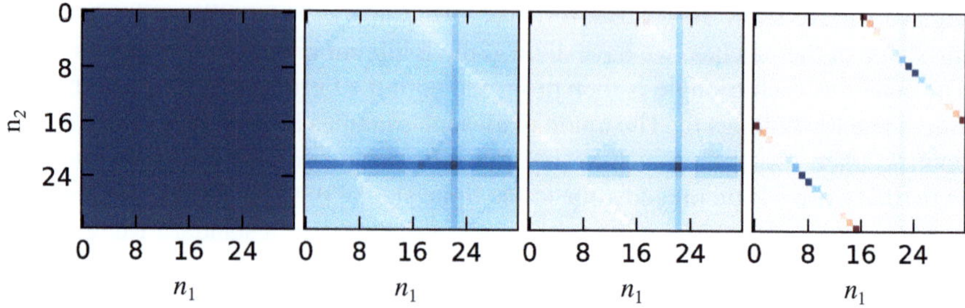

Fig. 10.12 Same as Fig. 10.11 slightly away from half filling for $\mu/t = -0.03$. From left to right, the RG time is given by $l = 0.00, 4.90, 9.56, 11.86$

coupling expansion, the approach breaks down as the couplings become large and the integration needs to be stopped. However, the nature of the diverging couplings can hint at the dominant correlations and the symmetry of the order parameter. Following Honerkamp et al. (2001), in Fig. 10.11 the relative strength of the couplings $u(n_1, n_2; n_3)$ is color coded as a function of n_1 and n_2 with n_3 fixed at $n_3 = 0$. As the flow progresses and we approach the critical cutoff scale, a cross-like pattern emerges in Fig. 10.11. The dominant couplings that form this cross are scatterings of the Umklapp and nesting type. It is well known that at half filling and strong coupling, the Hubbard model can be mapped onto the spin-$\frac{1}{2}$ Heisenberg model with nearest-neighbor antiferromagnetic exchange interaction $J \approx 4t^2/U$. On a square lattice the latter model has an ordered Néel ground state which suggests that the flow to strong coupling of the Umklapp and nesting interactions should be taken as an indication for such an ordered state.

In Fig. 10.12 the flow of the couplings is shown for a band that is slightly less than half filled ($t' = 0$). Initially, a cross-like pattern emerges again which is, however, superseded by a diagonal pattern consisting of pairing interactions with $\boldsymbol{k}_{n_2} = -\boldsymbol{k}_{n_1}$ at later stages in the flow. Note that the pairing interactions occur with different signs. Moving along the diagonal pattern in Fig. 10.12 the sign changes four times as $\boldsymbol{k}_{n_1}$ crosses the diagonals of the Brillouin zone. This suggests an ordered ground state of the BCS type with a d-wave symmetry and nodes in the order parameter along the diagonals of the Brillouin zone ($d_{x^2-y^2}$ symmetry). Thus a repulsive microscopic interaction can lead to strong pairing correlations or even an ordered state with d-wave symmetry. This is reminiscent of the Luttinger–Kohn mechanism that leads to pairing in higher angular momentum channels via feedback from higher energy scales. However, for the Hubbard model near half filling the RG flow is particularly strong due to Van Hove points, so that nesting and Umklapp scatterings lead to much higher critical energy scales.

In Fig. 10.13, we show the qualitative evolution of the strong coupling pattern as a function of the band filling. The flow is stopped at a critical scale $\Lambda_c = \Lambda_0 e^{-l_c}$, which we define by the condition that the largest flowing coupling reaches $160t$, i.e., twenty times the total band width of $8t$. One observes a crossover between a regime at and very close to half filling (where Umklapp and nesting interactions diverge) to

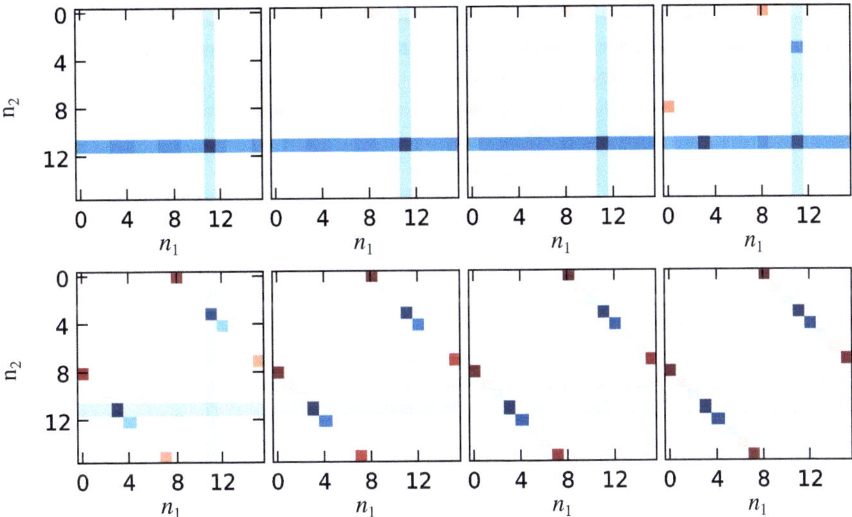

Fig. 10.13 Pattern of coupling strength color-coded as in Figs. 10.11 and 10.12 at the RG scale when the maximum occurring magnitude of the coupling strength reaches $160t$. Results are shown for 16 patches and from upper left to lower right are for $-\mu/t = 0.0, 0.005, 0.01, 0.015, 0.02, 0.025, 0.03, 0.035$

a regime with dominant pairing interactions. In the crossover regime, both types of couplings remain strong.

In Fig. 10.14, we show the dependence of the critical scale Λ_c on the chemical potential. The critical scale Λ_c is often interpreted as a critical temperature. With this in mind, Fig. 10.14 shows some qualitative similarities with the phase diagrams of high-T_c cuprate materials. At half filling there is antiferromagnetic order with an s-wave gap at the Fermi surface signaling an insulating state. Upon doping away from half filling by introducing holes into the system, the critical scale is reduced

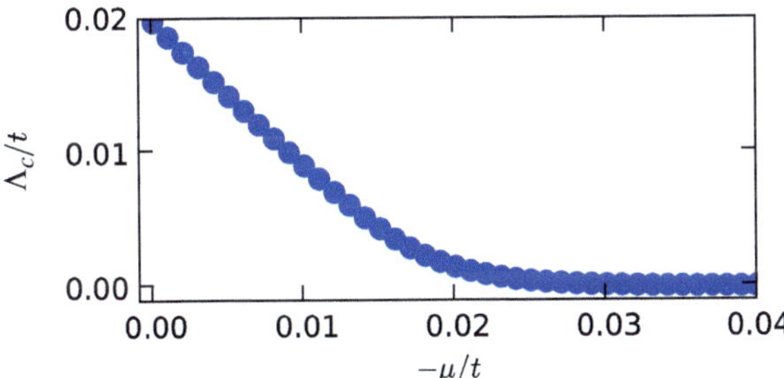

Fig. 10.14 Critical scale Λ_c as a function of the chemical potential for a Hubbard model on the square lattice with the parameters $t' = 0$ and $U/t = 1$. Numerically, the critical scale is determined as $\Lambda_c = \Lambda_0 e^{-l_c}$, where l_c is the renormalization group time at which the maximum magnitude of any of the couplings $u(n_1, n_2; n_3)$ reaches $160t$

and a d-wave paired state emerges. With further doping the critical scale eventually drops to zero. It is also tempting to associate the crossover regime with the mysterious pseudogap phase, although no real physical picture for this regime emerges from the RG analysis. Note that besides the global scale the width of the crossover in Fig. 10.14 also depends on the coupling strength U and vanishes for $U \to 0$.

References

Balents, L. and M. P. A. Fisher (1996), *Weak-coupling phase diagram of the two-chain Hubbard model*, Phys. Rev. B **53**, 12133.

Bourke, P. (1987), CONREC: *A contouring subroutine*, Byte: The Small Systems Journal **12**, 143, http://local.wasp.uwa.edu.au/_pbourke/papers/conrec.

Brazovskii, S. A. (1975), *Phase transition of an isotropic system to the heterogeneous state*, Zh. Eksp. Teor. Fiz **68**, 175.

Busche, T., L. Bartosch, and P. Kopietz (2002), *Dynamic scaling in the vicinity of the Luttinger liquid fixed point*, J. Phys.: Condens. Matter **14**, 8513.

Chitov, G. Y. and D. Sénéchal (1998), *Fermi liquid as a renormalization group fixed point: The role of interference in the Landau channel*, Phys. Rev. B **57**, 1444.

Dupuis, N. and G. Y. Chitov (1996), *Renormalization group approach to Fermi liquid theory*, Phys. Rev. B **54**, 3040.

Feldman, J., M. Salmhofer, and E. Trubowitz (1996), *Pertubation theory around non-nested Fermi surfaces I. Keeping the Fermi surface fixed*, J. Stat. Phys. **84**, 1209.

Fetter, A. L. and J. D. Walecka (1971), *Quantum Theory of Many Particle Systems*, McGraw-Hill, New York.

Furukawa, N. and T. M. Rice (1998), *Instability of a Landau Fermi liquid as the Mott insulator is approached*, J. Phys. Cond. Mat. **10**, L381.

Furukawa, N., T. M. Rice, and M. Salmhofer (1998), *Truncation of a two-dimensional Fermi surface due to quasiparticle gap formation at the saddle points*, Phys. Rev. Lett. **81**, 3195.

Galitskii, V. (1958), *The energy spectrum of a non-ideal Fermi gas*, Sov. Phys. JETP **7**, 104.

Halboth, C. J. and W. Metzner (2000), *Renormalization group analysis of the two-dimensional Hubbard model*, Phys. Rev. B **61**, 7364.

Haldane, F. D. M. (1981), *'Luttinger liquid theory' of one-dimensional quantum fluids. I. Properties of the Luttinger model and their extension to the general 1D interacting spinless Fermi gas*, J. Phys. C: Solid State Phys. **14**, 2585.

Hohenberg, P. C. and J. B. Swift (1995), *Metastability in fluctuation-driven first-order transitions: Nucleation of lamellar phases*, Phys. Rev. E **52**, 1828.

Honerkamp, C. and M. Salmhofer (2001), *Temperature-flow renormalization group and the competition between superconductivity and ferromagnetism*, Phys. Rev. B **64**, 184516.

Honerkamp, C., M. Salmhofer, N. Furukawa, and T. M. Rice (2001), *Breakdown of the Landau-Fermi liquid in two dimensions due to umklapp scattering*, Phys. Rev. B **63**, 035109.

Houghton, A. and J. B. Marston (1993), *Bosonization and fermion liquids in dimensions greater than one*, Phys. Rev. B **48**, 7790.

Katanin, A. A. (2009), *The two-loop functional renormalization group approach to the one- and two-dimensional Hubbard model*, Phys. Rev. B **79**, 235119.

Kopietz, P. and T. Busche (2001), *Exact renormalization group flow equations for nonrelativistic fermions: Scaling toward the Fermi surface*, Phys. Rev. B **64**, 155101.

Lederer, P., G. Montambaux, and D. Poilblanc (1987), *Antiferromagnetism and superconductivity in a quasi two-dimensional electron gas. Scaling theory of a generic Hubbard model*, J. Physique **48**, 1613.

Ledowski, S. and P. Kopietz (2003), *An exact integral equation for the renormalized Fermi surface*, J. Phys.: Condens. Matter **15**, 4779.

Ledowski, S. and P. Kopietz (2007), *Fermi surface renormalization and confinement in two coupled metallic chains*, Phys. Rev. B **75**, 045134.

Ledowski, S., P. Kopietz, and A. Ferraz (2005), *Self-consistent Fermi surface renormalization of two coupled Luttinger liquids*, Phys. Rev. B **71**, 235106.

Lifshitz, E. M. and L. P. Pitaevskii (1980), *Statistical Physics, Part 2*, Pergamon Press, Oxford.

Luther, A. and V. J. Emery (1974), *Backward scattering in the one-dimensional electron gas*, Phys. Rev. Lett. **33**, 589.

Luttinger, J. M. (1960), *Fermi surface and some simple equilibrium properties of a system of interacting fermions*, Phys. Rev. **119**, 1153.

Metzner, W. (2005), *Functional renormalization group computation of interacting Fermi systems*, Prog. Theor. Phys. Suppl. **160**, 58.

Metzner, W., C. Castellani, and C. Di Castro (1998), *Fermi systems with strong forward interaction*, Adv. Phys. **47**, 317.

Negele, J. W. and H. Orland (1988), *Quantum Many-Particle Systems*, Addison-Wesley, Redwood City.

Nozières, P. (1964), *Theory of Interacting Fermi Systems*, Benjamin, New York.

Pines, D. and P. Nozières (1966), *The Theory of Quantum Liquids, Vol I: Normal Fermi Liquids*, Benjamin, New York.

Press, W. H., S. A. Teukolsky, W. T. Vetterling, and B. P. Flannery (2007), *Numerical Recipes: The Art of Scientific Computing*, 3rd ed., Cambridge University Press, New York.

Sachdev, S. (1999), *Quantum Phase Transitions*, Cambridge University Press, Cambridge.

Sakurai, J. J. (1994), *Modern Quantum Mechanics*, Addison-Wesley, Reading, MA.

Salmhofer, M. (1998), *Continuous renormalization for fermions and Fermi liquid theory*, Comm. Math. Phys. **194**, 249.

Salmhofer, M. and C. Honerkamp (2001), *Fermionic renormalization group flows*, Prog. Theor. Phys. **105**, 1.

Sauli, F. and P. Kopietz (2006), *Low-density expansion for the two-dimensional electron gas*, Phys. Rev. B **74**, 193106.

Schönhammer, K. (2003), *Luttinger Liquids: The Basic Concepts*, in D. Baeriswyl and L. De Giorgi, editors, *Strong Interactions in Low Dimensions*, Kluwer Academic Publishers, Berlin.

Shankar, R. (1994), *Renormalization group approach to interacting fermions*, Rev. Mod. Phys. **66**, 129.

Shiwa, Y. (2006), *Exact renormalization group for the Brazovskii model of striped patterns*, J. Stat. Phys. **124**, 1207.

Sólyom, J. (1979), *The Fermi gas model of one-dimensional conductors*, Adv. Phys. **28**, 201.

Sudakov, V. V. (1956), *Vertex parts at very high energies in quantum electrodynamics*, Sov. Phys. JETP **3**, 65.

Tam, K.-M., S.-W. Tsai, and D. K. Campbell (2006), *Functional renormalization group analysis of the half-filled one-dimensional extended Hubbard model*, Phys. Rev. Lett. **96**, 036408.

Tsai, S.-W. and J. B. Marston (2001), κ-$(BEDT$-$TTF)_2X$ *organic crystals: Superconducting versus anti-ferromagnetic instabilities in the Hubbard model on an anisotropic triangular lattice*, Can. J. Phys. **79**, 1463.

Zanchi, D. and H. J. Schulz (2000), *Weakly correlated electrons on a square lattice: Renormalization group theory*, Phys. Rev. B **61**, 13609.

Chapter 11
Normal Fermions: Partial Bosonization in the Forward Scattering Channel

The advantage of the one-loop patching approximation discussed in Sect. 10.5 is that this approach is completely unbiased, because it retains on equal footing the RG flow of all marginal couplings related to the four-point vertex. An obvious disadvantage of the one-loop patching approximation is the fact that it breaks down when at least one of the marginal couplings becomes large, which usually happens when the RG cutoff Λ reaches a finite scale Λ_c. For the two-dimensional Hubbard model with nearest-neighbor hopping we have explicitly calculated the scale Λ_c in Sect. 10.5.4, see Fig. 10.14. Although it seems reasonable to assume that the coupling which diverges at the smallest RG scale indicates the dominant instability of the system, there is no proof that this is always the case. It would certainly be more convincing to detect the instabilities within a method which does not break down as soon as the RG flow leaves the weak coupling regime. Note also that wave function renormalization effects which are neglected within the one-loop patching approximation usually slow down the growth of the coupling constants and in certain cases completely remove the strong-coupling instabilities predicted by the one-loop patching approximation (Ferraz 2003, Freire et al. 2005, 2008). Unfortunately, within the framework of the purely fermionic FRG it is rather difficult to systematically include two-loop corrections responsible for the wave function renormalization into the analysis of the FRG flow equations (Katanin 2009).

In this chapter we shall describe an alternative FRG approach to interacting fermions which is based on the introduction of collective bosonic fluctuations via suitable Hubbard–Stratonovich transformations. Keeping in mind that the low-lying excitations of interacting fermions consist not only of fermionic quasiparticles but also of bosonic collective excitations (Pines and Nozières 1966), it is natural to introduce bosonic fields representing the relevant collective fluctuations into the FRG equations for interacting fermions. Of course, as will be discussed in more detail in Sect. 12.6, a given fermionic interaction can be decoupled in infinitely many ways by means of Hubbard–Stratonovich transformations (Hamann 1969, Wang et al. 1969, Castellani and Di Castro 1979, Schulz 1990, Macêdo and Coutinho-Filho 1991, Dupuis 2002, 2005, Borejsza and Dupuis 2003, Bartosch et al. 2009a), so that the choice of a particular Hubbard–Stratonovich decoupling implies an assumption about the nature of the dominant collective fluctuations in the system. The partially

Kopietz, P. et al.: *Normal Fermions: Partial Bosonization in the Forward Scattering Channel*. Lect. Notes Phys. **798**, 305–326 (2010)
DOI 10.1007/978-3-642-05094-7_11

bosonized FRG for fermions introduced in this chapter is therefore not as unbiased as the purely fermionic FRG discussed in Sect. 10.5. On the other hand, for special types of interactions a certain scattering channel might be singled out already in the bare action, suggesting a Hubbard–Stratonovich decoupling of the two-body interaction in this channel. A simple truncation in the bosonic sector of the partially bosonized theory corresponds then to an infinite resummation of the interaction in fermionic language, so that in this way the strong coupling regime is accessible. Another advantage of the partially bosonized FRG is that this method is very convenient to describe the symmetry-broken phase, because the order parameter can usually be constructed in terms of the expectation value of a suitably defined Hubbard–Stratonovich field, as will be discussed in Chap. 12.

The exact hierarchy of FRG flow equations for the one-line irreducible vertices of general Bose-Fermi theories has first been derived by Schütz et al. (2005) and can be obtained as a special case of the general FRG flow equations given in Chap. 7. A related approach using partially bosonized effective actions for interacting fermions has been developed by Correia et al. (2002), who applied a gradient expansion to a functional version of the Callan-Symanzik equation, and by Wetterich (2007), who truncated the exact FRG flow equation for the generating functional of the irreducible vertices using the derivative expansion. For recent applications of this method see (Baier et al. 2004, 2005, Ledowski and Kopietz 2007, Strack et al. 2008, Floerchinger et al. 2008, Jakubczyk et al. 2008, Bartosch et al. 2009a,b).

We consider in this chapter a system of fermions which interact via long-range density–density forces. The Euclidean action of our model is of the form (10.1), with the Gaussian action $S_0[\bar{\psi}, \psi]$ given in Eq. (10.2) and the interaction is now

$$S_1[\bar{\psi}, \psi] = \frac{1}{2} \sum_{\sigma\sigma'} \int_{\bar{K}} f_{\bar{k}}^{\sigma\sigma'} \bar{\rho}_{\bar{K}\sigma} \rho_{\bar{K}\sigma'} \,, \tag{11.1}$$

where the composite field $\rho_{\bar{K}\sigma}$ represents the Fourier components of the density,

$$\rho_{\bar{K}\sigma} = \int_K \bar{\psi}_{K\sigma} \psi_{K+\bar{K},\sigma} \,. \tag{11.2}$$

Note that the $\rho_{\bar{K}\sigma}$ have the symmetry $\bar{\rho}_{\bar{K}\sigma} = \rho_{-\bar{K}\sigma}$. The discrete flavor index σ is formally written as a spin projection, but includes all other flavor labels. Throughout this chapter, labels with a bar such as $\bar{K} = (i\bar{\omega}, \bar{k})$ refer to bosonic frequencies and momenta, while labels without a bar refer to fermionic ones. We assume that the momentum-dependent interaction parameters $f_{\bar{k}}^{\sigma\sigma'}$ are dominated by small momentum transfers $\bar{k}$ (forward scattering), so that $f_{\bar{k}}^{\sigma\sigma'}$ is negligibly small as soon as $|\bar{k}|$ exceeds a certain characteristic scale $q_c \ll k_F$. In real space the corresponding density–density interaction is then long range with characteristic length scale q_c^{-1}.

11.1 Hubbard–Stratonovich Transformation in the Forward Scattering Channel

For bare interactions whose Fourier transform is strongly enhanced for small momentum transfers, the effective interaction is strongly renormalized by particle-hole forward scattering processes. It is then natural to decouple the density–density interaction by means of a real Hubbard–Stratonovich field $\varphi_{\bar{K}\sigma}$ which couples to the Fourier components $\rho_{\bar{K}\sigma}$ of the density. As already discussed in Sect. 2.2.1, Hubbard–Stratonovich transformations are based on the formula (2.25) for multidimensional Gaussian integrals and its complex analogue. Setting $s = -iy$ in Eq. (2.25), we have

$$\left(\prod_{i=1}^{N} \int_{-\infty}^{\infty} \frac{dx_i}{\sqrt{2\pi}} \right) e^{-\frac{1}{2}x^T \mathbf{A} x - i y^T x} = [\det \mathbf{A}]^{-1/2} e^{-\frac{1}{2} y^T \mathbf{A}^{-1} y} \ . \tag{11.3}$$

Let us also give the complex version of this identity,

$$\left(\prod_{i=1}^{N} \int_{-\infty}^{\infty} \frac{d\mathrm{Re}z_i \, d\mathrm{Im}z_i}{\pi} \right) e^{-z^\dagger \mathbf{A} z - i a^\dagger z - i z^\dagger b} = [\det \mathbf{A}]^{-1} e^{-a^\dagger \mathbf{A}^{-1} b} \ , \tag{11.4}$$

where z, a and b are complex N-component vectors. We now read the identity (11.4) from right to left to decouple the density–density interaction (11.1) in terms of a real Hubbard–Stratonovich field φ_σ, whose Fourier components satisfy $\varphi_{-\bar{K}\sigma} = \varphi_{\bar{K}\sigma}^*$. To avoid double counting of the components with labels $\bar{K}$ and $-\bar{K}$, we write the interaction (11.1) in terms of just one of them denoted by $\bar{K} > 0$ and treat the $\bar{K} = 0$ term separately,

$$S_1[\bar{\psi}, \psi] = \sum_{\sigma\sigma'} \int_{\bar{K}>0} f_{\bar{k}}^{\sigma\sigma'} \bar{\rho}_{\bar{K}\sigma} \rho_{\bar{K}\sigma'} + \frac{1}{2\beta V} \sum_{\sigma\sigma'} f_0^{\sigma\sigma'} \bar{\rho}_{0\sigma} \rho_{0\sigma'} \ . \tag{11.5}$$

The first term can now be decoupled using Eq. (11.4) once for every $\bar{K}$, while for the second term we use the real Gaussian integral (11.3). The result of both contributions can be recombined to write the interaction in the form (Kopietz 1997, Schütz 2005)

$$e^{-S_1[\bar{\psi}, \psi]} = e^{-\frac{1}{2}\sum_{\sigma\sigma'} \int_{\bar{K}} f_{\bar{k}}^{\sigma\sigma'} \bar{\rho}_{\bar{K}\sigma} \rho_{\bar{K}\sigma'}} = \frac{\int \mathcal{D}[\varphi] e^{-S_0[\varphi] - S_1[\bar{\psi}, \psi, \varphi]}}{\int \mathcal{D}[\varphi] e^{-S_0[\varphi]}} \ , \tag{11.6}$$

where the free bosonic part is given by

$$S_0[\varphi] = \frac{1}{2} \sum_{\sigma\sigma'} \int_{\bar{K}} \left[f_{\bar{k}}^{-1} \right]^{\sigma\sigma'} \varphi_{\bar{K}\sigma}^* \varphi_{\bar{K}\sigma'} \ , \tag{11.7}$$

and the coupling between Fermi and Bose fields is

$$S_1[\bar{\psi}, \psi, \varphi] = i \sum_\sigma \int_{\bar{K}} \bar{\rho}_{\bar{K}\sigma} \varphi_{\bar{K}\sigma} = i \sum_\sigma \int_K \int_{\bar{K}} \bar{\psi}_{K+\bar{K},\sigma} \psi_{K\sigma} \varphi_{\bar{K}\sigma} \,. \tag{11.8}$$

The integration measure in Eq. (11.6) is

$$\mathcal{D}[\varphi] = \prod_\sigma \left[\frac{d\varphi_{0\sigma}}{\sqrt{2\pi}} \prod_{\bar{K}>0} \frac{d\mathrm{Re}\varphi_{\bar{K}\sigma}\, d\mathrm{Im}\varphi_{\bar{K}\sigma}}{\pi} \right] \,. \tag{11.9}$$

In field theory language, the real field φ represents the bosonic particle whose exchange mediates the interaction between the fermions.

Next, we rewrite the above Bose–Fermi theory in the superfield notation introduced in Chap. 6. For the ratio of the partition functions with and without interaction we write

$$\frac{\mathcal{Z}}{\mathcal{Z}_0} = \frac{\int \mathcal{D}[\bar{\psi}, \psi] e^{-S_0[\bar{\psi},\psi] - S_1[\bar{\psi},\psi]}}{\int \mathcal{D}[\bar{\psi}, \psi] e^{-S_0[\bar{\psi},\psi]}}$$

$$= \frac{\int \mathcal{D}[\bar{\psi}, \psi, \varphi] e^{-S_0[\bar{\psi},\psi] - S_0[\varphi] - S_1[\bar{\psi},\psi,\varphi]}}{\int \mathcal{D}[\bar{\psi}, \psi, \varphi] e^{-S_0[\bar{\psi},\psi] - S_0[\varphi]}} \equiv \frac{\int \mathcal{D}[\Phi] e^{-S[\Phi]}}{\int \mathcal{D}[\Phi] e^{-S_0[\Phi]}} \,, \tag{11.10}$$

where $\Phi = [\psi_\sigma, \bar{\psi}_\sigma, \varphi_\sigma]$ is a three-component superfield with one bosonic and two fermionic components,[1] and the superfield action is

$$S[\Phi] = S_0[\Phi] + S_1[\Phi] = S_0[\bar{\psi}, \psi] + S_0[\varphi] + S_1[\bar{\psi}, \psi, \varphi] \,. \tag{11.11}$$

As in Eq. (6.3), we write the quadratic part $S_0[\Phi]$ of our action in the symmetric form

$$S_0[\Phi] = S_0[\bar{\psi}, \psi] + S_0[\varphi] = -\frac{1}{2} \left(\Phi, [\mathbf{G}_0]^{-1} \Phi \right)$$

$$= -\frac{1}{2} \int_\alpha \int_{\alpha'} \Phi_\alpha [\mathbf{G}_0]^{-1}_{\alpha\alpha'} \Phi_{\alpha'} \,, \tag{11.12}$$

where $\mathbf{G}_0$ is now a matrix in frequency, momentum, spin, and field-type indices, and α is a "superfield label" for all of these indices, as explained after Eq. (6.3) in Chap. 6. For our theory, the matrix $\mathbf{G}_0^{-1}$ has the block structure

---

[1] It is understood that each component of $\Phi = [\psi_\sigma, \bar{\psi}_\sigma, \varphi_\sigma]$ consists of several flavors labeled by σ.

$$\mathbf{G}_0^{-1} = \begin{pmatrix} 0 & \zeta[\hat{G}_0^{-1}]^T & 0 \\ \hat{G}_0^{-1} & 0 & 0 \\ 0 & 0 & -\hat{F}_0^{-1} \end{pmatrix} , \tag{11.13}$$

where $\zeta = -1$ and $\hat{G}_0$ and $\hat{F}_0$ are infinite matrices in frequency, momentum, and spin space, with matrix elements

$$[\hat{G}_0]_{K\sigma,K'\sigma'} = \delta_{K,K'}\delta_{\sigma\sigma'}G_{0,\sigma}(K) , \tag{11.14a}$$

$$[\hat{F}_0]_{\bar{K}\sigma,\bar{K}'\sigma'} = \delta_{\bar{K}+\bar{K}',0}F_{0,\sigma\sigma'}(\bar{K}) , \tag{11.14b}$$

where

$$G_{0,\sigma}(K) = [i\omega - \xi_{k\sigma}]^{-1} , \tag{11.15a}$$

$$F_{0,\sigma\sigma'}(\bar{K}) = f_{\bar{k}}^{\sigma\sigma'} . \tag{11.15b}$$

Note that the bare interaction plays the role of a free bosonic Green function. The inverse of the matrix in Eq. (11.13) is the free superfield propagator,

$$\mathbf{G}_0 = \begin{pmatrix} 0 & \hat{G}_0 & 0 \\ \zeta\hat{G}_0^T & 0 & 0 \\ 0 & 0 & -\hat{F}_0 \end{pmatrix} , \tag{11.16}$$

which satisfies $\mathbf{G}_0^T = \mathbf{Z}\mathbf{G}_0 = \mathbf{G}_0\mathbf{Z}$, in agreement with Eq. (6.4). The superfield self-energy $\boldsymbol{\Sigma}$ is related to the exact superfield propagator $\mathbf{G}$ via the Dyson equation (6.34) and contains the fermionic irreducible self-energy $\Sigma_\sigma(K)$ and the one-interaction-line irreducible polarization $\Pi_\sigma(\bar{K})$ in the following blocks,

$$\boldsymbol{\Sigma} = \begin{pmatrix} 0 & \zeta[\hat{\Sigma}]^T & 0 \\ \hat{\Sigma} & 0 & 0 \\ 0 & 0 & \hat{\Pi} \end{pmatrix} , \tag{11.17}$$

where

$$[\hat{\Sigma}]_{K\sigma,K'\sigma'} = \delta_{K,K'}\delta_{\sigma\sigma'}\Sigma_\sigma(K') , \tag{11.18a}$$

$$[\hat{\Pi}]_{\bar{K}\sigma,\bar{K}'\sigma'} = \delta_{\bar{K}+\bar{K}',0}\delta_{\sigma\sigma'}\,\Pi_\sigma(\bar{K}') . \tag{11.18b}$$

These matrices are flavor-diagonal because the bare coupling $S_1[\bar{\psi}, \psi, \varphi]$ between Fermi and Bose fields in Eq. (11.8) is diagonal in the flavor index σ. The exact superfield Green function $\mathbf{G} = \left[\mathbf{G}_0^{-1} - \boldsymbol{\Sigma}\right]^{-1}$ has then the same block structure as the free propagator (11.16),

$$\mathbf{G} = \begin{pmatrix} 0 & \hat{G} & 0 \\ \zeta\hat{G}^T & 0 & 0 \\ 0 & 0 & -\hat{F} \end{pmatrix} , \tag{11.19}$$

where the blocks contain the exact single-particle Green function $G_\sigma(K) = [G_{0,\sigma}^{-1}(K) - \Sigma_\sigma(K)]^{-1}$ and the effective (screened) interaction $F_{\sigma\sigma'}(\bar{K})$,

$$[\hat{G}]_{K\sigma,K'\sigma'} = \left[\hat{G}_0^{-1} - \hat{\Sigma}\right]_{K\sigma,K'\sigma'}^{-1} = \delta_{K,K'}\delta_{\sigma\sigma'}G_\sigma(K)\,, \qquad (11.20a)$$

$$[\hat{F}]_{\bar{K}\sigma,\bar{K}'\sigma'} = \left[\hat{F}_0^{-1} + \hat{\Pi}\right]_{\bar{K}\sigma,\bar{K}'\sigma'}^{-1} = \delta_{\bar{K}+\bar{K}',0}\,F_{\sigma\sigma'}(\bar{K})\,. \qquad (11.20b)$$

11.2 Exact FRG Flow Equations

To derive exact FRG flow equations for the vertices of our mixed Bose–Fermi theory given above, we introduce a cutoff Λ into the Gaussian propagator, $\mathbf{G}_0 \to \mathbf{G}_{0,\Lambda}$, as discussed in Sect. 7.1. Since our theory contains both bosonic and fermionic fields, we have the freedom of introducing the cutoff into both bosonic and fermionic sectors, or only into one of them. In the latter case the structure of the FRG flow equations simplifies due to the absence of single-scale propagators associated with the field without cutoff, but we should impose a nontrivial initial condition on the FRG flow (Schütz and Kopietz 2006). In practice, it can be advantageous to introduce a cutoff only in the bosonic sector of the theory (Schütz et al. 2005, Ledowski and Kopietz 2007, Bartosch et al. 2009a,b). Since in the model considered here the bosonic field φ mediates the interaction between the fermions, the reduction of a momentum cutoff in the corresponding bosonic propagator amounts to the elimination of scattering processes involving large energy–momentum transfers. We therefore refer to this cutoff procedure as the *momentum transfer cutoff scheme*.

To begin with, let us first write down the exact FRG flow equations for a general cutoff procedure and subsequently discuss the simplifications in the momentum transfer cutoff scheme. The action given in Eqs. (11.8), (11.11), and (11.12) is a special case of the general class of actions considered in Chap. 7, so that the FRG flow equations of our model can be obtained from the flow equations derived there. For simplicity, we ignore the renormalization of the vacuum expectation value[2] of the bosonic field φ, so that we may use the FRG flow equations without vacuum expectation values given in Sect. 7.3. To classify the various diagrams, it is useful to switch from the general notation used in Chap. 7 to a more explicit notation which exhibits the different field types. We define the partially symmetrized flowing irreducible vertices $\Gamma_\Lambda^{(2n,m)}$ with $2n$ external fermion legs and m external boson legs by writing the functional Taylor expansion of the generating functional $\Gamma[\bar{\psi}, \psi, \varphi]$ of the irreducible vertices of our cutoff-dependent theory in the form

---

[2] The vacuum expectation value of our Hubbard–Stratonovich field φ is related to the Hartree correction to the fermionic self-energy, which can be eliminated by a shift of the chemical potential. Alternatively, we may assume that $f_{\bar{k}=0}^{\sigma\sigma'} = 0$, implying that also the vacuum expectation value $\langle\varphi_{K=0}\rangle$ vanishes.

$$\Gamma[\bar{\psi}, \psi, \varphi] = \sum_{n,m=0}^{\infty} \frac{1}{(n!)^2 m!} \int_{K_1'\sigma_1'} \cdots \int_{K_n'\sigma_n'} \int_{K_1\sigma_1} \cdots \int_{K_n\sigma_n} \int_{\bar{K}_1\bar{\sigma}_1} \cdots \int_{\bar{K}_m\bar{\sigma}_m}$$

$$\times \delta_{K_1'+\ldots+K_n', K_1+\ldots+K_n+\bar{K}_1+\ldots+\bar{K}_m}$$

$$\times \Gamma_\Lambda^{(2n,m)} \left(K_1'\sigma_1', \ldots, K_n'\sigma_n'; K_n\sigma_n, \ldots, K_1\sigma_1; \bar{K}_1\bar{\sigma}_1, \ldots, \bar{K}_m\bar{\sigma}_m \right)$$

$$\times \bar{\psi}_{K_1'\sigma_1'} \cdots \cdots \bar{\psi}_{K_n'\sigma_n'} \psi_{K_n\sigma_n} \cdots \cdots \psi_{K_1\sigma_1} \varphi_{\bar{K}_1\bar{\sigma}_1} \cdots \cdots \varphi_{\bar{K}_m\bar{\sigma}_m} .$$

$$(11.21)$$

Because our theory is characterized by three types of fields, we represent the vertices $\Gamma_\Lambda^{(2n,m)}$ graphically by triangles whose sides are associated with the different field types, as shown in Fig. 11.1. Note that in Eq. (11.21) we have factored out the momentum and frequency conserving δ-functions in the definition of the vertices $\Gamma_\Lambda^{(2n,m)}$. Apart from this, the totally symmetric vertices defined by the expansion (6.60) coincide with the corresponding partially symmetrized momentum conserving ones in Eq. (11.21) for the same order of the indices, see Eq. (10.29). As a consequence, we can obtain the flow equations for the vertices $\Gamma_\Lambda^{(2n,m)}$ by choosing in the corresponding flow equations for the completely symmetrized vertices derived in Sect. 7.3 a definite realization of the external legs and by carrying out the intermediate sums over the different field species, i.e., by drawing all possible lines in the intermediate loop (two possible orientations of solid lines or one wiggly line). On the right-hand side one then has to appropriately order all the legs on the vertices keeping track of signs for the interchange of two neighboring fermion legs. Having done so, we can use the pictorial dictionary in Fig. 11.1 to obtain diagrams involving the partially symmetrized vertices $\Gamma^{(2n,m)}$ appearing in Eq. (11.21). In this way, we obtain from the diagram representing the FRG flow of the completely symmetric two-point vertex shown in Fig. 7.2 the corresponding diagrams for the fermionic self-energy in Fig. 11.2 as well as for the irreducible polarization shown in Fig. 11.3.

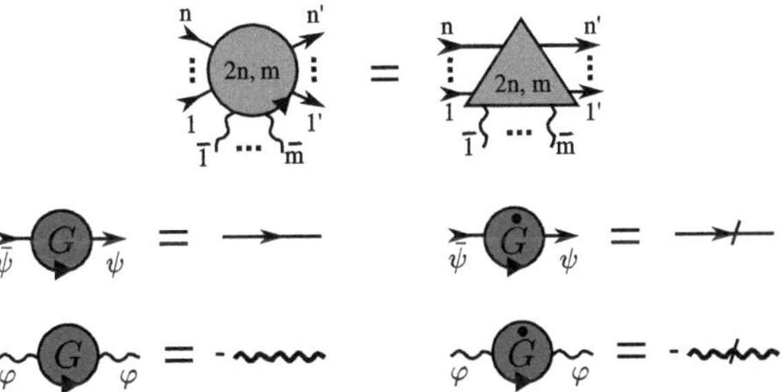

Fig. 11.1 Graphical notation for the partially symmetrized vertices $\Gamma_\Lambda^{(2n,m)}$ with $2n$ fermion legs and m boson legs defined via Eq. (11.21). We also show the corresponding vertex in the superfield notation used in Chaps. 6 and 7. The diagrams on the right-hand sides of the last two lines represent the exact propagators G, $\dot{G}$, F and $\dot{F}$ respectively. The wavy line is associated with the bosonic field φ, while for fermions we use the same symbols as in Fig. 10.1. Recall that our Bose field is real because it couples to the density, so that it should be represented graphically by an undirected line

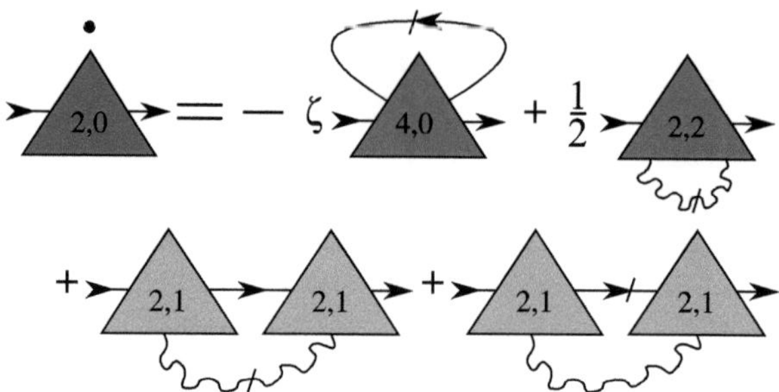

Fig. 11.2 Flow of the irreducible fermionic self-energy. The diagrams are obtained from the superfield diagrams shown in Fig. 7.2 by specifying the external legs to be one outgoing and one incoming fermion leg

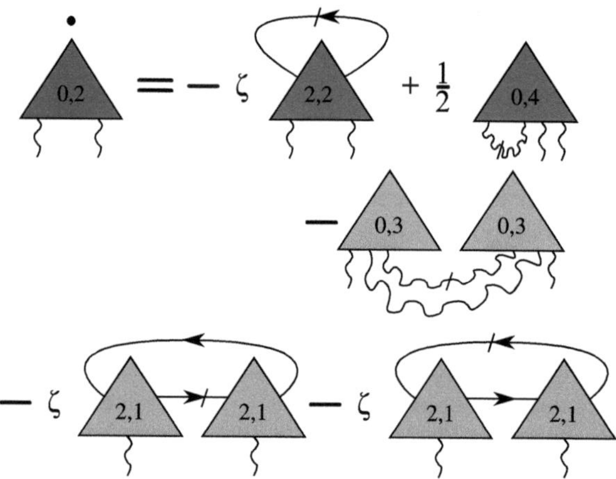

Fig. 11.3 Flow of the irreducible polarization, obtained from the superfield diagrams in Fig. 7.2 by setting both external legs equal to boson legs. Note that each closed fermion loop gives rise to an additional factor of $\zeta = -1$

Moreover, if we specify the external legs in the diagram for the totally symmetrized three-legged vertex shown in Fig. 7.3 to be two fermion legs and one boson leg, we obtain the diagram representing the FRG flow equation for the three-legged vertex shown in Fig. 11.4. Obviously all diagrams shown in Figs. 11.2, 11.3, and 11.4 can be subdivided into two classes: those involving fermionic single-scale propagators (where the slash appears on internal fermion lines), and those with bosonic single-scale propagators (where the slash appears on internal boson lines).

Let us now adopt the momentum transfer cutoff scheme, where only the bosonic propagator is regularized via a momentum transfer cutoff. Then all diagrams in Figs. 11.2, 11.3, and 11.4 involving slashed fermion lines should be simply omitted. Explicitly, the exact FRG flow equations for the fermionic self-energy $\Sigma_\sigma(K)$ and the irreducible polarization $\Pi_\sigma(\bar{K})$ are then

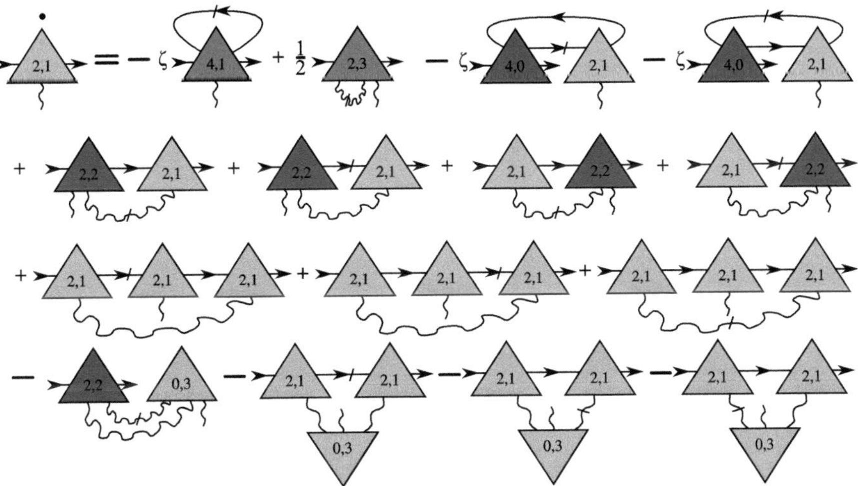

Fig. 11.4 Flow of the three-legged vertex with two fermion legs and one boson leg, obtained as a special case of the diagram in Fig. 7.3

$$\partial_\Lambda \Sigma_\sigma(K) = \frac{1}{2} \int_{\bar{K}} \dot{F}_{\sigma\sigma}(\bar{K}) \Gamma^{(2,2)}(K\sigma; K\sigma; \bar{K}\sigma, -\bar{K}\sigma)$$

$$+ \int_{\bar{K}} \dot{F}_{\sigma\sigma}(\bar{K}) G_\sigma(K + \bar{K}) \Gamma^{(2,1)}(K + \bar{K}\sigma; K\sigma; \bar{K}\sigma)$$

$$\times \Gamma^{(2,1)}(K\sigma; K + \bar{K}\sigma; -\bar{K}\sigma), \tag{11.22}$$

$$\partial_\Lambda \Pi_\sigma(\bar{K}) = \frac{1}{2} \int_{\bar{K}'} \dot{F}_{\sigma\sigma}(\bar{K}') \Gamma^{(0,4)}(\bar{K}'\sigma, -\bar{K}'\sigma, \bar{K}\sigma, -\bar{K}\sigma)$$

$$- \int_{\bar{K}'} \dot{F}_{\sigma\sigma}(\bar{K}') F_{\sigma\sigma}(\bar{K} + \bar{K}') \Gamma^{(0,3)}(-\bar{K}\sigma, \bar{K} + \bar{K}'\sigma, -\bar{K}'\sigma)$$

$$\times \Gamma^{(0,3)}(\bar{K}'\sigma, -\bar{K} - \bar{K}'\sigma, \bar{K}\sigma). \tag{11.23}$$

The simpler structure of the FRG flow equations in the momentum transfer cutoff scheme comes at the price of a nontrivial initial condition: at the initial scale $\Lambda = \Lambda_0$ all interaction lines are effectively turned off while fermion propagator lines are fully functional. In addition to the bare three-legged interaction vertex, the only one-line irreducible diagrams that can be drawn under these conditions are closed loops of fermionic propagators. These loops have to be symmetrized with respect to the exchange of external bosonic legs as shown in Fig. 11.5. The initial condition for the purely bosonic vertices is therefore given by the symmetrized fermion loops,

$$\Gamma^{(0,m)}_{\Lambda_0}(\bar{K}_1\sigma, \ldots, \bar{K}_m\sigma) = \frac{i^m}{m} \sum_P \int_K G_{0,\sigma}(K) G_{0,\sigma}(K + \bar{K}_{P(1)})$$

$$\ldots G_{0,\sigma}(K + \bar{K}_{P(1)} + \ldots + \bar{K}_{P(m-1)}). \tag{11.24}$$

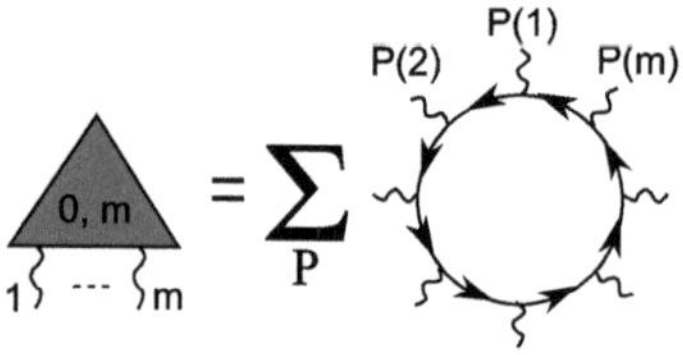

Fig. 11.5 Initial condition for the pure boson vertices in the momentum transfer cutoff scheme. The sum is taken over the $m!$ permutations of the labels of the external legs

A more formal derivation of the initial conditions in the momentum transfer cutoff scheme can be found in Schütz and Kopietz (2006). For fermions in D dimensions with a quadratic dispersion the fermion loops of the type (11.24) with $m > (D+1)$ can be reduced to the more elementary loop with $D+1$ external legs (Neumayr and Metzner 1998, 1999, Kopper and Magnen 2001, Pirooznia et al. 2008). In particular, for $D = 1$ the loops with $n > 2$ external legs can be expressed in terms of the two-loop, i.e., the noninteracting polarization. Explicit expressions for the symmetrized loops with three and four external legs for one-dimensional fermions with quadratic dispersion can be found in Pirooznia et al. (2008).

In order to assign scaling dimensions to the vertices which can be used to classify the vertices according to their relevance in the RG sense, we have to define how we rescale momenta, frequencies and fields under the RG transformation. Such a rescaling is not unique and depends on the nature of the fixed point under consideration. As discussed in Sect. 10.4 (see Eqs. (10.67), (10.76), and Fig. 10.7), in the presence of a sharp Fermi surface the fermionic momenta should be rescaled such that only the projection $\delta k_\| = \hat{\boldsymbol{v}}_F \cdot (\boldsymbol{k} - \boldsymbol{k}_F)$ is modified by the rescaling. But the component $\delta k_\|$ is defined with respect to a given point $\boldsymbol{k}_F$ on the Fermi surface, so that a global rescaling of all fermionic momenta in the vicinity of the Fermi surface leads to rather complicated geometric constructions, as discussed in Sect. 10.4.2. The rescaling problem simplifies for forward scattering problems where the maximal momentum transfer q_c is small compared with k_F, so that one can subdivide the Fermi surface into patches whose size is still larger than q_c but small compared with the typical k_F. Then the patches effectively decouple and it is sufficient to consider only a fixed reference point $\boldsymbol{k}_F$ on the Fermi surface (Schütz et al. 2005).

While the rescaling of purely bosonic vertices follows from straightforward power counting, the proper rescaling of mixed boson–fermion vertices in theories where the fermionic and bosonic sectors are characterized by different dynamic exponents is determined by the field with the largest dynamic exponent, as discussed by Schütz et al. (2005). See also the recent work by Yamamoto and Si (2009) for a detailed analysis of the rescaling problem in mixed Bose–Fermi theories. In this chapter we shall work with unrescaled FRG flow equations and thus avoid possible subtleties associated with the rescaling procedure in mixed Bose–Fermi theories.

11.3 Dyson–Schwinger and Skeleton Equations

As discussed in Sect. 6.3.3, the invariance of the functional integral under infinitesimal shifts of the integration variables implies the so-called Dyson–Schwinger equations (also called skeleton equations), which are relations between vertex functions of different order. These relations are valid for any value of the running cutoff Λ and can be used to truncate the hierarchy of flow equations (Bartosch et al. 2009a,b). For the general class of theories considered in Chaps. 6 and 7 the functional version of the Dyson–Schwinger equation is given in Eq. (6.132). For our coupled Fermi–Bose system with Euclidean action $S[\bar\psi, \psi, \varphi]$ given by Eqs. (11.8), (11.11), and (11.12) involving three types of fields Eq. (6.132) is actually equivalent with the following three equations (Schütz et al. 2005),

$$\left(J_{-\bar K\sigma} - \sum_{\sigma'} [f_{\bar k}^{-1}]^{\sigma'\sigma} \frac{\delta}{\delta J_{\bar K\sigma'}} \right) \mathcal{G} - i\zeta \int_K \frac{\delta^2 \mathcal{G}}{\delta j_{K+\bar K\sigma}\delta \bar j_{K\sigma}} = 0 \,, \quad (11.25a)$$

$$\left(\zeta \bar j_{K\sigma} + [i\omega - \xi_{k\sigma}] \frac{\delta}{\delta j_{K\sigma}} \right) \mathcal{G} - i \int_{\bar K} \frac{\delta^2 \mathcal{G}}{\delta j_{K+\bar K\sigma}\delta J_{-\bar K\sigma}} = 0 \,, \quad (11.25b)$$

$$\left(j_{K\sigma} + [i\omega - \xi_{k\sigma}] \frac{\delta}{\delta \bar j_{K\sigma}} \right) \mathcal{G} - i \int_{\bar K} \frac{\delta^2 \mathcal{G}}{\delta \bar j_{K-\bar K\sigma}\delta J_{-\bar K\sigma}} = 0 \,. \quad (11.25c)$$

Here, the sources are defined by writing

$$(J, \Phi) = (\bar j, \psi) + (\bar\psi, j) + (J^*, \varphi) =$$

$$\sum_\sigma \int_K \bar j_{K\sigma}\psi_{K\sigma} + \sum_\sigma \int_K \bar\psi_{K\sigma} j_{K\sigma} + \sum_\sigma \int_{\bar K} J^*_{\bar K\sigma}\varphi_{\bar K\sigma} \,, \quad (11.26)$$

which amounts to identifying the components of the supersource J_α by $(J_\alpha) = (\bar j, \zeta j, J^*)$. Expressing these equations in terms of the generating functionals $\mathcal{G}_c[\bar j, j, J]$ of the connected Green functions and the corresponding generating functional $\Gamma[\bar\psi, \psi, \varphi]$ of the irreducible vertices defined in Eqs. (6.19) and (6.61), we can alternatively write the Dyson–Schwinger equations in the following form,

$$\frac{\delta \Gamma}{\delta \varphi_{\bar K\sigma}} - i \int_K \left[\bar\psi_{K+\bar K,\sigma}\psi_{K\sigma} + \frac{\delta^2 \mathcal{G}_c}{\delta \bar j_{K\sigma}\delta j_{K+\bar K,\sigma}} \right] = 0 \,, \quad (11.27a)$$

$$\frac{\delta \Gamma}{\delta \psi_{K\sigma}} - i \int_{\bar K} \left[\zeta \bar\psi_{K+\bar K,\sigma}\varphi_{\bar K\sigma} + \frac{\delta^2 \mathcal{G}_c}{\delta j_{K+\bar K,\sigma}\delta J_{-\bar K\sigma}} \right] = 0 \,, \quad (11.27b)$$

$$\frac{\delta \Gamma}{\delta \bar\psi_{K\sigma}} - i \int_{\bar K} \left[\psi_{K-\bar K,\sigma}\varphi_{\bar K\sigma} + \frac{\delta^2 \mathcal{G}_c}{\delta \bar j_{K-\bar K,\sigma}\delta J_{-\bar K\sigma}} \right] = 0 \,. \quad (11.27c)$$

The second functional derivatives of $\mathcal{G}_c$ can be expressed in terms of the irreducible vertices using the tree expansion (6.82). Taking derivatives of Eqs. (11.27a),

(11.27b), and (11.27c) with respect to the fields and then setting the fields equal to zero, we obtain skeleton equations of the irreducible vertices of the underlying fermion model which are different from the skeleton equations obtained in the purely fermionic parametrization in Sect. 6.3.3. Let us begin by deriving a skeleton equation relating the irreducible self-energy to the three-legged boson-fermion vertex. To this end we simply differentiate Eq. (11.27c) with respect to $\psi_{K'\sigma}$. Using the fact that by definition

$$\left. \frac{\delta^2 \Gamma}{\delta \psi_{K'\sigma} \delta \bar{\psi}_{K\sigma}} \right|_{\text{fields}=0} = \delta_{K,K'} \Sigma_\sigma(K) \,, \tag{11.28}$$

we obtain

$$\delta_{K,K'} \Sigma_\sigma(K) = i \int_{\bar{K}} \left. \frac{\delta^3 \mathcal{G}_c}{\delta \psi_{K'\sigma} \delta \bar{J}_{K-\bar{K},\sigma} \delta J_{-\bar{K}\sigma}} \right|_{\text{fields}=0} . \tag{11.29}$$

On the other hand, from the $\nu = 1$ term in the expansion (6.82) it is easy to show that

$$\left. \frac{\delta^3 \mathcal{G}_c}{\delta \psi_{K'\sigma} \delta \bar{J}_{K-\bar{K},\sigma} \delta J_{-\bar{K}\sigma}} \right|_{\text{fields}=0}$$
$$= \delta_{K,K'} F_{\sigma\sigma}(\bar{K}) G_\sigma(K + \bar{K}) \Gamma^{(2,1)}(K + \bar{K}\sigma; K\sigma; \bar{K}\sigma) \,, \tag{11.30}$$

so that we finally obtain the skeleton equation

$$\boxed{\Sigma_\sigma(K) = i \int_{\bar{K}} F_{\sigma\sigma}(\bar{K}) G_\sigma(K + \bar{K}) \Gamma^{(2,1)}(K + \bar{K}\sigma; K\sigma; \bar{K}\sigma) \,,} \tag{11.31}$$

which is shown graphically in Fig. 11.6(a). Recall that in Sect. 6.3.3 we have derived an alternative skeleton equation for the irreducible self-energy involving the irreducible vertex with four fermionic external legs, see Eq. (6.139) and Fig. 6.7.

Similarly, we obtain the skeleton equation of the irreducible polarization by differentiating Eq. (11.27a) with respect to $\varphi_{-\bar{K}\sigma}$,

$$\boxed{\begin{aligned} \Pi_\sigma(\bar{K}) &= i \int_K \left. \frac{\delta^3 \mathcal{G}_c}{\delta \varphi_{-\bar{K}\sigma} \delta \bar{J}_{K,\sigma} \delta j_{K+\bar{K}\sigma}} \right|_{\text{fields}=0} \\ &= -i\zeta \int_K G_\sigma(K) G_\sigma(K + \bar{K}) \Gamma^{(2,1)}(K + \bar{K}\sigma; K\sigma; \bar{K}\sigma) \,, \end{aligned}} \tag{11.32}$$

which is shown diagrammatically in Fig. 11.6(b). Finally, applying the operator $\frac{\delta^2}{\delta \bar{\psi}_{K+\bar{K}\sigma} \delta \psi_{K\sigma}}$ to Eq. (11.27a) and subsequently setting the fields equal to zero we obtain the skeleton equation for the three-legged vertex shown in Fig. 11.6 (c),

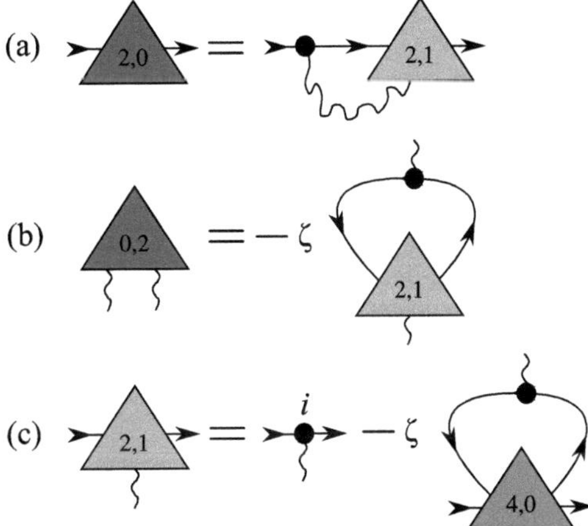

Fig. 11.6 Skeleton diagrams for (**a**) the one-particle irreducible fermionic self-energy; (**b**) the one-interaction-line irreducible polarization; and (**c**) the three-legged vertex with two fermion legs and one boson leg. The small *black circle* denotes the bare three-legged vertex. *Thin lines* denote external legs. The other graphical elements are the same as in Fig. 11.1

$$
\Gamma^{(2,1)}(K + \bar{K}\sigma; K\sigma; \bar{K}\sigma) = i
$$
$$
-i\zeta \int_{K'} G_\sigma(K')G_\sigma(K' + \bar{K})\Gamma^{(4,0)}(K + \bar{K}\sigma, K'\sigma; K' + \bar{K}\sigma, K\sigma)\,. \tag{11.33}
$$

Skeleton equations for higher-order vertices can be obtained analogously from the appropriate functional derivatives of Eqs. (11.27a), (11.27b), and (11.27c). We emphasise that the above skeleton equations are valid for any value of the running cutoff Λ, so that they can be used to close the hierarchy of FRG flow equations for theories involving both bosonic and fermionic fields (Bartosch et al. 2009a,b).

11.4 Ward Identities

Ward identities are relations between vertex functions of different order which follow from the symmetries of a given model. For normal Fermi systems with density-density interactions involving only small momentum transfers $|\boldsymbol{q}| \ll k_F$ the number of fermions with momenta in a given patch on the Fermi surface is approximately conserved. More precisely, the patch must be sufficiently small so that the variation of the corresponding Fermi velocity $\boldsymbol{v}_F$ within this patch can be neglected. Then we may associate with this patch an emergent $U(1)$-symmetry (Haldane 1992, 1994) corresponding to the conservation of the number of fermions with momenta in this patch. In one dimension, where the Fermi surface consists only of two points, the Ward identities associated with this $U(1) \times U(1)$-symmetry have been employed by Dzyaloshinskii and Larkin (1974) to calculate the Green function of the Tomonaga-Luttinger model exactly. A generalization of this approach to higher dimensions has been developed in Castellani et al. (1994) and Metzner et al. (1998). The enhanced symmetry of interacting Fermi systems with dominant forward scattering also forms the basis of the method of higher-dimensional bosonization (Haldane 1992, 1994, Houghton and Marston 1993, Houghton et al. 1994, 2000, Castro-Neto and Fradkin 1995, Kopietz et al. 1995, Kopietz and Schönhammer 1996, Kopietz 1997, Bartosch and Kopietz 1999).

The advantage of our FRG approach with momentum transfer cutoff is that it does not violate the Ward identities associated with the emergent $U(1)$-symmetries. Let us now derive these Ward identities within the framework of our functional integral approach. Consider the generating functional $\mathcal{G}[\bar{j}, j, J]$ of the Green function of our mixed Bose–Fermi theory, which according to Eqs. (6.13) and (11.10) can be written as

$$\mathcal{G}[\bar{j}, j, J] = \frac{1}{\mathcal{Z}} \int \mathcal{D}[\bar{\psi}, \psi, \varphi] e^{-S[\bar{\psi}, \psi, \varphi] + (\bar{j}, \psi) + (\bar{\psi}, j) + (J^*, \varphi)} \, , \tag{11.34}$$

where the action $S[\bar{\psi}, \psi, \varphi] \equiv S[\Phi]$ is defined in Eqs. (11.8), (11.11), and (11.12). If we rewrite the parts of the action involving the fermionic fields $\bar{\psi}$ and ψ in real space and imaginary time, we have

$$S[\bar{\psi}, \psi, \varphi] = S_0[\bar{\psi}, \psi] + S_0[\varphi] + S_1[\bar{\psi}, \psi, \varphi] \, , \tag{11.35}$$

where the bosonic part $S_0[\varphi]$ of the Gaussian action is given in Eq. (11.7), and

$$S_0[\bar{\psi}, \psi] = \sum_{\sigma} \int_X \bar{\psi}_\sigma(X) \partial_\tau \psi_\sigma(X)$$
$$+ \sum_{\sigma} \int d\tau \int d^D r \int d^D r' \, \bar{\psi}_\sigma(\tau, r) \xi_\sigma(r - r') \psi_\sigma(\tau, r') \, , \tag{11.36}$$
$$S_1[\bar{\psi}, \psi, \varphi] = i \sum_{\sigma} \int_X \bar{\psi}_\sigma(X) \psi_\sigma(X) \varphi_\sigma(X) \, . \tag{11.37}$$

Here, $X = (\tau, r)$, $\int_X = \int d\tau \int d^D r$, and the Fourier transform of the dispersion is defined by

$$\xi_\sigma(r) = \int \frac{d^D k}{(2\pi)^D} e^{ik \cdot r} \xi_{k\sigma} \, . \tag{11.38}$$

Suppose now we perform a local gauge transformation on the fermion fields, defining new fields ψ' and $\bar{\psi}'$ via

$$\psi_\sigma(X) = e^{i\alpha_\sigma(X)} \psi'_\sigma(X) \, , \quad \bar{\psi}_\sigma(X) = e^{-i\alpha_\sigma(X)} \bar{\psi}'_\sigma(X) \, , \tag{11.39}$$

where $\alpha_\sigma(X)$ is an arbitrary real function. It is easy to show that, to linear order in $\alpha_\sigma(X)$, our action $S[\bar{\psi}, \psi, \varphi]$ transforms as follows,

$$S[e^{-i\alpha}\bar{\psi}', e^{i\alpha}\psi', \varphi] = S[\bar{\psi}', \psi', \varphi] + i\sum_\sigma \int_X \bar{\psi}'_\sigma(X)[\partial_\tau \alpha_\sigma(X)]\psi'_\sigma(X)$$

$$- i\sum_\sigma \int d\tau \int d^D r \int d^D r'\, \bar{\psi}'_\sigma(\tau, \boldsymbol{r})[\alpha_\sigma(\tau, \boldsymbol{r}) - \alpha_\sigma(\tau, \boldsymbol{r}')]\xi_\sigma(\boldsymbol{r} - \boldsymbol{r}')\psi'_\sigma(\tau, \boldsymbol{r}')\,.$$

$$(11.40)$$

Using this relation, we see that the invariance of the generating functional (11.34) with respect to the change of integration variables defined by Eq. (11.39) implies, to linear order in $\alpha_\sigma(X)$,

$$\int \mathcal{D}[\bar{\psi}, \psi, \varphi] e^{-S[\bar{\psi},\psi,\varphi]+(\bar{j},\psi)+(\bar{\psi},j)+(J^*,\varphi)} \left\{ - \sum_\sigma \int_X \bar{\psi}_\sigma(X)[\partial_\tau \alpha_\sigma(X)]\psi_\sigma(X) \right.$$

$$+ \sum_\sigma \int d\tau \int d^D r \int d^D r'\, \bar{\psi}_\sigma(\tau, \boldsymbol{r})[\alpha_\sigma(\tau, \boldsymbol{r}) - \alpha_\sigma(\tau, \boldsymbol{r}')]\xi_\sigma(\boldsymbol{r} - \boldsymbol{r}')\psi_\sigma(\tau, \boldsymbol{r}')$$

$$\left. + (\bar{j}, \alpha\psi) - (\bar{\psi}\alpha, j) \right\} = 0\,.$$

$$(11.41)$$

Taking a functional derivative of this equation with respect to $\alpha_\sigma(X)$ we obtain in Fourier space

$$\int_K \left\{ [i\bar{\omega} - \xi_{k+\bar{k},\sigma} + \xi_{k\sigma}] \frac{\delta^2 \mathcal{G}}{\delta \bar{j}_{K\sigma} \delta j_{K+\bar{K}\sigma}} + \bar{j}_{K+\bar{K}\sigma} \frac{\delta \mathcal{G}}{\delta \bar{j}_{K\sigma}} - j_{K\sigma} \frac{\delta \mathcal{G}}{\delta j_{K+\bar{K}\sigma}} \right\} = 0.$$

$$(11.42)$$

Introducing the generating functional $\mathcal{G}_c$ of the connected Green functions as in Eq. (6.19), and expressing the last two terms in Eq. (11.42) in terms of the generating functional $\Gamma[\bar{\psi}, \psi, \varphi]$ of the irreducible vertices as defined in Eq. (6.61), we obtain

$$\boxed{\begin{aligned} \int_K \Big\{ &[i\bar{\omega} - \xi_{k+\bar{k},\sigma} + \xi_{k\sigma}] \frac{\delta^2 \mathcal{G}_c}{\delta \bar{j}_{K\sigma} \delta j_{K+\bar{K}\sigma}} \\ &+ \psi_{K\sigma} \frac{\delta\Gamma}{\delta\psi_{K+\bar{K}\sigma}} - \bar{\psi}_{K+\bar{K}\sigma} \frac{\delta\Gamma}{\delta\bar{\psi}_{K\sigma}} \Big\} = 0\,. \end{aligned}}$$

$$(11.43)$$

Alternatively, using the Dyson–Schwinger equation (11.27a), we may rewrite this as

$$\boxed{\begin{aligned} i\bar{\omega} &\left[\frac{\delta\Gamma}{\delta\varphi_{\bar{K}\sigma}} - i\int_K \bar{\psi}_{K+\bar{K}\sigma}\psi_{K\sigma} \right] - i\int_K (\xi_{k+\bar{k},\sigma} - \xi_{k\sigma}) \frac{\delta^2 \mathcal{G}_c}{\delta \bar{j}_{K\sigma} \delta j_{K+\bar{K}\sigma}} \\ &+ i\int_K \left[\psi_{K\sigma} \frac{\delta\Gamma}{\delta\psi_{K+\bar{K}\sigma}} - \bar{\psi}_{K+\bar{K}\sigma} \frac{\delta\Gamma}{\delta\bar{\psi}_{K\sigma}} \right] = 0\,. \end{aligned}}$$

$$(11.44)$$

From the functional Ward identities (11.43) and (11.44) we may derive Ward identities for the vertices by taking functional derivatives. For example, taking the derivative $\frac{\delta}{\delta\varphi_{-\bar{K}\sigma}}$ of Eq. (11.44) we obtain

$$i\bar{\omega}\Pi_\sigma(\bar{K}) - \Pi_\sigma^c(\bar{K}) = 0 \,, \tag{11.45}$$

where the irreducible polarization $\Pi_\sigma(\bar{K})$ has the skeleton expansion (11.32), and we have defined

$$\Pi_\sigma^c(\bar{K}) = -i\zeta \int_K (\xi_{k+\bar{k},\sigma} - \xi_{k\sigma})G_\sigma(K)G_\sigma(K+\bar{K})$$
$$\times \Gamma^{(2,1)}(K+\bar{K}\sigma; K\sigma; \bar{K}\sigma) \,. \tag{11.46}$$

Equation (11.45) is a relation between response functions, which follows more directly from the equation of continuity.

If we are interested in vertices involving at least one fermionic momentum our functional Ward identities (11.43) and (11.44) can be further simplified if we assume that the momentum transfered by the interaction is small. Then all fermionic momenta lie close to a given point $\boldsymbol{k}_{F\sigma}$ on the Fermi surface so that we may replace under the integral sign in Eqs. (11.43) and (11.44),

$$\xi_{k+\bar{k},\sigma} - \xi_{k\sigma} \to \boldsymbol{v}_{F\sigma} \cdot \bar{\boldsymbol{k}} \,. \tag{11.47}$$

This approximation amounts to the linearization of the energy dispersion relative to the point $\boldsymbol{k}_{F\sigma}$ on the Fermi surface. Using again the Dyson–Schwinger equation (11.27a), our master Ward identity (11.44) reduces to

$$(i\bar{\omega} - \boldsymbol{v}_{F\sigma} \cdot \bar{\boldsymbol{k}})\left[\frac{\delta\Gamma}{\delta\varphi_{\bar{K}\sigma}} - i\int_K \bar{\psi}_{K+\bar{K}\sigma}\psi_{K\sigma}\right]$$
$$+ i\int_K \left[\psi_{K\sigma}\frac{\delta\Gamma}{\delta\psi_{K+\bar{K}\sigma}} - \bar{\psi}_{K+\bar{K}\sigma}\frac{\delta\Gamma}{\delta\bar{\psi}_{K\sigma}}\right] = 0 \,. \tag{11.48}$$

Differentiating this simplified functional Ward identity with respect to the fields using the relation (11.28) as well as

$$\left.\frac{\delta^3\Gamma}{\delta\varphi_{\bar{K}\sigma}\delta\psi_{K\sigma}\delta\bar{\psi}_{K+\bar{K}\sigma}}\right|_{\text{fields}=0} = \Gamma^{(2,1)}(K+\bar{K}\sigma; K\sigma; \bar{K}\sigma) \,, \tag{11.49}$$

$$\left.\frac{\delta^4\Gamma}{\delta\varphi_{\bar{K}_1\sigma}\delta\varphi_{\bar{K}_2\sigma}\delta\psi_{K\sigma}\delta\bar{\psi}_{K+\bar{K}_1+\bar{K}_2\sigma}}\right|_{\text{fields}=0}$$
$$= \Gamma^{(2,2)}(K+\bar{K}_1+\bar{K}_2\sigma; K\sigma; \bar{K}_1\sigma, \bar{K}_2\sigma) \,, \tag{11.50}$$

and so on, we obtain the following Ward identities for the irreducible vertices,

$$G(K + \bar{K})\Gamma^{(2,1)}(K + \bar{K}; K; \bar{K})G(K) = \frac{-i}{i\bar{\omega} - \boldsymbol{v}_{F\sigma} \cdot \bar{\boldsymbol{k}}}\Big[G(K + \bar{K}) - G(K)\Big],$$
(11.51)

$$\begin{aligned}
&\Gamma^{(2,2)}\Big(K + \bar{K}_1 + \bar{K}_2; K; \bar{K}_1, \bar{K}_2\Big) \\
&= \frac{-i}{i\bar{\omega}_1 - \boldsymbol{v}_{F\sigma} \cdot \bar{\boldsymbol{k}}_1}\Big[\Gamma^{(2,1)}\Big(K + \bar{K}_1 + \bar{K}_2; K + \bar{K}_1; \bar{K}_2\Big) \\
&\qquad\qquad\qquad\qquad -\Gamma^{(2,1)}\Big(K + \bar{K}_2; K; \bar{K}_2\Big)\Big] \\
&= \frac{-i}{i\bar{\omega}_2 - \boldsymbol{v}_{F\sigma} \cdot \bar{\boldsymbol{k}}_2}\Big[\Gamma^{(2,1)}\Big(K + \bar{K}_1 + \bar{K}_2; K + \bar{K}_2; \bar{K}_1\Big) \\
&\qquad\qquad\qquad\qquad -\Gamma^{(2,1)}\Big(K + \bar{K}_1; K; \bar{K}_1\Big)\Big],
\end{aligned}$$
(11.52)

and for a general number $m \geq 2$ of external bosonic legs,

$$\begin{aligned}
&\Gamma^{(2,m)}\Big(K'; K; \bar{K}_1, \ldots, \bar{K}_m\Big) \\
&= \frac{-i}{i\bar{\omega}_l - \boldsymbol{v}_{F\sigma} \cdot \bar{\boldsymbol{k}}_l}\Big[\Gamma^{(2,m-1)}\Big(K'; K + \bar{K}_l; \bar{K}_1, \ldots, \bar{K}_{l-1}, \bar{K}_{l+1}, \ldots, \bar{K}_m\Big) \\
&\qquad\qquad\qquad -\Gamma^{(2,m-1)}\Big(K' - \bar{K}_l; K; \bar{K}_1, \ldots, \bar{K}_{l-1}, \bar{K}_{l+1}, \ldots, \bar{K}_m\Big)\Big],
\end{aligned}$$
(11.53)

where $1 \leq l \leq m$. In one dimension, the Ward identity (11.51) has been used by Dzyaloshinskii and Larkin (1974) to close the skeleton equation for the self-energy and thus obtain the exact Green function of the Tomonaga-Luttinger model. This strategy can also be generalized to higher dimensions if the interaction is dominated by small momentum transfers (Castellani et al. 1994, Metzner et al. 1998). Note that the Ward identities (11.51), (11.52), and (11.53) are all based on the simplified functional Ward identity (11.48) which relies on the linearization (11.47) of the fermionic energy dispersion. If this approximation is not made, we should start from the more general functional Ward identity (11.43) or (11.44). Then the Ward identities (11.51), (11.52), and (11.53) for the vertices acquire correction terms which have been studied in a mathematically rigorous way by Benfatto and Mastropietro (2005).

11.5 Exact Solution of the FRG Flow Equations for Fermions with Linear Dispersion via Ward Identities

In the momentum transfer cutoff scheme, the Ward identities (11.51) and (11.52) derived for models with linear energy dispersion are valid for any value of the running cutoff Λ. Substituting these identities into the exact FRG flow equation (11.22) for the self-energy $\Sigma_\sigma(K)$ in the momentum transfer cutoff scheme, we may eliminate the flowing vertices $\Gamma^{(2,2)}$ and $\Gamma^{(2,1)}$ on the right-hand side of Eq. (11.22) in favor of $\Sigma_\sigma(K)$ and thus obtain the following closed integro-differential equation for the flowing self-energy (Schütz et al. 2005),

$$\partial_\Lambda \Sigma_\sigma(K) = G_\sigma^{-2}(K) \int_{\bar{K}} \frac{\dot{F}_{\sigma\sigma}(\bar{K})}{(i\bar{\omega} - \boldsymbol{v}_{F\sigma} \cdot \boldsymbol{k})^2} \left[G_\sigma(K) - G_\sigma(K + \bar{K}) \right] . \quad (11.54)$$

Here, the index σ labels not only the different spin species, but also the different patches of the sectorized Fermi surface (Kopietz 1997). For example, for the spinless case in $D = 1$, $\sigma = \pm k_F$. Using the fact that in the momentum transfer cutoff scheme $G^2 \partial_\Lambda \Sigma = \partial_\Lambda G$ we can alternatively write Eq. (11.54) as a *linear* integro-differential equation for the fermionic Green function,

$$\partial_\Lambda G_\sigma(K) = \int_{\bar{K}} \frac{\dot{F}_{\sigma\sigma}(\bar{K})}{(i\bar{\omega} - \boldsymbol{v}_{F\sigma} \cdot \boldsymbol{k})^2} \left[G_\sigma(K) - G_\sigma(K + \bar{K}) \right] . \quad (11.55)$$

If we had simply set the vertex $\Gamma^{(2,2)}$ equal to zero in Eq (11.22) and had then closed this equation by means of the Ward identity (11.51), we would have obtained a nonlinear equation. Thus, the linearity of Eq. (11.55) is the result of a cancellation of nonlinear terms arising from both Ward identities (11.51) and (11.52). Because the second term on the right hand side of Eq. (11.55) is a convolution, we can easily solve this equation by means of a Fourier transformation to imaginary time and real space. Defining

$$G_\sigma(X) = \int_K e^{i(\boldsymbol{k}\cdot\boldsymbol{r} - \omega\tau)} G_\sigma(K) , \quad (11.56)$$

$$H_{\Lambda,\sigma}(X) = \int_{\bar{K}} e^{i(\bar{\boldsymbol{k}}\cdot\boldsymbol{r} - \bar{\omega}\tau)} \frac{\dot{F}_{\sigma\sigma}(\bar{K})}{(i\bar{\omega} - \boldsymbol{v}_{F\sigma} \cdot \bar{\boldsymbol{k}})^2} , \quad (11.57)$$

the flow equation (11.55) is transformed to

$$\left[\partial_\Lambda + H_{\Lambda\sigma}(X) - H_{\Lambda\sigma}(0) \right] G_\sigma(X) = 0 . \quad (11.58)$$

This implies the conservation law

$$\partial_\Lambda \left[\exp \left\{ \int_0^\Lambda d\Lambda' \left[H_{\Lambda'\sigma}(X) - H_{\Lambda'\sigma}(0) \right] \right\} G_\sigma(X) \right] = 0 . \quad (11.59)$$

Integrating from $\Lambda = 0$ to $\Lambda = \Lambda_0$, we obtain

$$G_\sigma(X) = G_{0,\sigma}(X)\, \exp\left[Q_\sigma(X)\right] , \tag{11.60}$$

with

$$Q_\sigma(X) = S_\sigma(0) - S_\sigma(X) , \tag{11.61}$$

and

$$
\begin{aligned}
S_\sigma(X) &= -\int_0^{\Lambda_0} d\Lambda'\, H_{\Lambda',\sigma}(X) \\
&= \int_{\bar{K}} \frac{\Theta(\Lambda_0 - |\bar{k}|) F_{\sigma\sigma}(\bar{K})}{(i\bar{\omega} - v_{F\sigma}\cdot\bar{k})^2}\, \cos(\bar{k}\cdot r - \bar{\omega}\tau) ,
\end{aligned}
\tag{11.62}
$$

where we have used the invariance of the effective interaction $F_{\sigma\sigma}(\bar{K})$ under $\bar{K} \to -\bar{K}$.

Another important consequence of the linearized energy dispersion considered here is the vanishing of all symmetrized closed fermion loops given in Eq. (11.24) with more than two external legs.[3] As a consequence, all interaction corrections to the irreducible polarization cancel, so that the bosonic self-energy defined in Eq. (11.18b) is simply given by the noninteracting polarization for a fixed flavor index σ,

$$\Pi_\sigma(\bar{K}) = \Pi_{0,\sigma}(\bar{K}) \equiv \zeta \int_K G_{0,\sigma}(K) G_{0,\sigma}(K + \bar{K}) . \tag{11.63}$$

The exact effective interaction $F_{\sigma\sigma'}(\bar{K})$ defined in Eq. (11.20b), which can be identified with the propagator of our Hubbard–Stratonovich field φ, is then simply given by the random phase approximation. This is most transparent in the functional integral approach to higher-dimensional bosonization developed in (Kopietz et al. 1995, Kopietz and Schönhammer 1996, Kopietz 1997), where one finds that for a linear dispersion the effective action of the Hubbard-Stratonovich field φ is Gaussian. For example, if the bare interaction $f_{\bar{k}}^{\sigma\sigma'} = f_{\bar{k}}$ is independent of the flavor labels σ and σ', this is also true for the effective interaction $F_{\sigma\sigma}(\bar{K}) = F(\bar{K})$ in Eq. (11.62), which is then explicitly given by

$$F(\bar{K}) = \frac{f_{\bar{k}}}{1 + f_{\bar{k}} \sum_\sigma \Pi_{0,\sigma}(\bar{K})} . \tag{11.64}$$

---

[3] This so-called *closed loop theorem* or *loop cancellation theorem* has been discussed by Bohr (1981) in the context of the one-dimensional Tomonaga–Luttinger model, and has later been generalized for higher-dimensional fermions with dominant forward scattering (Kopietz et al. 1995, Kopietz 1997, Metzner et al. 1998)

The solution in Eqs. (11.60), (11.61), and (11.62) is well known from the functional integral approach to bosonization (Fogedby 1976, Lee and Chen 1988, Kopietz et al. 1995, Kopietz and Schönhammer 1996, Kopietz 1997) where $Q_\sigma(X)$ arises as a Debye-Waller factor from Gaussian averaging over the distribution of the Hubbard–Stratonovich field. In one dimension, Eqs. (11.60), (11.61), and (11.62) can be shown to be equivalent to the exact solution for the Green function of the Tomonaga–Luttinger model obtained via conventional bosonization (Kopietz 1997). Once the exact single-particle Green function is known, we may substitute the result back into the Ward identities (11.51) and (11.53) and iteratively calculate the vertices $\Gamma^{(2,m)}$ with two fermion legs and an arbitrary number m of boson legs. In principle, this method can also be applied to vertices with more than two fermion legs. For example, the right-hand sides of the FRG flow equation for $\Gamma^{(4,m)}$ contain only vertices with no more than four fermion legs. Ward identities for these vertices obtained from our simplified functional Ward identity (11.48) would again yield a solution of this complete hierarchy, once the vertices $\Gamma^{(2,m)}$ are known. This procedure can be iterated to obtain vertices with an arbitrary number of external legs using at each step the complete flow of vertices with two fewer fermion legs than obtained in the previous step. We can thus obtain all correlation functions of our model within the framework of the FRG.

References

Baier, T., E. Bick, and C. Wetterich (2004), *Temperature dependence of antiferromagnetic order in the Hubbard model*, Phys. Rev. B **70**, 125111.

Baier, T., E. Bick, and C. Wetterich (2005), *Antiferromagnetic gap in the Hubbard model*, Phys. Lett. B **605**, 144.

Bartosch, L. and P. Kopietz (1999), *Correlation functions of higher dimensional Luttinger liquids*, Phys. Rev. B **59**, 5377.

Bartosch, L., H. Freire, J. J. Ramos Cardenas, and P. Kopietz (2009a), *Functional renormalization group approach to the Anderson impurity model*, J. Phys.: Condens. Matter **21**, 305602.

Bartosch, L., P. Kopietz, and A. Ferraz (2009b), *Renormalization of the BCS-BEC crossover by order parameter fluctuations*, Phys. Rev. B **80**, 104514.

Benfatto, G. and V. Mastropietro (2005), *Ward identities and chiral anomaly in the Luttinger liquid*, Commun. Math. Phys. **258**, 609.

Bohr, T. (1981), *Lectures on the Luttinger Model*, Nordita-preprint 81/4, Unpublished.

Borejsza, K. and N. Dupuis (2003), *Antiferromagnetism and single-particle properties in the two-dimensional half-filled Hubbard model: Slater vs Mott-Heisenberg*, Europhys. Lett. **63**, 722.

Castellani, C. and C. Di Castro (1979), *Arbitrariness and symmetry properties of the functional formulation of the hubbard hamiltonian*, Phys. Lett. **70A**, 37.

Castellani, C., C. Di Castro, and W. Metzner (1994), *Dimensional crossover from Fermi to Luttinger liquid*, Phys. Rev. Lett. **72**, 316.

Castro Neto, A. H. and E. H. Fradkin (1995), *Exact solution of the Landau fixed point via bosonization*, Phys. Rev. B **51**, 4084.

Correia, S., J. Polonyi, and J. Richert (2002), *The functional Callan-Symanzik equation for the Coulomb gas*, Ann. Phys. (New York) **296**, 214.

Dupuis, N. (2002), *Spin fluctuations and pseudogap in the two-dimensional half-filled Hubbard model at weak coupling*, Phys. Rev. B **65**, 245118.

Dupuis, N. (2005), *Effective action for superfluid Fermi systems in the strong coupling limit*, Phys. Rev. A **72**, 013606.

Dzyaloshinskii, I. E. and A. I. Larkin (1974), *Correlation functions for a one-dimensional Fermi system with long-range interaction (Tomonaga model)*, Sov. Phys. JETP **38**, 202.

Ferraz, A. (2003), *Non-Fermi liquid in a truncated two-dimensional Fermi surface*, Phys. Rev. B **68**, 075115.

Floerchinger, S., M. Scherer, D. S., and C. Wetterich (2008), *Particle-hole fluctuations in the BCS-BEC crossover*, Phys. Rev. B **78**, 174528.

Fogedby, H. C. (1976), *Correlation functions for the Tomonaga model*, J. Phys. C: Solid State Phys. **9**, 3757.

Freire, H., E. Corrêa, and A. Ferraz (2005), *Field-theoretical renormalization group for a flat two-dimensional Fermi surface*, Phys. Rev. B **71**, 165113.

Freire, H., E. Corrêa, and A. Ferraz (2008), *Breakdown of the Fermi-liquid regime in the 2D Hubbard model from a two-loop field-theoretical renormalization group approach*, Phys. Rev. B **78**, 125114.

Haldane, F. D. M. (1992), *Varenna Lectures (1992)*, Helv. Phys. Acta **65**, 152.

Haldane, F. D. M. (1994), *Luttinger's Theorem and Bosonization of the Fermi surface*, in J. R. Schrieffer and A. Broglia, editors, *Proceedings of the International School of Physics "Enrico Fermi"*, Course 121, North Holland, New York.

Hamann, D. R. (1969), *Fluctuation theory of dilute magnetic alloys*, Phys. Rev. Lett. **23**, 95.

Houghton, A. and J. B. Marston (1993), *Bosonization and fermion liquids in dimensions greater than one*, Phys. Rev. B **48**, 7790.

Houghton, A., H. J. Kwon, and J. B. Marston (1994), *Stability and single-particle properties of bosonized Fermi liquids*, Phys. Rev. B **50**, 1351.

Houghton, A., H. J. Kwon, and J. B. Marston (2000), *Multidimensional bosonization*, Adv. Phys. **49**.

Jakubczyk, P., P. Strack, A. A. Katanin, and W. Metzner (2008), *Renormalization group for phases with broken discrete symmetry near quantum critical points*, Phys. Rev. B **77**, 195120.

Katanin, A. A. (2009), *The two-loop functional renormalization group approach to the one- and two-dimensional Hubbard model*, Phys. Rev. B **79**, 235119.

Kopietz, P. (1997), *Bosonization of Interacting Fermions in Arbitrary Dimensions*, Springer, Berlin.

Kopietz, P. and K. Schönhammer (1996), *Functional bosonization of interacting fermions in arbitrary dimensions*, Z. Phys. B **100**, 259.

Kopietz, P., J. Hermisson, and K. Schönhammer (1995), *Bosonization of interacting fermions in arbitrary dimension beyond the Gaussian approximation*, Phys. Rev. B **52**, 10877.

Kopper, C. and J. Magnen (2001), *Singularity cancellation in fermion loops through Ward identities*, in *Annales Henri Poincaré*, volume 2, pages 513–524, Springer.

Ledowski, S. and P. Kopietz (2007), *Fermi surface renormalization and confinement in two coupled metallic chains*, Phys. Rev. B **75**, 045134.

Lee, D. K. K. and Y. Chen (1988), *Functional bosonisation of the Tomonaga-Luttinger model*, J. Phys. A: Math. Gen. **21**, 4155.

Macêdo, C. A. and M. D. Coutinho-Filho (1991), *Hubbard model: Functional integral approach and diagrammatic perturbation theory*, Phys. Rev. B **43**, 13515.

Metzner, W., C. Castellani, and C. Di Castro (1998), *Fermi systems with strong forward scattering*, Adv. Phys. **47**, 317.

Neumayr, A. and W. Metzner (1998), *Fermion loops, loop cancellation, and density correlations in two-dimensional Fermi systems*, Phys. Rev. B **58**, 15449.

Neumayr, A. and W. Metzner (1999), *Reduction formula for fermion loops and density correlations of the 1D Fermi gas*, J. Stat. Phys. **96**, 613.

Pines, D. and P. Nozières (1966), *The Theory of Quantum Liquids, Vol I: Normal Fermi Liquids*, Benjamin, New York.

Pirooznia, P., F. Schütz, and P. Kopietz (2008), *Dynamic structure factor of Luttinger liquids with quadratic energy dispersion and long-range interactions*, Phys. Rev. B **78**, 075111.

Schulz, H. J. (1990), *Effective action for strongly correlated fermions from functional integrals*, Phys. Rev. Lett. **65**, 2462.

Schütz, F. (2005), *Aspects of Strong Correlations in Low Dimensions*, Doktorarbeit, Goethe-Universität Frankfurt.

Schütz, F. and P. Kopietz (2006), *Functional renormalization group with vacuum expectation values and spontaneous symmetry breaking*, J. Phys. A: Math. Gen. **39**, 8205.

Schütz, F., L. Bartosch, and P. Kopietz (2005), *Collective fields in the functional renormalization group for fermions, Ward identities, and the exact solution of the Tomonaga-Luttinger model*, Phys. Rev. B **72**, 035107.

Strack, P., R. Gersch, and W. Metzner (2008), *Renormalization group flow for fermionic superfluids at zero temperature*, Phys. Rev. B **78**, 014522.

Wang, S. Q., W. E. Evenson, and J. R. Schrieffer (1969), *Theory of itinerant ferromagnets exhibiting localized-moment behavior above the curie point*, Phys. Rev. Lett. **23**, 92.

Wetterich, C. (2007), *Bosonic effective action for interacting fermions*, Phys. Rev. B **75**, 085102.

Yamamoto, S. J. and Q. Si (2009), *Renormalization group for mixed fermion-boson systems*, arXiv:0906.0014v1 [cond-mat.str-el].

Chapter 12
Superfluid Fermions: Partial Bosonization in the Particle–Particle Channel

While the decoupling of the fermionic two-body interaction in the forward scattering channel described in Chap. 11 is natural if the interaction involves only small momentum transfers, such a procedure is not appropriate for other types of interactions. For example, the superconducting instability of a normal metal is triggered by particle–particle scattering processes with vanishing total momentum, so that in this case a Hubbard–Stratonovich decoupling in the particle–particle channel is more natural. Of course, if the resulting mixed Bose–Fermi theory could be solved exactly, then the choice of the Hubbard–Stratonovich field should not matter. However, in practice one has to rely on approximations, so that it is important to introduce the physically relevant collective degrees of freedom from the beginning by means of a proper choice of the Hubbard–Stratonovich field. In fact, it is a priori not clear whether the physical properties of a given system can be simply described by means of a decoupling involving only a single Hubbard–Stratonovich field (Bartosch et al. 2009a). We shall further comment on multicomponent Hubbard–Stratonovich transformations in Sect. 12.6.

In this chapter, we shall focus on the superfluid state of a Fermi gas with a short-range attractive two-body interaction. As first pointed out by Eagles (1969), the superfluid state of such a system exhibits an interesting crossover as a function of the dimensionless parameter $1/(k_F a_s)$, where a_s is the s-wave scattering length in vacuum:[1] while for $1/(k_F a_s) \ll -1$, where the scattering length is small and negative, the paired state consists of weakly bound spatially extended Cooper pairs (BCS limit), in the opposite limit $1/(k_F a_s) \gg 1$ the superfluid state can be viewed as a Bose–Einstein condensate of tightly bound fermions (BEC limit). The qualitative features of the BCS–BEC crossover are correctly described by the mean-field approximation (Eagles 1969). However, in the vicinity of the unitary point, where $a_s = \infty$ and hence $1/(k_F a_s) = 0$, quantitatively accurate calculations are difficult because there is no small parameter to justify approximations. Due to experimental progress in the field of ultra-cold atoms (Bartenstein et al. 2004, Bourdel et al. 2004, Kinast et al. 2005, Bloch et al. 2008) it is now possible to directly examine

---

[1] The relation between the s-wave scattering length a_s and the bare interaction of our model is given in Eqs. (12.22), (12.25), and (12.27) below.

Kopietz, P. et al.: *Superfluid Fermions: Partial Bosonization in the Particle–Particle Channel.*
Lect. Notes Phys. **798**, 327–368 (2010)
DOI 10.1007/978-3-642-05094-7_12

the unitary point experimentally, so that it is important to have reliable quantitative results in the entire range of the BCS–BEC crossover. Many authors studied this crossover using either the T-matrix approximation to improve on the mean-field approximation (Leggett 1980, Nozières and Schmitt-Rink 1985, Drechsler and Zwerger 1992, Randeria 1995, Engelbrecht et al. 1997) or other field theoretical many-body techniques (Nishida and Son 2006, 2007, Veillette et al. 2007, Nikolić and Sachdev 2007, Haussmann et al. 2007, Diener et al. 2008). The problem has also been investigated using FRG methods, using either the derivative expansion (Birse et al. 2005, Krippa et al. 2005, Krippa 2007, 2009, Diehl et al. 2007a,b, 2009, Floerchinger et al. 2008) or truncated vertex expansions (Strack et al. 2008, Bartosch et al. 2009b).

The FRG approach described in Sect. 12.5 of this chapter is based on the truncation of the vertex expansion developed by Bartosch et al. (2009b), which combines skeleton equations and a Ward identity to close the infinite hierarchy of FRG flow equations; furthermore, we impose a cutoff only in the propagator of the bosonic Hubbard–Stratonovich field. This *total momentum cutoff scheme* is in a sense the particle–particle version of the momentum transfer cutoff scheme introduced in Sect. 11.2. The advantage of this scheme is that the usual BCS results (which are nonperturbative in the coupling constant) define the initial conditions for the RG flow, whereas in alternative FRG schemes using a cutoff in the fermionic sector (Salmhofer et al. 2004, Strack et al. 2008) one has to integrate the RG flow in order to recover the BCS results. We shall further comment on the advantages of the total momentum cutoff scheme in Sect. 12.5.

12.1 Hubbard–Stratonovich Transformation in the Particle–Particle Channel

Having discussed the particle-hole forward scattering channel in Chap. 11, we consider in this section a two-body interaction which singles out the particle–particle channel. Our model is defined in terms of the following fermionic action,

$$S[\bar{\psi}, \psi] = -\sum_{\sigma} \int_K (i\omega - \xi_{k\sigma})\bar{\psi}_{K\sigma}\psi_{K\sigma} - \int_P g_P \bar{b}_P b_P \,, \qquad (12.1)$$

where we allow at this point for spin-dependent energy dispersions $\xi_{k\sigma} = \epsilon_{k\sigma} - \mu_\sigma$ and chemical potentials μ_σ, and the composite fermionic fields b_P and $\bar{b}_P$ are defined by

$$b_P = \int_K \psi_{-K+\frac{P}{2}\downarrow}\psi_{K+\frac{P}{2}\uparrow} = \int_K \psi_{-K\downarrow}\psi_{K+P\uparrow} \,, \qquad (12.2a)$$

$$\bar{b}_P = \int_K \bar{\psi}_{K+\frac{P}{2}\uparrow}\bar{\psi}_{-K+\frac{P}{2}\downarrow} = \int_K \bar{\psi}_{K+P\uparrow}\bar{\psi}_{-K\downarrow} \,. \qquad (12.2b)$$

Here, $K = (i\omega, \boldsymbol{k})$ is again a collective label consisting of the fermionic Matsubara frequency $i\omega$ and the momentum $\boldsymbol{k}$ associated with a fermion, while $P = (i\bar{\omega}, \boldsymbol{p})$ denotes the bosonic Matsubara frequency $i\bar{\omega}$ and the total momentum $\boldsymbol{p}$ of a fermion pair. The $\boldsymbol{p}$-integration in Eq. (12.1) contains an implicit ultraviolet cutoff Λ_0, which is the maximally allowed total momentum of two particles in a scattering process. For an attractive interaction to be considered here the coupling g_p is positive.

We now decouple the interaction in the spin-singlet particle–particle channel by means of a complex bosonic Hubbard–Stratonovich field χ_P depending on the total energy–momentum P of a fermion pair with opposite spin. The ratio of the partition functions with and without interactions can then be written as

$$\frac{\mathcal{Z}}{\mathcal{Z}_0} = \frac{\int \mathcal{D}[\bar{\psi}, \psi, \bar{\chi}, \chi]\, e^{-S_0[\bar{\psi},\psi] - S_0[\bar{\chi},\chi] - S_1[\bar{\psi},\psi,\bar{\chi},\chi]}}{\int \mathcal{D}[\bar{\psi}, \psi, \bar{\chi}, \chi]\, e^{-S_0[\bar{\psi},\psi] - S_0[\bar{\chi},\chi]}} \,, \tag{12.3}$$

where the Gaussian part of our bare action consists of a fermionic and a bosonic part,

$$S_0[\bar{\psi}, \psi, \bar{\chi}, \chi] = S_0[\bar{\psi}, \psi] + S_0[\bar{\chi}, \chi] \,, \tag{12.4}$$

with

$$S_0[\bar{\psi}, \psi] = -\sum_\sigma \int_K (i\omega - \xi_{\boldsymbol{k}\sigma}) \bar{\psi}_{K\sigma} \psi_{K\sigma} \,, \tag{12.5}$$

$$S_0[\bar{\chi}, \chi] = \int_P g_p^{-1} \bar{\chi}_P \chi_P \,. \tag{12.6}$$

The boson–fermion interaction can be written as

$$\begin{aligned} S_1[\bar{\psi}, \psi, \bar{\chi}, \chi] &= \int_P [\bar{b}_P \chi_P + b_P \bar{\chi}_P] \\ &= \int_P \int_K [\bar{\psi}_{K+P\uparrow} \bar{\psi}_{-K\downarrow} \chi_P + \psi_{-K\downarrow} \psi_{K+P\uparrow} \bar{\chi}_P] \,. \end{aligned} \tag{12.7}$$

It should be noted that, in contrast to the corresponding Bose–Fermi coupling (11.8) describing a repulsive interaction in the forward scattering channel, there is no factor of i in the Bose–Fermi coupling (12.7), because here our interaction is attractive. Defining a two-component Fermi field

$$\psi_K = \begin{pmatrix} \psi_{K\uparrow} \\ \bar{\psi}_{-K\downarrow} \end{pmatrix} \,, \tag{12.8}$$

the fermionic part of the action in the numerator of Eq. (12.3) can be written as

$$S_0[\bar{\psi}, \psi] + S_1[\bar{\psi}, \psi, \bar{\chi}, \chi] = - \int_K \int_{K'} \psi_K^\dagger [\hat{G}^{-1}]_{KK'} \psi_{K'} \, ,$$

$$(12.9)$$

where $\hat{G}^{-1}$ is a matrix in spin, momentum, and frequency space, with matrix elements

$$[\hat{G}^{-1}]_{KK'} = \begin{pmatrix} \delta_{K,K'}(i\omega - \xi_{k\uparrow}) & -\chi_{K-K'} \\ -\bar{\chi}_{K'-K} & \delta_{K,K'}(i\omega + \xi_{-k\downarrow}) \end{pmatrix} \, . \qquad (12.10)$$

Before attacking this mixed Bose–Fermi field theory in Sect. 12.5 by means of the FRG machinery, it is instructive to examine our model first using simple mean-field theory and the Gaussian approximation, as discussed in Chap. 2.

12.2 Mean-field Approximation and BCS–BEC Crossover

The mean-field approximation amounts to performing the integration over the collective field χ in Eq. (12.3) in saddle point approximation. Therefore we simply replace the χ-integration by the value of the integrand at a single point, which is obtained by replacing

$$\chi_P \rightarrow \delta_{P,0} \Delta_0 \, . \qquad (12.11)$$

The real parameter Δ_0 is fixed by minimizing the free energy. The inverse Green function matrix in Eq. (12.10) is then approximated by

$$\hat{G}^{-1} \approx [\hat{G}_1^{-1}]_{KK'} = \delta_{K,K'} \begin{pmatrix} i\omega - \xi_{k\uparrow} & -\Delta_0 \\ -\Delta_0 & i\omega + \xi_{-k\downarrow} \end{pmatrix} \, . \qquad (12.12)$$

Assuming from now on that $\xi_{k\sigma} = \xi_k = \epsilon_k - \mu$ is independent of the spin projection σ, Eq. (12.3) reduces in this approximation to

$$\frac{\mathcal{Z}}{\mathcal{Z}_0} \approx \frac{\mathcal{Z}_1}{\mathcal{Z}_0} = e^{-\beta(\Omega_1 - \Omega_0)} \, , \qquad (12.13)$$

where the change of the grand canonical potential due to interactions is

$$\Omega_1 - \Omega_0 = -2T \sum_k \ln \left[\frac{\cosh(\beta E_k/2)}{\cosh(\beta \xi_k/2)} \right] + V \frac{\Delta_0^2}{g_0} \, . \qquad (12.14)$$

Here, the dispersion of the fermionic quasiparticles is

$$E_k = \sqrt{\xi_k^2 + \Delta_0^2} \, , \qquad (12.15)$$

and

$$\Omega_0 = -2T \sum_k \ln\left[1 + e^{-\beta \xi_k}\right] \tag{12.16}$$

is the grand canonical potential in the absence of interactions. The mean-field approximation to the grand canonical potential can be written as

$$\Omega_1 = -\sum_k (E_k - \xi_k) - 2T \sum_k \ln\left[1 + e^{-\beta E_k}\right] + V\frac{\Delta_0^2}{g_0}. \tag{12.17}$$

Minimizing this with respect to Δ_0 yields the usual BCS gap equation

$$\frac{1}{g_0} = \frac{1}{V}\sum_k \frac{1 - 2f(E_k)}{2E_k} = \frac{1}{V}\sum_k \frac{\tanh(\beta E_k/2)}{2E_k}, \tag{12.18}$$

where

$$f(E_k) = \frac{1}{e^{\beta E_k} + 1} \tag{12.19}$$

is the Fermi function. If we work at constant density $\rho = N/V$, then the chemical potential should be expressed in terms of ρ using the thermodynamic relation $N = -\partial\Omega/\partial\mu$. Within the mean-field approximation this yields

$$\boxed{\rho = \frac{1}{V}\sum_k \left[1 - \frac{\xi_k}{E_k}\tanh(\beta E_k/2)\right].} \tag{12.20}$$

The thermodynamics should then be derived from the free energy $F = \Omega + \mu N$, which reduces to the ground state energy in the zero temperature limit. In mean-field approximation we obtain from Eq. (12.17) for $T = 0$,

$$F \approx \Omega_1 + \mu N = -\sum_k \left[E_k - \xi_k - \mu\left(1 - \frac{\xi_k}{E_k}\right)\right] + V\frac{\Delta_0^2}{g_0}$$

$$= \sum_k \epsilon_k \left[1 - \frac{\xi_k}{E_k}\right] - V\frac{\Delta_0^2}{g_0}. \tag{12.21}$$

Without ultraviolet cutoff the gap equation (12.18) is ultraviolet divergent in dimensions $D \geq 2$. To regularize this divergence, we eliminate the bare interaction g_0 in favor of the two-body T-matrix at zero energy and vanishing total momentum in vacuum (corresponding to $\mu = 0$). The T-matrix g is related to the bare interaction g_0 via

$$\frac{1}{g_0} = \frac{1}{g} + \frac{1}{V}\sum_k \frac{1}{2\epsilon_k} \,. \tag{12.22}$$

Using this relation to eliminate the bare interaction g_0 from the gap equation (12.18), we obtain the regularized gap equation

$$\frac{1}{g} = \frac{1}{V}\sum_k \left[\frac{\tanh(\beta E_k/2)}{2E_k} - \frac{1}{2\epsilon_k}\right]. \tag{12.23}$$

The subtraction regularizes the ultraviolet divergence in dimensions $2 < D < 4$. In $D = 3$ the two-body T-matrix is related to the s-wave scattering length a_s via $g = -4\pi a_s/m$. Note that for $D > 2$ we may define the s-wave scattering length via (see e.g., Sauli and Kopietz 2006)

$$g = -\gamma_D \,\mathrm{sign}(a_s)\,|a_s|^{D-2}/m \,, \tag{12.24}$$

where γ_D is a numerical constant of the order of unity ($\gamma_3 = 4\pi$). For given values of the interaction g and the density ρ, Eqs. (12.20) and (12.23) determine the chemical potential μ and the gap Δ_0. For a numerical evaluation of these equations, it is convenient to express them in terms of dimensionless parameters. We therefore define the relevant dimensionless interaction

$$\tilde{g} = \nu_0 g \,, \tag{12.25}$$

where $\nu_0 = \nu(\epsilon_F)$ is the density of states (per spin projection) of the noninteracting system at the Fermi energy ϵ_F. Neglecting lattice effects, the energy dispersion is $\epsilon_k = k^2/(2m)$ and $\epsilon_F = k_F^2/(2m) = \frac{m}{2}v_F^2$. In D dimensions the energy-dependent density of states is

$$\nu(\epsilon) = \frac{1}{V}\sum_k \delta(\epsilon - \epsilon_k) = K_D m k_F^{D-2}(\epsilon/\epsilon_F)^{\frac{D-2}{2}} \,, \quad V \to \infty \,, \tag{12.26}$$

where the numerical constant K_D is given in Eq. (2.86). Our dimensionless coupling (12.25) can then be written as

$$\tilde{g} = -\tilde{\gamma}_D \,\mathrm{sign}(a_s)\,|k_F a_s|^{D-2} \,, \tag{12.27}$$

where $\tilde{\gamma}_D = K_D \gamma_D$. In $D = 3$ this reduces to $\tilde{g} = -2k_F a_s/\pi$.

For simplicity, we focus on the zero-temperature limit from now on. Introducing the dimensionless parameters

$$\tilde{\mu} = \frac{\mu}{\epsilon_F} \,, \quad \tilde{\Delta} = \frac{\Delta_0}{\epsilon_F} \,, \tag{12.28}$$

the regularized gap equation (12.23) can be written as

$$\frac{1}{\tilde{g}} = \frac{1}{2} \int_0^\infty dx\, x^{\frac{D-2}{2}} \left[\frac{1}{\sqrt{(x-\tilde{\mu})^2 + \tilde{\Delta}^2}} - \frac{1}{x} \right], \qquad (12.29)$$

while the particle number equation (12.20) reduces to the following relation between $\tilde{\mu}$ and $\tilde{\Delta}$,

$$1 = \frac{D}{4} \int_0^\infty dx\, x^{\frac{D-2}{2}} \left[1 - \frac{x-\tilde{\mu}}{\sqrt{(x-\tilde{\mu})^2 + \tilde{\Delta}^2}} \right], \qquad (12.30)$$

where we have used the fact that k_F is related to the density via $\rho = \frac{2K_D}{D} k_F^D$. Introducing the ground state energy per particle in units of the Fermi energy,

$$\tilde{\varepsilon} \equiv \frac{\varepsilon}{\epsilon_F} = \lim_{T \to 0} \frac{F}{N\epsilon_F}, \qquad (12.31)$$

we obtain from Eq. (12.21) for $N \equiv \rho V \to \infty$ with the help of the BCS gap equation (12.18)

$$\tilde{\varepsilon} = \frac{D}{4} \int_0^\infty dx\, x^{\frac{D}{2}} \left[1 - \frac{x-\tilde{\mu}}{\sqrt{(x-\tilde{\mu})^2 + \tilde{\Delta}^2}} - \frac{\tilde{\Delta}^2}{2x\sqrt{(x-\tilde{\mu})^2 + \tilde{\Delta}^2}} \right]. \qquad (12.32)$$

The term in the square braces vanishes as x^{-3} for large x, so that the integral is ultraviolet convergent for $D < 4$. For a given value of $\tilde{g}$, Eqs. (12.29) and (12.30) impose two conditions which can be used to determine the two unknowns $\tilde{\Delta}$ and $\tilde{\mu}$. It is easy to solve these equations numerically.[2] In Figs. 12.1 and 12.2 we show numerical results for the dimensionless mean-field gap $\tilde{\Delta} = \Delta_0/\epsilon_F$ and chemical potential $\tilde{\mu} = \mu/\epsilon_F$ in units of ϵ_F. As first discussed by Eagles (1969), in three dimensions the above mean-field equations give a qualitatively correct description of the crossover from a weakly coupled BCS superconductor with small negative scattering length to a Bose–Einstein condensate of tightly bound fermion pairs, characterized by a small positive scattering length. Of particular interest is the unitary point, where the scattering length diverges and hence $\tilde{g}^{-1} = 0$. The mean-field results are in this case

$$\tilde{\mu} = 0.5906 \quad , \quad \tilde{\Delta} = 0.6864 \quad , \quad \tilde{\varepsilon} = 0.3543 \quad \text{(mean field)}. \qquad (12.33)$$

---

[2] As pointed out by Marini et al. (1998), for $D = 3$ it is possible to express the integrals in Eqs. (12.29) and (12.30) in terms of complete elliptic integrals of the first and second kind, $K(\kappa)$ and $E(\kappa)$ (Abramowitz and Stegun 1965). This is also possible for the integrals appearing in the expression (12.32) for the ground state energy, which simplifies the numerical calculations.

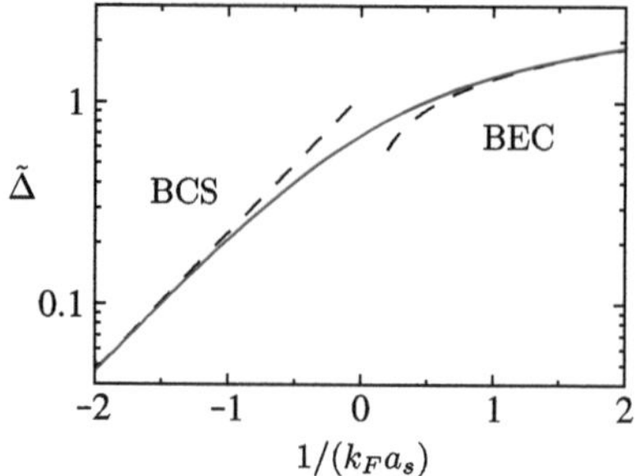

Fig. 12.1 Mean-field result for the dimensionless gap $\tilde{\Delta} = \Delta_0/\epsilon_F$ in three dimensions as a function of $1/(k_F a_s) = -2/(\pi \tilde{g})$. The *dashed lines* represent the asymptotic behavior in the BCS limit, $\ln \tilde{\Delta} \sim \pi/(2k_F a_s) + \ln(8/e^2)$, and in the BEC limit $\tilde{\Delta} \sim (3\pi^3/64)^{1/2} \, 1/(k_F a_s)^{1/2}$

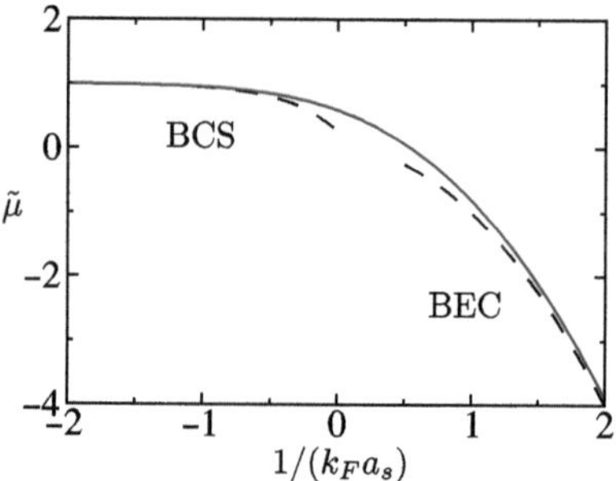

Fig. 12.2 Mean-field result for $\tilde{\mu} = \mu/\epsilon_F$ in three dimensions as a function of $1/(k_F a_s)$. The *dashed lines* are the asymptotic behavior in the BCS limit $\tilde{\mu} \sim 1 - (16/e^4)[5/2 - \pi/(2k_F a_s)]e^{\pi/(k_F a_s)}$, and in the BEC limit $\tilde{\mu} \sim -1/(k_F a_s)^2$

Experimentally, for $\tilde{\varepsilon} = \varepsilon/\epsilon_F$ values in the range between 0.19 and 0.31 have been reported (Bartenstein et al. 2004, Bourdel et al. 2004, Kinast et al. 2005). The mean-field results (12.33) can also be compared with quantum Monte Carlo simulations (Carlson et al. 2003, Chang et al. 2004, Astrakharchik et al. 2004, Carlson and Reddy 2005), which give at the unitary point

$$\tilde{\Delta} = 0.50 \pm 0.03 \quad , \quad \tilde{\varepsilon} = 0.25 \pm 0.01 \quad \text{(Monte Carlo)}. \tag{12.34}$$

From the universality of the free energy density at the unitary point it is easy to show (see e.g., Veillette et al. 2007) that the chemical potential is related to the ground state energy per particle via

$$\mu = \frac{5}{3}\varepsilon \, . \tag{12.35}$$

This relation is of course also satisfied by the mean-field results (12.33), but it is an exact relation at the unitary point in three dimensions. The Monte Carlo results (12.34) imply $\tilde{\mu} = \frac{5}{3}\tilde{\varepsilon} = 0.42$. Both the mean-field results for the gap and the mean-field energy per particle are significantly larger than the corresponding Monte Carlo results. It is therefore important to investigate fluctuation corrections to the mean-field approximation.

12.3 Gaussian Fluctuations

Corrections to the mean-field results (12.33) at the unitary point due to Gaussian fluctuations of the order-parameter field have been calculated by Veillette et al. (2007) and by Diener et al. (2008), but the results obtained by these two groups do not completely agree. Fluctuation corrections to the mean-field approximation have also been calculated by means of other many-body techniques (Nikolić and Sachdev 2007, Haussmann et al. 2007). To set the stage for the FRG calculation in Sect. 12.5, let us focus in this section on the calculation of the corrections to the mean-field results due to Gaussian fluctuations of our Hubbard–Stratonovich field.

12.3.1 Gaussian Effective Action

Integrating in Eq. (12.3) over the Fermi fields we obtain the formally exact expression

$$\frac{\mathcal{Z}}{\mathcal{Z}_0} = e^{-\beta(\Omega_1 - \Omega_0)} \frac{\int \mathcal{D}[\delta\bar{\chi}, \delta\chi] e^{-S_{\mathrm{eff}}[\delta\bar{\chi}, \delta\chi]}}{\int \mathcal{D}[\bar{\chi}, \chi] e^{-S_0[\bar{\chi}, \chi]}} , \tag{12.36}$$

where in the numerator we have shifted the fields according to $\chi_P = \delta_{P,0}\Delta_0 + \delta\chi_P$, and the effective action for the bosonic fluctuations is

$$\begin{aligned}
S_{\mathrm{eff}}[\delta\bar{\chi}, \delta\chi] &= S_0[\Delta_0 + \delta\bar{\chi}, \Delta_0 + \delta\chi] - \beta V g_0^{-1}\Delta_0^2 - \mathrm{Tr}\ln[1 - \hat{G}_1\hat{V}] \\
&= \int_P g_p^{-1}\delta\bar{\chi}_P\delta\chi_P + g_0^{-1}\Delta_0(\delta\chi_0 + \delta\bar{\chi}_0) + \sum_{n=1}^{\infty}\frac{\mathrm{Tr}[\hat{G}_1\hat{V}]^n}{n} .
\end{aligned} \tag{12.37}$$

Here, $\hat{G}_1$ is the mean-field fermionic Green function defined in Eq. (12.12), and the matrix $\hat{V}$ is given by

$$[\hat{V}]_{KK'} = \begin{pmatrix} 0 & \delta\chi_{K-K'} \\ \delta\bar{\chi}_{K'-K} & 0 \end{pmatrix} . \tag{12.38}$$

Within the Gaussian approximation, we expand $S_{\mathrm{eff}}[\delta\bar{\chi}, \delta\chi]$ to second order in the fluctuations,

$$\begin{aligned}
S_{\mathrm{eff}}[\delta\bar{\chi}, \delta\chi] &\approx \int_P g_p^{-1}\delta\bar{\chi}_P\delta\chi_P + \frac{1}{2}\mathrm{Tr}[\hat{G}_1\hat{V}]^2 \\
&\quad + g_0^{-1}\Delta_0(\delta\chi_0 + \delta\bar{\chi}_0) + \mathrm{Tr}[\hat{G}_1\hat{V}] .
\end{aligned} \tag{12.39}$$

To evaluate the traces, it is convenient to use the spectral representation of the mean-field Green function $\hat{G}_1$, which amounts to writing the inverse of the matrix in Eq. (12.12) in the form

$$[\hat{G}_1]_{KK'} = -\delta_{K,K'} \frac{1}{\omega^2 + E_k^2} \begin{pmatrix} i\omega + \xi_k & \Delta_0 \\ \Delta_0 & i\omega - \xi_k \end{pmatrix}$$

$$= \delta_{K,K'} \begin{pmatrix} B_0(K) & A_0(K) \\ A_0(K) & -B_0(-K) \end{pmatrix} = \delta_{K,K'} \sum_{\sigma=\pm} \frac{\mathbf{w}_{k\sigma} \mathbf{w}_{k\sigma}^T}{i\omega - \sigma E_k} , \tag{12.40}$$

where

$$A_0(K) = -\frac{\Delta_0}{\omega^2 + E_k^2} = u_k v_k \left[\frac{1}{i\omega - E_k} - \frac{1}{i\omega + E_k} \right] , \tag{12.41}$$

$$B_0(K) = -\frac{i\omega + \xi_k}{\omega^2 + E_k^2} = \frac{u_k^2}{i\omega - E_k} + \frac{v_k^2}{i\omega + E_k} , \tag{12.42}$$

and the components of the normalized eigenvectors $\mathbf{w}_{k\sigma}$ are the usual Bogoliubov coefficients u_k and v_k, i.e.,

$$\mathbf{w}_{k+} = \begin{pmatrix} u_k \\ v_k \end{pmatrix} , \quad \mathbf{w}_{k-} = \begin{pmatrix} -v_k \\ u_k \end{pmatrix} , \tag{12.43}$$

with

$$u_k^2 = \frac{1}{2} \left(1 + \frac{\xi_k}{E_k} \right) , \quad v_k^2 = \frac{1}{2} \left(1 - \frac{\xi_k}{E_k} \right) . \tag{12.44}$$

Note that by construction $\mathbf{w}_{k\sigma}^T \mathbf{w}_{k\sigma'} = \delta_{\sigma\sigma'}$, which for $\sigma = \sigma'$ reduces to $u_k^2 + v_k^2 = 1$. Moreover,

$$u_k^2 - v_k^2 = \frac{\xi_k}{E_k} , \quad u_k v_k = \frac{\Delta_0}{2E_k} . \tag{12.45}$$

For the first-order term in Eq. (12.39) we obtain

$$\text{Tr}[\hat{G}_1 \hat{V}] = \int_K \sum_\sigma \frac{\mathbf{w}_{k\sigma}^T \begin{pmatrix} 0 & \delta\chi_0 \\ \delta\bar{\chi}_0 & 0 \end{pmatrix} \mathbf{w}_{k\sigma}}{i\omega - \sigma E_k}$$

$$= -(\delta\chi_0 + \delta\bar{\chi}_0) \frac{\Delta_0}{V} \sum_k \frac{1 - 2f(E_k)}{2E_k} . \tag{12.46}$$

Using the BCS self-consistency condition (12.18), we see that this term cancels the other linear term $g_0^{-1} \Delta_0 (\delta\chi_0 + \delta\bar{\chi}_0)$ on the right-hand side of Eq. (12.39), so that in Gaussian approximation only the first line of the effective action in Eq. (12.39) survives.

An explicit expression for the second-order term $\mathrm{Tr}[\hat{G}_1\hat{V}]^2$ in terms of the complex fields $\delta\chi$ and $\delta\bar{\chi}$ has recently been derived by Veillette et al. (2007). For our purpose, it is more convenient to parameterize the Gaussian fluctuations in terms of two real fields φ_ℓ and φ_t, corresponding to the real and imaginary part of the complex field $\delta\chi$ (Engelbrecht et al. 1997). In frequency–momentum space, we set[3]

$$\delta\chi_P = \frac{1}{\sqrt{2}}[\varphi_{P\ell} + i\varphi_{Pt}]\,, \tag{12.47a}$$

$$\delta\bar{\chi}_P = \frac{1}{\sqrt{2}}[\varphi_{-P\ell} - i\varphi_{-Pt}] = \frac{1}{\sqrt{2}}\left[\varphi_{P\ell}^* - i\varphi_{Pt}^*\right]\,. \tag{12.47b}$$

The field φ_ℓ describes longitudinal fluctuations of the superfluid order parameter; the corresponding collective mode can be viewed as the condensed matter analogue of the Higgs boson in particle physics (Littlewood and Varma 1982, Varma 2002, Barankov and Levitov 2007, Barankov 2008). On the other hand, the field φ_t describes transverse fluctuations associated with the phase dynamics of the superfluid order parameter; the corresponding collective mode is the gapless Goldstone boson associated with the spontaneous breaking of the $U(1)$-phase symmetry in a superfluid. This mode is usually called the Bogoliubov–Anderson mode (Bogoliubov 1958, Anderson 1958, Schrieffer 1964). Expressing the Gaussian effective action (12.39) in terms of the fields φ_ℓ and φ_t, we obtain

$$\begin{aligned}
S_{\mathrm{eff}}[\varphi_\ell, \varphi_t] &= \frac{1}{2}\int_P g_p^{-1}(\varphi_{-P\ell}\varphi_{P\ell} + \varphi_{-Pt}\varphi_{Pt}) + \frac{1}{2}\mathrm{Tr}[\hat{G}_1\hat{V}]^2 \\
&= \frac{1}{2}\int_P \left\{ \left[g_p^{-1} + \Pi_0^{\ell\ell}(P)\right]\varphi_{-P\ell}\varphi_{P\ell} + \left[g_p^{-1} + \Pi_0^{tt}(P)\right]\varphi_{-Pt}\varphi_{Pt} \right. \\
&\quad \left. + \Pi_0^{\ell t}(P)\varphi_{-P\ell}\varphi_{Pt} + \Pi_0^{t\ell}(P)\varphi_{-Pt}\varphi_{P\ell}\right\}\,,
\end{aligned} \tag{12.48}$$

where the polarization functions are given by

$$\begin{aligned}
\Pi_0^{\ell\ell}(P) &= -\frac{1}{2}\int_K \left[B_0(K)B_0(-K+P) - A_0(K)A_0(K+P) + (P\to -P)\right] \\
&= \frac{1}{V}\sum_k \left\{ (u_k v_{k+p} + v_k u_{k+p})^2 \frac{E_k - E_{k+p}}{(E_k - E_{k+p})^2 + \bar{\omega}^2}\left[f(E_k) - f(E_{k+p})\right] \right. \\
&\quad \left. - (u_k u_{k+p} - v_k v_{k+p})^2 \frac{E_k + E_{k+p}}{(E_k + E_{k+p})^2 + \bar{\omega}^2}\left[1 - f(E_k) - f(E_{k+p})\right]\right\}\,,
\end{aligned} \tag{12.49a}$$

---

[3] The fields $\varphi_\ell(\tau, \boldsymbol{r})$ and $\varphi_t(\tau, \boldsymbol{r})$ are real functions of space and imaginary time. In momentum–frequency space this implies $\varphi_{P\ell}^* = \varphi_{-P\ell}$ and $\varphi_{Pt}^* = \varphi_{-Pt}$.

$$\Pi_0^{tt}(P) = -\frac{1}{2} \int_K [B_0(K)B_0(-K+P) + A_0(K)A_0(K+P) + (P \to -P)]$$

$$= \frac{1}{V} \sum_k \left\{ (u_k v_{k+p} - v_k u_{k+p})^2 \frac{E_k - E_{k+p}}{(E_k - E_{k+p})^2 + \bar{\omega}^2} \left[f(E_k) - f(E_{k+p}) \right] \right.$$

$$\left. - (u_k u_{k+p} + v_k v_{k+p})^2 \frac{E_k + E_{k+p}}{(E_k + E_{k+p})^2 + \bar{\omega}^2} \left[1 - f(E_k) - f(E_{k+p}) \right] \right\},$$

$$(12.49b)$$

$$\Pi_0^{\ell t}(P) = -\Pi_0^{t\ell}(P)$$

$$= -\frac{i}{2} \int_K [B_0(K)B_0(-K+P) - A_0(K)A_0(K+P) - (P \to -P)]$$

$$= -\frac{\bar{\omega}}{V} \sum_k \left\{ \left(u_k^2 v_{k+p}^2 - v_k^2 u_{k+p}^2 \right) \frac{f(E_k) - f(E_{k+p})}{(E_k - E_{k+p})^2 + \bar{\omega}^2} \right.$$

$$\left. - \left(u_k^2 u_{k+p}^2 - v_k^2 v_{k+p}^2 \right) \frac{1 - f(E_k) - f(E_{k+p})}{(E_k + E_{k+p})^2 + \bar{\omega}^2} \right\}. \qquad (12.49c)$$

Substituting the Gaussian effective action (12.48) into Eq. (12.36) and performing the Gaussian integrations, we obtain for the grand canonical potential in Gaussian approximation $\Omega \approx \Omega_1 + \Omega_2$, where Ω_1 is the mean-field result given in Eq. (12.17), and the correction due to Gaussian fluctuations is[4]

$$\Omega_2 = \frac{V}{2} \int_P \ln \left\{ \left[1 + g_p \Pi_0^{\ell\ell}(P) \right] \left[1 + g_p \Pi_0^{tt}(P) \right] + g_p^2 \left[\Pi_0^{\ell t}(P) \right]^2 \right\}. \qquad (12.50)$$

For an evaluation of Ω_2 and the resulting corrections to the thermodynamics at the unitary point, see (Veillette et al. 2007, Diener et al. 2008). Let us here only quote the more recent results by Diener et al. (2008), who found

$$\tilde{\Delta} = 0.47 \quad , \quad \tilde{\varepsilon} = 0.24 \quad \text{(Gaussian fluctuations)}. \qquad (12.51)$$

The corresponding value of the chemical potential in units of ϵ_F is according to Eq. (12.35) given by $\tilde{\mu} = \frac{5}{3}\tilde{\varepsilon} = 0.40$. Comparing these numbers with the mean-

---

[4] Using the coupling constant integration trick (see e.g., Schwiete and Efetov 2006), the Gaussian correction (12.50) to the grand canonical potential can be expressed in terms of correlation functions $\Pi_{\lambda g}^{\ell\ell}(P)$ and $\Pi_{\lambda g}^{tt}(P)$ obtained by substituting $g_p \to \lambda g_p$ in Eqs. (12.59a), and (12.59b),

$$\Omega_2 = \frac{V}{2} \int_P \int_0^1 d\lambda \frac{d}{d\lambda} \ln \left\{ \left[1 + \lambda g_p \Pi_0^{\ell\ell}(P) \right] \left[1 + \lambda g_p \Pi_0^{tt}(P) \right] + \lambda^2 g_p^2 \left[\Pi_0^{\ell t}(P) \right]^2 \right\}$$

$$= \frac{V}{2} \int_0^1 \frac{d\lambda}{\lambda} \int_P \left[\Pi_{\lambda g}^{\ell\ell}(P) + \Pi_{\lambda g}^{tt}(P) \right].$$

field results (12.33) and the Monte Carlo results (12.34) we conclude that Gaussian fluctuations account for most of the difference between the exact ground state energy and the mean-field energy.

12.3.2 Bosonic Propagators and Order-Parameter Correlations

To derive the correlation functions of the bosonic Hubbard–Stratonovich fields in Gaussian approximation, let us write our Gaussian effective action defined in Eq. (12.48) in matrix form,

$$S_{\text{eff}}[\varphi_\ell, \varphi_t] = \frac{1}{2} \int_P \int_{P'} (\varphi_{P\ell}, \varphi_{Pt})[\hat{F}^{-1}]_{PP'} \begin{pmatrix} \varphi_{P'\ell} \\ \varphi_{P't} \end{pmatrix} , \qquad (12.52)$$

where $\hat{F}^{-1}$ (which is matrix in the energy–momentum and field-type labels) consists of two contributions,

$$\hat{F}^{-1} = \hat{F}_0^{-1} + \hat{\Pi}_0 , \qquad (12.53)$$

with the first term given by the inverse bare interaction,

$$[\hat{F}_0^{-1}]_{PP'} = \delta_{P,-P'} \begin{pmatrix} g_{p'}^{-1} & 0 \\ 0 & g_{p'}^{-1} \end{pmatrix} , \qquad (12.54)$$

while the second term is the polarization matrix,

$$[\hat{\Pi}_0]_{PP'} = \delta_{P,-P'} \begin{pmatrix} \Pi_0^{\ell\ell}(P') & \Pi_0^{\ell t}(P') \\ \Pi_0^{t\ell}(P') & \Pi_0^{tt}(P') \end{pmatrix} . \qquad (12.55)$$

The Gaussian propagator of our Hubbard–Stratonovich fields is thus

$$[\hat{F}]_{PP'} = \begin{pmatrix} \langle \varphi_{P\ell}\varphi_{P'\ell} \rangle & \langle \varphi_{P\ell}\varphi_{P't} \rangle \\ \langle \varphi_{Pt}\varphi_{P'\ell} \rangle & \langle \varphi_{Pt}\varphi_{P't} \rangle \end{pmatrix} = \delta_{P,-P'} \begin{pmatrix} F_P^{\ell\ell} & F_P^{\ell t} \\ F_P^{t\ell} & F_P^{tt} \end{pmatrix}$$

$$= \delta_{P,-P'} \frac{g_p}{N(P)} \begin{pmatrix} 1 + g_p\Pi_0^{tt}(P) & -g_p\Pi_0^{\ell t}(P) \\ -g_p\Pi_0^{t\ell}(P) & 1 + g_p\Pi_0^{\ell\ell}(P) \end{pmatrix} , \qquad (12.56)$$

where

$$N(P) = \left[1 + g_p\Pi_0^{\ell\ell}(P) \right]\left[1 + g_p\Pi_0^{tt}(P) \right] + g_p^2[\Pi_0^{\ell t}(P)]^2 , \qquad (12.57)$$

and we have used $\Pi_0^{\ell t}(P) = -\Pi_0^{t\ell}(P)$. Keeping in mind that our Hubbard–Stratonovich fields φ_ℓ and φ_t are conjugate to the order parameter, the corresponding correlation functions in Eq. (12.56) should not be confused with the correlation functions of the superfluid order parameter (Kopietz 1997, De Palo et al. 1999). The latter can easily be obtained from the matrix

$$\hat{\Pi} = \left[\hat{\Pi}_0^{-1} + \hat{F}_0\right]^{-1} = \hat{\Pi}_0 \left[1 + \hat{F}_0 \hat{\Pi}_0\right]^{-1} . \tag{12.58}$$

Explicitly, we obtain for the order-parameter correlation functions in Gaussian approximation

$$[\hat{\Pi}]_{PP'}^{\ell\ell} = \delta_{P,-P'} \frac{\Pi_0^{\ell\ell}(P)\left[1 + g_p \Pi_0^{tt}(P)\right] + g_p \left[\Pi_0^{\ell t}(P)\right]^2}{N(P)} , \tag{12.59a}$$

$$[\hat{\Pi}]_{PP'}^{tt} = \delta_{P,-P'} \frac{\Pi_0^{tt}(P)\left[1 + g_p \Pi_0^{\ell\ell}(P)\right] + g_p \left[\Pi_0^{\ell t}(P)\right]^2}{N(P)} , \tag{12.59b}$$

$$[\hat{\Pi}]_{PP'}^{\ell t} = [\hat{\Pi}]_{P'P}^{t\ell} = \delta_{P,-P'} \frac{\Pi_0^{\ell t}(P)}{N(P)} , \tag{12.59c}$$

which is equivalent to the ladder approximation in the particle–particle channel.

For small momenta and frequencies, the bosonic correlation functions exhibit a pole with linear dispersion on the real frequency axis, which can be identified with the Goldstone boson associated with the broken $U(1)$-symmetry in the superfluid state, the Bogoliubov–Anderson (BA) mode (Bogoliubov 1958, Anderson 1958, Schrieffer 1964). To derive the dispersion of the BA mode, we expand the polarization functions in powers of momenta and frequencies. For our purpose it is sufficient to approximate

$$\Pi_0^{\ell\ell}(\boldsymbol{p}, i\bar{\omega}) \approx \Pi_0^{\ell\ell}(0, i0) = -v_0 I_0 + v_0 I_1 , \tag{12.60a}$$

$$\Pi_0^{tt}(\boldsymbol{p}, i\bar{\omega}) \approx -v_0 I_0 + \frac{v_0}{(2\Delta_0)^2}\left[I_1\bar{\omega}^2 + I_3(v_F \boldsymbol{p})^2\right], \tag{12.60b}$$

$$\Pi_0^{\ell t}(\boldsymbol{p}, i\bar{\omega}) \approx \Pi_0^{\ell t}(0, i\bar{\omega}) \approx \frac{v_0}{2\Delta_0} I_2 \bar{\omega} . \tag{12.60c}$$

The dimensionless coefficients I_n are at zero temperature given by

$$I_0 = \frac{1}{v_0 V} \sum_k \frac{1}{2E_k} = \frac{1}{v_0 g_0} , \tag{12.61a}$$

$$I_1 = \frac{1}{v_0 V} \sum_k \frac{\Delta_0^2}{2E_k^3} , \tag{12.61b}$$

$$I_2 = \frac{1}{v_0 V} \sum_k \frac{\Delta_0 \xi_k}{2E_k^3} , \tag{12.61c}$$

$$I_3 = \frac{1}{v_0 V} \sum_k \frac{\Delta_0^2}{2E_k^3}\left[\frac{\epsilon_k}{D\epsilon_F}\left(2 - \frac{3\xi_k^2}{E_k^2}\right) + \frac{\xi_k}{2\epsilon_F}\right] = \frac{1}{D} , \tag{12.61d}$$

where in the second equality of Eq. (12.61a) we have used the mean-field gap equation (12.18) at zero temperature to express the momentum sum in terms of the bare coupling g_0. The identity $I_3 = 1/D$ follows for $V \to \infty$ with the help of the relation (12.30) between $\tilde{\mu}$ and $\tilde{\Delta}$ after an integration by parts.[5] Note that for $D > 2$ the integral I_0 defined in Eq. (12.61a) is ultraviolet divergent. However, all physical quantities involve the combinations $g_0^{-1} + \Pi_0^{\ell\ell}(P)$ or $g_0^{-1} + \Pi_0^{tt}(P)$, so that the ultraviolet divergence can be absorbed into the bare coupling g_0. Therefore we introduce the corresponding renormalized coupling g as in Eq. (12.22) and write

$$g_0^{-1} - \Pi_0^{tt}(0, i0) = g_0^{-1} - v_0 I_0 = g^{-1} - v_0 \tilde{I}_0 \, , \tag{12.62}$$

where the integral $\tilde{I}_0$ is now ultraviolet convergent,

$$\tilde{I}_0 = \frac{1}{v_0 V} \sum_k \left[\frac{1}{2E_k} - \frac{1}{2\epsilon_k} \right] . \tag{12.63}$$

From Eqs. (12.60b) and (12.61a) we conclude that

$$g_0^{-1} + \Pi_0^{tt}(0, i0) = 0 \, , \tag{12.64}$$

which guarantees that the BA mode is gapless. Substituting the expansions (12.60a), 12.60b), and (12.60c) into Eq. (12.56) we obtain the leading long-wavelength and low-frequency behavior of the bosonic propagators in Gaussian approximation,

$$F_P^{\ell\ell} \approx \frac{DZ_c^2}{v_0} \frac{\bar{\omega}^2 + Z_1 c^2 \boldsymbol{p}^2}{\bar{\omega}^2 + c^2 \boldsymbol{p}^2} \, , \tag{12.65a}$$

$$F_P^{tt} \approx \frac{DZ_c^2}{v_0} \frac{(2\Delta_0)^2}{\bar{\omega}^2 + c^2 \boldsymbol{p}^2} \, , \tag{12.65b}$$

$$F_P^{\ell t} \approx -\frac{Z_2}{v_0} \frac{2\Delta_0 \bar{\omega}}{\bar{\omega}^2 + c^2 \boldsymbol{p}^2} . \tag{12.65c}$$

Here, the dimensionless renormalization factors Z_c, Z_1, and Z_2 are given by

$$Z_c^2 = \frac{c^2}{v_F^2} = \frac{I_1 I_3}{I_1^2 + I_2^2} \, , \quad Z_1 = \frac{I_1^2 + I_2^2}{I_1^2} \, , \quad Z_2 = \frac{I_2}{I_1^2 + I_2^2} . \tag{12.66}$$

After analytic continuation to real frequencies ($i\bar{\omega} \to \bar{\omega} + i0$) the propagators (12.65a), and (12.65b) exhibit poles at $\bar{\omega} = \pm c|\boldsymbol{p}|$ corresponding to the gapless BA mode, which is the Goldstone boson associated with the broken $U(1)$-symmetry in the superfluid state. Note that the poles in the transverse propagator F_P^{tt} have the largest residue, so that at long wavelength it is justified to retain only the contribution from transverse correlations. Using the definitions (12.61a), (12.61b), (12.61c), and (12.61d) as well as the mean-field gap equation (12.18), we can write the Gaussian

---

[5] In fact, the integrals I_1, I_2, as well as $\tilde{I}_0$ defined in Eq. (12.63) can be calculated analytically for all physically relevant dimensions $D = 1, 2, 3$; for $D = 3$ and $D = 1$ these integrals can be expressed in terms of complete elliptic integrals.

approximation for the velocity of the BA mode at zero temperature in the following simple form,

$$\frac{c}{v_F} = Z_c = \frac{1}{\sqrt{D}} \sqrt{\frac{\tilde{\mu} - \frac{D-2}{2\tilde{g}} \tilde{\Delta}^2}{1 + \left(\frac{D-2}{2\tilde{g}}\right)^2 \tilde{\Delta}^2}} \, , \tag{12.67}$$

which is valid in dimensions $2 < D < 4$. Recall that $\tilde{\mu} = \mu/\epsilon_F$, $\tilde{\Delta} = \Delta_0/\epsilon_F$, and $\tilde{g} = \nu_0 g$, see Eqs. (12.25), and (12.28). Note that at the unitary point where $\tilde{g}^{-1} = 0$ the relation (12.67) reduces to $Z_c = \sqrt{\tilde{\mu}/D}$, which gives $Z_c = 0.4437$ in three dimensions.

Finally, let us point out that the feedback of the Gaussian fluctuations onto the properties of the *fermionic* quasiparticles has not received much attention in the literature. It turns out that in dimensions $D \leq 3$ Gaussian fluctuations have a rather drastic effect on the fermionic single-particle excitations: the coupling of the gapless BA mode to the fermionic single-particle excitations leads to the breakdown of the quasiparticle picture for real frequencies ω close to the gap energy (Lerch et al. 2008).

12.4 Dyson–Schwinger Equations and Ward Identities

Before using the FRG to go beyond the Gaussian approximation, it is useful to derive exact relations between vertex functions of our mixed Bose–Fermi theory using the same functional methods as in Chaps. 11.3 and 11.4, where we have derived skeleton equations and Ward identities for the forward scattering model with interaction given in Eq. (11.1). Here, we derive the analogous relations for our attractive Fermi gas model defined in Eq. (12.1). In Sect. 12.5 we shall then use these relations to close the FRG flow in the bosonic sector.

12.4.1 Dyson–Schwinger and Skeleton Equations

The invariance of the functional integral under infinitesimal shifts in the integration variables implies functional relations between vertex functions of different order which have been derived in general form in Sect. 6.3.3, see Eq. (6.132). Let us now write down these equations for the special case of our mixed Bose–Fermi theory with Euclidean action

$$S[\bar{\psi}, \psi, \bar{\chi}, \chi] = S_0[\bar{\psi}, \psi] + S_0[\bar{\chi}, \chi] + S_1[\bar{\psi}, \psi, \bar{\chi}, \chi] \, , \tag{12.68}$$

where the fermionic and bosonic Gaussian parts $S_0[\bar{\psi}, \psi]$ and $S_0[\bar{\chi}, \chi]$ are given in Eqs. (12.5) and (12.6), while the interaction $S_1[\bar{\psi}, \psi, \bar{\chi}, \chi]$ is defined in Eq. (12.7). The generating functional of the Green functions now depends on four Grassmann sources $\bar{j}_\sigma$, j_σ as well as on two complex bosonic sources $\bar{J}$, J,

$$\mathcal{G}[\bar{j}, j, \bar{J}, J] = \frac{1}{\mathcal{Z}} \int \mathcal{D}[\bar{\psi}, \psi, \bar{\chi}, \chi] e^{-S[\psi,\psi,\bar{\chi},\chi]+(\bar{j},\psi)+(\psi,j)+(\bar{J},\chi)+(\bar{\chi},J)} . \quad (12.69)$$

For our Bose–Fermi theory involving four different types of fields, we can write down four different Dyson–Schwinger equations, which can be obtained from our general Dyson–Schwinger equation (6.132) by specifying the label α to refer to $\bar{\psi}$, ψ, $\bar{\chi}$, or χ. With the action given in Eq. (12.68) we obtain the Dyson–Schwinger equations

$$\left(\bar{J}_P - g_p^{-1} \frac{\delta}{\delta J_P} \right) \mathcal{G} - \int_K \frac{\delta^2 \mathcal{G}}{\delta j_{K+P\uparrow} \delta j_{-K\downarrow}} = 0 , \quad (12.70a)$$

$$\left(J_P - g_p^{-1} \frac{\delta}{\delta \bar{J}_P} \right) \mathcal{G} - \int_K \frac{\delta^2 \mathcal{G}}{\delta \bar{j}_{-K\downarrow} \delta \bar{j}_{K+P\uparrow}} = 0 , \quad (12.70b)$$

$$\left(\zeta \bar{j}_{K\sigma} + [i\omega - \xi_{k\sigma}] \frac{\delta}{\delta j_{K\sigma}} \right) \mathcal{G} - \zeta_\sigma \int_P \frac{\delta^2 \mathcal{G}}{\delta \bar{J}_{P-K,-\sigma} \delta J_P} = 0 , \quad (12.70c)$$

$$\left(j_{K\sigma} + [i\omega - \xi_{k\sigma}] \frac{\delta}{\delta \bar{j}_{K\sigma}} \right) \mathcal{G} - \zeta_\sigma \int_P \frac{\delta^2 \mathcal{G}}{\delta j_{P-K,-\sigma} \delta \bar{J}_P} = 0 , \quad (12.70d)$$

where $\zeta_\uparrow = \zeta$ and $\zeta_\downarrow = 1$. The analogous relations for the forward scattering model discussed in Chap. 11 are given in Eqs. (11.25a), (11.25b), and (11.25c). As in Sect. 11.3, we may alternatively express Eqs. (12.70a), (12.70b), and (12.70c) in terms of the generating functional $\mathcal{G}_c[\bar{j}, j, \bar{J}, J]$ of the connected Green functions and the corresponding generating functional $\Gamma[\bar{\psi}, \psi, \bar{\chi}, \chi]$ of the irreducible vertices,

$$\frac{\delta \Gamma}{\delta \chi_P} = \int_K \left[\bar{\psi}_{K+P\uparrow} \bar{\psi}_{-K\downarrow} + \frac{\delta^2 \mathcal{G}_c}{\delta j_{K+P\uparrow} \delta j_{-K\downarrow}} \right] , \quad (12.71a)$$

$$\frac{\delta \Gamma}{\delta \bar{\chi}_P} = \int_K \left[\psi_{-K\downarrow} \psi_{K+P\uparrow} + \frac{\delta^2 \mathcal{G}_c}{\delta \bar{j}_{-K\downarrow} \delta \bar{j}_{K+P\uparrow}} \right] , \quad (12.71b)$$

$$\frac{\delta \Gamma}{\delta \psi_{K\sigma}} = \zeta_\sigma \int_K \left[\psi_{P-K,-\sigma} \bar{\chi}_P + \frac{\delta^2 \mathcal{G}_c}{\delta \bar{J}_{P-K,-\sigma} \delta J_P} \right] , \quad (12.71c)$$

$$\frac{\delta \Gamma}{\delta \bar{\psi}_{K\sigma}} = \zeta_\sigma \int_K \left[\zeta \bar{\psi}_{P-K,-\sigma} \chi_P + \frac{\delta^2 \mathcal{G}_c}{\delta J_{P-K,-\sigma} \delta \bar{J}_P} \right] . \quad (12.71d)$$

The above relations may now be used to derive exact relations between correlation functions and vertex functions of different order. For example, if we set the sources equal to zero in Eq. (12.70b) and keep in mind that in the superfluid phase

$$\frac{\delta \mathcal{G}}{\delta \bar{J}_P} \bigg|_{J=0} = \langle \chi_P \rangle_{J=0} = \delta_{P,0} \langle \chi \rangle , \quad (12.72)$$

we obtain the following exact generalized gap equation relating the order parameter $\langle \chi \rangle$ and the anomalous component $A(K)$ of the exact fermionic propagator,

$$\boxed{\langle \chi \rangle = -g_0 \int_K A(K) \, .}$$

(12.73)

Note that at the mean-field level we approximate $\langle \chi \rangle \approx \Delta_0$ and replace $A(K)$ by $A_0(K)$ given in Eq. (12.41); then Eq. (12.73) reduces to the mean-field gap equation (12.18).

Equation (12.73) establishes a relation between the expectation value of our bosonic Hubbard–Stratonovich field and a purely fermionic correlation function. Similar relations can also be derived for bosonic correlation functions involving two and more powers of the fields χ and $\bar{\chi}$. For example, taking the functional derivative of Eq. (12.71a) with respect to $\bar{\chi}_{P'}$ and then setting all fields equal to zero we obtain,

$$\delta_{P,P'} \Pi^{\bar{\chi}\chi}(P) = \frac{\delta^2 \Gamma}{\delta \bar{\chi}_{P'} \delta \chi_P} \bigg|_{\text{fields}=0} = \int_K \frac{\delta^3 \mathcal{G}_c}{\delta \bar{\chi}_{P'} \delta j_{K+P\uparrow} \delta j_{-K\downarrow}} \bigg|_{\text{fields}=0} \, .$$

(12.74)

Similar to Eq. (11.30), the functional derivative on the right-hand side is determined by the $\nu = 1$ term in the tree expansion (6.82) of the second derivative of $\mathcal{G}_c$ in powers of irreducible vertices,

$$\frac{\delta^3 \mathcal{G}_c}{\delta \bar{\chi}_{P'} \delta j_{K+P\uparrow} \delta j_{-K\downarrow}} \bigg|_{\text{fields}=0} = - \frac{\delta^3 \mathcal{G}_c}{\delta j_{-K\downarrow} \delta j_{K+P\uparrow} \delta \bar{\chi}_{P'}} \bigg|_{\text{fields}=0}$$

$$= -\delta_{P,P'} B(-K) B(K+P) \Gamma^{\psi_\downarrow \psi_\uparrow \bar{\chi}}(-K, K+P; P) \, ,$$

(12.75)

where $B(K)$ is the normal component of the exact fermionic propagator, and $\Gamma^{\psi_\downarrow \psi_\uparrow \bar{\chi}}(-K, K+P; P)$ is the irreducible vertex with one bosonic and two fermionic legs of the type indicated by the superscripts. Substituting Eq. (12.75) into Eq. (12.74) and renaming $K \to -K$ we obtain the skeleton equation

$$\boxed{\Pi^{\bar{\chi}\chi}(P) = - \int_K B(K) B(-K+P) \Gamma^{\psi_\downarrow \psi_\uparrow \bar{\chi}}(K, -K+P; P) \, .}$$

(12.76)

By taking the functional derivative of the Dyson–Schwinger equation (12.71b) with respect to $\chi_{P'}$ we obtain an equation similar to Eq. (12.76) with the vertex $\Gamma^{\psi_\downarrow \psi_\uparrow \bar{\chi}}(K, P-K; P)$ replaced by $\Gamma^{\bar{\psi}_\uparrow \bar{\psi}_\downarrow \chi}(P-K, K; P)$. Skeleton equations for the anomalous bosonic correlation functions $\Pi^{\chi\chi}(P)$ and $\Pi^{\bar{\chi}\bar{\chi}}(P)$ can be derived analogously. Graphical representations of these relations are shown in Fig. 12.3.

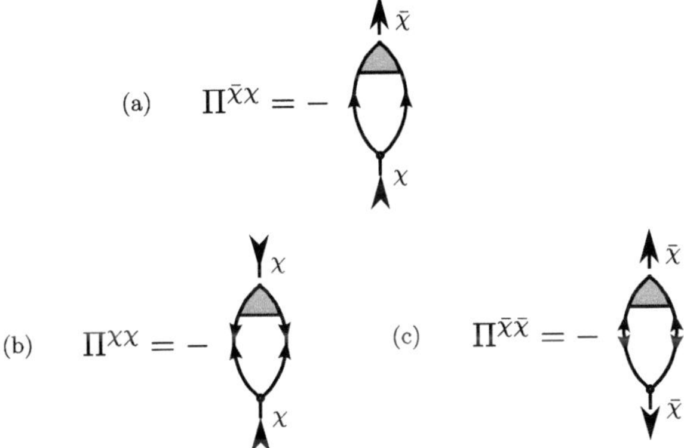

Fig. 12.3 Diagrammatic representation of the skeleton equations for the irreducible bosonic polarizations: (**a**) $\Pi^{\bar{\chi}\chi}(P)$, (**b**) $\Pi^{\chi\chi}(P)$, (**c**) $\Pi^{\bar{\chi}\bar{\chi}}(P)$. Here, *incoming dashed arrows* represent external legs associated with the Hubbard–Stratonovich field χ, while *outgoing arrows* represent its complex conjugate $\bar{\chi}$

12.4.2 Ward Identities

Using the same method as in Sect. 11.4 it is straightforward to derive Ward identities associated with the U(1)-symmetry of the action $S[\bar{\psi}, \psi, \bar{\chi}, \chi]$ defined in Eqs. (12.5), (12.6), (12.7), and (12.68). Therefore we introduce new fermionic fields,

$$\psi_\sigma(X) = e^{i\alpha_\sigma(X)}\psi'_\sigma(X) \quad , \quad \bar{\psi}_\sigma(X) = e^{-i\alpha_\sigma(X)}\bar{\psi}'_\sigma(X) , \tag{12.77}$$

and new bosonic fields,

$$\chi(X) = e^{i\alpha_\uparrow(X)+i\alpha_\downarrow(X)}\chi'(X) \quad , \quad \bar{\chi}(X) = e^{-i\alpha_\uparrow(X)-i\alpha_\downarrow(X)}\bar{\chi}'(X) , \tag{12.78}$$

with an arbitrary real function $\alpha_\sigma(X)$. Note that these transformations leave the interaction $S_1[\bar{\psi}, \psi, \bar{\chi}, \chi]$ in Eq. (12.7) invariant. Using the invariance of the integration measure $\mathcal{D}[\bar{\psi}, \psi, \bar{\chi}, \chi]$ in the functional integral representation (12.69) of the generating functional $\mathcal{G}[\bar{j}, j, \bar{J}, J]$ of the Green functions, we obtain after a simple calculation analogous to Sect. 11.4,

$$\int_K \left\{ \left[-G_0^{-1}(K) + G_0^{-1}(K+\bar{K}) \right] \frac{\delta^2 \mathcal{G}}{\delta \bar{j}_{K\sigma} \delta j_{K+\bar{K}\sigma}} \right.$$
$$\left. + \bar{j}_{K+\bar{K}\sigma} \frac{\delta \mathcal{G}}{\delta \bar{j}_{K\sigma}} - j_{K\sigma} \frac{\delta \mathcal{G}}{\delta j_{K+\bar{K}\sigma}} \right\}$$
$$+ \int_P \left\{ \left[g_p^{-1} - g_{p+k}^{-1} \right] \frac{\delta^2 \mathcal{G}}{\delta \bar{J}_P \delta J_{P+\bar{K}}} + \bar{J}_{P+\bar{K}} \frac{\delta \mathcal{G}}{\delta \bar{J}_P} - J_P \frac{\delta \mathcal{G}}{\delta J_{P+\bar{K}\sigma}} \right\} = 0 . \tag{12.79}$$

Finally, we express $\mathcal{G}$ in Eq. (12.79) in terms of the generating functional $\mathcal{G}_c[\bar{j}, j, \bar{J}, J]$ of the connected Green functions and the corresponding generating functional $\Gamma[\bar{\psi}, \psi, \bar{\chi}, \chi]$ of the irreducible vertices and obtain

$$
\begin{aligned}
&\int_K \Bigg\{ \left[-G_0^{-1}(K) + G_0^{-1}(K+\bar{K})\right] \frac{\delta^2 \mathcal{G}_c}{\delta \bar{j}_{K\sigma} \delta j_{K+\bar{K}\sigma}} \\
&\quad + \psi_{K\sigma} \frac{\delta \Gamma}{\delta \psi_{K+\bar{K}\sigma}} - \bar{\psi}_{K+\bar{K}\sigma} \frac{\delta \Gamma}{\delta \bar{\psi}_{K\sigma}} \Bigg\} \\
&+ \int_P \Bigg\{ \left[g_p^{-1} - g_{p+\bar{k}}^{-1}\right] \frac{\delta^2 \mathcal{G}_c}{\delta \bar{J}_P \delta J_{P+\bar{K}}} + \chi_P \frac{\delta \Gamma}{\delta \chi_{P+\bar{K}}} - \bar{\chi}_{P+\bar{K}} \frac{\delta \Gamma}{\delta \bar{\chi}_P} \Bigg\} = 0 \, .
\end{aligned}
$$

$$\tag{12.80}$$

From this functional Ward identity we may again obtain Ward identities relating irreducible vertices with different numbers of external legs by expanding the generating functionals Γ and $\mathcal{G}_c$ in powers of the fields and then comparing the coefficients of properly symmetrized monomials in the fields. Of particular interest is the Ward identity relating the anomalous self-energy $\Delta(K)$ to the three-legged vertices with two fermionic and one bosonic external leg. To derive this, let us set $\bar{K} = 0$ in Eq. (12.80), so that the terms involving the second derivatives of $\mathcal{G}_c$ drop out,

$$
\int_K \left\{ \psi_{K\sigma} \frac{\delta \Gamma}{\delta \psi_{K\sigma}} - \bar{\psi}_{K\sigma} \frac{\delta \Gamma}{\delta \bar{\psi}_{K\sigma}} \right\} = - \int_P \left\{ \chi_P \frac{\delta \Gamma}{\delta \chi_P} - \bar{\chi}_P \frac{\delta \Gamma}{\delta \bar{\chi}_P} \right\} \, . \tag{12.81}
$$

Comparing the terms involving the combinations $\bar{\psi}_{K\uparrow} \bar{\psi}_{-K\downarrow}$ or $\psi_{-K\downarrow}\psi_{K\uparrow}$ on both sides of Eq. (12.81) and keeping in mind that in the superfluid phase the bosonic field has a finite vacuum expectation value, $\chi_P = \delta_{P,0}\langle \chi \rangle + \delta\chi_P$, we find the exact Ward identities

$$
\langle \chi \rangle \Gamma^{\bar{\psi}_\uparrow \bar{\psi}_\downarrow \chi}(K, -K; 0) = \Delta(K) \, , \tag{12.82a}
$$
$$
\langle \bar{\chi} \rangle \Gamma^{\psi_\downarrow \psi_\uparrow \bar{\chi}}(K, -K; 0) = \bar{\Delta}(-K) \, . \tag{12.82b}
$$

In the following section we shall use these identities to close the hierarchy of FRG flow equations for our mixed Bose–Fermi theory.

12.5 FRG Approach with Total Momentum Cutoff

In this section we shall use our general superfield FRG formalism developed in Chap. 7 to calculate fluctuation corrections beyond the Gaussian approximation for the attractive Fermi gas model defined in Eq. (12.1). We thereby also show how the vertex expansion with partial bosonization works in practice when the bosonic field has a finite vacuum expectation value.

12.5.1 Superfield Notation

To begin with, we rewrite the Hubbard–Stratonovich transformed action of our
model given in Eqs. (12.4), (12.5), (12.6), and (12.7) in superfield notation. As in
Sect. 12.3, we parametrize the complex Hubbard–Stratonovich field χ in terms of
its real and imaginary parts describing longitudinal and transverse fluctuations,[6]

$$\chi_P = \frac{1}{\sqrt{2}}[\chi_{P\ell} + i\chi_{Pt}] \quad , \quad \bar{\chi}_P = \frac{1}{\sqrt{2}}[\chi_{-P\ell} - i\chi_{-Pt}] \,. \tag{12.83}$$

Comparing Eq. (12.83) with the corresponding mean-field expression (12.11), it
is clear that within mean-field theory the vacuum expectation value $\langle \chi \rangle$ of our
Hubbard–Stratonovich field can be identified with the mean-field gap Δ_0 appearing in the fermionic quasiparticle dispersion. However, such an identification is not
valid when fluctuation corrections to the mean-field approximation for the fermionic
Green functions are taken into account. Our mixed Bose–Fermi theory involves a
six-component superfield

$$\Phi = \begin{pmatrix} \psi_\uparrow \\ \psi_\downarrow \\ \bar{\psi}_\uparrow \\ \bar{\psi}_\downarrow \\ \chi_\ell \\ \chi_t \end{pmatrix} \,, \tag{12.84}$$

where the field component χ_ℓ has a finite expectation value. The ratio of the partition
functions with and without interactions can then be written as

$$\frac{Z}{Z_0} = \frac{\int \mathcal{D}[\bar{\psi}, \psi, \chi_\ell, \chi_t] e^{-S_0[\bar{\psi},\psi]-S_0[\chi_\ell,\chi_t]-S_1[\bar{\psi},\psi,\chi_\ell,\chi_t]}}{\int \mathcal{D}[\bar{\psi}, \psi, \chi_\ell, \chi_t] e^{-S_0[\bar{\psi},\psi]-S_0[\chi_\ell,\chi_t]}} \,, \tag{12.85}$$

where the fermionic part $S_0[\bar{\psi}, \psi]$ of the Gaussian action is given by Eq. (12.5), the
bosonic Gaussian part is

$$S_0[\chi_\ell, \chi_t] = \frac{1}{2} \int_P g_p^{-1} \left[\chi_{-P\ell}\chi_{P\ell} + \chi_{-Pt}\chi_{Pt} \right] \,, \tag{12.86}$$

and for the interaction we obtain from Eq. (12.7)

---

[6] We assume that the expectation value of our field χ is real, so that $\langle \chi \rangle = \langle \chi_\ell \rangle / \sqrt{2}$ and $\langle \chi_t \rangle = 0$.
The fields φ_ℓ and φ_t defined in Eqs. (12.47a) and (12.47b) are then given by $\varphi_\ell = \delta\chi_\ell = \chi_\ell - \langle \chi_\ell \rangle$
and $\varphi_t = \delta\chi_t = \chi_t$.

$$S_1[\bar{\psi}, \psi, \chi_\ell, \chi_t] = \frac{1}{\sqrt{2}} \int_K \int_P \left\{ [\psi_{K+P\uparrow}\bar{\psi}_{-K\downarrow} + \psi_{-K\downarrow}\psi_{K-P\uparrow}]\chi_{P\ell} \right.$$

$$\left. +i[\bar{\psi}_{K+P\uparrow}\bar{\psi}_{-K\downarrow} - \psi_{-K\downarrow}\psi_{K-P\uparrow}]\chi_{Pt} \right\} . \tag{12.87}$$

The Bose–Fermi theory defined above is a special case of the general class of theories discussed in Chaps. 6 and 7, so that the exact FRG flow equations for the order parameter and the irreducible vertices can be obtained as a special case of the general flow equations derived in Sect. 7.4. To do this, it is again necessary to rewrite our action in symmetrized superfield notation, taking into account that in the presence of symmetry breaking it is advantageous to define the bare matrix propagator $\mathbf{G}_0$ such that it satisfies Eq. (7.92). We therefore subtract the $p = 0$ component of the condensed field χ_ℓ from the Gaussian action, so that the Gaussian part of our superfield action involves only fields with vanishing expectation values,

$$S_0[\Phi] = S_0[\bar{\psi}, \psi] + S_0[\chi_\ell, \chi_t] - \frac{\chi_{0\ell}^2}{2\beta V g_0}$$

$$= S_0[\bar{\psi}, \psi] + \frac{1}{2} \int_P g_P^{-1} \left[(1 - \frac{\delta_{P,0}}{\beta V})\varphi_{-P\ell}\varphi_{P\ell} + \varphi_{-Pt}\varphi_{Pt} \right]$$

$$= -\frac{1}{2}(\delta\Phi, [\mathbf{G}_0^{-1}]\delta\Phi) , \tag{12.88}$$

where $\delta\Phi = \Phi - \langle\Phi\rangle$ and we have set again $\varphi_\ell = \delta\chi_\ell = \chi_\ell - \langle\chi_\ell\rangle$ and $\varphi_t = \delta\chi_t = \chi_t$ (see Eqs. (12.47a), and (12.47b)). Here, $\mathbf{G}_0^{-1}$ is a 6×6 matrix in field-type space of the same form as Eq. (11.13), but with the bosonic block now given by

$$[\hat{F}_0^{-1}]_{P\bar{\sigma}, P'\bar{\sigma}'} = \delta_{P,-P'}\delta_{\bar{\sigma},\bar{\sigma}'} \left(g_{P'}^{\bar{\sigma}'} \right)^{-1} , \tag{12.89}$$

where $\bar{\sigma} = \ell, t$ labels the two independent real components of our bosonic field, and longitudinal and transverse parts of the bare interaction are

$$g_P^\ell = [1 - (\beta V)^{-1}\delta_{P,0}]g_p \quad , \quad g_P^t = g_p . \tag{12.90}$$

The subtraction in Eqs. (12.88), and (12.90) guarantees that the conditions (7.92) are satisfied. The subtracted term is added again to the interaction in Eq. (12.87), so that after proper anti-symmetrization of the three-legged boson–fermion vertices with respect to the interchange of the fermionic labels the total interaction part of our superfield action can be written as

$$S_1[\Phi] = \frac{\chi_{0\ell}^2}{2\beta V g_0} + \langle \chi \rangle \int_K [\bar{\psi}_{K\uparrow}\bar{\psi}_{-K\downarrow} + \psi_{-K\downarrow}\psi_{K\uparrow}]$$

$$+ \frac{1}{2!} \int_{K_1\sigma_1} \int_{K_2\sigma_2} \int_{P\bar{\sigma}} \delta_{K_1+K_2,P} \Gamma_0^{\bar{\psi}_{\sigma_1}\bar{\psi}_{\sigma_2}\varphi_{\bar{\sigma}}}(K_1, K_2; P)\bar{\psi}_{K_1\sigma_1}\bar{\psi}_{K_2\sigma_2}\varphi_{P\bar{\sigma}}$$

$$+ \frac{1}{2!} \int_{K_1\sigma_1} \int_{K_2\sigma_2} \int_{P\bar{\sigma}} \delta_{K_1+K_2+P,0} \Gamma_0^{\psi_{\sigma_1}\psi_{\sigma_2}\varphi_{\bar{\sigma}}}(K_1, K_2; P)\psi_{K_1\sigma_1}\psi_{K_2\sigma_2}\varphi_{P\bar{\sigma}} \; ,$$

$$\tag{12.91}$$

where we have introduced the notation

$$\langle \chi \rangle = \frac{1}{\beta V}\langle \chi_{P=0}\rangle = \frac{1}{\beta V}\frac{\langle \chi_{P=0,\ell}\rangle}{\sqrt{2}} \; . \tag{12.92}$$

The bare three-legged vertices are

$$\Gamma_0^{\bar{\psi}_{\sigma_1}\bar{\psi}_{\sigma_2}\varphi_\ell}(K_1, K_2; P) = -\Gamma_0^{\psi_{\sigma_1}\psi_{\sigma_2}\varphi_\ell}(K_1, K_2; P) = \frac{\epsilon_{\sigma_1\sigma_2}}{\sqrt{2}} \; , \tag{12.93a}$$

$$\Gamma_0^{\bar{\psi}_{\sigma_1}\bar{\psi}_{\sigma_2}\varphi_t}(K_1, K_2; P) = \Gamma_0^{\psi_{\sigma_1}\psi_{\sigma_2}\varphi_t}(K_1, K_2; P) = i\frac{\epsilon_{\sigma_1\sigma_2}}{\sqrt{2}} \; , \tag{12.93b}$$

where

$$\epsilon_{\sigma_1\sigma_2} = \delta_{\sigma_1\uparrow}\delta_{\sigma_2\downarrow} - \delta_{\sigma_1\downarrow}\delta_{\sigma_2\uparrow} \tag{12.94}$$

is the antisymmetric ϵ-tensor. A graphical representation of the four nonzero bare vertices given in Eqs. (12.93a), and (12.93b) is shown in Fig. 12.4.

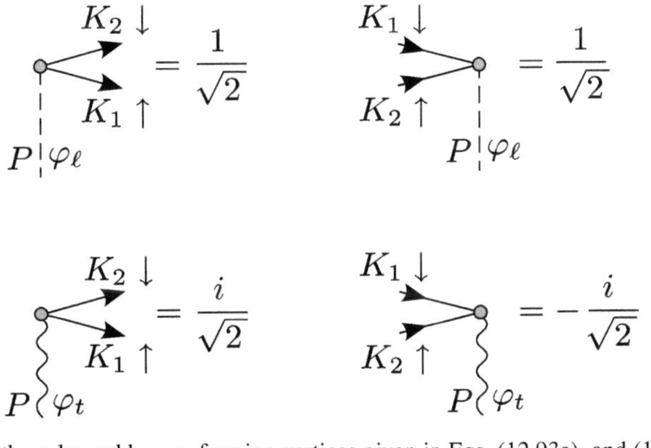

Fig. 12.4 Bare three-legged boson–fermion vertices given in Eqs. (12.93a), and (12.93b). *Outgoing arrows* represent $\bar{\psi}$, *incoming arrows* represent ψ, *dashed lines* represent the longitudinal part φ_ℓ of the order-parameter field, while *wavy lines* represent its transverse part φ_t

Let us examine the structure of the exact matrix Green function of the interacting system, which is related to the bare Green function via the Dyson equation (6.34), $\mathbf{G}^{-1} = \mathbf{G}_0^{-1} - \mathbf{\Sigma}$, where the superfield self-energy $\mathbf{\Sigma}$ has the block-structure

$$\mathbf{\Sigma} = \begin{pmatrix} \hat{\Delta}^\dagger & -\hat{\Sigma} & 0 \\ \hat{\Sigma} & \hat{\Delta} & 0 \\ 0 & 0 & \hat{\Pi} \end{pmatrix} . \tag{12.95}$$

The fermionic blocks in Eq. (12.95) can be parametrized in terms of two functions $\Sigma(K)$ and $\Delta(K)$ as follows,

$$[\hat{\Sigma}]_{K\sigma,K'\sigma'} = \delta_{K,K'}\delta_{\sigma\sigma'}\,\Sigma(K') , \tag{12.96a}$$

$$[\hat{\Delta}]_{K\sigma,K'\sigma'} = \delta_{K,-K'}\epsilon_{\sigma\sigma'}\,\Delta(\sigma'K') . \tag{12.96b}$$

The bosonic block $\hat{\Pi}$ contains the exact irreducible polarization functions $\Pi^{\bar{\sigma}\bar{\sigma}'}(P')$ associated with the collective bosonic fields,

$$[\hat{\Pi}]_{P\bar{\sigma},P'\bar{\sigma}'} = \delta_{P,-P'}\,\Pi^{\bar{\sigma}\bar{\sigma}'}(P') . \tag{12.97}$$

To lowest order in perturbation theory these functions are approximated by the noninteracting polarizations $\Pi_0^{\ell\ell}(P)$, $\Pi_0^{tt}(P)$, and $\Pi_0^{\ell t}(P)$ given in Eqs. (12.49a), (12.49b), and (12.49c). The exact inverse matrix propagator $\mathbf{G}^{-1}$ has therefore the block structure

$$\mathbf{G}^{-1} = \mathbf{G}_0^{-1} - \mathbf{\Sigma} = \begin{pmatrix} -\hat{\Delta}^\dagger & -\hat{G}^{-1} & 0 \\ \hat{G}^{-1} & -\hat{\Delta} & 0 \\ 0 & 0 & -\hat{F}^{-1} \end{pmatrix} , \tag{12.98}$$

where

$$\hat{G}^{-1} = \hat{G}_0^{-1} - \hat{\Sigma} , \tag{12.99a}$$

$$\hat{F}^{-1} = \hat{F}_0^{-1} + \hat{\Pi} . \tag{12.99b}$$

The matrix elements of the diagonal matrix $\hat{G}^{-1}$ are explicitly given by

$$[\hat{G}^{-1}]_{K\sigma,K'\sigma'} = \delta_{K,K'}\delta_{\sigma\sigma'}G^{-1}(K) = \delta_{K,K'}\delta_{\sigma\sigma'}[G_0^{-1}(K) - \Sigma(K)] . \tag{12.100}$$

The inverse of the infinite matrix in Eq. (12.98) is

$$\mathbf{G} = \begin{pmatrix} -\hat{\Delta}\hat{D}^{-1} & \hat{G}_-^{-1}\hat{D}^{-1} & 0 \\ -\hat{G}_-^{-1}\hat{D}^{-1} & -\hat{\Delta}^\dagger\hat{D}^{-1} & 0 \\ 0 & 0 & -\hat{F} \end{pmatrix} , \tag{12.101}$$

where the time-reversed block $\hat{G}_-^{-1}$ is obtained from $\hat{G}^{-1}$ by replacing $K \to -K$,

$$[\hat{G}_-^{-1}]_{K\sigma,K'\sigma'} = \delta_{K,K'}\delta_{\sigma\sigma'}G^{-1}(-K) , \tag{12.102}$$

and the diagonal matrix $\hat{D}$ is defined by

$$[\hat{D}]_{K\sigma,K'\sigma'} = \delta_{K,K'}\delta_{\sigma\sigma'}D(K)\,,\qquad(12.103)$$

with

$$D(K) = G^{-1}(K)G^{-1}(-K) + |\Delta(K)|^2\,.\qquad(12.104)$$

Spin-rotational invariance implies that the off-diagonal blocks in Eq. (12.101) involving the combination $\hat{G}_-^{-1}\hat{D}^{-1}$ must be proportional to the unit matrix, implying $D(K) = D(-K)$ and hence $|\Delta(K)| = |\Delta(-K)|$. Our superfield Green function (12.101) can therefore be written as

$$\mathbf{G} = \begin{pmatrix} \hat{A} & \hat{B} & 0 \\ -\hat{B} & \hat{A}^\dagger & 0 \\ 0 & 0 & -\hat{F} \end{pmatrix}\,,\qquad(12.105)$$

with the anomalous and normal blocks

$$[\hat{A}]_{K\sigma,K'\sigma'} = \delta_{K,-K'}\epsilon_{\sigma\sigma'}A(\sigma K)\,,\qquad(12.106a)$$

$$[\hat{B}]_{K\sigma,K'\sigma'} = \delta_{K,K'}\delta_{\sigma\sigma'}B(K)\,,\qquad(12.106b)$$

where

$$A(K) = -\frac{\Delta(K)}{D(K)}\quad,\quad B(K) = \frac{G^{-1}(-K)}{D(K)}\,.\qquad(12.107)$$

Within the Hartree–Fock approximation we set in the fermionic sector $\Delta(K) \approx \Delta_0$ and $\Sigma(K) \approx 0$, so that $G^{-1}(K) \approx G_0^{-1}(K) = i\omega - \xi_k$ and $D(K) \approx \omega^2 + E_k^2$. Moreover, in Gaussian approximation the bosonic self-energies $\Pi^{\bar{\sigma}\bar{\sigma}'}(P)$ are given by the bare polarization functions $\Pi_0^{\bar{\sigma}\bar{\sigma}'}(P)$ defined in Eqs. (12.49a), (12.49b), and (12.49c). In this approximation the inverse propagator (12.98) is given by

$$\mathbf{G}_1^{-1} = \begin{pmatrix} -\hat{\Delta}_0^\dagger & -\hat{G}_0^{-1} & 0 \\ \hat{G}_0^{-1} & -\hat{\Delta}_0 & 0 \\ 0 & 0 & -[\hat{F}_0^{-1} + \hat{\Pi}_0] \end{pmatrix}\,.\qquad(12.108)$$

12.5.2 Truncation of the Vertex Expansion

Let us now derive exact FRG flow equations for the one-line irreducible vertices of our mixed Bose–Fermi theory. Therefore we modify the free superfield propagator $\mathbf{G}_0$ by introducing an infrared cutoff Λ which suppresses the low-energy fluctuations such that for $\Lambda \to 0$ we recover our original model. All correlation functions and

self-energies are then Λ-dependent. Our aim is to derive an approximate closed system of FRG flow equations for the two fermionic components $\Sigma_\Lambda(K)$ and $\Delta_\Lambda(K)$
of the self-energy and for the vacuum expectation value $\langle \chi \rangle_\Lambda$ of the order-parameter
field. The exact flow equations for these quantities follow as a special case of the
general hierarchy of functional RG flow equations given in Sect. 7.4, which explicitly include the possibility that one or more components of the superfield Φ have a
finite expectation value. A short summary of the results presented in this section can
be found in Bartosch et al. (2009b).

We shall use here a cutoff procedure where only the bosonic sector of our theory is regularized via a cutoff Λ which suppresses bosonic fluctuations with total
momentum $|\boldsymbol{p}| < \Lambda$. This total momentum cutoff scheme is in a sense the particle–
particle analogue of the momentum transfer cutoff scheme introduced in Chap. 11.
For simplicity we use a sharp total momentum cutoff and modify the bare interactions as

$$g_P^{\bar{\sigma}} \to \Theta(\Lambda < |\boldsymbol{p}| < \Lambda_0)g_P^{\bar{\sigma}} . \tag{12.109}$$

where $\Theta(X) = 1$ if the logical expression X is true and $\Theta(X) = 0$ if X is wrong.
With this cutoff procedure the single-scale superfield propagator $\dot{\mathbf{G}}$ defined in general in Eq. (7.64) is of the form

$$\dot{\mathbf{G}} = - \begin{pmatrix} 0 & 0 & 0 \\ 0 & 0 & 0 \\ 0 & 0 & \dot{\hat{F}} \end{pmatrix} , \tag{12.110}$$

with

$$[\dot{\hat{F}}]_{P\bar{\sigma},P'\bar{\sigma}'} = -\delta(\Lambda - |\boldsymbol{p}|)[\hat{F}]_{P\bar{\sigma},P'\bar{\sigma}'} = \delta_{P,-P'}\dot{F}_P^{\bar{\sigma}\bar{\sigma}'} , \tag{12.111}$$

and

$$\dot{F}_P^{\bar{\sigma}\bar{\sigma}'} = -\delta(\Lambda - |\boldsymbol{p}|)F_P^{\bar{\sigma}\bar{\sigma}'} , \tag{12.112}$$

where it is understood that in the bosonic propagator $F_P^{\bar{\sigma}\bar{\sigma}'}$ on the right-hand side of
Eq. (12.112) we should replace the Θ-function by unity.

Consider now the exact FRG flow equation (7.94) for the order parameter, which
for general Bose–Fermi theories is shown graphically in Fig. 7.7. Keeping in mind
that according to Eq. (6.75) we may identify $\Gamma_{\alpha_1\alpha_2}^{(2)}$ with the self-energy $[\boldsymbol{\Sigma}]_{\alpha_1\alpha_2}$,
Eq. (7.94) can be written as

$$\int_{\beta_1} [\boldsymbol{\Sigma}]_{\alpha_1\beta_1} \partial_\Lambda \bar{\Phi}_{\beta_1}^0 = \frac{1}{2} \int_{\beta_1} \int_{\beta_2} [\dot{\mathbf{G}}]_{\beta_1\beta_2} \Gamma_{\beta_1\beta_2\alpha_1}^{(3)} . \tag{12.113}$$

This equation relates the RG flow of the vacuum expectation value $\bar{\Phi}_\alpha^0 = \langle \Phi_\alpha \rangle$
of the superfield in the absence of sources to the irreducible self-energy $\boldsymbol{\Sigma}$ and

the irreducible vertices $\Gamma^{(3)}$ with three external legs. To keep track of the various processes, it is at this point useful to introduce a more explicit graphical notation adopted to our specific model where the various types of fields are represented by different graphical symbols, as defined in Fig. 12.5. With this notation the exact flow equation (12.113) for the vacuum expectation value $\langle \chi_{P=0,\ell} \rangle / (\beta V) \equiv \sqrt{2}\langle \chi \rangle$ in the total momentum cutoff scheme can be written as

$$
\left[g_0^{-1} + \Pi^{\ell\ell}(0) \right] \sqrt{2}\partial_\Lambda \langle \chi \rangle = -\frac{1}{2} \int_P \left[\Gamma^{\ell\ell\ell}(P, -P, 0)\dot{F}^{\ell\ell}(P) \right.
$$
$$
\left. + \Gamma^{\ell tt}(P, -P, 0)\dot{F}^{tt}(P) + 2\Gamma^{\ell\ell t}(P, -P, 0)\dot{F}^{\ell t}(P) \right], \quad (12.114)
$$

where $\Gamma^{\ell\ell\ell}$, $\Gamma^{\ell\ell t}$, and $\Gamma^{\ell tt}$ are the irreducible vertices with three bosonic external legs of the type indicated by the superscripts. A graphical representation of the exact FRG flow equation (12.114) is shown in Fig. 12.6. The exact FRG flow equations for the vertices $\Gamma^{\ell\ell\ell}$, $\Gamma^{\ell\ell t}$, and $\Gamma^{\ell tt}$ can in turn be obtained as a special case of the general FRG flow equation (7.74), which is shown graphically in Fig. 7.3. However, instead of explicitly considering these rather complicated flow equations, we will use the skeleton equation (12.73) and the Ward identities (12.82a) and (12.82b) to

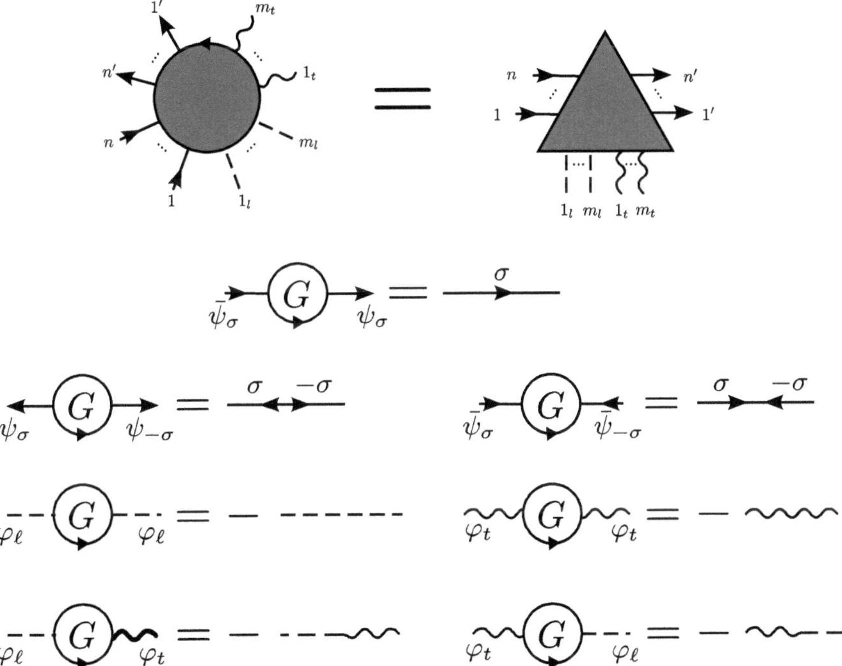

Fig. 12.5 Dictionary for the fermionic and bosonic propagators and vertices. The single-scale propagators are represented by similar symbols with *slashed lines*. See also Fig. 11.1 for a similar dictionary

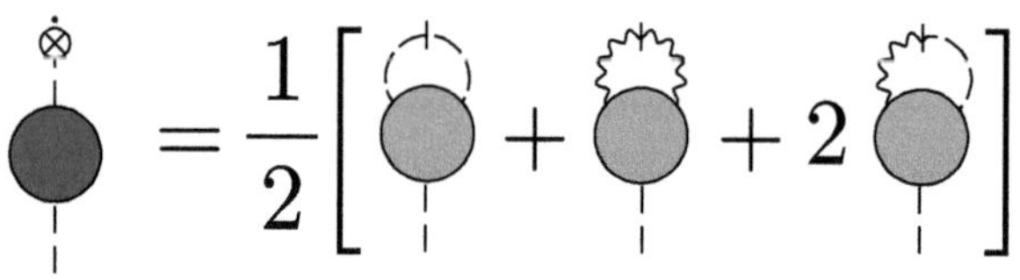

Fig. 12.6 Exact FRG flow equation (12.114) for the order parameter using physical vertices and propagators, as defined in Fig. 12.5. This is a decompressed version of the general order-parameter flow equation shown in Fig. 7.7 in the total momentum cutoff scheme

relate the order parameter $\langle \chi \rangle$ to the anomalous fermionic propagator $A(K)$ and other quantities which are more easily accessible, as will be explained below.

Next, consider the RG flow of the irreducible two-point vertices. For $n = 2$ our exact FRG flow equation (7.91) for the irreducible n-point vertices in the presence of fields with vacuum expectation values reduces to

$$
\partial_\Lambda [\boldsymbol{\Sigma}]_{\alpha_1 \alpha_2} = \int_{\beta_1} (\partial_\Lambda \bar{\Phi}^0_{\beta_1}) \Gamma^{(3)}_{\beta_1 \alpha_1 \alpha_2} - \frac{1}{2} \mathrm{Tr} \left[\mathbf{Z} \dot{\mathbf{G}} \Gamma^{(4)}_{\alpha_1 \alpha_2} \right] + \mathrm{Tr} \left[\mathbf{Z} \dot{\mathbf{G}} \Gamma^{(3)}_{\alpha_1} \mathbf{G} \Gamma^{(3)}_{\alpha_2} \right]
$$

$$
= \int_{\beta_1} (\partial_\Lambda \bar{\Phi}^0_{\beta_1}) \Gamma^{(3)}_{\beta_1 \alpha_1 \alpha_2} - \frac{1}{2} \int_{\beta_1} \int_{\beta_2} [\dot{\mathbf{G}}]_{\beta_1 \beta_2} \Gamma^{(4)}_{\beta_1 \beta_2 \alpha_1 \alpha_2}
$$

$$
+ \int_{\beta_1} \int_{\beta_2} \int_{\beta_3} \int_{\beta_4} [\dot{\mathbf{G}}]_{\beta_1 \beta_4} \Gamma^{(3)}_{\beta_1 \beta_2 \alpha_1} [\mathbf{G}]_{\beta_2 \beta_3} \Gamma^{(3)}_{\beta_3 \beta_4 \alpha_2} . \tag{12.115}
$$

If we specify the external labels α_1 and α_2 to refer to fermion fields, we obtain the exact FRG flow equations for the normal and anomalous fermionic self-energies $\Sigma(K)$ and $\Delta(K)$, while for bosonic external labels Eq. (12.115) gives the FRG flow of the irreducible bosonic polarizations $\Pi^{\ell\ell}(P)$, $\Pi^{tt}(P)$, and $\Pi^{\ell t}(P)$. Let us now truncate the exact FRG flow equation (12.115) by neglecting the bosonic energy–momentum in the vertices with one bosonic and two fermionic legs. In the complex χ-$\bar{\chi}$-basis this amounts to the approximation

$$
\Gamma^{\bar{\psi}_\uparrow \bar{\psi}_\downarrow \chi}(K, P - K; P) \approx \Gamma^{\bar{\psi}_\uparrow \bar{\psi}_\downarrow \chi}(K, -K; 0) = \gamma(K) , \tag{12.116a}
$$

$$
\Gamma^{\psi_\downarrow \psi_\uparrow \bar{\chi}}(K, P - K; P) \approx \Gamma^{\psi_\downarrow \psi_\uparrow \bar{\chi}}(K, -K; 0) = \bar{\gamma}(-K) , \tag{12.116b}
$$

where the flowing vertex function $\gamma(K)$ is normalized such that it reduces to unity in the noninteracting limit. In the real basis spanned by the longitudinal (φ_ℓ) and transverse (φ_t) components of the field one should take the different normalization of the vertices given in Eqs. (12.93a) and (12.93b) into account. As a further approximation, we shall neglect all other vertices with four and more external legs. These higher-order vertices vanish at the initial scale and we assume that their effect remains small throughout the entire RG flow. Specifying the external legs in Eq. (12.115) to be fermionic, and using the total momentum cutoff scheme in the bosonic sector, we then obtain the following two approximate FRG flow equations for the anomalous and normal self-energy,

$$\partial_\Lambda \Delta(K) = \gamma(K)\partial_\Lambda\langle\chi\rangle + \frac{1}{2}\gamma^2(K)\int_P \left[\dot{F}_P^{\ell\ell} - \dot{F}_P^{tt}\right]A(P-K)\,, \quad (12.117)$$

$$\partial_\Lambda \Sigma(K) = -\frac{1}{2}\gamma^2(K)\int_P \left[\dot{F}_P^{\ell\ell} + \dot{F}_P^{tt} - 2i\,\dot{F}_P^{\ell t}\right]B(P-K)\,. \quad (12.118)$$

Graphical representations of these equations are shown in Fig. 12.7.

Obviously, we also need an equation for the vertex $\gamma(K)$; instead of writing down a FRG flow equation for this vertex, we use the Ward identity (12.82a), which implies in combination with Eq. (12.116a) that our flowing vertex function $\gamma(K)$ can be expressed in terms of the flowing anomalous self-energy and the flowing order parameter as follows,

$$\Delta(K) = \gamma(K)\langle\chi\rangle\,. \quad (12.119)$$

Substituting this Ward identity into our flow equation (12.117) for the anomalous self-energy, the latter transmutes into the following flow equation for the vertex function $\gamma(K)$,

$$\partial_\Lambda \ln\gamma(K) = \frac{\gamma^2(K)}{2\Delta(K)}\int_P \left[\dot{F}_P^{\ell\ell} - \dot{F}_P^{tt}\right]A(P-K)\,. \quad (12.120)$$

Note that $\partial_\Lambda\gamma(K)$ is proportional to $\gamma^3(K)$; a similar term is also contained in the general FRG flow equation (7.74) for vertices with three external legs. For simplicity, we shall from now on neglect the momentum dependence of the vertex function, approximating $\gamma(K) \approx \gamma(0) \equiv \gamma_\Lambda$.

To obtain a closed system of flow equations, we also need the FRG flow of the irreducible bosonic polarizations $\Pi^{\ell\ell}(P)$, $\Pi^{tt}(P)$, and $\Pi^{\ell t}(P)$. In our total momentum cutoff scheme where only the bosonic components of the single-scale propagator are nonzero, the FRG flow of these functions is driven by the purely bosonic

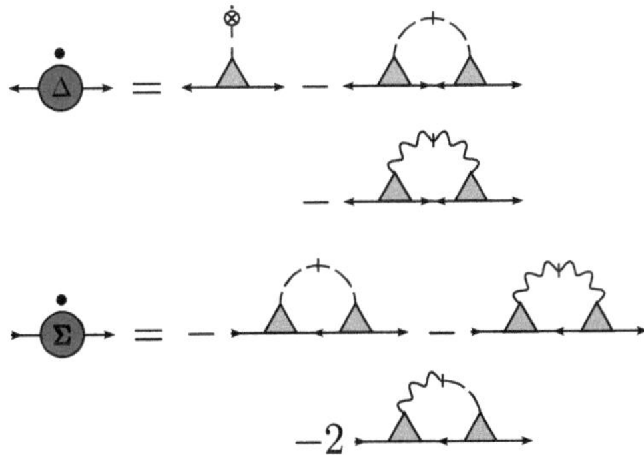

Fig. 12.7 Truncated FRG flow equation for the fermionic self-energies in the total momentum cutoff scheme, where only the bosonic propagators are regularized via a cutoff. The *upper diagram* represents the flow equation (12.117) for the anomalous self-energy $\Delta(K)$, while the *lower diagram* represents the flow equation (12.118) for the normal self-energy $\Sigma(K)$

vertices with three and four external legs appearing on the right-hand side of the flow equation (12.115). However, the skeleton equations discussed in Sect. 12.4 provide an alternative way of closing the hierarchy of FRG flow equations without explicitly considering the flow of bosonic vertices with more than two external legs: The key observation is that the purely bosonic correlation functions $\Pi^{\bar{\chi}\chi}(P)$, $\Pi^{\chi\chi}(P)$, and $\Pi^{\bar{\chi}\bar{\chi}}(P)$ of our mixed Bose–Fermi theory can be expressed in terms of the three-legged vertices with one bosonic and two fermionic legs and purely fermionic propagators. The skeleton equations for the irreducible bosonic polarizations $\Pi^{\ell\ell}(P)$, $\Pi^{tt}(P)$ and $\Pi^{\ell t}(P)$ can be obtained via a simple change of basis from the skeleton equations for the polarizations $\Pi^{\bar{\chi}\chi}(P)$, $\Pi^{\chi\chi}(P)$, and $\Pi^{\bar{\chi}\bar{\chi}}(P)$ discussed in Sect. 12.4.1, see Fig. 12.3. It should be noted that the skeleton equations are valid for any given value of the cutoff parameter Λ. Replacing the three-legged vertices with two fermionic legs and one bosonic leg by γ_Λ times their initial values given in Eqs. (12.93a) and (12.93b) we obtain the approximate skeleton equations for the irreducible bosonic polarization functions,

$$\Pi^{\ell\ell}(P) = -\frac{\gamma_\Lambda}{2} \int_K \left[B(K)B(-K+P) - A(K)A(K+P) + (P \to -P) \right],$$

(12.121a)

$$\Pi^{tt}(P) = -\frac{\gamma_\Lambda}{2} \int_K \left[B(K)B(-K+P) + A(K)A(K+P) + (P \to -P) \right],$$

(12.121b)

$$\Pi^{\ell t}(P) = -\frac{i\gamma_\Lambda}{2} \int_K \left[B(K)B(-K+P) - A(K)A(K+P) - (P \to -P) \right].$$

(12.121c)

These expressions resemble the corresponding noninteracting polarizations in Eqs. (12.49a), (12.49b), and (12.49c), with the important difference that the fermionic propagators on the right-hand sides are now renormalized by fermionic self-energy corrections, as determined by the FRG flow equations (12.117) and (12.118). In addition, one of the bare three-legged vertices with two fermionic legs and one bosonic leg is multiplied by the vertex renormalization factor γ_Λ, whose frequency and momentum dependence we ignore.

Finally, let us also write down the exact FRG flow equation for the grand canonical potential Ω_Λ, which follows from our general flow equation (7.70) for the irreducible vertex $\Gamma_\Lambda^{(0)}$ without external legs. Keeping in mind that in our cutoff scheme only the bosonic block of $\mathbf{G}_{0,\Lambda}$ is nonzero, and carefully taking the sharp cutoff limit of Eq. (7.70) using the relation (8.49), we obtain for the flow of the grand canonical potential

$$\partial_\Lambda \Omega_\Lambda = -\frac{V}{2} \int_P \delta(|\boldsymbol{p}| - \Lambda) \ln\left\{ [1 + g_0 \Pi^{\ell\ell}(P)][1 + g_0 \Pi^{tt}(P)] + g_0^2 [\Pi^{\ell t}(P)]^2 \right\},$$

(12.122)

which should be compared with the corresponding correction Ω_2 in Gaussian approximation given in Eq. (12.50). The formally exact flow equation (12.122) should be integrated with the initial condition $\Omega_{\Lambda_0} = \Omega_1$, where Ω_1 is the mean-field result for the grand canonical potential given in Eq. (12.17).

Keeping in mind that our flow equation (12.117) for the anomalous self-energy $\Delta(K)$ has turned into a flow equation for the vertex function $\gamma(K)$, we still need one additional condition to close the system of flow equations (12.118), (12.119), (12.120), (12.121a), (12.121b), and (12.121c). Fortunately, this can be achieved in a very simple and physically transparent way without explicitly considering the FRG flow of the bosonic three-legged vertices (Bartosch et al. 2009b). The crucial observation is that in the superfluid state the bosonic propagators must exhibit a pole on the real frequency axis associated with the gapless Bogoliubov–Anderson mode. From the requirement that our approximation (12.121a), (12.121b), and (12.121c) for the bosonic self-energies is consistent with the existence of a gapless BA mode we obtain an additional constraint which uniquely fixes the flowing order parameter $\langle \chi \rangle$. The anomalous self-energy $\Delta(K)$ can then be obtained from the Ward identity (12.119). To derive the corresponding FRG flow equation for $\langle \chi \rangle$, we note that, for a given value of the cutoff Λ, the propagator of the bosonic fields is of the form (12.56) but with the noninteracting polarizations $\Pi_0^{\ell\ell}(P)$, $\Pi_0^{tt}(P)$, and $\Pi_0^{\ell t}(P)$ replaced by the flowing irreducible polarizations $\Pi^{\ell\ell}(P)$, $\Pi^{tt}(P)$, and $\Pi^{\ell t}(P)$, which in our truncation are given by Eqs. (12.121a), (12.121b), and (12.121c). The existence of the gapless BA mode is therefore guaranteed if

$$g_0^{-1} + \Pi^{tt}(0, i0) = 0 . \tag{12.123}$$

Obviously, this condition is satisfied at the initial RG scale $\Lambda = \Lambda_0$, where the irreducible transverse polarization $\Pi_{\Lambda_0}^{tt}(0, i0) = \Pi_0^{tt}(0, i0)$ is given by the non-interacting one and Eq. (12.123) reduces to Eq. (12.64). In order to guarantee that Eq. (12.123) is valid for arbitrary Λ, we impose the condition

$$\partial_\Lambda \left[g_0^{-1} + \Pi^{tt}(0, i0) \right] = 0 . \tag{12.124}$$

With $\Pi^{tt}(0, i0)$ given by the skeleton equation (12.121b), the condition (12.124) implies a functional relation between $\partial_\Lambda \Delta(K)$ and $\partial_\Lambda \Sigma(K)$. By demanding that this functional relation is consistent with our flow equations for the fermionic self-energies, we obtain the desired flow equation of the order parameter $\langle \chi \rangle$. Eqs. (12.118), (12.119), (12.120), (12.121a), (12.121b), (12.121c), and (12.124) form a closed system of flow equations for the fermionic self-energies $\Delta(K)$ and $\Sigma(K)$, which implicitly takes the effect of the bosonic three-legged vertices in the exact order-parameter flow equation (12.114) into account.

12.5.3 Truncation with Momentum-Independent Self-Energies

We now further simplify our system of flow equations by neglecting the momentum dependence of the fermionic self-energies and by keeping only the leading frequency dependence of the normal self-energy,

$$\Delta(K) \approx \Delta(0) \equiv \Delta_\Lambda \,, \tag{12.125}$$

$$\Sigma(K) \approx \Sigma(0) + \left.\frac{\partial \Sigma(i\omega)}{\partial(i\omega)}\right|_{\omega=0} i\omega \equiv \Sigma_\Lambda - \left(\frac{1}{Z_\Lambda} - 1\right) i\omega \,, \tag{12.126}$$

where

$$Z_\Lambda = \frac{1}{1 - \left.\frac{\partial \Sigma(i\omega)}{\partial(i\omega)}\right|_{\omega=0}} \tag{12.127}$$

is the flowing wave function renormalization factor. The fermionic Green functions resemble then the mean-field Green functions given in Eqs. (12.41) and (12.42), and it is useful to define

$$A(K) = Z_\Lambda \tilde{A}(K) \quad, \quad B(K) = Z_\Lambda \tilde{B}(K) \,, \tag{12.128}$$

with

$$\tilde{A}(K) = -\frac{\tilde{\Delta}_\Lambda}{\omega^2 + \tilde{E}_{k\Lambda}^2} \quad, \quad \tilde{B}(K) = -\frac{i\omega + \tilde{\xi}_{k\Lambda}}{\omega^2 + \tilde{E}_{k\Lambda}^2} \,. \tag{12.129}$$

The renormalized anomalous self-energy $\tilde{\Delta}_\Lambda$ and the energy dispersions $\tilde{E}_{k\Lambda}$ and $\tilde{\xi}_{k\Lambda}$ appearing here are given by

$$\tilde{E}_{k\Lambda} = \sqrt{\tilde{\xi}_{k\Lambda}^2 + \tilde{\Delta}_\Lambda^2} \,, \tag{12.130a}$$

$$\tilde{\xi}_{k\Lambda} = \tilde{\epsilon}_{k\Lambda} - \tilde{\mu}_\Lambda \,, \tag{12.130b}$$

$$\tilde{\epsilon}_{k\Lambda} = Z_\Lambda \epsilon_k \,, \tag{12.130c}$$

$$\tilde{\Delta}_\Lambda = Z_\Lambda \Delta_\Lambda \,, \tag{12.130d}$$

$$\tilde{\mu}_\Lambda = Z_\Lambda(\mu - \Sigma_\Lambda) \,. \tag{12.130e}$$

The frequency integrations in Eqs. (12.121a), (12.121b), and (12.121c) can now be done analytically. Defining the dimensionless polarizations $\tilde{\Pi}^{\bar{\sigma}\bar{\sigma}'}(P)$ via

$$\Pi^{\bar{\sigma}\bar{\sigma}'}(P) = Z_\Lambda^2 \gamma_\Lambda v_0 \tilde{\Pi}^{\bar{\sigma}\bar{\sigma}'}(P) \,, \tag{12.131}$$

we can obtain the dimensionless functions $\tilde{\Pi}^{\bar{\sigma}\bar{\sigma}'}(P)$ from Eqs. (12.49a), (12.49b), and (12.49c) via the substitution $E_k \to \tilde{E}_{k\Lambda}$, $\xi_k \to \tilde{\xi}_{k\Lambda}$, $\Delta \to \tilde{\Delta}_\Lambda$, and an overall division by v_0. Imposing the gaplessness of the BA mode, Eq. (12.123) implies at $T = 0$ the constraint

$$\frac{1}{g_0} = Z_\Lambda^2 \gamma_\Lambda \frac{1}{V} \sum_k \frac{1}{2\tilde{E}_{k\Lambda}} \,. \tag{12.132}$$

The divergencies appearing on both sides of this equation can again be cured by eliminating the bare interaction g_0 in favor of the two-body T-matrix g,

$$\frac{1}{g_0} = \frac{1}{g} + Z_\Lambda^2 \gamma_\Lambda \frac{1}{V} \sum_k \frac{1}{2\tilde{\epsilon}_{k\Lambda}} \ . \tag{12.133}$$

This leads to the regularized gap equation

$$\frac{1}{Z_\Lambda^2 \gamma_\Lambda g} = \frac{1}{V} \sum_k \left[\frac{1}{2\tilde{E}_{k\Lambda}} - \frac{1}{2\tilde{\epsilon}_{k\Lambda}} \right] \ . \tag{12.134}$$

It is satisfying to see that exactly the same equation follows directly from the skeleton equation (12.73) without explicitly imposing the gaplessness of the BA mode. The gapless BA mode is therefore a natural consequence of our truncation scheme.

At this point, it is useful to summarize again our system of FRG flow equations derived so far for a sharp total momentum cutoff. The rescaled shifted chemical potential $\tilde{\mu}_\Lambda$ defined in Eq. (12.130e) and the vertex renormalization factor γ_Λ satisfy

$$\Lambda \partial_\Lambda \tilde{\mu}_\Lambda = \eta_\Lambda \tilde{\mu}_\Lambda - \gamma_\Lambda \left(\frac{\Lambda}{k_{F\Lambda_0}} \right)^D \epsilon_{F\Lambda_0} \int \frac{d\bar{\omega}}{2\pi} \left[\tilde{F}_P^{\ell\ell} + \tilde{F}_P^{tt} - 2i \tilde{F}_P^{\ell t} \right] \tilde{B}(P) \ , \tag{12.135}$$

$$\Lambda \partial_\Lambda \ln \gamma_\Lambda = -\frac{\gamma_\Lambda}{\tilde{\Delta}_\Lambda} \left(\frac{\Lambda}{k_{F\Lambda_0}} \right)^D \epsilon_{F\Lambda_0} \int \frac{d\bar{\omega}}{2\pi} \left[\tilde{F}_P^{\ell\ell} - \tilde{F}_P^{tt} \right] \tilde{A}(P) \ . \tag{12.136}$$

Here, the values of the Fermi energy $\epsilon_{F\Lambda_0}$ and the Fermi momentum $k_{F\Lambda_0}$ are determined by the noninteracting system at the beginning of the RG flow. The flowing wave function renormalization factor satisfies

$$\Lambda \partial_\Lambda Z_\Lambda = \eta_\Lambda Z_\Lambda \ , \tag{12.137}$$

with the flowing anomalous dimension

$$\eta_\Lambda = \gamma_\Lambda \left(\frac{\Lambda}{k_{F\Lambda_0}} \right)^D \epsilon_{F\Lambda_0} \int \frac{d\bar{\omega}}{2\pi} \left[\tilde{F}_P^{\ell\ell} + \tilde{F}_P^{tt} - 2i \tilde{F}_P^{\ell t} \right] \frac{\tilde{E}_\Lambda^2 - \bar{\omega}^2 + 2i\bar{\omega}\tilde{\xi}_\Lambda}{(\bar{\omega}^2 + \tilde{E}_\Lambda^2)^2} \ , \tag{12.138}$$

where $\tilde{E}_\Lambda$ and $\tilde{\xi}_\Lambda$ are obtained from $\tilde{E}_{p\Lambda}$ and $\tilde{\xi}_{p\Lambda}$ by setting $|p| = \Lambda$. The rescaled gap parameter $\tilde{\Delta}_\Lambda$ is determined by the generalized gap equation

$$\frac{1}{\tilde{g}_\Lambda} = \frac{1}{\nu_0 V} \sum_k \left[\frac{1}{2\tilde{E}_{k\Lambda}} - \frac{1}{2\tilde{\epsilon}_{k\Lambda}} \right] \ , \tag{12.139}$$

where the dimensionless renormalized coupling $\tilde{g}_\Lambda$ is defined by

$$\tilde{g}_\Lambda = Z_\Lambda^2 \gamma_\Lambda \nu_0 g \ . \tag{12.140}$$

From the solution of the above system of equations we obtain the flowing order parameter using the Ward identity

$$\langle \chi \rangle = \frac{\tilde{\Delta}_\Lambda}{Z_\Lambda \gamma_\Lambda} \ . \tag{12.141}$$

Finally, the rescaled bosonic propagators are given by

$$\begin{pmatrix} \tilde{F}_P^{\ell\ell} & \tilde{F}_P^{\ell t} \\ \tilde{F}_P^{t\ell} & \tilde{F}_P^{tt} \end{pmatrix} = \frac{1}{\tilde{N}(P)} \begin{pmatrix} \tilde{g}_\Lambda^{-1} + \tilde{\Pi}_{\Lambda,r}^{tt}(P) & -\tilde{\Pi}_\Lambda^{\ell t}(P) \\ -\tilde{\Pi}_\Lambda^{t\ell}(P) & \tilde{g}_\Lambda^{-1} + \tilde{\Pi}_{\Lambda,r}^{\ell\ell}(P) \end{pmatrix} \ , \tag{12.142}$$

where

$$\tilde{N}(P) = \left[\tilde{g}_\Lambda^{-1} + \tilde{\Pi}_{\Lambda,r}^{\ell\ell}(P) \right] \left[\tilde{g}_\Lambda^{-1} + \tilde{\Pi}_{\Lambda,r}^{tt}(P) \right] + \left[\tilde{\Pi}_\Lambda^{\ell t}(P) \right]^2 \ , \tag{12.143}$$

and the regularized dimensionless diagonal polarizations are defined by

$$\tilde{\Pi}_{\Lambda,r}^{\bar{\sigma}\bar{\sigma}}(P) = \tilde{\Pi}_\Lambda^{\bar{\sigma}\bar{\sigma}}(P) + \frac{1}{\nu_0 V} \sum_k \frac{1}{2\tilde{\epsilon}_{k\Lambda}} \ . \tag{12.144}$$

Note that by adding and subtracting the divergent quantity $\frac{1}{\nu_0 V}\sum_k \frac{1}{2\tilde{\epsilon}_{k\Lambda}}$ we have removed the divergencies from the inverse interaction and the polarizations. Due to the sharp momentum cutoff and the rotational symmetry we can carry out all momentum integrations analytically such that the collective label P reduces to $P = (i\bar{\omega}, \Lambda)$. By construction, in the total momentum cutoff scheme the initial values of the order parameter and the fermionic self-energies are given by the mean-field values discussed in Sect. 12.2, $\Delta_{\Lambda_0} = \langle \chi \rangle_{\Lambda_0} = \Delta_0 = 0.6864 \, \epsilon_F$ and $\Sigma_{\Lambda_0} = 0$. To compare the flowing order parameter and self-energies with the mean-field results, it is important to keep in mind that within the FRG one works at fixed chemical potential μ. As a consequence, the density of fermions and hence the Fermi energy $\epsilon_{F\Lambda}$ of a noninteracting system which has exactly the same density as the interacting system are functions of the flow parameter Λ. We should therefore use the thermodynamic relation between the density and the chemical potential to eliminate the chemical potential in favor of the density which is a function of Λ. Within our approximations the renormalized particle number corresponding to Eq. (12.20) at $T = 0$ can either be obtained from the grand canonical potential Ω_Λ using the thermodynamic relation $N = -\partial\Omega/\partial\mu$, or from the normal part of the single particle Green function. Using the latter approach we obtain as an approximate result for the flowing renormalized particle number density,

$$\rho_\Lambda = \frac{Z_\Lambda}{V} \sum_k \left[1 - \frac{\tilde{\xi}_{k\Lambda}}{\tilde{E}_{k\Lambda}} \right] \ . \tag{12.145}$$

Typical RG flows at the unitary point $(1/(k_F a_s) = 0)$ in three dimensions are shown in Fig. 12.8 for the vertex renormalization factor γ_Λ and the quasiparticle residue Z_Λ. Moreover, in Fig. 12.9 we show the flow of the single particle gap $\tilde{\Delta}_\Lambda$, the order parameter $\chi_\Lambda = \langle \chi \rangle$, and the chemical potential in units of the Fermi energy $\epsilon_{F\Lambda}$ of a noninteracting system, which has exactly the same density as our interacting system,

$$\epsilon_{F\Lambda} = \frac{k_{F\Lambda}^2}{2m} = \epsilon_{F\Lambda_0} \left(\frac{\rho_\Lambda}{\rho_{\Lambda_0}} \right)^{2/D} . \tag{12.146}$$

Note that while mean-field theory does not distinguish between $\tilde{\Delta}_\Lambda$ and $\langle \chi \rangle$, conceptually the renormalized single-particle gap and the order parameter are different quantities, so that it is not surprising that our FRG calculation shows that these

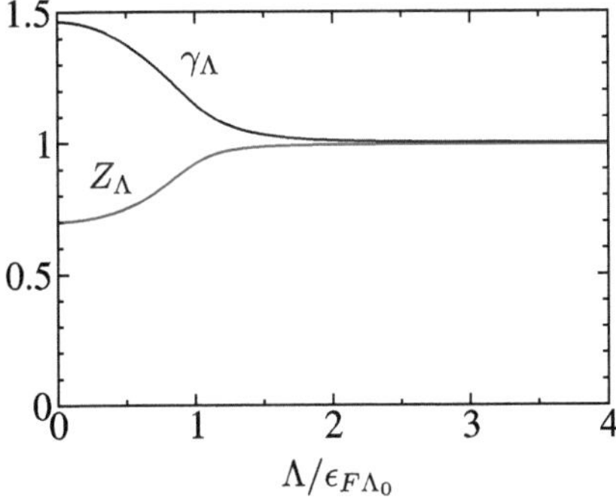

Fig. 12.8 Numerical solution of the FRG flow equations (12.135–12.144) at the unitary point $1/(k_F a_s) = 0$ in three dimensions. The graph shows the flowing wave function renormalization factor Z_Λ and the flowing vertex renormalization factor γ_Λ associated with the vertex with one boson and two fermion legs

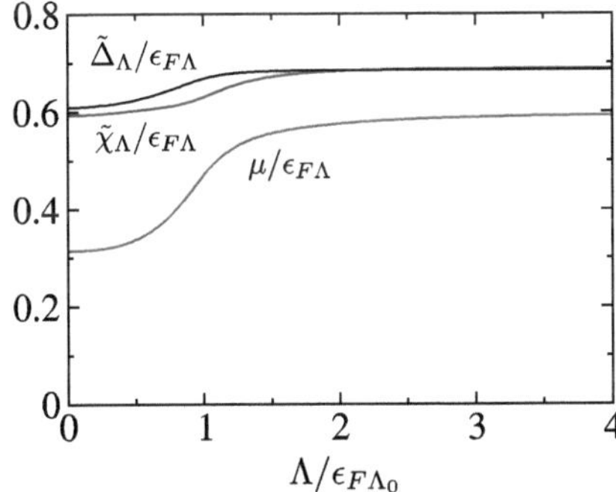

Fig. 12.9 Graph of the flowing single particle gap $\tilde{\Delta}_\Lambda$, the flowing order parameter χ_Λ, and the chemical potential μ at the unitary point $1/(k_F a_s) = 0$ in three dimensions divided by the flowing Fermi energy $\epsilon_{F\Lambda}$ of a noninteracting system, which has exactly the same density as our interacting system

quantities are renormalized differently. For a more detailed analysis of Eqs. (12.135–12.144) we refer the reader to Bartosch et al. (2009b). Here, we only quote the renormalized parameters at the unitary point in units of the physical Fermi energy $\epsilon_F = \lim_{\Lambda \to 0} \epsilon_{F\Lambda}$,

$$\mu/\epsilon_F = 0.32 \, , \quad \tilde{\Delta}/\epsilon_F = 0.61 \, , \quad \langle \chi \rangle / \epsilon_F = 0.59 \quad \text{(FRG)} \, . \qquad (12.147)$$

Comparing these FRG results with the mean-field results (12.33) and the recent Monte Carlo results (12.34) by Carlson and Reddy (2005), we conclude that the fluctuation corrections included in the FRG lead to a significant improvement, although our FRG calculation does not produce numerically reliable values for μ and $\tilde{\Delta}$. However, in view of the approximations inherent in our truncation (which is based on a low-energy expansion of the vertex functions and completely neglects particle-hole fluctuations), our FRG calculation yields satisfactory numerical values. Note that the main purpose of this chapter was not to produce numerically accurate results at the unitary point, but to show how the partially bosonized FRG in the symmetry-broken phase works in practice. Curiously, our result $\mu/\epsilon_F = 0.32$ agrees precisely with the experimentally determined value given by Bartenstein et al. (2004). Moreover, the results of the Gaussian approximation given by Diener et al. (2008) are amazingly close to the most recent Monte Carlo results Carlson and Reddy (2005). In view of the errors inherent in the experiments as well as in the Gaussian approximation, this agreement might be accidental.

12.6 Outlook

We hope that we have convinced the reader that the FRG is a powerful new formulation of the Wilsonian RG which provides a unified and aesthetically appealing framework of formulating Wilsonian RG transformations for any problem which can be formulated in terms of functional integrals. Our goal in this book was to introduce readers with no previous contact with the RG to the main ideas of the FRG and to show in detail how the FRG works in practice. The three parts of this book reflect the three basic steps which are necessary to achieve this goal: one first has to learn the main ideas of the renormalization group from Part I, then one has to learn the FRG from Part II, and finally one should see how this method can be used to solve problems and eventually calculate numbers which can be compared with experiments. In Part III we have focussed on fermionic many-body systems, but it should now be obvious that FRG methods are also useful to study interacting bosons. In fact, several authors have used FRG methods to study interacting Bose gases in various dimensions (Dupuis and Sengupta 2007, Wetterich 2008, Floerchinger and Wetterich 2008, 2009a,b, Sinner et al. 2009, Dupuis 2009a,b, Eichler 2009, Eichler et al. 2009). Due to limitations in space and time, we do not discuss these interesting applications of the FRG in this book and refer the interested reader to the literature cited above. Our main message to the reader is that the FRG formalism as described in this book is a very powerful tool which can be used to tackle a great variety of research problems of current interest.

To conclude this book, let us mention three promising recent developments in the theory of the FRG, which in our opinion will probably lead to further interesting new results in the future:

(a) **Partial bosonization in several competing channels:**
In Chaps. 11 and 12 we have shown that in some cases it can be useful to decouple the interaction between fermions with the help of a suitable bosonic Hubbard–Stratonovich (HS) field and consider the FRG flow equations of the resulting coupled Bose–Fermi theory. For the simplified models considered in Chaps. 11 and 12 the proper choice of the HS decoupling was obvious from the special form of the interaction. However, HS decouplings can be introduced in many different ways and in more realistic models it is a priori not clear which HS decoupling is most appropriate. This ambiguity has been discussed for many decades in the literature (Hamann 1969, Wang et al. 1969, Castellani and Di Castro 1979, Schulz 1990, Macêdo and Coutinho-Filho 1991, Dupuis 2002, Borejsza and Dupuis 2003, Dupuis 2005, Bartosch et al. 2009a). In exceptional cases the resulting mixed Bose–Fermi theory can be solved exactly, such as the forward-scattering model discussed in Chap. 11. However, in general one has to rely on rather simple approximations to solve the coupled Bose–Fermi theory, so that the proper choice of the HS decoupling is crucial. In fact, in some cases it might be advantageous to decouple the interaction in terms of two or more HS fields to take into account the existence of competing instabilities (Bartosch et al. 2009a). In this context it should also be mentioned that recently Husemann and Salmhofer (2009a,b) have proposed a new decomposition of the fermionic two-body interaction of the Hubbard model, which amounts to a decoupling of the flowing effective interaction in terms of three different HS fields associated with different scattering channels. They have shown that this decomposition reduces the ambiguities inherent in the introduction of boson fields, although it does not completely eliminate all ambiguities. We believe that this and similar techniques based on multicomponent HS transformations are promising developments whose potential has not yet been fully explored.

(b) **Scale-dependent Hubbard–Stratonovich transformations:**
In general, HS transformations can be used to eliminate two-body interactions (which are represented by quartic interaction terms in the original fermion fields) in favor of cubic boson–fermion couplings involving one boson and two fermion fields. In this way direct two-body interactions between the original fermion fields are completely eliminated in the bare action. However, terms involving four fermion fields are again generated as the RG is iterated. These terms contain also scattering processes in channels which are different from the channel which has been singled out by the HS field. Although the effect of these terms can indeed be negligible if the physics is dominated by the scattering channel singled out by the HS field, in problems with several competing channels or in situations where the dominant channel is a priori not known it

may be important to take these terms into account. An elegant way to achieve this within the framework of the FRG is based on scale-dependent HS transformations (Gies and Wetterich 2002, Jaeckel and Wetterich 2003, Gies 2006). The idea is to choose the HS field as a scale-dependent nonlinear functional of certain fields of the system. This functional is chosen such that the direct two-body interactions are eliminated at any RG scale. This technique has been successfully used to study certain aspects of quantum chromo-dynamics (Gies and Wetterich 2004, Braun 2009) and has also been used to study ultracold atomic gases (see e.g. Floerchinger et al. 2008). It seems to us that this is a promising technique which should be further developed and applied to the models describing strongly correlated electrons in condensed matter.

(c) **Nonequilibrium FRG:**

While enormous research efforts have been devoted to the understanding of quantum mechanical many-body systems in equilibrium, the behavior of these systems under nonequilibrium conditions has received much less attention. Of course, the nonequilibrium problem is more difficult, but the physics of many-body systems far from equilibrium is expected to be very rich. We believe that the FRG approach is a promising tool to study nonperturbative aspects of many-body systems under nonequilibrium conditions. For example, some classes of nonequilibrium problems can be mapped onto classical field theories, which can then be studied nonperturbatively using the FRG (Canet et al. 2004, 2005), revealing new nonequilibrium fixed points. But also in quantum field theories stationary nonequilibrium states can be characterized by nonthermal fixed points (Berges and Hoffmeister 2009).

For interacting bosons or fermions, the basic FRG equations for the irreducible vertices under nonequilibrium conditions can be easily written down: starting point is the functional integral formulation of the Keldysh formalism (see e.g., Kamenev 2004), where the number of field components has to be doubled in order to take into account that out of equilibrium the forward and the backward propagation in time are independent. All nonequilibrium correlation functions can be expressed in terms of functional averages of a suitably defined effective action. One then modifies the theory by introducing a suitable cutoff either in time (Gasenzer and Pawlowski 2008, Schoeller 2009, Schoeller and Reininghaus 2009) or in the frequency domain (Gezzi et al. 2007, Jakobs et al. 2007). The problem is then formally identical to the corresponding functional integral formulation of the equilibrium problem, so that the resulting flow equations are formally identical to the flow equations derived in Chap. 7. Note that the FRG flow equations in Chap. 7 have been derived for field theories involving arbitrary multicomponent fields, so that all flow equations of Chap. 7 are also valid for the nonequilibrium Keldysh action for bosons, fermions, or mixtures thereof; the superfield label α simply has to be enlarged by another index which keeps track of the Keldysh componets of the fields. The difficult and largely unexplored problem is now to find sensible truncations of the formally exact hierarchy of FRG flow equations, which do not violate essential properties such as the causality conditions (Kamenev 2004). We believe that in the near future the nonequilibrium quantum many-body problem will attract more and more attention, and that the FRG will be a competitive alternative to other nonperturbative tools, such as the method of continuous unitary transformations (Kehrein 2005, 2006) or the time-dependent density matrix renormalization group (Schollwöck 2005, Manmana et al. 2007, 2009).

References

Abramowitz, M. and I. A. Stegun (1965), *Handbook of Mathematical Functions*, Dover, New York.

Anderson, P. W. (1958), *Coherent excited states in the theory of superconductivity: Gauge invariance and the Meissner effect*, Phys. Rev. **110**, 827.

Astrakharchik, G. E., J. Boronat, J. Casulleras, and S. Giorgini (2004), *Equation of state of a Fermi gas in the BEC-BCS crossover: A Quantum Monte Carlo study*, Phys. Rev. Lett. **93**, 200404.

Barankov, R. (2008), *The Higgs resonance in fermionic pairing*, arXiv:0812.4575 [cond-mat.supr-con].

Barankov, R. A. and L. S. Levitov (2007), *Excitation of the Dissipationless Higgs Mode in a Fermionic Condensate*, arXiv:0704.1292 [cond-mat.supr-con].

Bartenstein, M., A. Altmeyer, S. Riedl, S. Jochim, C. Chin, J. H. Denschlag, and R. Grimm (2004), *Crossover from a Molecular Bose-Einstein Condensate to a Degenerate Fermi Gas*, Phys. Rev. Lett. **92**, 120401.

Bartosch, L., H. Freire, J. J. Ramos Cardenas, and P. Kopietz (2009a), *Functional renormalization group approach to the Anderson impurity model*, J. Phys.: Condens. Matter **21**, 305602.

Bartosch, L., P. Kopietz, and A. Ferraz (2009b), *Renormalization of the BCS–BEC crossover by order-parameter fluctuations*, Phys. Rev. B **80**, 104514.

Berges, J. and G. Hoffmeister (2009), *Nonthermal fixed points and the functional renormalization group*, Nucl. Phys. B **813**, 383.

Birse, M. C., B. Krippa, J. A. McGovern, and N. R. Walet (2005), *Pairing in many-fermion systems: An exact renormalisation group treatment*, Phys. Lett. B **605**, 287.

Bloch, I., J. Dalibard, and W. Zwerger (2008), *Many-body physics with ultracold gases*, Rev. Mod. Phys. **80**, 885.

Bogoliubov, N. N. (1958), *On a new method in the theory of superconductivity*, Nuovo Cimento **7**, 794.

Borejsza, K. and N. Dupuis (2003), *Antiferromagnetism and single-particle properties in the two-dimensional half-filled Hubbard model: Slater vs Mott-Heisenberg*, Europhys. Lett. **63**, 722.

Bourdel, T., L. Khaykovich, J. Cubizolles, J. Zhang, F. Chevy, M. Teichmann, L. Tarruell, S. Kokkelmans, and C. Salomon (2004), *Experimental study of the BEC-BCS crossover region in lithium 6*, Phys. Rev. Lett. **93**, 050401.

Braun, J. (2009), *The QCD phase boundary from quark-gluon dynamics*, Eur. Phys. J. C **64**, 459.

Canet, L., B. Delamotte, O. Deloubrière, and N. Wschebor (2004), *Nonperturbative renormalization-group study of reaction-diffusion processes*, Phys. Rev. Lett. **92**, 195703.

Canet, L., H. Chaté, B. Delamotte, I. Dornic, and M. A. Muñoz (2005), *Nonperturbative fixed point in a nonequilibrium phase transition*, Phys. Rev. Lett. **95**, 100601.

Carlson, J., S. Y. Chang, V. R. Pandharipande, and K. E. Schmidt (2003), *Superfluid Fermi gases with large scattering length*, Phys. Rev. Lett. **91**, 050401.

Carlson, J. and S. Reddy (2005), *Asymmetric two-component fermion systems in strong soupling*, Phys. Rev. Lett. **95**, 060401.

Castellani, C. and C. Di Castro (1979), *Arbitrariness and symmetry properties of the functional formulation of the Hubbard hamiltonian*, Phys. Lett. A **70**, 37.

Chang, S. Y., V. R. Pandharipande, J. Carlson, and K. E. Schmidt (2004), *Quantum Monte Carlo studies of superfluid Fermi gases*, Phys. Rev. A **70**, 043602.

De Palo, S., C. Castellani, C. Di Castro, and B. Chakraverty (1999), *Effective action for superconductors and BCS-Bose crossover*, Phys. Rev. B **60**, 564.

Diehl, S., H. Gies, J. M. Pawlowski, and C. Wetterich (2007a), *Flow equations for the BCS-BEC crossover*, Phys. Rev. A **76**, 021602(R).

Diehl, S., H. Gies, J. M. Pawlowski, and C. Wetterich (2007b), *Renormalization flow and universality for ultracold fermionic atoms*, Phys. Rev. A **76**, 053627.

Diehl, S., S. Floerchinger, H. Gies, J. M. Pawlowski, and C. Wetterich (2009), *Functional renormalization group approach to the BCS-BEC crossover*, arXiv:0907.2193 [cond-mat.quant-gas].

Diener, R. B., R. Sensarma, and M. Randeria (2008), *Quantum fluctuations in the superfluid state of the BCS-BEC crossover*, Phys. Rev. A **77**, 023626.

Drechsler, M. and W. Zwerger (1992), *Crossover from BCS-superconductivity to Bose-condensation*, Ann. Phys. (Leipzig) **504**, 15.

Dupuis, N. (2002), *Spin fluctuations and pseudogap in the two-dimensional half-filled Hubbard model at weak coupling*, Phys. Rev. B **65**, 245118.

Dupuis, N. (2005), *Effective action for superfluid Fermi systems in the strong coupling limit*, Phys. Rev. A **72**, 013606.

Dupuis, N. (2009a), *Infrared behavior and spectral function of a Bose superfluid at zero temperature*, Phys. Rev. A **80**, 043627.

Dupuis, N. (2009b), *Unified picture of superfluidity: From Bogoliubov's approximation to Popov's hydrodynamic theory*, Phys. Rev. Lett. **102**, 190401.

Dupuis, N. and K. Sengupta (2007), *Non-perturbative renormalization group approach to zero-temperature Bose systems*, Europhys. Lett. **80**, 50007.

Eagles, D. M. (1969), *Possible pairing without superconductivity at low carrier concentrations in bulk and thin-film superconducting semiconductors*, Phys. Rev. **186**, 456.

Eichler, C. (2009), *Anwendung der Funktionalen Renormierungsgruppe auf wechselwirkende Bosonen*, Diplomarbeit, Goethe-Universität Frankfurt.

Eichler, C., N. Hasselmann, and P. Kopietz (2009), *Condensate density of interacting bosons: A functional renormalization group approach*, Phys. Rev. E **80**, 051129.

Engelbrecht, J. R., M. Randeria, and C. A. R. Sáde Melo (1997), *BCS to bose crossover: Broken-symmetry state*, Phys. Rev. B **55**, 15153.

Floerchinger, S., M. Scherer, D. S., and C. Wetterich (2008), *Particle-hole fluctuations in the BCS-BEC Crossover*, Phys. Rev. B **78**, 174528.

Floerchinger, S. and C. Wetterich (2008), *Functional renormalization for Bose-Einstein condensation*, Phys. Rev. A **77**, 053603.

Floerchinger, S. and C. Wetterich (2009a), *Nonperturbative thermodynamics of an interacting Bose gas*, Phys. Rev. A **79**, 063602.

Floerchinger, S. and C. Wetterich (2009b), *Superfluid Bose gas in two dimensions*, Phys. Rev. A **79**, 013601.

Gasenzer, T. and J. M. Pawlowski (2008), *Towards far-from-equilibrium quantum field dynamics: A functional renormalisation-group approach*, Phys. Lett. B **670**, 135.

Gezzi, R., T. Pruschke, and V. Meden (2007), *Functional renormalization group for nonequilibrium quantum many-body problems*, Phys. Rev. B **75**, 045324.

Gies, H. (2006), *Introduction to the functional RG and applications to gauge theories*, arXiv:hep-ph/0611146.

Gies, H. and C. Wetterich (2002), *Renormalization flow of bound states*, Phys. Rev. D **65**, 065001.

Gies, H. and C. Wetterich (2004), *Universality of spontaneous chiral symmetry breaking in gauge theories*, Phys. Rev. D **69**, 025001.

Hamann, D. R. (1969), *Fluctuation theory of dilute magnetic alloys*, Phys. Rev. Lett. **23**, 95.

Haussmann, R., W. Rantner, S. Cerrito, and W. Zwerger (2007), *Thermodynamics of the BCS-BEC crossover*, Phys. Rev. A **75**, 023610.

Husemann, C. and M. Salmhofer (2009a), *Efficient Fermionic One-Loop RG for the 2D-Hubbard Model at Van Hove Filling*, arXiv:0902.1651v1 [cond-mat.str-el].

Husemann, C. and M. Salmhofer (2009b), *Efficient parametrization of the vertex function, Ω scheme, and the (t, t') Hubbard model at van Hove filling*, Phys. Rev. B **79**, 195125.

Jaeckel, J. and C. Wetterich (2003), *Flow equations without mean field ambiguity*, Phys. Rev. D **68**, 025020.

Jakobs, S. G., V. Meden, and H. Schoeller (2007), *Nonequilibrium functional renormalization group for interacting quantum systems*, Phys. Rev. Lett. **99**, 150603.

Kamenev, A. (2004), *Many-Body Theory of Non-equilibrium Systems*, in H. Bouchiat, Y. Gefen, S. Guèron, G. Montambaux and J. Dalibard, editors, *Les Houches*, volume LX, Elesvier, North-Holland, Amsterdam.

Kehrein, S. (2005), *Scaling and decoherence in the nonequilibrium Kondo model*, Phys. Rev. Lett. **95**, 056602.

Kehrein, S. (2006), *The Flow Equation Approach to Many-Particle Systems*, Springer, Berlin.

Kinast, J., A. Turlapov, J. E. Thomas, Q. Chen, J. Stajic, and K. Levin (2005), *Heat capacity of a strongly interacting Fermi gas*, Science **307**, 1296.

Kopietz, P. (1997), *Bosonization of Interacting Fermions in Arbitrary Dimensions*, Springer, Berlin.

Krippa, B. (2007), *Superfluidity in many fermion systems: Exact renormalization group treatment*, Eur. Phys. J. A **31**, 734.

Krippa, B. (2009), *Exact renormalization group flow for ultracold Fermi gases in the unitary limit*, J. Phys. A: Math. Theor. **42**, 465002.

Krippa, B., M. C. Birse, N. R. Walet, and J. A. McGovern (2005), *Exact renormalisation group and pairing in many-fermion systems*, Nucl. Phys. A **749**, 134.

Leggett, A. J. (1980), *Modern Trends in the Theory of Condensed Matter*, in A. Pekalski and R. Przystawa, editors, *Lecture Notes in Physics*, volume 115, Springer, Berlin.

Lerch, N., L. Bartosch, and P. Kopietz (2008), *Absence of fermionic quasiparticles in the superfluid state of the attractive Fermi gas*, Phys. Rev. Lett. **100**, 050403.

Littlewood, P. B. and C. M. Varma (1982), *Amplitude collective modes in superconductors and their coupling to charge-density waves*, Phys. Rev. B **26**, 4883.

Macêdo, C. A. and M. D. Coutinho-Filho (1991), *Hubbard model: Functional-integral approach and diagrammatic perturbation theory*, Phys. Rev. B **43**, 13515.

Manmana, S. R., S. Wessel, R. M. Noack, and A. Muramatsu (2007), *Strongly correlated fermions after a quantum quench*, Phys. Rev. Lett. **98**, 210405.

Manmana, S. R., S. Wessel, R. M. Noack, and A. Muramatsu (2009), *Time evolution of correlations in strongly interacting fermions after a quantum quench*, Phys. Rev. B **79**, 155104.

Marini, M., F. Pistolesi, and G. C. Strinati (1998), *Evolution from BCS superconductivity to Bose condensation: analytic results for the crossover in three dimensions*, Eur. Phys. J. B **1**, 151.

Nikolić, P. and S. Sachdev (2007), *Renormalization group fixed points, universal phase diagram, and 1/N expansion for quantum liquids with interactions near the unitarity limit*, Phys. Rev. A **75**, 033608.

Nishida, Y. and D. T. Son (2006), *ε-Expansion for a Fermi gas at infinite scattering length*, Phys. Rev. Lett. **97**, 050403.

Nishida, Y. and D. T. Son (2007), *Fermi gas near unitarity around four and two spatial dimensions*, Phys. Rev. A **75**, 063617.

Nozières, P. and S. Schmitt-Rink (1985), *Bose condensation in an attractive fermion gas: From weak to strong coupling superconductivity*, J. Low Temp. Phys. **59**, 195.

Randeria, M. (1995), *Crossover from BCS Theory to Bose-Einstein Condensation*, in A. Griffin, D. Snorke, and S. Stringari, editors, *Bose-Einstein Condensation*, Cambridge University Press, Cambridge, England.

Salmhofer, M., C. Honerkamp, W. Metzner, and O. Lauscher (2004), *Renormalization group flows into phases with broken symmetry*, Progr. Theoret. Phys. **112**, 943.

Sauli, F. and P. Kopietz (2006), *Low-density expansion for the two-dimensional electron gas*, Phys. Rev. B **74**, 193106.

Schoeller, H. (2009), *A perturbative non-equilibrium renormalization group method for dissipative quantum mechanics: Real-time RG in frequency space*, Eur. Phys. J. Special Topics **168**, 179.

Schoeller, H. and F. Reininghaus (2009), *Real-time renormalization group in frequency space: A two-loop analysis of the nonequilibrium anisotropic Kondo model at finite magnetic field*, Phys. Rev. B **80**, 045117.

Schollwöck, U. (2005), *The density-matrix renormalization group*, Rev. Mod. Phys. **77**, 259.

Schrieffer, J. R. (1964), *Theory of Superconductivity*, W. A. Benjamin, Inc., New York.

Schulz, H. J. (1990), *Effective action for strongly correlated fermions from functional integrals*, Phys. Rev. Lett. **65**, 2462.

Schwiete, G. and K. B. Efetov (2006), *Temperature dependence of the spin susceptibility of a clean Fermi gas with repulsion*, Phys. Rev. B **74**, 165108.

Sinner, A., N. Hasselmann, and P. Kopietz (2009), *Spectral function and quasiparticle damping of interacting bosons in two dimensions*, Phys. Rev. Lett. **102**, 120601.

Strack, P., R. Gersch, and W. Metzner (2008), *Renormalization group flow for fermionic superfluids at zero temperature*, Phys. Rev. B **78**, 014522.

Varma, C. M. (2002), *Higgs boson in superconductors*, J. Low Temp. Phys. **126**, 901.

Veillette, M. Y., D. E. Sheehy, and L. Radzihovsky (2007), *Large-N expansion for unitary superfluid Fermi gases*, Phys. Rev. A **75**, 043614.

Wang, S. Q., W. E. Evenson, and J. R. Schrieffer (1969), *Theory of itinerant ferromagnets exhibiting localized-moment behavior above the Curie point*, Phys. Rev. Lett. **23**, 92.

Wetterich, C. (2008), *Functional renormalization for quantum phase transitions with nonrelativistic bosons*, Phys. Rev. B **77**, 064504.

Index

Kopietz, P. et al.: *Index.* Lect. Notes Phys. **798**, 369–375 (2010)
DOI 10.1007/978-3-642-05094-7

MIX
Papier aus verantwortungsvollen Quellen
Paper from responsible sources
FSC® C105338

If you have any concerns about our products,
you can contact us on
ProductSafety@springernature.com

In case Publisher is established outside the EU,
the EU authorized representative is:
Springer Nature Customer Service Center GmbH
Europaplatz 3, 69115 Heidelberg, Germany

Printed by Libri Plureos GmbH
in Hamburg, Germany